Lecture Notes in Computer Science   11159

Commenced Publication in 1973
Founding and Former Series Editors:
Gerhard Goos, Juris Hartmanis, and Jan van Leeuwen

Editorial Board

Advanced Research in Computing and Software Science
Subline of Lecture Notes in Computer Science

Subline Series Editors

Subline Advisory Board

More information about this series at http://www.springer.com/series/7407

Andreas Brandstädt · Ekkehard Köhler
Klaus Meer (Eds.)

Graph-Theoretic Concepts in Computer Science

44th International Workshop, WG 2018
Cottbus, Germany, June 27–29, 2018
Proceedings

 Springer

Editors
Andreas Brandstädt
Universität Rostock
Rostock
Germany

Ekkehard Köhler
Brandenburgische Technische Universität
Cottbus-Senftenberg
Cottbus
Germany

Klaus Meer
Brandenburgische Technische Universität
Cottbus-Senftenberg
Cottbus
Germany

ISSN 0302-9743 ISSN 1611-3349 (electronic)
Lecture Notes in Computer Science
ISBN 978-3-030-00255-8 ISBN 978-3-030-00256-5 (eBook)
https://doi.org/10.1007/978-3-030-00256-5

Library of Congress Control Number: 2018953364

LNCS Sublibrary: SL1 – Theoretical Computer Science and General Issues

This Springer imprint is published by the registered company Springer Nature Switzerland AG
The registered company address is: Gewerbestrasse 11, 6330 Cham, Switzerland

Preface

The 44th International Workshop on Graph-Theoretic Concepts in Computer Science, WG 2018, took place at the Schlosshotel Lübbenau, Lübbenau, Germany, June 27–29, 2018. It was organized by the Brandenburg University of Technology Cottbus-Senftenberg. There were approximately 70 participants from several countries all over the world, including Brazil, Canada, the Czech Republic, France, Germany, India, Italy, The Netherlands, Norway, Poland, Slovenia, Sweden, the UK, and the USA. WG 2018 continued the series of the 43 previous WG conferences. Including this year's meeting, since 1975 the workshop has taken place 24 times in Germany, five times in The Netherlands, three times in France, twice in Austria and the Czech Republic, and once in Greece, Israel, Italy, Norway, Slovakia, Switzerland, Turkey, and the UK. WG conferences aim to connect theory and applications by demonstrating how graph-theoretic concepts can be applied in various areas of computer science. The goal is to present recent research results and to identify and explore directions for future research.

The Program Committee of WG 2018 was chaired by Andreas Brandstädt (Rostock), Ekkehard Köhler (Cottbus), and Klaus Meer (Cottbus). It was responsible for the selection of three invited speakers and for running the reviewing process of all submitted contributions. We received 66 submitted papers. One submission was withdrawn by the author, another one had to be withdrawn by the PC-chairs because of simultaneous submission to another conference with published proceedings. The remaining 64 submissions were reviewed intensely by the Program Committee and many expert sub-reviewers. For almost all submissions at least four reviews were collected. Finally, 30 of the submitted papers were accepted for publication in this volume. We would like to mention that the competition this year was very tough and several good papers could not be accepted. Without the help of our expert referees, the production of the volume would have been impossible. We would like to thank all the sub-reviewers for their excellent work; their names are listed in the organization section of this preface.

There were three excellent invited talks given by Maria Chudnovsky (Princeton University, USA) on "4-Coloring Graphs with No Induced 6-Vertex Path"; by Martin Milanič (University of Primorska in Koper, Slovenia) on "Strong Cliques and Graph Classes"; and by Martin Skutella (TU Berlin, Germany) on "Flows over Time and Submodular Function Minimization".

In addition, there was the first WG Test-of-Time Award given for a highly influential paper that has been presented at a previous WG conference. The award committee consisted of H. L. Bodlaender, R. Möhring, and G. Woeginger. The winning paper was "Bounding the Bandwidth of NP-Complete Problems" from WG 1980 authored by Burkhard Monien and Ivan Hal Sudborough. The talk was given by Burkhard Monien.

Springer generously funded both a Best Paper Award (BPA) and a Best Student Paper Award (BSPA), which were given during the WG 2018 conference.

Édouard Bonnet and Paweł Rzążewski won the BPA for their paper "Optimality Program in Segment and String Graphs". Jelco M. Bodewes and Marieke van der Wegen won the BSPA for their paper "Recognizing Hyperelliptic Graphs in Polynomial Time", co-authored by Hans L. Bodlaender and Gunther Cornelissen. Congratulations to the winners!

Acknowledgments

The organizers of WG 2018 would like to acknowledge and thank the following entities for their financial and non-financial support (in alphabetic order): Brandenburg University of Technology Cottbus-Senftenberg, Deutsche Forschungsgemeinschaft, Springer, and an anonymous donor.

We thank Andrej Voronkov for his outstanding EasyChair system which facilitated the work of the Program Committee and the editors considerably.

July 2018   Andreas Brandstädt
                                                               Ekkehard Köhler
                                                                    Klaus Meer

Organization

Program Committee

Therese Biedl	University of Waterloo, Canada
Andreas Brandstädt (Co-chair)	Universität Rostock, Germany
Kathie Cameron	Wilfrid Laurier University, Waterloo, Canada
Steven Chaplick	Julius-Maximilians-Universität Würzburg, Germany
Derek Corneil	University of Toronto, Canada
Feodor Dragan	Kent State University, USA
Thomas Erlebach	University of Leicester, UK
Celina De Figueiredo	Universidade Federal do Rio de Janeiro, Brazil
Fedor Fomin	University of Bergen, Norway
Michel Habib	IRIF, CNRS and Université Paris-Diderot, France
Pinar Heggernes	University of Bergen, Norway
Ekkehard Köhler (Co-chair)	BTU Cottbus-Senftenberg, Germany
Jan Kratochvíl	Charles University, Prague, Czech Republic
Dieter Kratsch	Université de Lorraine, Metz, France
Ross Mcconnell	Colorado State University, Fort Collins, USA
Klaus Meer (Co-chair)	BTU Cottbus-Senftenberg, Germany
Petra Mutzel	University of Dortmund, Germany
Christophe Paul	LIRMM, CNRS, Université de Montpellier, France
Dieter Rautenbach	Universität Ulm, Germany
Ignasi Sau	LIRMM, CNRS, Université de Montpellier, France
Oliver Schaudt	RWTH Aachen, Germany
Dimitrios M. Thilikos	LIRMM, CNRS, Université de Montpellier, France and National and Kapodistrian University of Athens, Greece
Annegret Wagler	Université Clermont Auvergne, Clermont-Ferrand, France

Additional Reviewers

Adjiashvili, David
Barat, Janos
Beisegel, Jesse
Belmonte, Rémy
Bessy, Stéphane
Bianchi, Silvia
Biniaz, Ahmad
Bonamy, Marthe

Botler, Fábio
Bousquet, Nicolas
Campos, Victor
Chakraborty, Sankardeep
Chepoi, Victor
Cohen, Nathann
Coudert, David
Cunha, Luís

Da Fonseca, Guilherme D.
Dabrowski, Konrad Kazimierz
Dahn, Christine
Damaschke, Peter
Deligkas, Argyrios
Denkert, Carolin
Dos Santos, Vinícius F.

Droschinsky, Andre
Ducoffe, Guillaume
Dyer, Martin
Ehard, Stefan
Eschen, Elaine
Fagerberg, Rolf
Felsner, Stefan
Fernau, Henning
Fiala, Jiri
Fluschnik, Till
Foucaud, Florent
Gabow, Harold
Giannopoulou, Archontia
Gijswijt, Dion
Golovach, Petr
Gonçalves, Daniel
Gronemann, Martin
Guarnera, Heather
Gurvich, Vladimir
Gutin, Gregory
Henning, Michael
Hols, Eva-Maria
Huang, Shenwei
Jabrayilov, Adalat
Kamiński, Marcin
Kanté, Mamadou
    Moustapha
Kim, Eunjung
Klavžar, Sandi
Klemz, Boris
Knauer, Kolja
Knop, Dušan
Koster, Arie
Kothari, Nishad
Kratsch, Stefan
Kulik, Ariel
Kurz, Denis

Kwon, O-joung
Lampis, Michael
Le, Van Bang
Leitert, Arne
Liedloff, Mathieu
Lima, Carlos Vinicius
Lingas, Andrzej
Lopes, Raul
Maffray, Frederic
Makowsky, Johann A.
Maniatis, Spyridon
Mertzios, George B.
Milanič, Martin
Miranda, Alberto
Mitsou, Valia
Mohammed,
    Abdulhakeem
Mondal, Debajyoti
Mouatadid, Lalla
Müller, Haiko
Munaro, Andrea
Naves, Guyslain
Nelson, Peter
Nisse, Nicolas
Ochem, Pascal
Oettershagen, Lutz
Oliveira, Fabiano
Oum, Sang-il
Panda, B. S.
Papadopoulos, Charis
Papagelis, Manos
Perarnau, Guillem
Plummer, Mike
Pupyrev, Sergey
Ramanujan, M. S.
Raymond, Jean-Florent
Saettler, Aline

Saito, Akira
Salazar, Gelasio
Saulpic, David
Saurabh, Saket
Sawada, Joe
Scheffler, Robert
Schrezenmaier, Hendrik
Schulz, André
Shalom, Mordechai
Shi, Yongtang
Sidorowicz, Elzbieta
Siebertz, Sebastian
Simoes, Jefferson
Souza, Uéverton
Spoerhase, Joachim
Sritharan, R.
Strash, Darren
Strehler, Martin
Strømme, Torstein
Szigeti, Zoltan
Todinca, Ioan
Torres, Luis M.
Ulmer, Arthur
van Dijk, Thomas C.
van Leeuwen, Erik Jan
van Rooij, Johan M. M.
Vaxès, Yann
Viennot, Laurent
Watrigant, Rémi
Weller, Mathias
Wollan, Paul
Wood, David R.
Wrochna, Marcin
Yen, Hsu-Chun
Zeman, Peter
Zey, Bernd

Local Organization

Local organizers of WG 2018 were the Lehrstuhl für Diskrete Mathematik und Grundlagen der Informatik and the Lehrstuhl Theoretische Informatik of the Brandenburg University of Technology, Cottbus, Germany. The conference took place at the Schlosshotel Lübbenau.

Members of the Organizing Committee were Jesse Beisegel, Carolin Denkert, Romain Gengler, Diana Hübner, Uwe Jähnert, Karla Kersten, Ekkehard Köhler (co-chair), Klaus Meer (co-chair), Robert Scheffler, and Martin Strehler.

Contents

On Dispersable Book Embeddings

Jawaherul Md. Alam[1], Michael A. Bekos[2(✉)], Martin Gronemann[3],
Michael Kaufmann[2], and Sergey Pupyrev[1]

[1] Department of Computer Science, University of Arizona, Tucson, USA
jawaherul@gmail.com, spupyrev@gmail.com
[2] Institut für Informatik, Universität Tübingen, Tübingen, Germany
{bekos,mk}@informatik.uni-tuebingen.de
[3] Institut für Informatik, Universität zu Köln, Köln, Germany
gronemann@informatik.uni-koeln.de

Abstract. In a *dispersable book embedding*, the vertices of a graph G
are ordered along a line ℓ, called *spine*, and the edges of G are drawn
at different half-planes bounded by ℓ, called *pages*, such that: (i) no two
edges of the same page cross, and (ii) no two edges of the same page
share a common endvertex. The minimum number of pages needed in
a dispersable book embedding of G is called its *dispersable book thickness*, $dbt(G)$. Graph G is called *dispersable* if $dbt(G)$ equals the maximum
degree of G, $\Delta(G)$ (note that $dbt(G) \geq \Delta(G)$ always holds).

Back in 1979, Bernhart and Kainen conjectured that every k-regular
bipartite graph G is dispersable and showed that it holds for $k \in \{1,2\}$.
In this paper, we disprove the conjecture for the cases $k = 3$ (with a
computer-aided proof), and $k = 4$ (with a purely combinatorial proof).
In particular, we show that the bipartite 3-regular *Gray* graph has dispersable book thickness four, while the bipartite 4-regular *Folkman* graph
has dispersable book thickness five. On the positive side, we show that
every 3-connected 3-regular bipartite planar graph is dispersable.

1 Introduction

The book embedding problem is a well studied problem in graph theory due
to its numerous applications with early results dating back to early 1970s; see
e.g., [12]. The input in this problem is a graph G and the task is to find a *linear
order* of the vertices of G along a line ℓ, called the *spine of the book*, and an
assignment of the edges of G to different half-planes, called *pages of the book*,
delimited by the spine, such that no two edges of the same page cross; see Fig. 1a
for an illustration. The minimum number of pages that are required by any book
embedding of G is commonly referred to as its *book thickness* (but also as *stack
number* or *page number*) and is denoted by $bt(G)$.

For planar input graphs, the literature is really rich. The most notable result
is due to Yannakakis [26], who proved that the book thickness of a planar graph
is at most four. Better upper bounds are only known for restricted subclasses,
such as planar 3-trees [16] (which fit in books with three pages), subgraphs of

© Springer Nature Switzerland AG 2018
A. Brandstädt et al. (Eds.): WG 2018, LNCS 11159, pp. 1–14, 2018.
https://doi.org/10.1007/978-3-030-00256-5_1

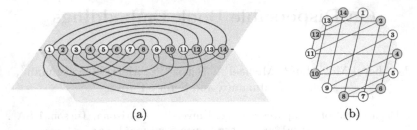

Fig. 1. (a) A dispersable book embedding with 3 pages of the 3-regular and bipartite Heawood graph [15], and (b) an equivalent circular embedding with a 3-edge-coloring, in which no two edges of the same color cross. (Color figure online)

planar Hamiltonian graphs [5], 4-connected planar graphs [22], planar graphs without separating triangles [19], Halin graphs [10], bipartite planar graphs [14], planar 2-trees [9], planar graphs of maximum degree 4 [3] (which fit in books with two pages), and outerplanar graphs [5] (which fit in single-page books). Note that, in general, the problem of testing, whether a maximal planar graph has book thickness two, is equivalent to determining whether it is Hamiltonian, and thus is NP-complete [25].

For non-planar graphs, the literature is significantly limited. It is known that the book thickness of a complete n-vertex graph is $\Theta(n)$ [5], while all graphs with subquadratic number of edges [21], subquadratic genus [20] or sublinear treewidth [11] have sublinear book thickness. The book thickness is known to be bounded, e.g., for bounded genus graphs [20] and, more generally, all minor-closed graph families [6]. The reader is referred to [12] for a survey.

In this paper, we focus on *dispersable book embeddings* [5], in which additionally no two edges of the same page share a common endvertex. The *dispersable book thickness* of a graph G, denoted by $dbt(G)$, is defined analogously to the book thickness as the minimum number of pages required by any dispersable book embedding of G. By definition $dbt(G) \geq \Delta(G)$ holds, where $\Delta(G)$ is the maximum degree of G. Finally, a graph G is called *dispersable* if and only if $dbt(G) = \Delta(G)$; see Fig. 1a. Note that any book embedding with k pages can be equivalently transformed into a circular embedding with a k-edge-coloring, in which no two edges of the same color cross, and vice versa [5,16]. In the dispersable case, the graphs induced by the edges of the same color must additionally be 1-regular; see Fig. 1b. The order in which the vertices appear in a circular embedding with $\Delta(G)$ colors (or, equivalently on the spine of a dispersable book embedding with $\Delta(G)$ pages), if any, is called *dispersable order*.

Dispersable book embeddings were first studied by Bernhart and Kainen [5], who back in 1979 proved that the book thickness of the graph formed by the Cartesian product of a dispersable bipartite graph B and an arbitrary graph H is upper bounded by the degree of B plus the book thickness of H (that is, $bt(B \times H) \leq bt(H) + \Delta(B)$), and posed the following conjecture (see also [18]):

Conjecture 1 (Bernhart and Kainen [5]). Every k-regular bipartite graph G is dispersable, that is, $dbt(G) = k$.

It is easy to verify that the conjecture holds for $k \leq 2$. As every k-regular bipartite graph admits a proper k-edge-coloring, Conjecture 1 implies that the dispersable book thickness of a regular bipartite graph equals its chromatic index. Overbay [23], who continued the study of dispersable embeddings in her Ph.D. thesis, observed that not every proper k-edge coloring yields a dispersable book embedding, and that bipartiteness is a necessary condition in the conjecture. She also proved that several classes of graphs are dispersable; among them are trees, binary cube graphs, and complete graphs.

Our contribution: In Sect. 2, we disprove Conjecture 1 for $k = 4$, by showing, with a purely combinatorial proof, that the Folkman graph [13], which is 4-regular and bipartite, has dispersable book thickness five. In Sect. 3, we first show how one can appropriately adjust a relatively recent SAT-formulation of the book embedding problem [4] for the dispersable case, and, using this formulation, we demonstrate that the 3-regular bipartite Gray graph [7] has dispersable book thickness four, thus, disproving Conjecture 1 for $k = 3$. Note that since both graphs are not planar, their (non-dispersable) book thickness is at least three. In [1], we demonstrate that it is exactly three. In Sect. 4, we show that 3-connected 3-regular bipartite planar graphs are dispersable. Our findings lead to a number of interesting research directions, which we list in Sect. 5, where we also conjecture that every (i.e., not necessarily 3-connected) 3-regular planar bipartite graph is dispersable.

2 The Dispersable Book Thickness of the Folkman Graph

The Folkman graph [13] can be constructed in two steps starting from K_5 as follows. First, we replace every edge by a path of length two to obtain a bipartite graph (see Fig. 2b). Then, we add for every vertex of the original K_5 a copy with the same neighborhood (see Fig. 2c). The resulting graph is the Folkman graph, which is clearly 4-regular and bipartite. We refer to a vertex of the original K_5 and to its copy as *twin* vertices. The remaining vertices of the Folkman, i.e., the ones obtained from the paths, are referred to as *connector vertices*. We denote the five pairs of twin vertices by A_1, A_2, B_1, B_2, C_1, C_2, D_1, D_2, E_1, E_2, and the ten connector vertices by ab, ac, ad, ae, bc, bd, be, cd, ce, de; see Fig. 2c.

To prove that the dispersable book thickness of the Folkman graph is five, it suffices to prove that its dispersable book thickness cannot be four, and that it admits a dispersable book embedding with five pages. For the latter, refer to [1]. For the former, we will assume for a contradiction that the Folkman graph admits a circular embedding with a 4-edge-coloring, in which (i) no two edges of the same color cross, and (ii) the graphs induced by the edges of the same color are 1-regular. Since by Property (ii) adjacent edges must have different colors, we name them "crossing" such that we can use Property (i) also for them. In the drawings, we use red, green, blue, and orange to indicate the four colors of the

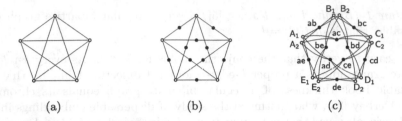

Fig. 2. Construction steps for the Folkman graph [13].

edges; black is used for an unknown (or not yet specified) color. For any subset of at least three twin or connector vertices of the Folkman graph, say A_1, ab and B_2, we denote the clockwise order in which they appear along the boundary of the circular embedding by $(\ldots A_1 \ldots ab \ldots B_2 \ldots)$. Every two vertices, say ab and A_1, form two intervals, $[ab, A_1]$ and $[A_1, ab]$, in the clockwise order that correspond to the two arcs on the circle.

Useful Lemmas: In the following, we investigate properties of a dispersable book embedding with four pages of the Folkman graph. We start with a property that was first observed by Overbay [23] and latter reproved by Hoske [17].

Lemma 1 (Overbay [23]). *For any regular bipartite graph, the vertices from both partitions are alternating in a dispersable order.*

For the Folkman graph, Lemma 1 implies alternating twin and connector vertices in a dispersable order. In the following, we adjust this implication.

Lemma 2. *Let A_1, A_2 be a pair of twins and $[A_1, A_2]$, $[A_2, A_1]$ be the two intervals defined by them in a dispersable order. Then one of the following holds:*

- *one of the intervals contains exactly one connector vertex corresponding to the twins, and another one contains all other connectors, that is, the order is* $(\ldots A_1 \ ax \ A_2 \ldots ay \ldots au \ldots av \ldots)$;
- *both intervals contain two connectors corresponding to the twins (and possibly other connectors), that is, the order is* $(\ldots A_1 \ldots ax \ldots ay \ldots A_2 \ldots au \ldots av \ldots)$.

Proof Sketch. Assuming that there is an interval with three or four of A's connectors, or with only one A's connector and some other connectors, we easily see that there exist five pairwise crossing edges; a contradiction. □

Denote the number of A's connectors in $[x, y]$ by $\delta_A(x, y)$. Lemma 2 defines two possible configurations for a pair of twins, A_1 and A_2. The first one, which we call *1–3 configuration*, is when $\delta_A(A_1, A_2) = 1$ and $\delta_A(A_2, A_1) = 3$, that is, the first interval contains one connector and another interval contains three connectors. In that case, the twins have to lie next to each other in the order (that is, there are no other twins in between); we call such twins *close*. In the second

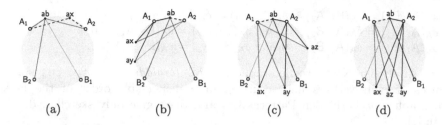

Fig. 3. Illustration for the proof of Lemma 3. (Color figure online)

configuration, called *2–2 configuration*, $\delta_A(A_1, A_2) = \delta_A(A_2, A_1) = 2$ holds. Here, the twins are called *far* (as there is at least one other twin in between).

Lemmas 3 and 4 describe properties of two pairs of twins that either do not alternate along the spine (*non-crossing twin-pairs*) or that do (*crossing twin-pairs*). We prove Lemma 3 in detail. The proof of Lemma 4 is analogous; see [1].

Lemma 3. *Let A_1, A_2 and B_1, B_2 be two non-crossing twin-pairs, i.e., the order is $(\ldots A_1 \ldots A_2 \ldots B_1 \ldots B_2 \ldots)$. For connector ab, one of the following holds:*

- *i. ab is in $[A_2, B_1]$, that is, $(\ldots A_1 \ldots A_2 \ldots ab \ldots B_1 \ldots B_2 \ldots)$;*
- *ii. ab is in $[B_2, A_1]$, that is, $(\ldots A_1 \ldots A_2 \ldots B_1 \ldots B_2 \ldots ab \ldots)$;*
- *iii. A_1 and A_2 are close twin vertices and the four twins are separated by A's connectors, that is, the order is $(\ldots A_1\ ab\ A_2 \ldots ax \ldots B_1 \ldots ay \ldots B_2 \ldots az \ldots)$;*
- *iv. B_1 and B_2 are close twin vertices and the four twins are separated by B's connectors, that is, the order is $(\ldots B_1\ ab\ B_2 \ldots bx \ldots A_1 \ldots by \ldots A_2 \ldots bz \ldots)$.*

Proof. If $ab \in [A_2, B_1] \cup [B_2, A_1]$, then the lemma holds. Let w.l.o.g. $ab \in [A_1, A_2]$. If A_1 and A_2 were far, then by Lemma 2 there would be a connector of A, say ax, in $[A_1, A_2]$. It follows that one of (A_1, ax), (A_2, ax) cannot be colored; see Fig. 3a. Thus, ab is the only connector of A in $[A_1, A_2]$. Hence, A_1 and A_2 are close twins.

Twins B_1 and B_2 define three sub-intervals on $[A_2, A_1]$. If two of A's connectors, say ax and ay, belong to the left- or rightmost one, say w.l.o.g. the former, then (A_2, ab), (A_2, ax), (A_2, ay), (B_1, ab) and (B_2, ab) pairwise cross (see Fig. 3b); a contradiction. Finally, if two of A's connectors, say ax and ay, are on the central sub-interval, then by symmetry we may assume that the fourth of A's connectors, say az, belongs either to $[A_2, B_1]$ or to $[B_1, B_2]$. In both cases, edge (B_2, ab) crosses (A_1, ax), (A_1, ay) and (A_1, az), which implies that all must have different colors; see Figs. 3c and d. Thus, (A_1, ab) needs a fifth color; a contradiction. We conclude that each of the intervals contain one connector. □

Corollary 1. *Let the order be $(\ldots A_1\ ab\ A_2 \ldots)$. Then B_1 and B_2 are not close.*

Lemma 4. *Let A_1, A_2 and B_1, B_2 be two crossing twin-pairs, i.e., the order is $(\ldots A_1 \ldots B_1 \ldots A_2 \ldots B_2 \ldots)$. Then one of the following holds:*

i. $\delta_A(A_1, B_1) = \delta_A(B_1, A_2) = \delta_A(A_2, B_2) = \delta_A(B_2, A_1) = 1;$
ii. $\delta_A(A_1, B_1) = \delta_A(A_2, B_2) = 2$ *and* $\delta_A(B_1, A_2) = \delta_A(B_2, A_1) = 0;$
iii. $\delta_A(A_1, B_1) = \delta_A(A_2, B_2) = 0$ *and* $\delta_A(B_1, A_2) = \delta_A(B_2, A_1) = 2.$

Case Analysis: Next, we determine several *forbidden patterns*, i.e., subsequences of twin vertices not occurring in a dispersable order of the Folkman graph. For Forbidden Patterns 3, 4 and 8 we give only sketches; details are in [1].

Forbidden Pattern 1 $(\ldots A_1 \cdot B_1 \cdot A_2 \ldots)$. *Between any twin pair, there is not exactly one single twin vertex.*

Proof. For a contradiction, let B_1 be the only twin in $[A_1, A_2]$. By Lemma 2, there are two of A's connectors in $[A_1, A_2]$, say ax and ay, and two A's connectors in $[A_2, A_1]$, say au and av; see Fig. 4a. The edges from B_1 to the three connectors different than ab cross both (A_1, ay) and (A_2, ax); a contradiction.   □

Forbidden Pattern 2 $(\ldots A_1 \cdot B_1 \cdot B_2 \cdot A_2 \ldots)$. *Between any twin pair, there are not exactly two same twin vertices, e.g., B_1 and B_2.*

Proof. For a contradiction, let B_1, B_2 be the only twins in $[A_1, A_2]$. By Lemma 1, we assume that the order is $(\ldots A_1 \, x \, B_1 \, y \, B_2 \, z \, A_2 \ldots)$, where x, y, z are connectors. By Lemma 2, two of them are connectors of A, including ab (by Lemma 3). If ab were y, then by Lemma 3.iv both x and z would be B's connectors, a contradiction since two of x, y, z are connectors of A. Hence, ab is not y, and so there exist two B's connectors, say bu and bv, in $[A_2, A_1]$; see Fig. 4b. But now (B_1, bv), (B_2, bu), (B_2, bv), (A_1, z), and (A_2, x) pairwise cross, a contradiction; see Fig. 4c.   □

Forbidden Pattern 3 $(\ldots A_1 \cdot B_1 \cdot C_1 \cdot A_2 \ldots)$. *Between any twin pair, there are not exactly two different twin vertices, e.g., B_1 and C_1.*

Proof Sketch. For a contradiction, let B_1, C_1 be the only two twins in $[A_1, A_2]$. By Lemma 2, A_1 and A_2 have two connectors in each of $[A_1, A_2]$ and $[A_2, A_1]$. By Lemma 3, ad, ae $\in [A_2, A_1]$, and thus ab, ac $\in [A_1, A_2]$; w.l.o.g. let ae be before ad in $[A_2, A_1]$. By Lemma 4, $\delta_B(A_1, A_2) = \delta_C(A_1, A_2) = 2$ holds. By symmetry, two cases bc $\in [A_1, B_1]$ and bc $\in [B_1, C_1]$ exist, which we lead to contradiction.   □

Forbidden Pattern 4 $(\ldots A_1 \cdot B_1 \ldots B_2 \cdot A_2 \ldots)$. *It is not possible to have a non-crossing pair of adjacent twins.*

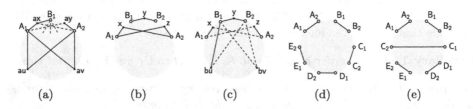

(a)   (b)   (c)   (d)   (e)

Fig. 4. Illustrations for (a–c) Forbidden Patterns 1 and 2, and (d–e) Theorem 1.

Proof Sketch. If A_1, A_2 and B_1, B_2 were non-crossing adjacent twins, then by Forbidden Pattern 2 neither A_1, A_2 nor B_1, B_2 are close. By Lemma 3, we assume w.l.o.g. $ab \in [B_2, A_2]$. By Lemma 1, there is a connector in $[A_1, B_1]$, which can be adjacent to one of A_1 or B_1 or not. A contradiction is obtained in each case. $\square$

Forbidden Pattern 5 $(\ldots A_1 \cdot B_1 \cdot C_1 \ldots A_2 \cdot B_2 \cdot C_2 \ldots)$. *It is not possible to have a* crossing triple, *i.e., a triple of consecutive twins that pairwise cross.*

Proof. For a contradiction, let the order be $(\ldots A_1 \times B_1 \text{ y } C_1 \ldots A_2 \text{ u } B_2 \text{ v } C_2 \ldots)$, where x, y, u, v are intermediate connectors. Since A_1, A_2, B_1, B_2, and C_1, C_2 form three pairs of crossing twins, by Lemma 4, the number of B's connectors in $[A_1, C_1]$ equals the number of B's connectors in $[A_2, C_2]$, which implies that four, two, or zero out of x, y, u, v are B's connectors. We refer to the first two cases as *non-zero crossing triple*, while to the third as *zero crossing triple*.

(i) B *has four connectors among* x, y, u, v. By symmetry, we may assume $x = ab$. This implies that there is no connector of A in $[A_2, C_2]$. Lemma 4, however, applied on A and C implies that there must exist an A's connector in $[A_2, C_2]$.

(ii) B *has two connectors among* x, y, u, v. Assume w.l.o.g. that x is a connector of B. By Lemma 4.i, u is also a connector of B. We prove by contradiction that $x \notin \{ab, bc\}$. Assume first that $x = bc$ and let w.l.o.g. the color of (C_1, bc) be blue. Since (C_1, bc) cannot be crossed by another blue edge, the edge (B_1, y) exists and is blue contradicting the fact that B has two connectors among x, y, u, v. Assume now that $x = ab$. Since $\delta_A(A_1, B_1) = 1$, by Lemma 4.i $\delta_A(A_2, B_2) = 1$. So, $u \in [A_2, B_2]$ is a connector of A. Since $ab \in [A_1, B_1]$ and u is a connector of B, we have a contradiction. It follows that either $x = bd$ or $x = be$ holds. By symmetry, either $u = bd$ or $u = be$ holds. By Lemma 2, there is a connector of B in each of $[C_1, A_2]$ and $[C_2, A_1]$. W.l.o.g. assume $ab \in [C_1, A_2]$ and $bc \in [C_2, A_1]$. Hence, (B_1, bc), (C_1, bc), (A_1, ab), (B_1, ab), and (u, B_1) pairwise cross; a contradiction.

(iii) B *has zero connectors among* x, y, u, v. By (i) and (ii), we may assume that no non-zero crossing triple exists. By Lemma 4, two connectors of B exist in each of $[C_1, A_2]$ and $[C_2, A_1]$. Note that x is not a connector of C, as otherwise the four edges incident to B_1 would cross (C_1, x). By symmetry, u is not a connector of C, and y and v are not connectors of A. Also, $ac \notin [A_1, C_1] \cup [A_2, C_2]$.

Let $\delta(C_1, A_2)$ and $\delta(C_2, A_1)$ be the number of twin vertices in $[C_1, A_2]$ and $[C_2, A_1]$. Clearly, $\delta(C_1, A_2) + \delta(C_2, A_1) \leq 4$ holds. Since there exist two B's connectors in each of $[C_1, A_2]$ and $[C_2, A_1]$, there exist at least one twin vertex in each of $[C_1, A_2]$ and $[C_2, A_1]$. Thus, $\delta(C_1, A_2), \delta(C_2, A_1) \geq 1$. Assume w.l.o.g. that D_1 is encountered first in $[C_1, A_2]$. The first twin vertex in $[C_2, A_1]$ cannot be D_2, as otherwise B_1, C_1, D_1, and B_2, C_2, D_2 would form a non-zero crossing triple containing C's connectors. By symmetry, let E_1 be the first twin vertex in $[C_2, A_1]$.

We claim that $\delta(C_1, A_2), \delta(C_2, A_1) \leq 2$. For a contradiction, let $\delta(C_1, A_2) = 3$ (the case $\delta(C_2, A_1) = 3$ is symmetric). Then, $[C_1, A_2]$ contains D_1, D_2, E_2. If D_2 precedes E_2 in $[C_1, A_2]$, then E_1, A_1, B_1, and E_2, A_2, B_2 form a non-zero crossing triple containing connectors of A. Otherwise, D_2 follows E_2 and thus D_1, E_2, D_2 form Forbidden Pattern 1. Hence, our claim holds.

Since $D_1 \in [C_1, A_2]$, $E_1 \in [C_2, A_1]$ and $\delta(C_1, A_2) \leq 2$, either $D_1, E_2 \in [C_1, A_2]$ or $D_1, D_2 \in [C_1, A_2]$ holds. In the former case, D and E form Forbidden Pattern 4. In the latter case, the order is $(A_1 \cdot B_1 \cdot C_1 \cdot D_1 \cdot D_2 \cdot A_2 \cdot B_2 \cdot C_2 \cdot E_1 \cdot E_2 \cdot)$. Recall that $ac \notin [A_1, C_1] \cup [A_2, C_2]$. By Lemma 2, $ac \notin [D_1, D_2]$ and $ac \notin [E_1, E_2]$. So, ac belongs to one of $[C_1, D_1]$, $[D_2, A_2]$, $[C_2, E_1]$, $[E_2, A_1]$. In the first case, (A_1, ac) is crossed by the four edges out of B_1; a contradiction. The other cases are similar. □

Forbidden Pattern 6 $(\ldots A_1 \cdot B_1 \ldots A_2 \cdot B_2 \ldots)$. *It is not possible to have a crossing pair of adjacent twins.*

Proof. For a contradiction, let A_1, A_2 and B_1, B_2 be a crossing pair of adjacent twins. Since by Forbidden Patterns 1 and 3 there are at least two twin vertices in each of $[B_1, A_2]$ and $[B_2, A_1]$, we may assume that the order is $(\ldots X \cdot A_1 \cdot B_1 \cdot Y \ldots U \cdot A_2 \cdot B_2 \cdot V \ldots)$. By Forbidden Pattern 3, X and Y are not twins; same with U and V. Let w.l.o.g. $X = D_1$, $Y = C_1$. We may also assume $U \neq E_1, E_2$. Thus, $U \in \{C_2, D_2\}$. If $U = D_2$, then D, A, B form Forbidden Pattern 5. Hence, $U = C_2$.

Since the remaining twins are D_2, E_1, and E_2, and since one of these is V, there are either zero, or one, or two twin vertices in $[C_1, C_2]$. One yields Forbidden Pattern 1, while two yield Forbidden Pattern 2 or 3. So, C_1 and C_2 are close.

It follows that D_2, E_1, E_2 are all in $[B_2, D_1]$; hence, their relative order is: $(E_1 \cdot D_2 \cdot E_2)$, or $(E_2 \cdot D_2 \cdot E_1)$, or $(E_1 \cdot E_2 \cdot D_2)$, or $(E_2 \cdot E_1 \cdot D_2)$, or $(D_2 \cdot E_1 \cdot E_2)$, or $(D_2 \cdot E_2 \cdot E_1)$. The first four yield Forbidden Patterns 1 or 2. By the symmetry of the last two cases, we may assume that the order is $(A_1 \; x \; B_1 \; y \; C_1 \; z \; C_2 \; u \; A_2 \; v \; B_2 \cdot E_1 \cdot E_2 \cdot D_2 \cdot D_1 \cdot)$, where x, y, z, u, v are intermediate connectors.

Since C_1, C_2, D_1, D_2 and E_1, E_2 are close, by Corollary 1, $z \in [C_1, C_2]$ is neither cd nor ce. By Lemma 2, z is either ac or bc. By symmetry, we may assume $z = ac$. Since $ac \in [C_1, C_2]$, by Lemma 3.iii and iv applied for C and A, there is a C's connector in each of $[C_1, C_2]$, $[C_2, A_2]$, $[A_2, A_1]$ and $[A_1, C_1]$. Thus, $u \in [C_2, A_2]$ and $y \in [B_1, C_1]$ are C's connectors. By Lemma 2, there are two A's connectors in $[A_1, A_2]$; thus, $x \in [A_1, B_1]$ is a connector of A^1. By symmetry, v is a B's connector. By Lemma 4, $x \in [A_1, B_1]$ is also a connector of B, which implies that $x = ab$ (recall that x is a connector of A). Since $x \in [A_1, B_1]$ is a connector of A, again by Lemma 4, $v \in [A_2, B_2]$ must be a connector of A. Since we have shown that v is a connector of B, $v = ab$; a contradiction as $ab \in [A_1, B_1]$. □

Forbidden Pattern 7 $(\ldots A_1 \cdot B_1 \cdot C_1 \cdot D_1 \cdot A_2 \ldots)$. *Between any twin pair, it is impossible to have exactly three pairwise different twins.*

Proof. For a contradiction, let the order be $(\ldots X \cdot A_1 \cdot B_1 \cdot C_1 \cdot D_1 \cdot A_2 \cdot Y \ldots)$, where X and Y are the twins preceding A_1 and following A_2. If $X = B_2$, then A and B form Forbidden Pattern 1; if $X = C_2$, then A and C form Forbidden

---

[1] $[A_1, A_2]$ is the union of $[A_1, B_1]$, $[B_1, C_1]$, $[C_1, C_2]$, $[C_2, A_2]$. As in the last three there are C's connectors including $ac \in [C_1, C_2]$, the second A's connector can be only in $[A_1, B_1]$.

Pattern 3; if $X = D_2$, then D and A form Forbidden Pattern 6. Thus, $X = E_1$. By symmetry, $Y = E_2$ holds. But then A and E form Forbidden Pattern 4.   □

Forbidden Pattern 8 $(\ldots A_1 \cdot B_1 \cdot C_1 \cdot C_2 \cdot A_2 \ldots)$. *Between any twin pair, it is impossible to have exactly three twins, such that two of them are same.*

Proof Sketch. Assume that only $B_1, C_1, C_2 \in [A_1, A_2]$. By Forbidden Pattern 1, C_1, C_2 are close. So, the order is $(A_1 \cdot B_1 \cdot C_1 \cdot C_2 \cdot A_2 \cdot U \cdot V \cdot X \cdot Y \cdot Z \cdot)$, where U, V, X, Y, Z are the remaining twins. If $B_2 = Z$, then A and B form Forbidden Pattern 1; if $B_2 = Y$, then A and B form Forbidden Pattern 3; if $B_2 = U$, then A and B form Forbidden Pattern 6. We lead the cases $B_2 = V$ and $B_2 = X$ to contradiction.   □

Theorem 1. *The dispersable book thickness of the Folkman graph is five.*

Proof. We argue that the dispersable book thickness of the Folkman graph is not four. Let $d[A_1, A_2]$ be the number of twin vertices in $[A_1, A_2]$ including A_1 and A_2, and let $d(A) = \min(d[A_1, A_2], d[A_2, A_1])$. By Forbidden Pattern 1, $d(A) \neq 3$; by Forbidden Patterns 2 and 3, $d(A) \neq 4$; by Forbidden Patterns 7 and 8, $d(A) \neq 5$. As $d(A) \in \{2, 6\}$, any pair of twins is close or opposite in a dispersable order.

We argue that at most one pair of twins is opposite. Indeed, let A_1, A_2 be one opposite twin pair. By Forbidden Pattern 6, the twin vertices next to A_1 and A_2 cannot be opposite. This directly implies that all remaining twin pairs are close.

Figures 4d and e shows the remaining two cases, in which no or one pair of twins is opposite. In the former case, by Lemma 2 there is an A's connector, say w.l.o.g. ab, in $[A_1, A_2]$. Then, by Corollary 1, twins B_1, B_2 must be far; a contradiction. For the latter case, let C_1, C_2 be the pair of opposite twins. By Lemma 2 and Corollary 1, the connectors between the four close pairs can only be C's connectors. Hence, the order is $(C_2 \, x \, A_1 \, ac \, A_2 \, y \, B_1 \, bc \, B_2 \, z \, C_1 \, u \, D_1 \, cd \, D_2 \, v \, E_1 \, ce \, E_2 \, w)$, where x, y, z, u, v, w are the remaining connectors.

Applying Lemma 3 on A and C, we conclude that there exist A's connectors in both $[A_2, C_1]$ and $[C_2, A_1]$. Thus, x is a connector of A. Similarly, we conclude that z is B's connector. Next observe that $z \neq ab$, as otherwise five edges, (ab, A_1), (ab, A_2), (ab, B_1), (bc, C_2), (bc, C_1), pairwise cross. Hence, $y = ab$. Symmetrically, u is D's connector, w is E's connector, and $v = de$. So, bd is either z or u. Both cases are impossible, as (D_1, bd) or (B_2, bd) would cross four C_1's edges.   □

Corollary 2. *The Folkman graph is not dispersable.*

3   The Dispersable Book Thickness of the Gray Graph

In this section, we study the dispersable book thickness of the Gray graph [7], which can be constructed in two steps starting from three copies of $K_{3,3}$. First, we subdivide every edge. Then, for each newly introduced vertex u in the first copy, with v and w being its counterparts in the other two copies, we add a new vertex connected to u, v and w. The resulting graph is the Gray graph, which is clearly 3-regular and bipartite; for an illustration refer to [1].

Our computer-aided proof is based on appropriately adjusting a relatively recent formulation of the (non-dispersable) book embedding problem as a SAT instance by Bekos et al. [4]. In their formulation, Bekos et al. use three different variables, denoted by σ, ϕ and χ, with the following meanings: (i) for a pair of vertices u and v, variable $\sigma(u, v)$ is true, if and only if u is to the left of v along the spine, (ii) for an edge e and a page i, variable $\phi_i(e)$ is true, if and only if edge e is assigned to page i of the book, and (iii) for a pair of edges e and e', variable $\chi(e, e')$ is true, if and only if e and e' are assigned to the same page. Hence, there exist in total $O(n^2 + m^2 + pm)$ variables, where n denotes the number of vertices of the graph, m its number of edges, and p the number of available pages. A set of $O(n^3 + m^2)$ clauses ensure that the underlying order is linear, and that no two edges of the same page cross; for details see [4].

For the dispersable case, we must additionally guarantee that no two adjacent edges are on the same page. This requirement can be easily encoded by introducing for every pair of edges e, e' with a common endvertex the clause $\neg\chi(e, e')$. Observe that there is no need to introduce new variables, and that the total number of constraints is not asymptotically affected. Using this adjustment, we proved that the dispersable book thickness of the Gray graph cannot be three, and that it admits a dispersable book embedding with four pages (see [1]). We summarize these findings in the following theorem.

Theorem 2. *The Gray graph is not dispersable. Its dispersable book thickness is four.*

4   3-Connected 3-Regular Bipartite Planar Graphs

The Gray graph, which is 3-connected, 3-regular and bipartite, is not dispersable. Since it contains $K_{3,3}$ as minor, it is not planar. We next prove that when adding planarity to the requirements, every such graph is dispersable. We refer to a 3-connected 3-regular bipartite planar graph as *Barnette graph* for short (due to Barnette's Conjecture [2] which states that every such graph is Hamiltonian).

Lemma 5. *Let $G = (V, E)$ be a Barnette graph with its dual $G^* = (V^*, E^*)$. Then, there exists a 3-edge coloring $E_r \sqcup E_g \sqcup E_b = E$ for G, and a 3-vertex coloring $V_r^* \sqcup V_g^* \sqcup V_b^* = V^*$ for G^* so that: (i) Every facial cycle of G is bichromatic, i.e., the edges on a facial cycle of G alternate between two colors. (ii) Every face of G is colored differently from its bounding edges. (iii) The edges of G^* that connect vertices of V_g^* to vertices of V_b^* are in one-to-one correspondence with the edges of E_r, and induce a connected subgraph.*

Proof. Since G is 3-regular and bipartite, G^* is maximal planar and every vertex has even degree. By the 3-color theorem, G^* has chromatic number 3 [24], and since it is maximal planar, G^* is *uniquely 3-colorable* [8], i.e., it has a unique 3-vertex coloring up to permutation of the colors, say V_r^*, V_g^* and V_b^*.

We first show (ii). Every edge e of G bounds two faces that are colored differently in G^*. Hence, we can assign to e the third color. Since every vertex v

of G is incident to three faces (which are colored differently in G^*), no two edges of v have the same color. The result is a proper 3-edge coloring E_r, E_g, E_b of G. Now (i) follows from (ii): On every facial cycle f of G, two adjacent edges have distinct colors, which are different from the color of f in G^*. Thus, every face of G is bichromatic. Next we show (iii). By (ii), any edge of G^* that corresponds to an edge of E_r has one endpoint in V_g^* and one in V_b^*. Conversely, by construction every edge of G^* in the induced subgraph of $V_g^* \cup V_b^*$ corresponds to an edge in E_r of G. Hence, the edges of G^* that connect vertices of V_g^* to vertices of V_b^* are in one-to-one correspondence with the edges of E_r. Property (iii) follows from [8], where it is proved that for any k-vertex coloring of a uniquely k-colorable graph, the subgraph induced by any two of the k colors is connected.   □

We now show that it is possible to determine a dispersable order for a Barnette graph G such that the coloring of Lemma 5 for G is a valid page assignment. Our construction is based on determining a *subhamiltonian cycle* C of an auxiliary two-page book embedding of G, i.e., a cyclic order of the vertices so that when adding missing edges between consecutive vertices planarity is preserved.

Theorem 3. *Each Barnette graph $G = (V, E)$ has a 3-edge coloring $E_r \sqcup E_g \sqcup E_b = E$ and a subhamiltonian cycle C so that edges of (i) E_r are in the interior of C or on C, (ii) E_b are in the exterior of C or on C, (iii) E_g are on C.*

Proof Sketch. In the proof, we assume that $E_r \sqcup E_g \sqcup E_b$ is a 3-edge coloring of G, and that $V_r^* \sqcup V_g^* \sqcup V_b^*$ is a 3-vertex coloring of the dual $G^* = (V^*, E^*)$ of G satisfying the Properties i–iii of Lemma 5. By Lemma 5.iii, the subgraph G_{bg}^* of G^* induced by $V_g^* \cup V_b^*$ is connected. Hence, we can construct a spanning tree T^* of G_{bg}^*. This tree and the one-to-one correspondence between the edges of E_r and the edges of G_{bg}^* yield a partition of E_r into two sets T_r and N_r, such that $T_r \sqcup N_r = E_r$, as follows. An edge $e \in E_r$ belongs to T_r, if the edge of G^* corresponding to e belongs to T^*. Otherwise, e belongs to N_r. We also assume T^* to be rooted at a leaf ρ, such w.l.o.g. $\rho \in V_b^*$.

The proof is given by a recursive geometric construction of the subhamiltonian cycle C. Consider an arbitrary edge $(u, v) \in T_r$ of G, and let p and q be the faces to its left and its right side, respectively, as we move along (u, v) from u to v. Then, (p, q) is an edge of T^*. Since T^* is a tree, the removal of (p, q) results in two trees T_p^* and T_q^*. W.l.o.g. we assume that ρ belongs to T_p^*. For the recursive step of our algorithm, we assume that we have already computed a *simple* and *plane* cycle C_p for the subgraph $G_p = (V_p, E_p)$ of G induced by the vertices of the faces of G in T_p^*, which satisfies the following additional invariants: **(I.1)** edge (u, v) is on C_p, **(I.2)** every edge $e \in T_r \cap E_p$ is in the interior of C_p or on C_p, **(I.3)** every edge $e \in E_b \cap E_p$ is in the exterior of C_p or on C_p, **(I.4)** every edge $e \in E_g \cap E_p$ is on C_p, and **(I.5)** every edge $e \in N_r$ that bounds two faces h, h', with $h \in T_p^*$ and $h' \notin T_p^*$, is such that: (i) if $h \in V_b^*$, then both endpoints of e are on C_p, (ii) if $h \in V_g^*$, then none of the endpoints of e is on C_p.

Let $G_q = (V_q, E_q)$ be the subgraph of G induced by the vertices of the faces of G in T_q^*. Let also $q_1, \ldots, q_k$, with $k \geq 0$, be the children of q in T^* (if any). We proceed by considering two cases; $q \in V_b^*$ and $q \in V_g^*$; see Figs. 5a and b,

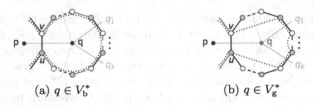

(a) $q \in V_b^*$ (b) $q \in V_g^*$

Fig. 5. The solid (dotted) gray edges belong to $\mathcal{T}^*$ ($G_{bg}^* \setminus \mathcal{T}^*$). The solid (dashed) red edges belong to T_r (N_r). Cycle C_q is drawn dotted black. (Color figure online)

respectively. Note that by Lemma 5.i and ii in the former case, the edges of q alternate between red and green, while in the latter case between red and blue.

Assume first that $q \in V_b^*$. We remove from C_p edge (u, v), which exists by **I.1** resulting in a path from u to v. Then, cycle C_q for $\mathcal{T}_p^* \cup \{q\}$ is obtained by this path and the path from u to v in face q. Assume now that $q \in V_g^*$. If q is a leaf in $\mathcal{T}^*$ (i.e., the only edge incident to q that belongs to T_r is edge (u, v)), then C_p is a (simple and plane) cycle also for $\mathcal{T}_p^* \cup \{q\}$, which clearly satisfies **I.1–I.5**. Let now q be w.l.o.g. not a leaf in $\mathcal{T}^*$. Thus, there exist edges of q, different from (u, v), that belong to T_r. Denote by $w_1, \ldots, w_\ell$ the endvertices of these edges as they appear in a clockwise traversal of q starting from u. We remove from C_p the edge (u, v), which exists by **I.1**. This results in a path from u to v. The cycle C_q that is obtained by this path and the path $u \to w_1 \to \ldots \to w_\ell \to v$ is a cycle for $\mathcal{T}_p^* \cup \{q\}$. We prove in [1] that in both cases **I.1–I.5** hold.

The base of our recursive algorithm corresponds to the face $\rho \in V_b^*$ that is the root of $\mathcal{T}^*$, where we choose C_ρ to be the facial cycle of ρ, which trivially satisfies **I.1–I.5**. Once we have traversed $\mathcal{T}^*$, we have computed a simple and plane cycle C, which by **I.2–I.4**, satisfies i–iii of our theorem. We show that C is a subhamiltonian cycle of G as follows. Since $\mathcal{T}^*$ is a spanning tree of G_{bg}^*, every green edge of G bounds a face that is in $\mathcal{T}^*$, and by **I.4** we may assume that both its endpoints are consecutive along C. As every vertex is incident to a green edge, it follows that C is a subhamiltonian cycle of G. $\square$

Assigning the green edges to a third page yields then the desired result.

Corollary 3. *Every Barnette graph is dispersable.*

5   Conclusions

There is a number of interesting questions raised by our work. Does there exist a non-dispersable bipartite graph for every $k \geq 5$? Is it possible to provide an upper bound on the dispersable book thickness of k-regular bipartite graphs (e.g., $k + 1$)? We conjecture that all (not necessarily 3-connected) 3-regular bipartite planar graphs are dispersable. Since Folkman and Gray graphs are not vertex-transitive, we ask whether all vertex-transitive regular bipartite graphs are dispersable.

Acknowledgment. Our work is partially supported by DFG grant KA812/18-1. We would like to thank Prof. Paul Kainen for bringing this problem to our attention. We also thank Jessica Wolz for discussions on experimental aspects.

References

1. Alam, J.M., Bekos, M.A., Gronemann, M., Kaufmann, M., Pupyrev, S.: On dispersable book embeddings. CoRR abs/1803.10030 (2018)
2. Barnette, D.W.: Conjecture 5. In: Tutte, W.T. (ed.) Recent Progress in Combinatorics, Proceedings of the Third Waterloo Conference on Combinatorics, pp. xiv+347. Academic Press, New York, London (1969)
3. Bekos, M.A., Gronemann, M., Raftopoulou, C.N.: Two-page book embeddings of 4-planar graphs. Algorithmica **75**(1), 158–185 (2016)
4. Bekos, M.A., Kaufmann, M., Zielke, C.: The book embedding problem from a SAT-Solving perspective. In: Di Giacomo, E., Lubiw, A. (eds.) GD 2015. LNCS, vol. 9411, pp. 125–138. Springer, Cham (2015)
5. Bernhart, F., Kainen, P.C.: The book thickness of a graph. J. Comb. Theory, Ser. B **27**(3), 320–331 (1979)
6. Blankenship, R.: Book embeddings of graphs. Ph.D. thesis, Louisiana State University (2003)
7. Bouwer, I.: On edge but not vertex transitive regular graphs. J. Comb. Theory, Ser. B **12**(1), 32–40 (1972)
8. Chartrand, G., Geller, D.P.: On uniquely colorable planar graphs. J. Comb. Theory **6**(3), 271–278 (1969)
9. Chung, F.R.K., Leighton, F.T., Rosenberg, A.L.: Embedding graphs in books: a layout problem with applications to VLSI design. SIAM J. Algebraic Discret. Methods **8**(1), 33–58 (1987)
10. Cornuéjols, G., Naddef, D., Pulleyblank, W.: Halin graphs and the travelling salesman problem. Math. Program. **26**(3), 287–294 (1983)
11. Dujmović, V., Wood, D.R.: Graph treewidth and geometric thickness parameters. Discret. Comput. Geom. **37**(4), 641–670 (2007)
12. Dujmović, V., Wood, D.R.: On linear layouts of graphs. Discret. Math. Theor. Comput. Sci. **6**(2), 339–358 (2004)
13. Folkman, J.: Regular line-symmetric graphs. J. Comb. Theory **3**(3), 215–232 (1967)
14. de Fraysseix, H., de Mendez, P.O., Pach, J.: A left-first search algorithm for planar graphs. Discret. Comput. Geom. **13**, 459–468 (1995)
15. Gerbracht, E.: Eleven unit distance embeddings of the Heawood graph. CoRR abs/0912.5395 (2009)
16. Heath, L.S.: Embedding planar graphs in seven pages. In: FOCS, pp. 74–83. IEEE Computer Society (1984)
17. Hoske, D.: Book embedding with fixed page assignments. Bachelor thesis, Karlsruhe Institute for Technology (2012)
18. Kainen, P.C.: Crossing-free matchings in regular outerplane drawings. In: Knots in Washington XXIX. George Washington Univ., Washington, DC, USA (2009). http://faculty.georgetown.edu/kainen/circLayouts.pdf
19. Kainen, P.C., Overbay, S.: Extension of a theorem of Whitney. Appl. Math. Lett. **20**(7), 835–837 (2007)
20. Malitz, S.: Genus g graphs have pagenumber $O(\sqrt{g})$. J. Algorithms **17**(1), 85–109 (1994)

21. Malitz, S.: Graphs with E edges have pagenumber $O(\sqrt{E})$. J. Algorithms **17**(1), 71–84 (1994)
22. Nishizeki, T., Chiba, N.: Planar Graphs: Theory and Algorithms. Elsevier, New York (1988)
23. Overbay, S.B.: Generalized book embeddings. Ph.D. thesis, Colorado State University (1998)
24. Steinberg, R.: The state of the three color problem. In: Gimbel, J., Kennedy, J.W., Quintas, L.V. (eds.) Quo Vadis, Graph Theory?. Elsevier, New York (1993). Ann. Discret. Math. **55**, 211–248
25. Wigderson, A.: The complexity of the Hamiltonian circuit problem for maximal planar graphs. Technical report TR-298, EECS Department, Princeton University (1982)
26. Yannakakis, M.: Embedding planar graphs in four pages. J. Comput. Syst. Sci. **38**(1), 36–67 (1989)

Characterising AT-free Graphs with BFS

Jesse Beisegel[✉]

Brandenburg University of Technology, Cottbus, Germany
beisegel@b-tu.de

Abstract. An asteroidal triple free graph is a graph such that for every independent triple of vertices no path between any two avoids the third. In a recent result from Corneil and Stacho, these graphs were characterised through a linear vertex ordering called an AT-free order. Here, we use techniques from abstract convex geometry to improve on this result by giving a vertex order characterisation with stronger structural properties and thus resolve an open question by Corneil and Stacho. These orderings are generated by a modification of BFS which runs in polynomial time. Furthermore, we give a linear time algorithm which employs multiple applications of (L)BFS to compute AT-free orders in claw-free AT-free graphs and a generalisation of these.

1   Introduction

In a classical paper of algorithmic graph theory by Lekkerkerker and Boland from the early 1960s [17] the authors used a forbidden substructure called an *asteroidal triple* to characterise interval graphs. An asteroidal triple is an independent triple of vertices, such that for any two of them there is a path that avoids the third. This definition gave rise to the introduction of the class of asteroidal triple free graphs (AT-free graphs) and due to the fact that these graphs form a superclass of both the interval and cocomparability graphs, there has been considerable research interest for the last two decades.

AT-free graphs are widely believed to exhibit a "linear structure" [14] akin to the interval graphs and two results in particular corroborate this claim: In [8] it was shown that every AT-free graph contains a *dominating pair*, i.e., a pair of vertices such that every path between them forms a dominating set for the whole graph. This result was strengthened in the same paper [8] which characterised AT-free graphs with the so-called *spine property*: A graph H has the spine property, if for every non-adjacent dominating pair s and t there exists a neighbour of t, say t', such that s and t' are a dominating pair in the connected component of $H - t$ that contains s. As shown in [8], a graph G is an asteroidal triple free graph if and only if every connected induced subgraph of G has the spine property. This can be seen as a generalisation of the fact that the maximal cliques of interval graphs form a chain.

An important algorithmic tool in the theory of interval graphs has been their characterising linear vertex ordering, the *interval order*. This is a linear

© Springer Nature Switzerland AG 2018
A. Brandstädt et al. (Eds.): WG 2018, LNCS 11159, pp. 15–26, 2018.
https://doi.org/10.1007/978-3-030-00256-5_2

ordering $\tau = (v_1, \ldots, v_n)$ of the vertices of a graph $G = (V, E)$ such that for $u \prec_\tau v \prec_\tau w$ and $uw \in E$ we have $uv \in E$. It was long conjectured that such a characterising linear vertex ordering must also exist for AT-free graphs and while in a recent result [11] this conjecture was answered in the positive, the notion of these orderings leaves quite a bit of freedom.

Ideally, such an ordering would somehow capture the structure given in the spine property in [8] (as it is in the case of interval orderings which immediately gives us the chain of maximal cliques). However, the so-called *LexComp* ordering that is constructed in [11] has one significant drawback: For some graphs the resulting ordering is "folded" in a way that seems to contradict our notion of linear behaviour. For example, given the path graph with $2n + 1$ vertices, the P_{2n+1}, where the vertices are numbered from left to right along the path, we would expect any viable linear vertex ordering to be $(1, 2, \ldots, 2n + 1)$ or its inversion. The algorithm in [11], on the other hand, might output $(n + 1, n, n + 2, n - 1, \ldots, 1, 2n + 1)$. In addition, this construction can even yield vertex orders $\tau := (v_1, \ldots, v_n)$ such that there are $i \in \{1, \ldots, n\}$ for which $G[v_1, \ldots, v_i]$ is not connected - for example the circuit in five vertices, i.e., C_5. More examples can be found in Fig. 1.

In an attempt to remedy this issue, the authors of [11] investigate whether it is possible to find AT-free orderings that coincide with search orders. After proving that there are graphs G such that no LBFS ordering of G is an AT-free order, they conjecture that every AT-free graph has an AT-free order that is a BFS order.

Conjecture 1. [11] Let $G = (V, E)$ be an AT-free graph. Then there exists a BFS ordering $\tau = (v_1, \ldots, v_n)$ that is an AT-free order.

We will prove an even stronger version of this conjecture, and show how such an order can be used to wed the notion of an AT-free ordering to the spine property. We will also give a polynomial time algorithm to compute such an order that takes approximately the same time as the previous best known algorithm to compute AT-free orders, i.e. $\mathcal{O}(nm)$ [11]. The best known algorithm to recognise AT-free graphs uses fast matrix multiplication and takes $\mathcal{O}(n^{2.82})$ time [16] and it can be shown that recognition of AT-free graphs is at least as hard as recognising graphs without an independent set of size three [20].

For the special case of claw-free AT-free graphs and a generalisation of these we give linear time algorithms to compute AT-free (L)BFS orders. This is a surprising result, as it was shown in [13] that the recognition of claw-free AT-free graphs is at least as hard as triangle recognition. This dichotomy is of striking resemblance to the case of comparability graphs, where a characterising linear ordering in the form of a transitive orientation can be found in linear time, while there is no known recognition algorithm that is faster than matrix multiplication [20]. Due to these facts, we conjecture that it is possible to compute AT-free orderings in linear time in the general case using some form of modified breadth-first-search. As is the case for comparability graphs, such a linear ordering might then be used for linear time optimisation algorithms that are robust for AT-free graphs, i.e. which can be applied without solving recognition first (for further information on robust algorithms see [20]).

Arbitrary AT-free: $(4, 5, 2, 7, 3, 6, 1, 8)$
LexComp: $(4, 5, 3, 6, 2, 7, 1, 8)$
$\text{BFS}^{\text{conv}}(G, 1)$: $(1, 2, 3, 4, 5, 6, 7, 8)$

Fig. 1. Graph with its various AT-free orders

2 Preliminaries

In the following, we will exclusively refer to simple connected graphs G with vertex set V and edge set E. The *neighbourhood* of v in G is the set $N_G(v) :=$ $\{w \; : \; vw \in E\}$ and $N[v] := N(v) + v$. A vertex with only one neighbour in G will be called a *pendant vertex*. A *walk* W of length k in G is a succession of vertices $(v_1, \ldots, v_{k+1})$ such that $v_i v_{i+1} \in E$ for all $i \in \{1, \ldots, k\}$. If a walk P has the additional property that all vertices are distinct, we call P a *path*. We say that a path P *avoids* a vertex v, if v does not have any neighbours on P, while a vertex v *intercepts* a path P if it has at least one neighbour on P.

The *distance* between two vertices s and t is the length of a shortest path between these vertices and will be denoted by $\text{dist}_G(s, t)$. The set of vertices that have distance k to a vertex s is called the *k-th distance layer* from s of G and is denoted by $L_G^k(s)$. For every vertex $v \in V$ we say that $N_s^k(v) := L_G^k(s) \cap N_G(v)$. A vertex x with largest distance from s is called *eccentric* with respect to s and its distance to s is the *eccentricity* $\text{ecc}_G(s)$ of s. The *eccentricity* of G is the largest such value among all vertices.

A subset $D \subseteq V$ is called a *dominating set* of G if every vertex in V has a neighbour in D. If the set D forms a path in G it is called a *dominating path*. Two vertices s and t of G form a *dominating pair*, if every path between them is dominating. A permutation $\tau := (v_1, \ldots, v_n)$ of the vertices of G will be called a *linear vertex ordering*.

Given a linear vertex ordering τ we can formulate a derivative of *Breadth First Search* called BFS+(τ). This algorithm is a breadth first search which prioritises vertices that are further to the right in τ, i.e. at any point of the search where neighbours of the current vertex are added to the queue, the vertices with highest τ-value are added first.

Lexicographic Breadth First Search (Algorithm 1) was introduced in [19] to recognise chordal graphs and has been an important ingredient in many recognition and optimisation algorithms since.

If two vertices have the same label in step 6, we say that they are *tied*. We call a set of tied vertices S encountered in step 6 of Algorithm 1 a *slice*. Given an LBFS order τ and two vertices u and v with $u \prec_\tau v$, we denote the vertex-minimal slice with respect to τ containing u and v as $\Gamma_{u,v}^\tau$.

Algorithm 1. LBFS

Input: Connected graph $G = (V, E)$ and a distinguished vertex $s \in V$
Output: A vertex ordering τ

1 **begin**
2 $\quad$ $label(s) \leftarrow n$;
3 $\quad$ **for** *each vertex* $v \in V - s$ **do**
4 $\quad\quad$ $\lfloor$ $label(v) \leftarrow \emptyset$;

5 $\quad$ **for** $i \leftarrow 1$ *to* n **do**
6 $\quad\quad$ pick an unnumbered vertex v with lexicographically largest label;
7 $\quad\quad$ $\tau(i) \leftarrow v$;
8 $\quad\quad$ **for** *each unnumbered vertex* $u \in N(v)$ **do**
9 $\quad\quad\quad$ $\lfloor$ append $(n - i)$ to $label(w)$;

As before with the BFS, given a linear vertex order τ, we can define an LBFS+(τ) in the following way: At any point in the search at which we encounter a slice, i.e. a set of tied vertices, the vertex of highest τ-value is chosen first.

There are many interesting properties and applications of LBFS, and some of these can be found in [7]. Here we will need one result in particular, which is a useful tool for the analysis of LBFS and LBFS+ orders.

Lemma 1 (Prior Path Lemma). [10] *Let τ be an arbitrary LBFS of a graph G and let $u, v \in V$ with $u \prec_\tau v$. Let w be the τ-first vertex of the connected component C_u of $\Gamma^\tau_{u,v}$ containing u. There exists a w-u-path in $\Gamma^\tau_{u,v}$ all of whose vertices, with the possible exception of u, are not adjacent to v. Moreover, all vertices on this path, other than u, occur before u in τ. Such a path is called a prior path.*

Finally, a graph will be called *claw-free*, if it does not contain a claw graph, i.e. the $K_{1,3}$, as an induced subgraph. We will call the three independent vertices the *prongs* and the fourth vertex the *base* of the claw.

3 Convex Geometries and AT-free Graphs

Definition 1. [12] *A set V and a family of subsets $\mathcal{C}$ of V form a* convexity space, *if $\emptyset, V \in \mathcal{C}$ and $\mathcal{C}$ is closed under intersection. The smallest convex set* $conv(X)$ *containing a set $X \subseteq V$ is called the* convex hull *of X. We say that a convexity space $(V, \mathcal{C})$ is a* convex geometry, *if for every convex set X and two points $p, q \in V \backslash X$:*

$$q \in conv(X + p) \Rightarrow p \notin conv(X + q).$$

This is sometimes referred to as the anti-exchange property. *A convex set whose complement is also convex is called a* halfspace.

The anti-exchange property motivates an ordering of the ground set V of a convex geometry: An ordering $\tau = (v_1, \ldots, v_n)$ is a *convexity ordering*, if $\{v_1, \ldots v_i\}$ is convex for every $i \in \{1, \ldots, n\}$. If $\{v_1, \ldots, v_i\}$ is a halfspace for every $i \in \{1, \ldots, n\}$, then we call τ a *halfspace ordering*.

One way to define a convexity space is through strict betweenness. Following [5] we say that a *strict betweenness* over a ground set V is a ternary relation $\mathcal{B} \subset V^3$ such that

$(a, b, c) \in \mathcal{B}$ implies that $(c, b, a) \in \mathcal{B}$ and a, b, and c are distinct.

The convexity space with regard to this betweenness is then defined to be the pair $(V, \mathcal{C}_\mathcal{B})$ where

$$\mathcal{C}_\mathcal{B} := \{C \subseteq V \ : \ \{a, c\} \subseteq C \text{ and } (a, b, c) \in \mathcal{B} \text{ implies } b \in C\}.$$

On graphs we can define just such a strict betweenness on the set of vertices and thus we can construct a convexity space in the following way:

Definition 2. *Given a graph $G = (V, E)$ we say that $(x, y, z) \in \mathcal{B}_D(G)$, if there is a chordless x-y-path that avoids z and a chordless y-z-path that avoids x. The set of vertices y with $(x, y, z) \in \mathcal{B}_D(G)$ is called the domination interval of x and z and is denoted by $I_D(x, z)$. The ternary relation $\mathcal{B}_D(G)$ is called the* domination betweenness *of G and it is easy to see that this is a strict betweenness. As a result, we obtain a convexity space $(V, \mathcal{C}_{\mathcal{B}_D}(G))$ which we will call the* domination convexity *of G.*

A vertex y is said to be *admissible*, if there are no two vertices x and z such that $(x, y, z) \in \mathcal{B}_D(G)$. An *AT-free ordering* is an ordering $\tau = (v_1, \ldots, v_n)$ of the vertices such that for any $(x, y, z) \in \mathcal{B}_D(G)$ we have $y \prec_\tau x$ or $y \prec_\tau z$. It is easy to see that for any such ordering $\{v_1, \ldots, v_i\}$ is domination convex for any $i \in \{1, \ldots, n\}$. If τ is such that for any $(x, y, z) \in \mathcal{B}_D(G)$ we have $x \prec_\tau y \prec_\tau z$ we say that it is a *bilateral AT-free ordering* of G.

The connection between convexity theory and AT-free graphs was recently made in [3,4] and it was furthermore shown that the convexity space thus defined is in fact a convex geometry. In the following we have bundled that result with a number of other characterising properties of AT-free graphs:

Theorem 1. [2–4,8,11,12,15] *Given a graph G, its domination betweenness $\mathcal{B}_D$ and its domination convexity $\mathcal{C}_{\mathcal{B}_D}$, the following statements are equivalent:*

(i) *G is AT-free.*
(ii) *If $(w, x, y) \in \mathcal{B}_D(G)$ and $(x, y, z) \in \mathcal{B}_D(G)$ then $(w, x, z) \in \mathcal{B}_D(G)$, i.e., $\mathcal{B}_D(G)$ is a transitive ternary relation.*
(iii) *Every connected induced subgraph of G has the spine property.*
(iv) *G has an AT-free order.*
(v) *$(V, \mathcal{C}_{\mathcal{B}_D}(G))$ is a convex geometry.*

4   AT-free BFS-Orders

Theorem 2. *Let G be a connected AT-free graph. Then for any vertex $s \in V$ there is a linear vertex order $\tau := (s = v_1, \ldots, v_n)$ that is an AT-free order and a BFS order.*

Proof. Let τ be a BFS order starting in an arbitrary vertex s of G with the following tie-break rule: At each step i choose the vertex v_i such that $\text{conv}(\{s = v_1, \ldots, v_i\})$ has smallest cardinality among all allowed choices at step i. We will show, that $\{s = v_1, \ldots, v_i\}$ is convex for $i \in \{1, \ldots, n\}$, which implies that τ is an AT-free order. The proof will be by induction on the BFS steps.

For $k = 1$ the claim is true, as every one element set is convex in $\mathcal{C}$.

We show the claim for step k, assuming it is true for $k - 1$. Suppose v_k is chosen. Then $\{v_1, \ldots, v_{k-1}\}$ is convex and v_k is such that $\text{conv}(\{v_1, \ldots, v_{k-1}\} + v_k)$ is smallest among all vertices that can be chosen by the search in step k. As we are conducting a BFS there is a vertex $y \in \{v_1, \ldots v_{k-1}\}$ that is adjacent to all possible choices, but no others. Assume that $\{v_1, \ldots, v_k\}$ is not convex. Then there is a vertex $p \in V \setminus \{v_1, \ldots, v_k\}$, such that $(v, p, v_k) \in \mathcal{B}_D$ for some vertex $v \in \{v_1, \ldots v_{k-1}\}$. As $(V, \mathcal{C}_{\mathcal{B}_D}(G))$ is a convex geometry, we can deduce that $\text{conv}(\{v_1, \ldots, v_{k-1}\} + p) \subsetneq \text{conv}(\{v_1, \ldots v_{k-1}\} + v_k)$. This implies that $yp \notin E$ due to the choice of v_k. Let w be the vertex that forced v into the BFS ordering (it may be that $y = w$). Due to the definition of BFS we see that $\text{dist}_G(s, w) \leq \text{dist}_G(s, y) < \text{dist}_G(s, p)$. We can assume that $wp \notin E$, as otherwise p would have been chosen before v_k. Therefore, the vertices $\{v, v_k, p\}$ form an asteroidal triple, due to the p-avoiding walk from v to v_k along w, s and y. This is a contradiction to fact that G is AT-free. $\square$

This theorem implies an algorithm for computing an AT-free BFS order which will be denoted by BFS^{conv}.

Any such ordering $\tau := (v_1, \ldots, v_n)$ obviously has the property that for every $i \in \{1, \ldots, n\}$ the induced subgraph $G[\{v_1, \ldots, v_i\}]$ is connected. This is already an improvement on the orders produced by the algorithm given in [11] and in Fig. 1 we compare orders computed by the different algorithms. On the other hand, returning to the example given in the introduction, the P_{2k+1} path graph, we can see that starting the BFS^{conv} in vertex $k + 1$ still yields an undesirable order.

Starting in an admissible vertex, which in the case of P_{2k+1} will be one of the endpoints or one of their neighbours, is an easy remedy of this problem. However, with a little modification to our search routine we can not only solve this issue, but make an intriguing link with the AT-free graphs characterisation through the spine property. We shall call a vertex ordering $\tau = (v_1, \ldots, v_n)$ a *monotone dominating pair order*, if for every $i \in \{1, \ldots, n\}$ the vertices v_1 and v_i form a dominating pair in the induced subgraph $G[v_1, \ldots, v_i]$.

Theorem 3. [9] *Let $G = (V, E)$ be a connected AT-free graph and suppose that s is an admissible vertex. Let $\tau = (v_1, \ldots, v_n)$ be a vertex order produced by LBFS*

Algorithm 2. BFS^{conv}

Input: Connected graph G and a distinguished vertex $s \in V$
Output: A vertex ordering σ
1 **begin**
2   Compute $I(v, w)$ for every pair of vertices $v, w \in V$;
3     $L \leftarrow \{s\}$;
4     $S \leftarrow \emptyset$;
5   **for** $i \leftarrow 1$ *to* n **do**
6   Choose the first vertex v from L such that there are no $u \in S$ and
       $z \in V - S$ with $z \in I(u, v)$;
7   Delete v from L;
8        $\sigma(i) \leftarrow v$;
9        $S \leftarrow S \cup \{v\}$;
10   **for** *each unnumbered vertex* w *adjacent to* v **do**
11   **if** $w \notin L$ **then**
12   Append w to end of L;

(G, s). *Then for any* $i \in \{1, \ldots, n\}$ *the vertices* v_1 *and* v_i *form a dominating pair of* $G[v_1, \ldots, v_n]$, *i.e.,* τ *is a monotone dominating pair order.*

In the following we will prove an analogous result for BFS^{conv}.

Lemma 2 ($\star^1$). *Let* $G = (V, E)$ *be an AT-free graph and let* s *be an admissible vertex of eccentricity* $k > 2$. *If* $\tau := (s = v_1, \ldots, v_n = t)$ *is the output of* $BFS^{\mathrm{conv}}(G, s)$, *then* s *and* t *form a dominating pair.*

However, applying a BFS^{conv} with an admissible start vertex must not always result in a monotone dominating pair order, as can be seen in Fig. 2.

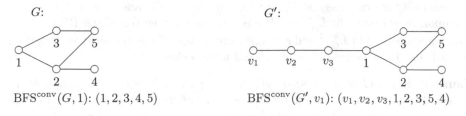

Fig. 2. Graph for which BFS^{conv} does not necessarily output a monotone dominating pair ordering and the graph G' constructed from G as in Theorem 4.

In [8] it is shown that for an AT-free graph G and an admissible vertex s the graph G' obtained by adding a pendant vertex v to s is also AT-free and v is admissible in G'. With this operation we can artificially raise the eccentricity of our starting vertex and generalise Lemma 2 to all AT-free graphs.

---

[1] The full proofs of theorems marked with ($\star$) can be found in [1].

Theorem 4. *Let G be a connected AT-free graph. For every admissible vertex s there is a vertex ordering τ beginning in s that is both AT-free and a monotone dominating pair ordering.*

Proof. We construct an auxiliary graph by adding a three vertex path to s in the following way: $G' = (V + \{v_1, v_2, v_3\}, E + \{v_1 v_2, v_2 v_3, v_3 s\})$. As s is admissible, the graph G' is again AT-free and v_1 is admissible in G' with $\mathrm{ecc}_{G'}(v_1) > 2$. The order $\tau' = (v_1, v_2, v_3, w_1, \ldots, w_n)$ that is generated by $\mathrm{BFS}^{\mathrm{conv}}(G', v_1)$ is an AT-free order and with Lemma 2 it is easy to see that $\tau = (w_1, \ldots, w_n)$ is a monotone dominating pair order for G. □

5   AT-free Orders in Claw-free AT-free Graphs

After having established the existence of AT-free BFS orders and a polynomial-time algorithm for their computation, we are interested in finding a simple linear time algorithm. In many graph classes, forbidding induced claw-graphs yields strong structural properties for BFS searches. For example, in [6,18] the authors use these structural properties to generate unit interval respectively minimal triangulation orderings. As in the papers cited above, we will use successive applications of BFS as well as LBFS.

Lemma 3. *Let G be claw-free and AT-free. Then the last vertex of a BFS is admissible.*

Proof. Let s be the first and z the last vertex of the BFS and let $k := \mathrm{dist}_G(s, z)$. Suppose there are $a, b \in V$ such that $(a, z, b) \in \mathcal{B}_D(G)$. As G is AT-free, at least one of a or b must be in the last layer $L_G^k(s)$ of the BFS, w.l.o.g. this is a. If $\mathrm{dist}_G(s, b) < \mathrm{dist}_G(s, z)$, then $N_s^{k-1}(a) \subseteq N_s^{k-1}(z)$, as otherwise there is a z-avoiding a-b-path. If $\mathrm{dist}_G(s, b) = \mathrm{dist}_G(s, a) = \mathrm{dist}_G(s, z)$, then either $N_s^{k-1}(a) \subseteq N_s^{k-1}(z)$ or $N_s^{k-1}(b) \subseteq N_s^{k-1}(z)$, as G is AT-free, and without loss of generality we can assume this to be true for a. Therefore, a and z have a common neighbour c in $L_G^{k-1}(s)$. If c is not the start vertex of the BFS, then c has a neighbour d in L_G^{k-2} and a, z, c, d form a claw. If c is the start vertex, then b must also be adjacent to c and a, b, c, d form a claw. □

Lemma 4. *Let G be a claw-free, AT-free graph and let $s \in V$ be admissible in G and t eccentric with respect to s. Then all but the first distance layers of s, i.e., $L_G^0(s), L_G^2(s), \ldots, L_G^k(s)$, with $k = \mathrm{ecc}_G(s)$, are cliques and s and t form a dominating pair.*

Proof. For $L_G^0(s)$ this is obvious. Let $i \geq 2$ and suppose there are $a, b \in L_G^i(s)$ with $ab \notin E$. As s is admissible, without loss of generality $N_s^{i-1}(a) \subseteq N_s^{i-1}(b)$. Therefore a and b have a common neighbour $c \in L_G^{i-1}$. This c in turn has a neighbour $d \in L_G^{i-2}$ and a, b, c, d form a claw, which is a contradiction to the assumption.

As any path P between s and t has one vertex from each distance layer $L_G^i(s)$ and s is adjacent to all vertices in $L_G^1(s)$ they must form a dominating pair. □

Theorem 5. *Let G be an AT-free, claw-free graph. Then a BFS starting in an admissible vertex yields an AT-free order that is a monotone dominating pair order.*

Proof. Let τ be such a BFS on G starting in an admissible vertex s. Suppose $(a, z, b) \in \mathcal{B}_D(G)$ and $a, b \prec_\tau z$. We can assume that a, b and z do not have the same distance to s (otherwise we can construct a claw as above). As G is AT-free, on the other hand, at least one of a or b must be in the same layer as z. W.l.o.g. we can assume that b and z are in the same layer $L_G^i(s)$ and a is in layer $L_G^j(s)$ with $j < i$. As b and z are independent of each other, they must be in the first layer of the BFS. As a cannot be the start vertex (it is not adjacent to the other two), this is a contradiction. Lemma 4 states that τ must be a monotone dominating pair order. □

Lemma 5 ($\star$). *Let $G = (V, E)$ be a connected graph with a dominating pair s and t. Let u and v be two vertices with $uv \notin E$ and $\mathrm{dist}_G(s, u) < \mathrm{dist}_G(s, v)$. Then $\mathrm{dist}_G(t, u) \geq \mathrm{dist}_G(t, v)$.*

Corollary 1 ($\star$). *Let G be a claw-free AT-free graph. Then G has a bilateral AT-free ordering and this order can be found in linear time.*

In the proof of Theorem 5 we can see that the main obstacles are triples of vertices $a, b, z \in V(G)$ with $(a, z, b) \in \mathcal{B}_D(G)$ that form the prongs of a claw. This justifies the following:

Definition 3. *Let G be a graph and let $a, b, z, c \in V$ induce a claw with base c. We will call such a claw a* bad claw, *if $(a, z, b) \in \mathcal{B}_D(G)$.*

It seems reasonable to expect that by forbidding such bad claws we will be able to get similar results to the ones above. On the other hand, there are examples of AT-free bad-claw-free graphs for which the above procedure does not yield either an AT-free order nor a bilateral AT-free ordering (see Fig. 3). In particular, Lemma 3 does not hold in general for these graphs. Therefore, we will use LBFS which guarantees us an admissible vertex as its end-vertex.

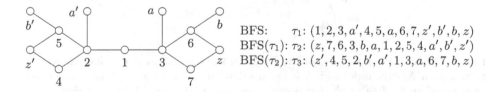

BFS: τ_1: $(1, 2, 3, a', 4, 5, a, 6, 7, z', b', b, z)$
BFS(τ_1): τ_2: $(z, 7, 6, 3, b, a, 1, 2, 5, 4, a', b', z')$
BFS(τ_2): τ_3: $(z', 4, 5, 2, b', a', 1, 3, a, 6, 7, b, z)$

Fig. 3. A bad-claw-free graph for which BFS does not yield an AT-free order

Lemma 6. [9] *Let $G = (V, E)$ be an AT-free graph and let τ be an ordering of V produced by an LBFS. Then the vertex $t := \tau(n)$ is admissible in G.*

In fact, the properties of LBFS even make up for the absence of the strong structural property of Lemma 4 and we can prove analogues to both Theorem 5 and Corollary 1.

Theorem 6. *Let G be AT-free and bad-claw-free. Then an LBFS starting in an admissible vertex yields an AT-free order that is a monotone dominating pair order.*

Proof. Let τ be an LBFS order starting in an admissible vertex s. Suppose $(a, z, b) \in \mathcal{B}_D(G)$ and $a, b \prec_\tau z$. Without loss of generality, we see that $i := \mathrm{dist}_G(s, b) = \mathrm{dist}_G(s, z)$, as G is AT-free. For that same reason either $N_s^{i-1}(b) \subseteq N_s^{i-1}(z)$ or $N_s^{i-1}(a) \subseteq N_s^{i-1}(z)$ or both.

Now suppose $\mathrm{dist}_G(s, a) = i$. As s is admissible, and a, b and z are independent, they must have a common neighbour c with $\mathrm{dist}_G(s, c) = i - 1$ and therefore a, b and z and c form a bad claw, which is a contradiction.

Therefore, we can assume that $j := \mathrm{dist}_G(s, a) < i$. With the above we see that $N_s^{i-1}(b) \subseteq N_s^{i-1}(z)$ and there is a b-avoiding a-z-path P. Let x be the τ-last vertex of P. As $b \prec_\tau z \preceq_\tau x$, due to Theorem 3 the vertex b must see every s-x-path and thus also every x-a-path, which is a contradiction. Thus, every LBFS starting in an admissible vertex yields an AT-free order.

Finally, Theorem 3 states that every LBFS order of an AT-free graph starting in an admissible vertex is a monotone dominating pair order.   □

Corollary 2 ($\star$). *Let G be an AT-free graph that does not have a bad claw as an induced subgraph. Then G has a bilateral AT-free ordering and such an order can be found in linear time.*

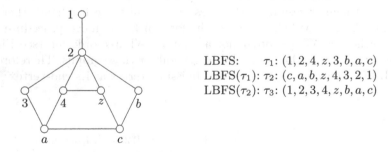

LBFS: τ_1: $(1, 2, 4, z, 3, b, a, c)$
LBFS(τ_1): τ_2: $(c, a, b, z, 4, 3, 2, 1)$
LBFS(τ_2): τ_3: $(1, 2, 3, 4, z, b, a, c)$

Fig. 4. Example of a graph with a bad claw. On the right, one can see that the second τ_2 is not an AT-free order and τ_3 is not a bilateral AT-free order. In fact, this is an example of an AT-free graph that does not possess a bilateral AT-free ordering.

These results indicate that a linear time algorithm to construct AT-free orders could also exist for the general case of AT-free graphs. However, none of the techniques used for the (bad-)claw-free graphs can be transferred. In [11] it was already shown that there are AT-free graphs which do not possess AT-free orders that are also LBFS orders. In addition, Fig. 4 shows a graph which does not possess a bilateral AT-free ordering. Therefore, it will be necessary to use a different search algorithm, possibly a BFS-derivative based on BFS^{conv}. We summarise these suppositions in the following:

Conjecture 2. Let $G = (V, E)$ be an AT-free graph. There is a linear time algorithm that computes an AT-free (BFS) order.

6   Conclusion

We resolved an open question from [11] by proving that any given AT-free graph has an AT-free order that coincides with a BFS order. The proof implied a polynomial time algorithm for the computation of such an order that is at least as fast as recognition. As a result, we were able to show that there is a close link between the vertex order characterisation of AT-free graphs, and their characterisation through the spine property. As checking whether a vertex order is an AT-free order is in fact of the same difficulty as recognising AT-free graphs, it should still be possible to find AT-free orders in linear time. This could be done by giving a linear time implementation of BFS^{conv} or by constructing another search scheme with similar structural properties.

For the special case of claw-free AT-free graphs we have shown that multiple applications of BFS yield AT-free orders with additional structural properties. In fact, if we exchange generic BFS with LexMinBFS, a derivative defined in [18], we can construct an AT-free, monotone dominating pair order that is also a minimal interval completion order. While claw-free AT-free graphs form a strongly restricted subclass of AT-free graphs, it is important to recall that their recognition has been shown to be at least as hard as triangle recognition, the same bound given to the recognition of general AT-free graphs. Furthermore, the results on bad-claw-free graphs can be seen as a first step toward a resolution of Conjecture 2, and give us a strong notion where the algorithmic difficulties lie.

Linear vertex orderings of other graph classes, such as interval orderings or cocomparability orderings, have found many applications in optimisation algorithms on these classes. To the best knowledge of the author, no such results are known with respect to AT-free orderings. By using AT-free BFS orderings such results might be easier to attain. Two of the most likely candidates are the independent set problem and the vertex colouring problem. However, in the case of vertex colouring even for cocomparability graphs there is no known algorithm that utilises the cocomparability ordering. Should it be possible to compute AT-free orders in linear time, it might even be possible to develop robust optimisation algorithms (see [20]) on AT-free graphs, similar to the maximum clique algorithm on comparability graphs.

Finally, it is still an open question whether every AT-free graph admits a DFS order whose reversal is AT-free [11].

References

1. Beisegel, J.: Characterising AT-free graphs with BFS. arXiv preprint (2018)
2. Broersma, H., Kloks, T., Kratsch, D., Müller, H.: Independent sets in asteroidal triple-free graphs. SIAM J. Discret. Math. **12**(2), 276–287 (1999)
3. Chang, J.M., Kloks, T., Wang, H.-L.: Gray codes for AT-free orders via antimatroids. In: Lipták, Z., Smyth, W.F. (eds.) IWOCA 2015. LNCS, vol. 9538, pp. 77–87. Springer, Cham (2016). https://doi.org/10.1007/978-3-319-29516-9_7
4. Chang, J.M., Kloks, T., Wang, H.-L.: Convex geometries on AT-free graphs and an application to generating the AT-free orders. arXiv preprint arXiv:1706.06336 (2017)
5. Chvátal, V.: Antimatroids, betweenness, convexity. In: Cook, W., Lovász, L., Vygen, J. (eds.) Research Trends in Combinatorial Optimization, pp. 57–64. Springer, Berlin (2009). https://doi.org/10.1007/978-3-540-76796-1_3
6. Corneil, D.G.: A simple 3-sweep LBFS algorithm for the recognition of unit interval graphs. Discret. Appl. Math. **138**(3), 371–379 (2004)
7. Corneil, D.G.: Lexicographic breadth first search – a survey. In: Hromkovič, J., Nagl, M., Westfechtel, B. (eds.) WG 2004. LNCS, vol. 3353, pp. 1–19. Springer, Heidelberg (2004). https://doi.org/10.1007/978-3-540-30559-0_1
8. Corneil, D.G., Olariu, S., Stewart, L.: Asteroidal triple-free graphs. SIAM J. Discret. Math. **10**(3), 399–430 (1997)
9. Corneil, D.G., Olariu, S., Stewart, L.: Linear time algorithms for dominating pairs in asteroidal triple-free graphs. SIAM J. Comput. **28**(4), 1284–1297 (1999)
10. Corneil, D.G., Olariu, S., Stewart, L.: The LBFS structure and recognition of interval graphs. SIAM J. Discret. Math. **23**(4), 1905–1953 (2009)
11. Corneil, D.G., Stacho, J.: Vertex ordering characterizations of graphs of bounded asteroidal number. J. Graph Theory **78**(1), 61–79 (2015)
12. Edelman, P.H., Jamison, R.E.: The theory of convex geometries. Geometriae Dedicata **19**(3), 247–270 (1985)
13. Hempel, H., Kratsch, D.: On claw-free asteroidal triple-free graphs. Discret. Appl. Math. **121**(1), 155–180 (2002)
14. Köhler, E.: Linear structure of graphs and the knotting graph. In: Schulz, A., Skutella, M., Stiller, S., Wagner, D. (eds.) Gems of Combinatorial Optimization and Graph Algorithms, pp. 13–27. Springer, Cham (2015). https://doi.org/10.1007/978-3-319-24971-1_2
15. Köhler, E.G.: Graphs Without Asteroidal Triples. Cuvillier, Göttingen (1999)
16. Kratsch, D., Spinrad, J.: Between $\mathcal{O}(nm)$ and $\mathcal{O}(n^\alpha)$. SIAM J. Comput. **36**(2), 310–325 (2006)
17. Lekkerkerker, C., Boland, J.: Representation of a finite graph by a set of intervals on the real line. Fundamenta Mathematicae **51**(1), 45–64 (1962)
18. Meister, D.: Recognition and computation of minimal triangulations for AT-free claw-free and co-comparability graphs. Discret. Appl. Math. **146**(3), 193–218 (2005)
19. Rose, D.J., Tarjan, R.E., Lueker, G.S.: Algorithmic aspects of vertex elimination on graphs. SIAM J. Comput. **5**(2), 266–283 (1976)
20. Spinrad, J.P.: Efficient Graph Representations. American Mathematical Society, Providence (2003)

Edge Partitions of Optimal 2-plane and 3-plane Graphs

Michael A. Bekos[1], Emilio Di Giacomo[2], Walter Didimo[2], Giuseppe Liotta[2], Fabrizio Montecchiani[2(✉)], and Chrysanthi Raftopoulou[3]

[1] Universität Tübingen, Tübingen, Germany
`bekos@informatik.uni-tuebingen.de`
[2] Università degli Studi di Perugia, Perugia, Italy
`{emilio.digiacomo,walter.didimo,`
`giuseppe.liotta,fabrizio.montecchiani}@unipg.it`
[3] National Technical University of Athens, Athens, Greece
`crisraft@mail.ntua.gr`

Abstract. A topological graph is a graph drawn in the plane. A topological graph is k-plane, $k > 0$, if each edge is crossed at most k times. We study the problem of partitioning the edges of a k-plane graph such that each partite set forms a graph with a simpler structure. While this problem has been studied for $k = 1$, we focus on *optimal* 2-plane and 3-plane graphs, which are 2-plane and 3-plane graphs with maximum density. We prove the following results. (i) It is not possible to partition the edges of a simple optimal 2-plane graph G into a 1-plane graph and a forest, while (ii) an edge partition of G formed by a 1-plane graph and two plane forests always exists and can be computed in linear time. (iii) We describe efficient algorithms to partition the edges of G into a 1-plane graph and a plane graph with maximum vertex degree 12, or with maximum vertex degree 8 if G is such that its crossing-free edges form a graph with no separating triangles. (iv) We exhibit an infinite family of simple optimal 2-plane graphs such that in any edge partition composed of a 1-plane graph and a plane graph, the plane graph has maximum vertex degree at least 6. (v) We show that every optimal 3-plane graph whose crossing-free edges form a biconnected graph can be decomposed into a 2-plane graph and two plane forests.

1 Introduction

Partitioning the edges of a graph such that each partite set induces a subgraph with a simpler structure is a fundamental problem in graph theory with various applications, including the design of graph drawing algorithms. For example, a classic result by Schnyder [17] states that the edge set of any maximal planar graph can be partitioned into three trees, which can be used to efficiently compute planar straight-line drawings on a grid of polynomial size. Edge partitions

Research funded in part by the project: "Algoritmi e sistemi di analisi visuale di reti complesse e di grandi dimensioni" - Ricerca di Base 2018, Dip. Ing. Univ. Perugia.

A. Brandstädt et al. (Eds.): WG 2018, LNCS 11159, pp. 27–39, 2018.
https://doi.org/10.1007/978-3-030-00256-5_3

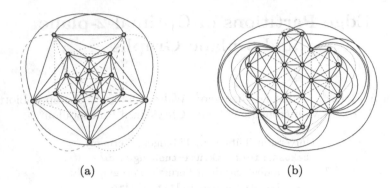

(a)   (b)

Fig. 1. An edge partition of: (a) an optimal 2-plane graph into a 1-plane graph (solid) and two plane forests (dashed and dotted); (b) an optimal 3-plane graph into a 2-plane graph (solid) and two plane forests (dashed and dotted).

of planar graphs have also been studied by Gonçalves [11], who proved that the edges of every planar graph can be partitioned into two outerplanar graphs, thus solving a conjecture by Chartrand et al. [4], and improving previous results (see, e.g., [8,12]). More in general, there exist various graph parameters based on edge partitions. For example, the *arboricity* of a graph G is the minimum number of forests needed to cover all edges of G, while G has *thickness* t if it is the union of t planar graphs. Durocher and Mondal [9] studied the interplay between the thickness t of a graph and the number of bends per edge in a drawing that can be partitioned into t planar sub-drawings.

Recently, edge partitions have been studied for the family of 1-*planar graphs*. A graph is k-*planar* ($k \geq 1$) if it can be drawn in the plane such that each edge is crossed at most k times [16]; a topological graph is k-*plane* if it has at most k crossings per edge. Ackerman [1] proved that the edges of a 1-plane graph can be partitioned into a plane graph (a topological graph with no crossings) and a plane forest, extending an earlier result by Czap and Hudáck [5]. A 1-planar graph with n vertices is *optimal* if it contains exactly $4n - 8$ edges, which attains the maximum density for 1-planar graphs. Lenhart et al. [14] proved that every optimal 1-plane graph can be partitioned into two plane graphs such that one has maximum vertex degree four, where the bound on the vertex degree is worst-case optimal. Di Giacomo et al. [7] proved that every triconnected 1-plane graph can be partitioned into two plane graphs such that one has maximum vertex degree six, which is also a tight bound. This result is exploited to show that every such a graph has a visibility representation in which the vertices are orthogonal polygons with few reflex corners each, while the edges are horizontal and vertical lines of sight between vertices. Additional results on edge partitions of various subclasses of 1-plane graphs are reported in [6].

While 1-planar graphs have been extensively studied (see also the survey by Kobourov et al. [13]), and their structure has been deeply understood, this is not the case for 2-planar and 3-planar graphs. These graphs have a more complex structure with at most $5n - 10$ and $5.5n - 11$ edges, respectively [16]. Similarly to 1-planar graphs, a 2-planar (respectively, 3-planar) graph with n vertices is *optimal* if it contains $5n - 10$ (respectively, $5.5n - 11$) edges (see also Sect. 2). Examples of optimal 2-plane and 3-plane graphs are shown in Figs. 1(a) and (b), respectively. Bekos et al. [3] recently characterized optimal 2-planar and optimal 3-planar graphs, and showed that these graphs have a regular structure; refer to Sect. 2 for details. In this paper, we build upon this characterization and we initiate the study of edge partitions of simple (i.e., with neither self-loops nor parallel edges) optimal 2-plane graphs. Figure 1(a) shows an edge partition of an optimal 2-plane graph into a 1-plane graph and two plane forests. We then extend some of our results to a subclass of optimal 3-plane graphs; an edge partition of an optimal 3-plane graph is shown in Fig. 1(b). Our contributions are as follows.

- We prove that it is not possible to partition the edges of a simple optimal 2-plane graph G into a 1-plane graph and a forest. Note that, by Nash-Williams formula [15], 2-planar graphs have arboricity at most five, while 1-planar graphs have arboricity at most four. Hence, our result implies that in a decomposition of G into five forests, it is not possible to pick four of them forming a 1-plane graph.
- On the positive side, every simple optimal 2-plane graph can be partitioned into a 1-plane graph and two plane forests. This result exploits some insights in the structure of optimal 2-plane graphs. Also, the edge partition can be computed in linear time.
- Additionally, we prove that the edges of a simple optimal 2-plane graph can always be partitioned into a 1-plane graph and a plane graph with maximum vertex degree 12. This upper bound on the vertex degree can be lowered to 8 if the optimal 2-plane graph is such that its crossing-free edges form a graph with no separating triangles. Both bounds are achieved with constructive techniques that work in polynomial time.
- Besides the above upper bound on the vertex degree, we establish a non-trivial lower bound. We exhibit an infinite family of simple optimal 2-plane graphs such that in any edge partition composed of a 1-plane graph and a plane graph, the plane graph has maximum vertex degree at least 6.
- We finally consider (non-simple) optimal 3-plane graphs and prove that any such a topological graph whose crossing-free edges form a biconnected graph can be partitioned into a 2-plane graph and two plane forests.

For space reasons, some proofs marked with * are omitted and can be found in [2].

2   Preliminaries and Notation

Drawings and Planarity. A graph is *simple* if it contains neither self-loops nor parallel edges. A *drawing* of a graph $G = (V, E)$ is a mapping of the vertices of V to points of the plane, and of the edges of E to Jordan arcs connecting their corresponding endpoints but not passing through any other vertex. We

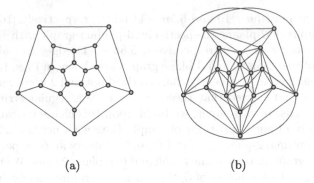

(a)   (b)

Fig. 2. (a) The pentangulation of the graph G in Fig. 1(a). (b) A 1-plane graph obtained from G by removing two adjacent chords from each filled pentagon.

only consider *simple* drawings, i.e., drawings such that two arcs representing two edges have at most one point in common, and this point is either a common endpoint or a common interior point where the two arcs properly cross each other. A graph drawn in the plane is also called a *topological graph*. The *crossing graph* $C(G)$ of a topological graph G has a vertex for each edge of G and an edge between two vertices if and only if the two corresponding edges of G cross each other. A topological graph is *plane* if it has no edge crossings. A plane graph subdivides the plane into topologically connected regions, called *faces*. The infinite region is the *outerface*. The *degree* of a face f is the number of vertices encountered in a closed walk along the boundary of f. If a vertex v is encountered $k > 0$ times, then v has *multiplicity* k in f. In a biconnected graph, all vertices have multiplicity one in the faces they belong to.

k-planar Graphs. A topological graph is *k-plane* if each edge is crossed at most k times. A *pentangulation* (resp., *hexangulation*) P (resp., H) is a plane graph such that all its faces are 5-cycles (resp., 6-cycles), which we call *pentagons* (resp., *hexagons*). Two parallel edges are *homotopic* if the interior or the exterior region bounded by their curves contain no vertices. A self loop is *homotopic* if the interior or the exterior region bounded by its curve contains no vertices. Bekos et al. [3] proved that an n-vertex graph G is *optimal 2-planar* if and only if it admits a drawing without homotopic self-loops and homotopic parallel edges, such that the graph formed by the crossing-free edges is a pentangulation $P(G)$ with n vertices, and each face of $P(G)$ has five crossing edges in its interior, which we call *chords* in the following. Also, each chord has exactly two crossings. A pentagon with its five chords routed as described above will be called a *filled pentagon*. Figure 2(a) shows the pentangulation $P(G)$ of the optimal 2-plane graph G of Fig. 1(a). Similarly, Bekos et al. proved that an n-vertex graph G is optimal 3-planar if and only if it admits a drawing without homotopic self-loops and parallel edges, such that the graph formed by the crossing-free edges is a hexangulation $H(G)$ with n vertices, and each face of $H(G)$ has eight crossing edges in its interior, which we call *chords* in the following. A hexagon with its eight chords routed as described above will be called a *filled hexagon*.

Arboricity and Orientations. The *arboricity* of a graph is the minimum number of forests into which its edges can be partitioned. Nash-Williams [15] proved that a graph G has arboricity $a \geq 1$ if and only if, $a = \max\{\lceil \frac{m_S}{n_S - 1} \rceil\}$ over all subgraphs S of G with $n_S \geq 2$ vertices and m_S edges. A *d-orientation* of a graph G is an orientation of the edges of G such that each vertex has at most d outgoing edges, for some integer $d \geq 1$. Note that if a graph has arboricity a, it admits an a-orientation (the converse may not be true). Given two vertices s and t of a graph G, an *st-orientation* of G is an orientation of its edges such that G becomes a directed acyclic graph with a single source s and a single sink t.

Edge Partitions. Given a topological graph $G = (V, E)$, an *edge partition* of G is denoted by $\langle E_1, \ldots, E_p \rangle$, for some $p > 1$, where $E = E_1 \cup \cdots \cup E_p$ and $E_i \cap E_j = \emptyset$ $(1 \leq i \neq j \leq p)$. We denote by $G[E_i]$ the topological graph obtained from G by removing all edges not in E_i and all the isolated vertices.

3 Edge Partitions of Optimal 2-plane Graphs

We begin by observing the following property, which will be useful in the remainder of this section.

Property 1 (*). Let $G' = (V, E \setminus R)$ be a topological graph obtained by removing a subset R of crossing edges from a simple optimal 2-plane graph $G = (V, E)$. G' is 1-plane if and only if R has two adjacent chords for each filled pentagon of G.

For example, the graph in Fig. 2(b) is obtained by removing two adjacent chords from each filled pentagon of an optimal 2-plane graph.

3.1 Edge Partitions with Acyclic Subgraphs

As already mentioned, the edge set of a 1-plane graph can always be partitioned into a plane graph and a plane forest [1]. One may wonder whether this result can be generalized to 2-plane graphs, that is, whether the edge set of every 2-plane graph can be partitioned into a 1-plane graph and a forest. Theorem 1 shows that this may not be possible. In particular, this is never the case for optimal 2-plane graphs. On the positive side, Theorem 2 gives a constructive technique to partition the edges of every optimal 2-plane graph into a 1-plane graph and two plane forests (rather than one).

Theorem 1. *Let G be a simple optimal 2-plane graph. G has no edge partition $\langle E_1, E_2 \rangle$ such that $G[E_1]$ is a 1-plane graph and $G[E_2]$ is a forest.*

Proof. Let $\langle E_1, E_2 \rangle$ be an edge partition such that $G[E_1]$ is a 1-plane graph. By Property 1, E_2 has at least two chords for each filled pentagon of G. By Euler's formula, if G has n vertices, the pentangulation $P(G)$ has $\frac{2}{3}(n - 2)$ faces, and thus G has $\frac{2}{3}(n-2)$ filled pentagons. Then E_2 contains at least $2 \times \frac{2}{3}(n-2)$ edges, and hence $G[E_2]$ can be a forest only if $n \leq 5$. On the other hand $n > 5$, because $P(G)$ has at least two internal faces (otherwise either $P(G)$ would be a 5-cycle and G would have two parallel chords or $P(G)$ would be drawn nonplanar). □

Lemma 1 (*). *The pentangulation $P(G)$ of a simple optimal 2-plane graph G is biconnected.*

Theorem 2. *Every n-vertex simple optimal 2-plane graph $G = (V, E)$ has an edge partition $\langle E_1, E_2, E_3 \rangle$, which can be computed in $O(n)$ time, such that $G[E_1]$ is a 1-plane graph and both $G[E_2]$ and $G[E_3]$ are plane forests.*

Proof. To construct the desired edge partition, we first guarantee that $E' = E \setminus E_1$ contains two adjacent chords for each filled pentagon of G, which implies that $G[E_1]$ is a 1-plane graph by Property 1. We then color the edges of E' with two colors, say green and red, so that each monochromatic set is a plane forest. The set of green edges will correspond to E_2, while the set of red edges to E_3.

We aim at computing an st-orientation of $P(G)$. Recall that, given a biconnected plane graph and two vertices s and t on its outerface, it is possible to construct an st-orientation of the graph in linear time (see, e.g., [10,18]). By Lemma 1, the pentangulation $P(G)$ of G is biconnected, and hence we can compute an st-orientation of $P(G)$ (with s and t that belong to the outerface). According to this orientation, all outgoing edges of any vertex $v \in P(G)$ appear consecutively around v, followed by all the incoming edges of v ([18, Lemma 2]). For any vertex $v \in P(G)$ distinct from s and t, this allows us to uniquely define the *leftmost* (*rightmost*) face of v as the face containing the last incoming and first outgoing edges (last outgoing and first incoming edges, respectively) of v in clockwise order around v. We use this fact to classify the internal faces of $P(G)$ in different types. By [18, Lemma 1], each internal face f of $P(G)$ has a source vertex $s(f)$ and a target vertex $t(f)$, and consists of two directed paths from $s(f)$ to $t(f)$, say $p_l(f)$ and $p_r(f)$. Since $P(G)$ is a pentangulation, we have $|p_l(f)| + |p_r(f)| = 5$, $|p_l(f)| \leq 4$, and $|p_r(f)| \leq 4$. We say that f is a face of type $i - j$ if $|p_l(f)| = i$ and $|p_r(f)| = j$. Hence, in total there exist exactly four different *types* of internal faces: $1 - 4$, $4 - 1$, $2 - 3$ and $3 - 2$; refer to Fig. 3. For the first two types of faces we select to be part of E' the two chords of f in G that are incident to the target vertex $t(f)$. In the other two types, we select the two edges that are incident to the middle vertex of the directed path with edge-length 2. If $i < j$ (resp., $i > j$) we color the selected edges red (resp., green). Note that we have not selected and colored any chord of the outerface; this selection will be made at the very end.

We now claim that each monochromatic subgraph induced by the red and green edges is a forest. We prove this claim for the red subgraph, symmetric arguments hold for the green one. We orient each pair of red edges of every interior face f of $P(G)$ towards their common end-vertex, and we observe that if (u, v) is a directed red edge in a face f from u to v, then f is the leftmost face of u. Since, the leftmost face of each vertex is unique, it follows that every vertex has at most one outgoing red edge. Hence, a cycle of red edges would be actually

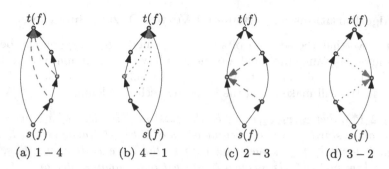

Fig. 3. Illustration for Theorem 2. Red (green) edges are dashed (dotted). (Color figure online)

a directed cycle (otherwise it would contain at least one vertex with out-degree two, contradicting the previous statement). Consider the plane subgraph G_{red} of G containing the edges of $P(G)$ (oriented according to the st-orientation defined above) and the red edges (each pair oriented towards the common end-vertex). We show that G_{red} does not contain directed cycles, which implies that the red subgraph is a forest. We actually prove a stronger property of G_{red}, namely, we show that the orientation of G_{red} is an st-orientation. The proof is by induction on the number $i \geq 0$ of internal faces of $P(G)$ having red chords in G_{red}. If $i = 0$, the statement trivially follows since G_{red} corresponds to $P(G)$. Assume the claim holds for $i \geq 0$, and suppose there are $i + 1$ internal faces of $P(G)$ having red chords in G_{red}. Consider any such face f of $P(G)$, and let G'_{red} be the graph obtained from G_{red} by removing the two red chords of f. G'_{red} is st-oriented by the inductive hypothesis. Obviously, reinserting the two removed chords in G'_{red} creates neither new sources nor new sinks. Moreover, reinserting the two chords cannot create a directed cycle, since each reinserted chord (u, v) connects either vertices on opposite paths of face f, which implies that there cannot be a directed path in G'_{red} from v to u [18, Lemma 4], or $v = t(f)$, which implies that there is already a directed path from u to v in G'_{red} and thus there cannot be a directed path from v to u because G'_{red} is acyclic.

It remains to select and color two chords of G from the outerface of $P(G)$. For each interior face f of $P(G)$, red or green edges are never incident to the source vertex $s(f)$. Hence, there is neither a red nor a green edge incident to s (which is the source of the graph). We arbitrarily select one of the two chords of G in the outerface of $P(G)$ that is incident to s to be red and the other one to be green. Since the degree of s in the red (green) subgraph is equal to one, it follows that no cycle is created. Moreover, since an st-orientation can be computed in $O(n)$ time, and since G has $O(n)$ faces and $O(n)$ edges, the theorem follows. Figure 1(a) shows an edge partition computed with the described algorithm. □

Theorem 2 together with the result by Ackerman [1] directly imply the following.

Corollary 1. *Every simple optimal 2-plane graph has an edge partition $\langle E_1, E_2, E_3, E_4 \rangle$ such that $G[E_1]$ is a plane graph, and $G[E_i]$ is a plane forest, for $i \geq 2$.*

3.2   Edge Partitions with Bounded Vertex Degree Subgraphs

We now prove that the edge set of a simple optimal 2-plane graph can be partitioned into a 1-plane graph and a plane graph whose maximum vertex degree is bounded by a small constant. An analogous result holds for optimal 1-plane graphs [14]. We will make use of the following technical lemma.

Lemma 2 (*). *Let v_0, v_1, v_2, v_3, v_4 be the (distinct) vertices of a 5-cycle C in clockwise order starting from v_0. Let the edges of C be arbitrarily oriented. There exists an index $0 \leq j \leq 4$ such that each of the three vertices v_j, v_{j+2}, v_{j+3} (indexes taken modulo 5) is incident to at least one outgoing edge of the 5-cycle.*

Theorem 3. *Every n-vertex simple optimal 2-plane graph $G = (V, E)$ has an edge partition $\langle E_1, E_2 \rangle$, which can be computed in $O(n)$ time, such that $G[E_1]$ is a 1-plane graph and $G[E_2]$ is a plane graph of maximum vertex degree 12.*

Proof. We construct the desired edge partition as follows. Remove three chords from every pentagon of $P(G)$ such that the resulting graph G' is plane and all its faces have degree three. Compute a 3-orientation of G' in linear time, by using the algorithm in [17]. From now on, we assume that the edges of $P(G)$ are directed according to this 3-orientation. For each filled pentagon of G we select three vertices that satisfy the conditions of Lemma 2, and we mark to be part of E_2 the two chords of the pentagon incident to the selected vertices. All other edges are part of E_1. Since each vertex has at most three outgoing edges in the 3-orientation of $P(G)$, and each of these edges is shared by exactly two pentagons (as otherwise G would be non-simple), we have that each vertex is selected for at most six pentagons and therefore is incident to at most 12 edges in E_2.

$G[E_1]$ is 1-plane by Property 1. $G[E_2]$ is a graph with maximum vertex degree 12 as shown above, and no two edges of $G[E_2]$ cross, because either they share an end-vertex or they are inside different pentagons of $P(G)$. $\square$

Theorem 3 can be improved if $P(G)$ has no separating triangles.

Theorem 4 (*). *Every n-vertex simple optimal 2-plane graph $G = (V, E)$ whose pentangulation $P(G)$ has no separating triangles has an edge partition $\langle E_1, E_2 \rangle$, which can be computed in $O(n^{1.5})$ time, such that $G[E_1]$ is a 1-plane graph and $G[E_2]$ is a plane graph of maximum vertex degree 8.*

The next corollary is a consequence of Theorems 3 and 4, together with the fact that every 3-connected 1-plane graph can be decomposed into a plane graph and a plane graph with maximum vertex degree 6 [7].

Corollary 2 (*). *Every n-vertex simple optimal 2-plane graph G has an edge partition $\langle E_1, E_2, E_3 \rangle$, which can be computed in $O(n)$ time, such that $G[E_1]$ is plane, $G[E_2]$ is plane with maximum vertex degree 12, and $G[E_3]$ is plane with maximum vertex degree 6. Also, if $P(G)$ has no separating triangles, then G has an edge partition $\langle E_1, E_2, E_3 \rangle$, which can be computed in $O(n^{1.5})$ time, such that $G[E_1]$ is plane, $G[E_2]$ is plane with maximum vertex degree 8, and $G[E_3]$ is plane with maximum vertex degree 6.*

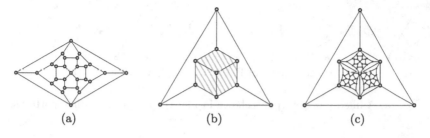

(a)   (b)   (c)

Fig. 4. Illustration for Theorem 5.

We conclude this section by proving a lower bound for the maximum vertex degree of an edge partition into a 1-plane graph and a plane graph.

Theorem 5. *There exists an infinite family $\mathcal{G}$ of simple optimal 2-plane graphs, such that in any edge partition $\langle E_1, E_2 \rangle$ of $G \in \mathcal{G}$ where $G[E_1]$ is 1-plane and $G[E_2]$ is plane, $G[E_2]$ has maximum vertex degree at least 6.*

Proof. For every $n \geq 12$, we construct a graph G_n as described in the following. Consider the plane graph G_1 in Fig. 4(a). Note that all faces of G_1 have degree five, except for the outer face which is a 4-cycle. Construct the graph G_2 by gluing the graph G_1 in the three gray quandrangular faces of the graph in Fig. 4(b). Note that all faces of G_2 have degree five, except for the outer face which is a 3-cycle. Then, starting from an n-vertex maximal plane graph M_n, identify each face of M_n (including its outer face) with the outer face of a copy of G_2. This results in a pentangulation P_n with $O(n)$ vertices. G_n is obtained by adding all five chords inside each pentagon of P_n. Graph G_n is optimal 2-plane as it satisfies the characterization in [3] (see also Sect. 2), and it is simple because P_n is simple and triconnected. Consider any edge partition $\langle E_1, E_2 \rangle$ of G_n, such that $G[E_1]$ is 1-plane. By Property 1, E_2 contains at least two chords of each filled pentagon of G_n. Therefore, for each face of M_n, there are at least three edges of E_2 having one end-vertex in M_n (at least one for each filled pentagon incident to the outer face of the copy of G_2 identified with this face). This means that E_2 contains at least $3(2n - 4) = 6n - 12$ edges incident to vertices of M_n. Let k be the maximum number of edges of E_2 that are incident to a single vertex of M_n, we have $kn \geq 6n - 12 \Rightarrow k \geq 6$ for $n \geq 12$.   □

4   Edge Partitions of Optimal 3-plane Graphs

In this section we study optimal 3-plane graphs and we aim at showing the existence of a decomposition into a 2-plane graph and two plane forests. It is known that no optimal 3-plane graph is simple [3], and hence its hexangulation may also be non-simple. We show that a similar strategy as the one used in the proof of Theorem 2 can be employed provided that the hexangulation of the graph is biconnected and hence each of its faces is a simple 6-cycle. Consider

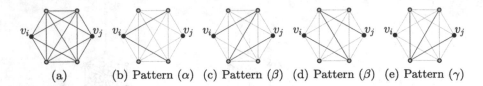

(a)   (b) Pattern (α)   (c) Pattern (β)   (d) Pattern (β)   (e) Pattern (γ)

Fig. 5. (a) A filled hexagon (poles shown in black). (b)–(d) The three patterns.

a filled hexagon h of an optimal 3-plane graph G. If $H(G)$ is biconnected, h contains six distinct vertices, which we denote by $v_0, v_1, \ldots, v_5$ following their clockwise order in a closed walk along the boundary of h. Refer to Fig. 5(a). We know that h contains 8 chords (see Sect. 2), and, in particular, there are only two vertices of h that are not connected by an edge of h; we call these two vertices the *poles* of h (black in Fig. 5(a)). Let v_i and v_j ($0 \le i < j \le 5$) be the poles of h. Note that $j - i = 3$, and that each chord of h is crossed at most twice after removing one of the following *patterns*: (α) the two chords of h incident to v_i or to v_j (see Fig. 5(b)); (β) one of the two *Z*-paths (v_i, v_{i+2}), (v_{i+2}, v_{j+2}), (v_{j+2}, v_j) and (v_i, v_{j+1}), (v_{j+1}, v_{i+1}), (v_{i+1}, v_j), where indexes are taken modulo 6 (see Figs. 5(c) and (d)); (γ) any three adjacent chords of h (see Fig. 5(e)).

Theorem 6. *Every n-vertex optimal 3-plane graph $G = (V, E)$ whose hexangulation $H(G)$ is biconnected has an edge partition $\langle E_1, E_2, E_3 \rangle$, which can be computed in $O(n)$ time, such that $G[E_1]$ is a 2-plane graph, and both $G[E_2]$ and $G[E_3]$ are plane forests.*

Proof. To construct the desired edge partition, we first guarantee that $E' = E \setminus E_1$ contains, for each filled hexagon of G, one of the three patterns described above, which implies that $G[E_1]$ is 2-plane. We then color the edges of E' with two colors, say green and red, so that each monochromatic set is a plane forest. The set of green edges will correspond to E_2, while the set of red edges to E_3.

We compute an *st*-orientation of $H(G)$ by choosing a pole of the outerface as vertex s. Recall that each internal face f of $H(G)$ has a source vertex $s(f)$ and a target vertex $t(f)$, and consists of two directed paths from $s(f)$ to $t(f)$, say $p_l(f)$ and $p_r(f)$. The number of edges $|p_l(f)|, |p_r(f)|$ of the two paths is at most 5, and in particular $|p_l(f)| + |p_r(f)| = 6$. We say that f is a face of type $i - j$ if $|p_l(f)| = i$ and $|p_r(f)| = j$. Hence, in total there exist exactly five different *types* of internal faces: $1 - 5$, $5 - 1$, $2 - 4$, $4 - 2$, $3 - 3$; refer to Fig. 6. For the first two types of faces we add to E' the two or three chords of f in G that are incident to the target vertex $t(f)$ (i.e., we remove either pattern (α) or (γ)). In the type $1 - 5$ ($5 - 1$) we color these edges red (green). For the types $2 - 4$ ($4 - 2$), we add to E' the two or three chords incident to the middle vertex of $p_l(f)$ ($p_r(f)$), and we color them red (green). For the type $3 - 3$, we distinguish a set of cases based on the position of the poles. Suppose first that the poles are $s(f)$ and $t(f)$, then we add to E' the two chords incident to $t(f)$ (pattern (α)), and we color red (green) the chord incident to a vertex of $p_r(f)$ ($p_l(f)$); see Fig. 6(e). Otherwise,

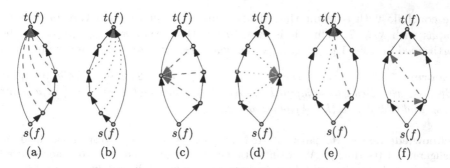

Fig. 6. Illustration for Theorem 6. Red (green) edges are dashed (dotted). (Color figure online)

among the two possible Z-paths, there is one that does not contain neither $s(f)$ nor $t(f)$ (pattern (β)), and we remove it; see Fig. 6(f).

We now claim that each monochromatic subgraph induced by the red and green edges is a forest. We prove this claim for the red subgraph, symmetric arguments hold for the green one. We orient the edges such that all red (green) edges are outgoing with respect to their end-vertex belonging to $p_r(f)$ ($p_l(f)$), note that there is always such an end-vertex. This orientation implies that each vertex has at most one outgoing red edge, hence a cycle of red edges would be actually a directed cycle. Consider the plane subgraph G_{red} of G containing the oriented edges of $H(G)$ and the oriented red edges. Since each red edge in a face f either connects a vertex of $p_r(f)$ to a vertex of $p_l(f)$, or it is incident to $t(f)$, a similar argument as the one used in the proof of Theorem 2 shows that the orientation of G_{red} is an st-orientation, and thus that there are no directed cycles.

It remains to select and color two chords of G from the outerface of $H(G)$. As in the proof of Theorem 2, there is neither a red nor a green edge incident to the vertex s of the outerface, which is a pole by construction. We color red one of the two chords of the outerface incident to s, and we color green the other one (i.e., we remove pattern (α) from the outerface). Since the degree of s in the red (green) subgraph is equal to one, no cycle is created. An example is shown in Fig. 1(b). Moreover, since an st-orientation can be computed in $O(n)$ time, and since G has $O(n)$ faces and $O(n)$ edges, the theorem follows.   □

5   Open Problems

A natural question is whether the edges of a (simple) optimal 2-plane graph can be partitioned into a plane graph and two forests. Moreover, the problem of partitioning the edges of an optimal 3-plane graph into a 2-plane graph and a forest is still open. Reducing the gap between the upper bound on the vertex degree of Theorem 3 and the lower bound of Theorem 5 is an interesting problem. Also, can we improve the time complexity of Theorem 4?

We conclude with a result that sheds some light on the structure of k-plane graphs with $k \geq 2$. While it is easy to see that every k-plane graph can be partitioned into $k + 1$ plane graphs, the next theorem shows a stronger property.

Theorem 7 (*). *Every n-vertex k-plane graph ($k \geq 2$) has an edge partition $\langle E_1, E_2 \rangle$, which can be computed in $O(k^{1.5}n)$ time, such that $G[E_1]$ is a plane graph and $G[E_2]$ is a $(k - 1)$-plane graph.*

Acknowledgments. Research started at the 2017 GNV Workshop, held in Heiligkreuztal (Germany). We wish to thank the organizers of the workshop and all the participants for the fruitful atmosphere and the useful discussions.

References

1. Ackerman, E.: A note on 1-planar graphs. Discret. Appl. Math. **175**, 104–108 (2014)
2. Bekos, M., Di Giacomo, E., Didimo, W., Liotta, G., Montecchiani, F., Raftopoulou, C.: Edge partitions of optimal 2-plane and 3-plane graphs. CoRR, abs/1802.10300 (2018)
3. Bekos, M.A., Kaufmann, M., Raftopoulou, C.N.: On optimal 2- and 3-planar graphs. In: SoCG 2017. LIPIcs, vol. 77, pp. 16:1–16:16. Schloss Dagstuhl - Leibniz-Zentrum fuer Informatik (2017)
4. Chartrand, G., Geller, D., Hedetniemi, S.: Graphs with forbidden subgraphs. J. Comb. Theory Ser. B **10**(1), 12–41 (1971)
5. Czap, J., Hudák, D.: On drawings and decompositions of 1-planar graphs. Electron. J. Comb. **20**(2), P54 (2013)
6. Di Giacomo, E., et al.: New results on edge partitions of 1-plane graphs. Theor. Comput. Sci. **713**, 78–84 (2018)
7. Di Giacomo, E., et al.: Ortho-polygon visibility representations of embedded graphs. Algorithmica **80**(8), 2345–2383 (2018)
8. Ding, G., Oporowski, B., Sanders, D.P., Vertigan, D.: Surfaces, tree-width, clique-minors, and partitions. J. Comb. Theory Ser. B **79**(2), 221–246 (2000)
9. Durocher, S., Mondal, D.: Relating graph thickness to planar layers and bend complexity. In: ICALP 2016. LIPIcs, vol. 55, pp. 10:1–10:13. Schloss Dagstuhl - Leibniz-Zentrum fuer Informatik (2016)
10. Even, S., Tarjan, R.E.: Computing an st-numbering. Theor. Comput. Sci. **2**(3), 339–344 (1976)
11. Gonçalves, D.: Edge partition of planar graphs into two outerplanar graphs. In: STOC 2005, pp. 504–512. ACM (2005)
12. Kedlaya, K.S.: Outerplanar partitions of planar graphs. J. Comb. Theory Ser. B **67**(2), 238–248 (1996)
13. Kobourov, S.G., Liotta, G., Montecchiani, F.: An annotated bibliography on 1-planarity. Comput. Sci. Rev. **25**, 49–67 (2017)
14. Lenhart, W.J., Liotta, G., Montecchiani, F.: On partitioning the edges of 1-plane graphs. Theor. Comput. Sci. **662**, 59–65 (2017)
15. Nash-Williams, C.S.A.: Edge-disjoint spanning trees of finite graphs. J. Lond. Math. Soc. **s1–36**(1), 445–450 (1961)
16. Pach, J., Tóth, G.: Graphs drawn with few crossings per edge. Combinatorica **17**(3), 427–439 (1997)

17. Schnyder, W.: Embedding planar graphs on the grid. In: SODA 1990, pp. 138–148. SIAM (1990)
18. Tamassia, R., Tollis, I.G.: A unified approach to visibility representations of planar graphs. Discret. Comput. Geom. **1**, 321–341 (1986)

On Minimum Connecting Transition Sets
in Graphs

Thomas Bellitto[1]([⊠]) and Benjamin Bergougnoux[2]([⊠])

[1] Université de Bordeaux, LABRI, CNRS, Bordeaux, France
thomas.bellitto@u-bordeaux.fr
[2] Université Clermont Auvergne, LIMOS, CNRS, Clermont-Ferrand, France
benjamin.bergougnoux@uca.fr

Abstract. A forbidden transition graph is a graph defined together with a set of permitted transitions *i.e.* unordered pair of adjacent edges that one may use consecutively in a walk in the graph. In this paper, we look for the smallest set of transitions needed to be able to go from any vertex of the given graph to any other. We prove that this problem is NP-hard and study approximation algorithms. We develop theoretical tools that help to study this problem.

1 Introduction

Graphs are the model of choice to solve routing problems in all sorts of networks. Depending on the applications, we sometimes need to express stronger constraints than what the standard definitions allow for. Indeed, in many practical cases, including optical networks, road networks or public transit systems among others, the set of possible walks a user can take is much more complex than the set of walks in a graph (see [1] or [2] for examples). To model a situation where a driver coming from a given road may not turn left while both the road he comes from and the road on the left exists, we have to define the permitted walks by taking into account not only the edges of the graph that a walk may use but also the transitions. A transition is a pair of adjacent edges and we call forbidden-transition graph a graph defined together with a set of permitted transitions.

Graphs with forbidden transitions have appeared in the literature in [3] and have received a lot of interest since, as well as other more specific models such as properly colored paths [4,5]. Many problems are harder in graphs with forbidden transitions, such as determining the existence of an elementary path (a path that does not use twice the same vertex) between two vertices which is a well-known polynomial problem in graph without forbidden transitions and has been proved NP-complete otherwise [6]. Algorithms for this problem have been studied in the general case [7] and also on some subclasses of graphs [8].

B. Bergougnoux—This work is supported by French Agency for Research under the GraphEN project (ANR-15-CE-0009).

A. Brandstädt et al. (Eds.): WG 2018, LNCS 11159, pp. 40–51, 2018.
https://doi.org/10.1007/978-3-030-00256-5_4

Forbidden transitions can also be used to measure the robustness of graph properties. In [9], Sudakov studies the Hamiltonicity of a graph with the idea that even Hamiltonian graphs can be more or less strongly Hamiltonian (an Hamiltonian graph is a graph in which there exists an elementary cycle that uses all the vertices). The number of transitions one needs to forbid for a graph to lose its Hamiltonicity gives a measure of its robustness: if the smallest set of forbidden transitions that makes a graph lose its Hamiltonicity has size 4, this means that this graph can withstand the failure of three transitions, no matter where the failures happen.

The notion we are interested in this paper is not Hamiltonicity but connectivity (the possibility to go from any vertex to any other), which is probably one of the most important properties we expect from any communication or transport network. However, our work differs from others in that we are not looking for the minimum number of transitions to forbid to disconnect the graph but for the minimum number of transitions to allow to keep the graph connected (the equivalent of minimum spanning trees for transitions). In other words, we are looking for the maximum number of transitions that can fail without disconnecting the graph, provided we get to choose which transitions still work. This does not provide a valid measure of the robustness of the network but measuring the robustness is only one part (the definition of the objective function) of the problem of robust network design. In most practical situations, robustness is achievable but comes at a cost and the optimization problem consists in creating a network as robust as possible for the minimum cost. In this respect, it makes sense to be able to choose where the failure are less likely to happen. Our problem highlights which transitions are the most important for the proper functioning of the network and this is where special attention must be paid in its design or maintenance. As long as those transitions work, connectivity is assured.

We also would like to point out that in practice, unusable transitions are not always the result of a malfunction. Consider a train network and imagine that there is a train going from a town A to a town B and one going from the town B to a town C. In the associated graph, there is an edge from A to B and one from B to C but if the second train leaves before the first one arrives, the transition is not usable and this kind of situation is clearly unavoidable in practice even if no special problem happens. Highlighting the most important transitions in the network thus helps design the schedule, even before the question of robustness arises.

Unlike Hamiltonicity or the existence of elementary path between two vertices, testing the connectivity is an easy task to perform even on graphs with forbidden transitions (note that a walk connecting two vertices does not have to be elementary). However, we prove that the problem of determining the smallest set of transitions that maintains the connectivity of the given graph is NP-hard even on co-planar graph which is the main contribution of the paper (see Sect. 3). Other notable contributions include a $O(|V|^2)$-time $\frac{3}{2}$-approximation (Theorem 3) and a reformulation of the problem (Theorem 2) which was of great help in the proofs of the other results and could hopefully be useful again in subsequent works.

Definitions and Notations. Throughout this paper, we only consider finite simple graphs, *i.e.* undirected graphs with a finite number of vertices, no multiple edges and no loop. Let G be a graph. The vertex set of G is denoted by $V(G)$ and its edge set by $E(G)$. The size of a set S is denoted by $|S|$. We denote by $d(v)$ the degree of a vertex v. We write xy to denote an edge $\{x, y\}$. We define a walk in G as a sequence $W = (v_1, \ldots, v_k)$ of vertices such that for all $i \leqslant k - 1$, $v_i v_{i+1} \in E(G)$ and we say that W uses the edge $v_i v_{i+1}$. Here, we say that the walk W leads from the vertex v_1 to v_k.

For $X \subseteq V(G)$, we denote by $G[X]$ the subgraph of G induced by X. We also denote by $G - X$ the subgraph of G induced by $V(G) \setminus X$. For $x \in V(G)$, we write $G - x$ instead of $G - \{x\}$. We denote by $\overline{G}$ the complement of G *i.e.* the graph such that $V(\overline{G}) = V(G)$ and $E(\overline{G}) = \{xy \in \binom{V(G)}{2} : xy \notin E(G)\}$. We say that a graph G is co-connected if and only if $\overline{G}$ is connected. We also call *co-connected components* (or *co-cc*) of G the connected components of $\overline{G}$.

Transitions. A transition is a set of two adjacent edges. We write abc for the transition $\{ab, bc\}$. If a walk uses the edges ab and bc consecutively (with $a \neq c$), we say that it uses the transition abc. For example, the walk (u, v, w, v, x) uses the transitions uvw and wvx. Let T be a set of transitions of G and $W = (v_1 \ldots v_k)$ be a walk on G. We say that W is T-compatible if and only if it only uses transitions of T *i.e.* for all $i \in [1, k - 2]$, we have $v_i v_{i+1} v_{i+2} \in T$ or $v_i = v_{i+2}$ (*i.e.* $v_i v_{i+1}$ and $v_{i+1} v_{i+2}$ are the same edge). Observe that a walk consisting of two vertices is always T-compatible. If for all vertices u and v of $V(G)$, there exists a T-compatible walk between u and v, then we say that G is T-connected and that T is a connecting transition set of G. The problem we study here is the following:

Minimum Connecting Transition Set (MCTS)

Input: A connected graph G.
Output: A minimum connecting transition set of G.

A more complete version of this paper is available on arXiv [10]. We refer the reader to the arXiv version for some of the proofs.

2   Polynomial Algorithms and Structural Results

In this section, we only consider graphs with at least 2 vertices. Our problem is trivial otherwise.

Lemma 1. *If G is a tree then a minimum connecting transition set of G has size $|V(G)| - 2$.*

Proof. We first prove that $|V(G)| - 2$ transitions are enough to connect G. For every vertex v of G, we pick a neighbor of v that we call $f(v)$. For every neighbor $u \neq f(v)$ of v, we allow the transition $uvf(v)$. We end up with the transition set $T = \{uvf(v) : v \in V(G), u \in N(v) \setminus \{f(v)\}\}$. Let u and v be vertices of G. Since G is connected, there exists a walk $(u, u_1, u_2, \ldots, u_k, v)$. The walk $(u, u_1, f(u_1), u_1, u_2, f(u_2), u_2, \ldots, u_k, f(u_k), u_k, v)$ is T-compatible and still leads from u to v. This proves that G is T-connected. The size of T is $|T| = \sum_{v \in V(G)} (d(v) - 1) = 2|E(G)| - |V(G)|$. Since G is a tree, $|E(G)| = |V(G)| - 1$ and thus, $|T| = |V(G)| - 2$.

Let us now prove by induction on the number n of vertices of G that at least $n - 2$ transitions are necessary to connect G. This is obvious for $n = 2$. Let us assume that it holds for n and let G be a tree with $n + 1$ vertices. Let T be a minimum connecting transition set of G. Let uv be an internal edge of T if any (*i.e.* an edge such that u and v are not leaves). Let a and b be two vertices from different connected components of $G - \{u, v\}$. Every walk leading from a to b in G therefore uses the edge uv and thus, two transitions containing uv. This proves that every internal edge of T belongs to at least two transitions of T. If every edge of G belongs to at least two transitions of T, T has size at least $|E(G)| = |V(G)| - 1$ which concludes the proof. Otherwise, let uv be an edge that belongs to at most one transition of T. This means that one of its vertices, say v, is a leaf. It is straightforward to check that uv must belong to one transition of T, otherwise G would not be T-connected. Let t be the transition in T containing uv. The graph $G - v$ is $T \setminus \{t\}$-connected and is a tree. By the induction hypothesis, this means that $|T \setminus \{t\}| \geq n - 3$ and $|T| \geq n - 2$. This concludes the proof of the lemma. $\square$

Let us also note that a linear-time algorithm to compute an optimal solution can be easily deduced from this proof. Since every connected graph contains a spanning tree, we have the following corollary.

Corollary 1. *Every connected graph G has a connecting transition set of size $|V(G)| - 2$.*

Note however that in the general case, this bound is far from tight. The most extreme case is the complete graph where every vertex can be connected to every other with a walk of one edge, that therefore uses no transition. Thus, the empty set is a connecting transition set of the complete graph. The following result aims at tightening the upper bound on the size of the minimum connecting transition set of a graph.

Theorem 1. *Every connected graph G has a connecting transition set of size $\tau(G)$ where*

$$\tau(G) = \sum_{\substack{C \text{ co-cc of } G \\ |C| \geq 2}} \begin{cases} |C| - 2 & \text{if the subgraph of } G \text{ induced by } C \text{ is connected} \\ |C| - 1 & \text{otherwise} \end{cases}$$

Proof. By definition, if u and v belong to different co-connected components of G, there is an edge $uv \in E(G)$ and there is therefore a walk between u and v is compatible with any transition set. We only have to find a transition set that connects all the vertices that belong to the same co-connected component.

Let C be a co-connected component of G with at least 2 vertices. If $G[C]$ is connected, Corollary 1 provides a transition set of size $|C| - 2$ that connects C. Otherwise, since G is connected, we know that $V(G) \neq C$ and there exists a vertex $v \notin C$. Hence, v is adjacent to every vertex of C and $C \cup \{v\}$ induces a connected subgraph of G. Corollary 1 provides a set of size $|C \cup \{v\}| - 2 = |C| - 1$ that connects C. By iterating this on every C, we build a connecting transition set T of size $\tau(G)$. □

Note that this bound can be computed in $O(|V(G)|^2)$. However, this bound is still not tight. Let us consider the graph $\overline{P_7}$ whose vertex set is $\{v_1, \ldots, v_7\}$ and where every vertex v_i, $2 \leqslant i \leqslant 6$ is connected to every vertex of the graph but v_{i-1} and v_{i+1}. Since the graph is connected and co-connected, $\tau(\overline{P_7}) = 5$ but the set $T = \{v_3 v_1 v_4, v_2 v_4 v_1, v_6 v_4 v_7, v_5 v_7 v_4\}$ is a connecting transition set of size only 4. To better understand this solution, let us consider the spanning tree of $\overline{P_7}$ depicted in Fig. 1:

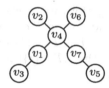

Fig. 1. A spanning tree of $\overline{P_7}$.

Note that the set T described above does not connect this spanning tree. Indeed, one can not go from v_1, v_2 or v_3 to v_5, v_6 or v_7 using a T-compatible walk in the tree. However, these vertices are already connected to each other by edges that do not belong to the spanning tree. The optimal solution here does not consist in connecting a spanning tree of G but in connecting a spanning tree of $G[\{v_1, v_2, v_3, v_4\}]$ and one of $G[\{v_4, v_5, v_6, v_7\}]$ and the cost is $(4-2)+(4-2) = 4$ instead of $7 - 2 = 5$.

In fact, we will prove that to each optimal connecting transition set T of a graph G corresponds a unique decomposition of G into subgraphs $G_1, G_2, \ldots, G_k$ such that T is the disjoint union of $T_1, T_2, \ldots, T_k$, where each T_i is the connecting transition set of some spanning tree of G_i. Observe that the size of T is uniquely determined by its correspondent decomposition, *i.e.*, $|T| = |V(G_1)| - 2 + \cdots + |V(G_k)| - 2$. Hence, finding an optimal connecting transition set is equivalent to finding its correspondent decomposition. In the following, we reformulate **MCTS** into this problem of graph decomposition which is easier to work with.

Definition 1. *Connecting Hypergraph*

Let G be a graph. A connecting hypergraph of G is a set H of subsets of $V(G)$, such that

- For all $E \in H$, we have $G[E]$ is connected and $|E| \geqslant 2$.
- For all $uv \notin E(G)$, there exists $E \in H$ such that $u, v \in E$ (we say that the hyperedge E connects u and v).

We define the problem of optimal connecting hypergraph as follows:

Optimal Connecting HyperGraph (OCHG)

Input: A connected graph G.
Output: A connecting hypergraph H that minimizes $\mathrm{cost}(H) = \sum_{E \in H} (|E| - 2)$.

In the next theorem, we prove that **OCHG** is a reformulation of **MCTS**.

Theorem 2. *Let G be a graph.*

- *The size of a minimum connecting transition set of G is the same as the cost of an optimal connecting hypergraph.*
- *A solution of one of these problems on G can be deduced in polynomial time from a solution of the other.*

Proof. Let G be a graph. This theorem is implied by the two following claims.

Claim. Let $H = \{E_1, \ldots, E_k\}$ be a connecting hypergraph of G. There exists a connecting transition set T of size at most $\mathrm{cost}(H)$.

By the definition of a connecting hypergraph, each E_i induces a connected graph and by Corollary 1, there exists a subset of transitions T_i of size $|E_i| - 2$ such that $G[E_i]$ is T_i-connected. Let $T = \bigcup_{i \leqslant k} T_i$. By definition, for all $uv \notin E(G)$, there exists i such that $u, v \in E_i$. Since $G[E_i]$ is T_i-connected and $T_i \subseteq T$, there is a T-compatible walk between u and v in G which means that G is T-connected. Since $T = \bigcup_{i \leqslant k} T_i$, $|T| \leqslant \sum_{i \leqslant k} |T_i| = \sum_{i \leqslant k} (|E_i| - 2) = \mathrm{cost}(H)$.

Claim. Let T be a connecting transition set of G. There exists a connecting hypergraph $H = \{E_1, \ldots, E_k\}$ of cost at most $|T|$.

Let $\sim$ be the relation on T such that $t \sim t'$ if t and t' share at least one common edge. We denote by $\mathcal{R}$ the transitive closure of $\sim$. Let $T_1, \ldots, T_k$ be the equivalence classes of $\mathcal{R}$. For all $i \leqslant k$, we denote by E_i the set of vertices induced by T_i. We claim that the hypergraph $\{E_1, \ldots, E_k\}$ is a connecting hypergraph and that, for all i, $|T_i| \geqslant |E_i| - 2$.

By construction, for all i, we have $|E_i| \geqslant 3$ since T_i contains at least one transition and thus, three vertices. Furthermore, since G is T-connected, there exists a T-compatible walk W between every pair $uv \notin E(G)$. All the transitions that W uses must be in T and are pairwise equivalent for $\mathcal{R}$. Thus, for all $uv \notin E(G)$, there exists i such that both u and v belong to E_i.

It remains to prove that for all i, $|E_i| - 2 \leqslant |T_i|$. We prove by induction on n that every set T of n pairwise equivalent transitions induces a vertex set of size at most $n + 2$. This property trivially holds for $n = 1$. Now, suppose that it is true for sets of size n and let T be a set of pairwise equivalent transitions of size $n + 1$. Let $P = t_1, \ldots, t_r$ be a maximal sequence of distinct transitions of T such that, for all $i \leqslant r - 1$, $t_i \sim t_{i+1}$. One can check that all the transitions of $T \setminus \{t_1\}$ are still pairwise equivalent (otherwise, P would not be maximal). By the induction hypothesis, $T \setminus \{t_1\}$ induces at most $n + 2$ vertices. Since t_1 shares an edge (and thus at least 2 vertices) with t_2, it induces at most one vertex not induced by $T \setminus \{t_1\}$. Thus T induces at most $n + 3$ vertices.   □

Let us note that the bound provided in Theorem 1 suggests a $O(|V|^2)$-time heuristic for **OCHG** which consists in building the set H as follows:

$$H = \bigcup_{\substack{C \text{ co-cc of } G \\ |C| \geqslant 2}} \begin{cases} C \text{ if the subgraph of } G \text{ induced by } C \text{ is connected} \\ C \cup \{v\} \text{ with } v \notin C \text{ otherwise} \end{cases}$$

We use the reformulation given by Theorem 2 to generalize Lemma 1:

Lemma 2 ([10]). *If G has a cut vertex, then a minimum connecting transition set of G has size $|V(G)| - 2$.*

The following lemma helps us prove that **MCTS** admits a $\frac{3}{2}$-approximation and its NP-hardness. It proves that if the graph is co-connected, we can restrict ourselves to some specific connecting hypergraph.

Lemma 3 ([10]). *Let G be a connected graph. If G is co-connected or G has a dominating vertex x and $G - x$ is connected and co-connected, then there exists an optimal connecting hypergraph $H = \{E_1, \ldots, E_k\}$ on G such that for all i, $G[E_i]$ is co-connected.*

We can also prove that **MCTS** has a polynomial $\frac{3}{2}$-approximation:

Theorem 3 ([10]). *For every connected graph G and optimal connecting transition set T of G, the size of T is at least $2/3\tau(G)$, where $\tau(G)$ is the function defined in Theorem 1.*

3   NP-hardness

In this section, we give a proof of NP-hardness of **OCHG** which involves very dense graphs. Hence, we prefer to work with the complementary graphs and therefore prove the NP-hardness of the following problem that we call **co-OCHG**:

Definition 2 *(co-Connecting Hypergraph).* *Let G be a graph. A co-connecting hypergraph is a collection of hyperedges $E_1, \ldots, E_r \subseteq V(G)$ such that*

- For all $i \leqslant r$, $G[E_i]$ is co-connected and $|E_i| \geqslant 2$.
- For all $uv \in E(G)$, there exists i such that $u, v \in E_i$ (we say that the hyperedge E_i covers the edge uv).

---

co-Optimal Connecting HyperGraph (co-OCHG)

Input: A co-connected graph G.
Output: A co-Connecting Hypergraph that minimizes $\text{cost}(H) = \sum_{E \in H} (|E| - 2)$.

We prove the NP-hardness of this problem by reducing 3-SAT to it. We restrict ourselves to the version of 3-SAT where each variable has at least one positive and one negative occurrence and each clause has exactly 3 literals that are associated to different variables. It is folklore that this restrictions of 3-SAT is NP-complete.

Let $\mathscr{F} = \{c_1, \ldots, c_m\}$ be an instance of 3-SAT with n variables. We will construct from $\mathscr{F}$ a graph $G_\mathscr{F}$ such that $\mathscr{F}$ is satisfiable if and only if $G_\mathscr{F}$ admits a co-covering hypergraph of cost $25\,m$.

We start by describing how to construct $G_\mathscr{F}$. To simplify the construction and the proofs, we give labels to some vertices and some edges. The set of labels we use are $\{c_i, T_{i,x}, F_{i,x} : i \leqslant m, x \text{ variable of } \mathscr{F}\}$. For each clause c_i and each variable x occurring in c_i, we create the gadget $g(x, c_i)$. If x occurs positively in c_i then $g(x, c_i)$ is the graph depicted in Fig. 2a, otherwise, if x occurs negatively in c_i then $g(x, c_i)$ is the graph depicted in Fig. 2b. Each gadget $g(x, c_i)$ contains a vertex labelled c_i and two edges labelled $T_{i,x}$ and $F_{i,x}$ (Fig. 2).

(a) The gadget $g(x, c_i)$ if x appears in c_i. (b) The gadget $g(x, c_i)$ if $\overline{x}$ appears in c_i.

Fig. 2.

We then create a new vertex for each clause c_i that we connect to the three vertices labelled c_i and to an additional vertex of degree 1. We thus have for each clause a graph like the one depicted in Fig. 3 that we call $g(c_i)$.

Finally, for each variable x, we do the following. Let $c_{i_1}, \ldots, c_{i_\ell}$ be the clause where x appears. Observe that $\ell \geqslant 2$ since every variable has a positive and a negative occurrence. For each $j \leqslant \ell$, we merge the edge labelled $T_{i_j, x}$ in $g(x, c_{i_j})$ with the edge labelled $F_{i_k, x}$ in $g(x, c_{i_k})$ (where $k = j + 1 \mod \ell$) such that the resulting edge has an extremity of degree one. We consider that this edge has both $T_{i_j, x}$ and $F_{i_k, x}$ as labels. For example, if a variable x appears positively in the clauses c_1 and c_4 and negatively in the clause c_3, the Fig. 4 depicts what the graph looks like around the gadget associated to the variable x.

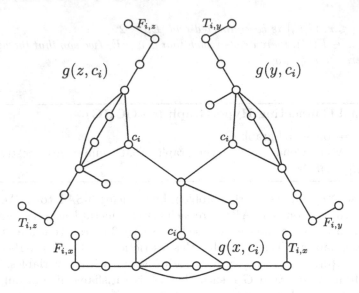

Fig. 3. The clause-gadget associated to the clause $c_i = (x \vee \neg y \vee \neg z)$.

By connecting all the gadgets $g(x, c_i)$ as described above, we obtain the gadget graph $G_{\mathscr{F}}$. We may assume that $G_{\mathscr{F}}$ is connected. Otherwise, this means that $\mathscr{F}$ is the conjunction of two formulas that share no common variables and $\mathscr{F}$ is satisfiable if and only if those two formulas are. Observe that $G_{\mathscr{F}}$ is trivially co-connected. Moreover, the size of $G_{\mathscr{F}}$ is polynomial in n and m.

Now, we prove that $\mathscr{F}$ is satisfiable if and only if $G_{\mathscr{F}}$ admits a co-covering hypergraph of cost $25\,m$. We start with the following lemma which proves the existence of an optimal co-covering hypergraph where every hyperedge is contained in the vertex set of some clause-gadget.

Lemma 4 ([10]). *There exists an optimal co-connecting hypergraph H of $G_{\mathscr{F}}$ such that $H = H_1 \cup H_2 \cup \cdots \cup H_m$ and for all $i \leqslant m$, we have $V(H_i) \subseteq V(g(c_i))$.*

Let $H = H_1 \cup \cdots \cup H_m$ be an optimal co-connecting hypergraph of G such that for all $i \leqslant m$, we have $V(H_i) \subseteq V(g(c_i))$.

Observe that the labelled edges are the only edges of $G_{\mathscr{F}}$ to belong to several clause-gadgets. Thus, for each $i \leqslant m$, the non-labelled edges of $g(c_i)$ must be covered by H_i. Consequently, the cost of H_i is fully determined by which labelled edges of $g(c_i)$ it covers. We want to prove that $G_{\mathscr{F}}$ is satisfiable if and only if the labelled edges can be covered in a way such that each H_j has cost 25.

Let c_i be a clause of $\mathscr{F}$ and let us study the cost of H_i in function of which labelled edges it covers. Let x be a variable of c_i. We recall that the gadget $g(x, c_i)$ differs depending on whether x appears positively or negatively in c_i but in both cases, the gadget has an edge labelled $F_{i,x}$ and one labelled $T_{i,x}$. The subgraph of $g(x, c_i)$ induced by H_i can take four values (up to isomorphims) depending on which of the following situations occurs:

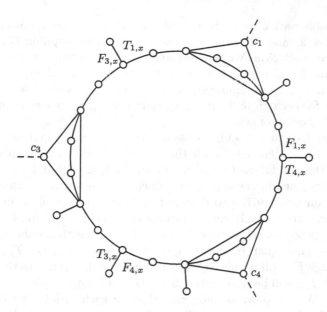

Fig. 4. The gadgets associated to the variable x.

- H_i covers neither $T_{i,x}$ nor $F_{i,x}$. We call this configuration N (for "none").
- H_i covers both $T_{i,x}$ and $F_{i,x}$. We call this configuration B (for "both").
- H_i covers $T_{i,x}$ and x appears positively in c_i or H_i covers $F_{i,x}$ and x appears negatively in c_i. We call this configuration S (for "satisfied").
- H_i covers $T_{i,x}$ and x appears negatively in c_i or H_i covers $F_{i,x}$ and x appears positively in c_i. We call this configuration U (for "unsatisfied").

Hence, the edges that H_i covers are determined (up to isomorphism) by the configurations encountered for each of the three variables that appear in c_i. Since the clause-gadget is symmetric, the order does not matter: the configuration SUN is exactly the same as the configuration NSU. Thus, we find that H_i can cover 20 different sets of edges up to isomorphisms. We determined the optimal values of $\text{cost}(H_i)$ for each case via a computer-assisted exhaustive search. The results are the following:

Configuration	Minimal cost	conf.	min.	conf.	min.	conf.	min.
BBB	28	BUS	26	UUU	26	UNN	25
BBU	27	BUN	26	UUS	25	SSS	25
BBS	27	BSS	26	UUN	25	SSN	25
BBN	27	BSN	26	USS	25	SNN	25
BUU	26	BNN	26	USN	25	NNN	25

The first observation we make is that the optimal value of $cost(H_i)$ is necessarily at least 25 and an optimal co-connecting hypergraph on $G_{\mathscr{F}}$ therefore always costs at least 25 m. We now investigate the case where the optimal cost is exactly 25 m. To this end, we suppose that H has a cost of 25 m.

We note that every configuration that contains a B costs at least 26. Thus, we know that for each H_i and each x appearing in c_i, H_i covers at most one of the two labelled edges of $g(x, c_i)$.

Let us now look at the gadgets associated to a variable x that appears in ℓ clauses (cf. Fig. 4). For all j such that x appears in c_j, the hypergraph H_j either covers the two labelled edges of $g(x, c_j)$ (B), one (S or U) or none (N). Since every edge must be covered at least once, this means that a solution where no configuration involves B also does not feature a configuration involving N. Hence, for every H_i, the only configurations that occur are S and U.

Let us suppose that H_i covers the edge $T_{i,x}$. Since the configuration B is impossible, we know that H_i does not cover the edge $F_{i,x}$. Let $T_{j,x}$ the other label of the edge $F_{i,x}$. Since this edge has to be covered, this means that H_j must cover the edge $T_{j,x}$ and because the configuration B is impossible, it cannot cover the edge $F_{j,x}$. We can prove by induction that for each variable x either, for all gadget $g(x, c_i)$, H_i covers the edge $T_{i,x}$ or for all gadget $g(x, c_i)$, H_i covers the edge $F_{i,x}$. In the first case, we say that the variable x is set to True, and in the second case, to False. If the variable x is set to True, this means all its positive occurrence will lead to a S configuration in the clause where it appears and conversely.

Finally, we notice that the cost of an optimal co-connecting hypergraph on the configurations SSS, SSU and SUU is 25 while it is 26 on the configuration UUU. Therefore, there exists a solution of cost 25 m if and only of there exists a way to affect all the variables to either True or False such that every clause is satisfied by at least one variable, which comes down to saying that the formula $\mathscr{F}$ is satisfiable.

This proves that **co-OCGH** and therefore **OCGH** and **MCTS** are all NP-hard. Moreover, Lichtenstein proved in [11] that 3-SAT remains NP-complete when restricted to formulas whose incidence graph is planar. The incidence graph of a formula $\mathscr{F}$ is the bipartite graph representing the relation of belonging between the variables and the clauses of $\mathscr{F}$. Clearly, if the incidence graph of $\mathscr{F}$ is planar then $G_{\mathscr{F}}$ is planar too. We conclude that **MCTS** is NP-hard even on co-planar graphs.

4   Conclusion

Our work proves that finding a minimum connecting transition set is NP-hard even on co-planar graphs. This notably implies the NP-hardness of other problems that generalizes this one such as finding a minimum connecting transition set in a graph that already has forbidden transitions.

A lot of our results suggest that the density of the graph has an impact on the complexity of **MCTS**. Consequently, it would be interesting to study the complexity of this problem on sparse graphs such as planar graphs or graphs with bounded treewith.

Further works could lead us to generalize this study to directed graphs, that are more suitable for many practical applications. Another interesting continuation of this work would also be the study of low-stretch connecting transition sets, a problem that is already well-studied for minimal spanning trees [12]. Intuitively, it consists in looking for a subset of transitions T such that the shortest T-compatible path between two vertices is not much longer than the shortest path in the graph with no forbidden transitions, which is also an important criteria of robustness.

Acknowledgments. The authors would like to thank Marthe Bonamy, Mamadou M. Kanté, Arnaud Pêcher, Théo Pierron and Xuding Zhu for the interest they showed for our work and for inspiring discussions.

References

1. Ahmed, M., Lubiw, A.: Shortest paths avoiding forbidden subpaths. In: 26th International Symposium on Theoretical Aspects of Computer Science, STACS 2009, Freiburg, Germany, Proceedings, 26–28 February 2009, pp. 63–74 (2009)
2. Bellitto, T.: Separating codes and traffic monitoring. Theor. Comput. Sci. **717**, 73–85 (2017)
3. Kotzig, A.: Moves without forbidden transitions in a graph. Matematický časopis **18**(1), 76–80 (1968)
4. Chen, C.C., Daykin, D.E.: Graphs with Hamiltonian cycles having adjacent lines different colors. J. Comb. Theory Ser. B **21**(2), 135–139 (1976)
5. Gutin, G., Kim, E.J.: Properly coloured cycles and paths: results and open problems. In: Graph Theory, Computational Intelligence and Thought, Essays Dedicated to Martin Charles Golumbic on the Occasion of His 60th Birthday. pp. 200–208 (2009)
6. Szeider, S.: Finding paths in graphs avoiding forbidden transitions. Discret. Appl. Math. **126**(2–3), 261–273 (2003)
7. Kanté, M.M., Laforest, C., Momège, B.: An exact algorithm to check the existence of (elementary) paths and a generalisation of the cut problem in graphs with forbidden transitions. In: van Emde Boas, P., Groen, F.C.A., Italiano, G.F., Nawrocki, J., Sack, H. (eds.) SOFSEM 2013. LNCS, vol. 7741, pp. 257–267. Springer, Heidelberg (2013). https://doi.org/10.1007/978-3-642-35843-2_23
8. Kanté, M.M., Moataz, F.Z., Momège, B., Nisse, N.: Finding paths in grids with forbidden transitions. In: Mayr, E.W. (ed.) WG 2015. LNCS, vol. 9224, pp. 154–168. Springer, Heidelberg (2016). https://doi.org/10.1007/978-3-662-53174-7_12
9. Sudakov, B.: Robustness of graph properties. arXiv (2016)
10. Bellitto, T., Bergougnoux, B.: On minimum connecting transition sets in graphs. arXiv (2018)
11. Lichtenstein, D.: Planar formulae and their uses. SIAM J. Comput. **11**(2), 329–343 (1982)
12. Peleg, D.: Low stretch spanning trees. In: Diks, K., Rytter, W. (eds.) MFCS 2002. LNCS, vol. 2420, pp. 68–80. Springer, Heidelberg (2002). https://doi.org/10.1007/3-540-45687-2_5

Recognizing Hyperelliptic Graphs
in Polynomial Time

Jelco M. Bodewes[1], Hans L. Bodlaender[1,3], Gunther Cornelissen[2],
and Marieke van der Wegen[1,2(✉)]

[1] Department of Information and Computing Sciences, Utrecht University,
P.O. Box 80.089, 3508 TB Utrecht, The Netherlands
M.vanderWegen@uu.nl
[2] Mathematical Institute, Utrecht University, P.O. Box 80.010,
3508 TA Utrecht, The Netherlands
[3] Department of Mathematics and Computer Science,
Eindhoven University of Technology, P.O. Box 513,
5600 MB Eindhoven, The Netherlands

Abstract. Based on analogies between algebraic curves and graphs,
Baker and Norine introduced *divisorial gonality*, a graph parameter for
multigraphs related to treewidth, multigraph algorithms and number
theory. We consider so-called *hyperelliptic graphs* (multigraphs of gonality 2) and provide a safe and complete set of reduction rules for such
multigraphs, showing that we can recognize hyperelliptic graphs in time
$O(n \log n + m)$, where n is the number of vertices and m the number of
edges of the multigraph. A corollary is that we can decide with the same
runtime whether a two-edge-connected graph G admits an involution σ
such that the quotient $G/\langle\sigma\rangle$ is a tree.

1 Introduction

Motivation. In this paper, we consider a graph theoretic problem that finds its
origin in algebraic geometry, and can be formulated in terms of a specific type
of graph search, namely *monotone chip firing*. The case with two chips is of
special interest in the application, and we show that we can decide this case in
$O(n \log n + m)$ time on a multigraph with n vertices and m edges.

In algebraic geometry, a special role is played by so-called *hyperelliptic* curves;
these are smooth projective algebraic curves possessing an involution, i.e. an
automorphism of order two, for which the quotient is the projective line. Such
curves can be described by an affine equation $y^2 = f(x)$, for some one-variable
polynomial $f(x)$ without repeated roots. They are widely studied and used, for
example in the study of moduli spaces of abelian surfaces, invariants of binary

---

H. L. Bodlaender—This author was partially supported by the NETWORKS project,
funded by the Netherlands Organisation for Scientific Research.

A. Brandstädt et al. (Eds.): WG 2018, LNCS 11159, pp. 52–64, 2018.
https://doi.org/10.1007/978-3-030-00256-5_5

quadratic forms, diophantine problems (finding integer or rational solutions to such equations), and in so-called "hyperelliptic curve cryptography" (see, e.g., [15,28]).

Recognizing hyperelliptic curves is an important, decidable problem in algorithmic algebraic geometry; an algorithm has been implemented when the curve is given by some set of polynomial equations, e.g., in the computer algebra package MAGMA [13]. No exact runtime analysis is available, but, the method being dependent on Gröbner basis computations, worst-case performance is expected to be more than exponential in the input size.

In recent work of Baker and Norine [5], the notion of a "hyperelliptic graph" was introduced, based on an analogy between algebraic curves and multigraphs. We show that the recognition problem for hyperelliptic graphs can be solved in quasilinear time. This can be applied to the recognition of certain hyperelliptic curves, since if an algebraic curve has a non-hyperelliptic stable reduction graph, the curve itself cannot be hyperelliptic (see [4, 3.5]).

Divisorial Gonality. Hyperelliptic graphs are graphs with *divisorial gonality* at most two. The notion of divisorial gonality has several equivalent definitions; intuitively, we use a chip firing game: we have a graph and some initial configuration that assigns a non-negative number of "chips" to each vertex. We can fire a subset of vertices by moving a chip along each outgoing edge of the subset, *if* every vertex has sufficiently many chips. We say that an initial configuration reaches a vertex if a sequence of firings results in that vertex having at least one chip. The divisorial gonality of a graph is the minimum number of chips needed for an initial configuration to reach each vertex of the graph. It actually suffices to consider a 'monotone' variant of the chip firing procedure, in which the sequence of subsets that are fired to reach a vertex is increasing; this is similar to several other graph search games, where the optimal number of searchers does not increase when we require the search to be monotone, see e.g., [7,25].

Known Results. The termination of similar Mancala-style games was discussed by Björner, Lovász and Shor [8]. In the guise of "abelian sandpile model", they play an important role in the study of self-organized criticality in statistical physics [3,19]. The chip firing game introduced by Baker and Norine is relevant for classical combinatorial problems about graphs, relating to spanning trees [14], the uniqueness of graph involutions [5], and potential theory on electrical network graphs [6]. A polynomial bound on the minimal number of required firings to terminate the Björner, Lovász and Shor-game was given by Tardos [30].

We study the divisorial gonality of graphs from the point of view of computational complexity. The analogous problem of computing the gonality of an algebraic curve is decidable [29].

The divisorial gonality of a graph G is related to treewidth, $tw(G)$, by an inequality [21]

$$\mathrm{dgon}(G) \geq \mathrm{tw}(G). \tag{1}$$

Since treewidth is insensitive to the presence of multiple edges while divisorial gonality is not, the parameters are different; actually, they are not "tied" in the sense of Norin [27]: there exists G with $\text{tw}(G) = 2$ but $\text{dgon}(G)$ arbitrarily high [24]. We know that treewidth is FPT, and that computing divisorial gonality is NP-hard and in XP [22], [20, Sect. 5].

Our Results. Our main result is the following.

Theorem A (= Theorem 1). *There is an algorithm that decides whether a graph G is hyperelliptic in $O(n \log n + m)$ time.*

To obtain our algorithm, we provide a safe and complete set of reduction rules. Similar to recognition algorithms for graphs of treewidth 2 or 3 (see [1]), in our algorithm the rules are applied to the graph until no further rule application is possible; we decide positively if and only if this results in the empty graph. One novelty is that some of the rules introduce constraints on pairs of vertices, which we model by colored edges. To deal with the fact that some of the rules are not local, we use a data structure that allows us to find an efficient way of applying these rules, leading to the stated running time. Omitted proofs and details can be found in [9], in which we also consider other variants of gonality.

The computational complexity of the problem "Does a graph admit a non-trivial automorphism" (solvable in quasi-polynomial time [2]) is very sensitive to alterations of the question. For example, deciding whether a graph has a *fixed point free* automorphism of order two is NP-complete (see Lubiw [26]). Our main result implies the following result as corollary.

Corollary A (= Corollary 1). *There is an algorithm that, given a two-edge-connected graph G, decides in $O(n \log n + m)$ time whether G admits an involution σ such that the quotient $G/\langle \sigma \rangle$ is a tree.*

2   Preliminaries

2.1   Definitions

Whenever we write "graph" we refer to a multigraph $G = (V, E)$, where V is the set of vertices and E is a multiset of edges.

There is a number of different definitions of divisorial gonality. The one we use is shown to be equivalent to the chip firing procedure without the 'monotonicity' property by [20]. The definition given here allows us to prove correctness of the reduction rules in our algorithm, and avoids more heavy algebraic terminology.

A *divisor* D in a graph $G = (V, E)$ is a mapping $D \colon V \to \mathbb{Z}$ (a divisor represents a distribution of chips, see Sect. 1). We call a divisor D *effective* (notation $D \geq 0$) if $D(v) \geq 0$ for all $v \in V$. The degree, $\deg(D)$, of a divisor D equals $\sum_{v \in V} D(v)$.

Given an effective divisor D and a set of vertices $W \subseteq V$, we call W *valid for* D, if for each $v \in W$, $D(v) \geq |E(v, V \setminus W)|$ (i.e., v has at least as many chips as it has neighbors in $V \setminus W$). If W is valid for D, we can *fire* W starting from D, this yields another divisor: for $v \in W$, $D(v)$ is decreased by the number of edges from v to $V \setminus W$, and for $x \in V \setminus W$, $D(x)$ is increased by the number of edges from W to x. Intuitively, firing W means moving a chip along all edges from W to $V \setminus W$. Note that the divisor obtained by firing is effective as well.

We call two effective divisors D and D' *equivalent*, in notation $D \sim D'$, if there is a sequence of subsets $A_1 \subseteq A_2 \subseteq \ldots \subseteq A_{k-1} \subset A_k = V$, such that for all i the set A_i can be fired when $A_1, \ldots, A_{i-1}$ are fired starting from D, and the divisor obtained by firing $A_1, \ldots, A_k$ is D'. This defines an equivalence relation on the set of effective divisors [20, Chap. 3]. For two equivalent effective divisors D and D', we call the difference of functions $D' - D$ the *transformation* from D to D', and the sequence $A_1 \subseteq A_2 \subseteq \ldots \subseteq A_{k-1} \subset A_k = V$ the *level set decomposition* of this transformation. This level set decomposition is unique [20, Remark 3.8].

We say that an effective divisor D *reaches* a vertex v, if there exists a D' such that $D \sim D'$ and $D'(v) \geq 1$. The *divisorial gonality*, dgon(G), of a graph G is the minimum degree of an effective divisor D that reaches each vertex of G.

Example 1. Let T be a tree. Then T has divisorial gonality 1. Let v be a vertex of T and consider the divisor D with $D(v) = 1$ and $D(x) = 0$ for all $x \neq v$. This divisor has degree 1 and reaches each vertex of T: Let w be a vertex of T. Let vu be the first edge on the unique path from v to w. Let A_v be the component that contains v of the cut induced by vu. Firing A_v yields the divisor $D(u) = 1$ and $D(x) = 0$ for all $x \neq u$, thus we moved a chip from v to u. Repeating this process yields a divisor with a chip on w.

Example 2. Let G be a cycle, then G has divisorial gonality 2. First note that every set of vertices of G induces a cut of size at least 2. Hence for all degree 1 divisors, there are no valid sets. Hence a degree 1 divisor does not reach every vertex. To see that there is a divisor with 2 chips that reaches every vertex, number the vertices $v_1, v_2, \ldots, v_n$ and consider the divisor D with a chip on v_1 and a chip on v_n. To reach a vertex v_k with $k \leq \frac{n}{2}$, fire the set $\{v_i \mid 1 \leq i \leq j\} \cup \{v_i \mid n - j + 1 \leq i \leq n\}$ for $j = 1, 2, \ldots, i - 1$. Analogous for a vertex v_k with $\frac{n}{2} \leq k \leq n$.

Example 3. Consider the graph G in Fig. 1. This graph has treewidth 1 and divisorial gonality 3. A divisor that reaches all vertices either has a chip on u and 2 more chips to reach both v and w, or has at least 3 chips to move along the three edges from v to u. See also [16, Table 3].

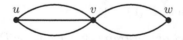

Fig. 1. Graph with divisorial gonality 3 and treewidth 1 (see Example 3).

Example 4. Consider the graph G in Fig. 2. This graph has treewidth 2 and divisorial gonality 3. A divisor that reaches all vertices needs two chips to traverse the left cycle and 2 chips to traverse the right cycle. But we cannot move two chips from u to v, so these two chips on the left side cannot be the same as the two on the right side. Hence we need at least three chips.

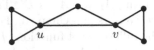

Fig. 2. Graph with divisorial gonality 3 and treewidth 2 (see Example 4).

2.2 Constraints

General Constraints. In the process of applying reduction rules to a graph, we will need to keep track of certain restrictions otherwise lost by removal of vertices and edges. We will maintain these restrictions in the form of a set of pairs of vertices, called constraints, and then extend the notion of divisorial gonality to graphs with constraints.

Definition 1. *Given a graph $G = (V, E)$, a constraint on G is an unordered pair of vertices $v, w \in V$, usually denoted as (v, w), where v and w can be the same vertex.*

Constraints are, like edges, pairs of vertices, so we can consider them as an extra set of edges. We will use $\mathcal{C}$ to represent this set.

Checking whether a graph has gonality two or lower is the same as checking whether there exists a divisor on our graph with degree two that reaches all vertices. Our constraints place restrictions on what divisors we consider, as well as what sets we are allowed to fire.

Definition 2. *Given a set of constraints $\mathcal{C}$, and two equivalent effective divisors D and D'. We call D and D' $\mathcal{C}$-equivalent (in notation $D \sim_{\mathcal{C}} D'$), if for every set A_i of the level set decomposition of $D' - D$ and every constraint $(u, v) \in \mathcal{C}$, either $u, v \in A_i$ or $u, v \notin A_i$.*

Note that this defines a finer equivalence relation. Now we can extend the definition of *reach* using $\mathcal{C}$-equivalence: a divisor D *reaches* a vertex v, if there exists a D' such that $D \sim_{\mathcal{C}} D'$ and $D'(v) \geq 1$.

Definition 3. *Given a set of constraints $\mathcal{C}$. A divisor D satisfies $\mathcal{C}$ if for every constraint $(u, v) \in \mathcal{C}$ there is a divisor $D' \sim_{\mathcal{C}} D$ such that $D'(u) \geq 1$ and $D'(v) \geq 1$ if $u \neq v$ and $D'(u) \geq 2$ if $u = v$.*

Definition 4. *Given a graph $G = (V, E)$ with constraints $\mathcal{C}$, we call a divisor D suitable if it is effective, has degree 2, reaches all vertices using the $\mathcal{C}$-equivalence relation and satisfies all constraints in $\mathcal{C}$.*

Definition 5. *We will say that a graph with constraints has divisorial gonality 2 or lower if it admits a suitable divisor. Note that for a graph with no constraints this is equivalent to the usual definition of divisorial gonality 2 or lower. We will denote the class of graphs with constraints that have divisorial gonality two or lower as $\mathcal{G}_2$.*

Constraints and Cycles. It will be useful to determine when constraints are non-conflicting locally:

Definition 6. *Let C be a cycle in a graph G with constraints $\mathcal{C}$. Let $\mathcal{C}_C \subseteq \mathcal{C}$ be the subset of the constraints that contain a vertex in C. We call the constraints $\mathcal{C}_C$ compatible if the following hold.*

(i) *If $(v, w) \in \mathcal{C}_C$ then both $v \in C$ and $w \in C$.*
(ii) *For each $(v, w) \in \mathcal{C}_C$ and $(v', w') \in \mathcal{C}_C$, the divisor given by assigning a chip to v and w must be equivalent to the one given by assigning a chip to v' and w' on the subgraph consisting of C.*

2.3 Reduction Rules, Safeness and Completeness

A *reduction rule* is a rule that can be applied to a graph to produce a smaller graph. Our final goal with the set of reduction rules is to show that it can be used to characterize the graphs in a certain class, that of the graphs with divisorial gonality two, by reduction to the empty graph. For this we need to make sure that membership of the class is invariant under our reduction rules.

Definition 7. *Let U be a rule and S be a set of reduction rules. Let $\mathcal{A}$ be a class of graphs. We call U safe for $\mathcal{A}$ if for all graphs G and H such that H can be produced by applying rule U to G it follows that $H \in \mathcal{A} \Longleftrightarrow G \in \mathcal{A}$. We call S safe for $\mathcal{A}$ if every rule in S is safe for $\mathcal{A}$.*

Apart from our rule sets being safe, we also need to know that, if a graph is in our class, it is always possible to reduce it to the empty graph.

Definition 8. *Let S be a set of reduction rules and $\mathcal{A}$ be a class of graphs. We call S complete for $\mathcal{A}$ if for any graph $G \in \mathcal{A}$ it holds that G can be reduced to the empty graph by applying some finite sequence of rules from S.*

For any rule set that is both complete and safe for $\mathcal{A}$ the rule set is suitable for characterizing $\mathcal{A}$: a graph G can be reduced to the empty graph if and only if G is in $\mathcal{A}$. Additionally it is not possible to make a wrong choice early on that would prevent the graph from being reduced to the empty graph: if $G \in \mathcal{A}$ and G can be reduced to H, then H can be reduced to the empty graph.

These properties ensure that we can use the set of reduction rules to create an algorithm for recognition of the graph class.

3   Reduction Rules for Divisorial Gonality

We will now show that there exists a set of reduction rules that is safe and complete for the class of graphs with divisorial gonality at most two. We will assume that our graph is loopless and connected. Loops can simply be removed from the graph since they never impact the divisorial gonality and a disconnected graph has divisorial gonality two or lower exactly when it consists of two trees, which can easily be checked in linear time. All reduction rules below maintain connectedness.

The Reduction Rules

We are given a connected loopless graph $G = (V, E)$ and a yet empty set of constraints $\mathcal{C}$. The following rules are illustrated in Fig. 3, where a constraint is represented by a red dashed edge.

We start by covering the two possible end states of our reduction:

Rule E_1. *Given a graph consisting of exactly one vertex, remove that vertex.*

Rule E_2. *Given a graph consisting of exactly two vertices, u and v, connected to each other by a single edge, and $\mathcal{C} = \{(u, v)\}$, remove both vertices.*

Next are the reduction rules to get rid of vertices with degree one. These rules are split by what constraint applies to the vertex:

Rule T_1. *Let v be a leaf, such that v has no constraints in $\mathcal{C}$. Remove v.*

Rule T_2. *Let v be a leaf, such that its only constraint in $\mathcal{C}$ is (v, v). Let u be its neighbor. Remove v and add the constraint (u, u) if it does not exist yet.*

Rule T_3. *Let v_1 be a leaf, such that its only constraint in $\mathcal{C}$ is (v_1, v_2), where v_2 is another leaf, whose only constraint is also (v_1, v_2). Let u_1 be the neighbor of v_1 and u_2 be the neighbor of v_2 (these can be the same vertex). Then remove v_1 and v_2 and add the constraint (u_1, u_2) if it does not exist yet.*

Finally we have a set of reduction rules that apply to cycles containing at most 2 vertices with degree greater than two. The rules themselves are split by the number of vertices with degree greater than two.

Rule C_1. *Let C be a cycle of vertices with degree two. If the set of constraints $\mathcal{C}_C$ on C is compatible, then replace C by a new single vertex.*

Rule C_2. *Let C be a cycle with one vertex v with degree greater than two. If the set of constraints $\mathcal{C}_C$ on C plus the constraint (v, v) is compatible, then remove all vertices except v in C and add the constraint (v, v) if it does not exist yet.*

Rule C_3. *Let C be a cycle with two vertices v and u of degree greater than two. If there exists a path from v to u that does not share any edges with C and the set of constraints $\mathcal{C}_C$ on C plus the constraint (v, u) is compatible, then remove all vertices of C except v and u, remove all edges in C and add the constraint (v, u) if it does not exist yet.*

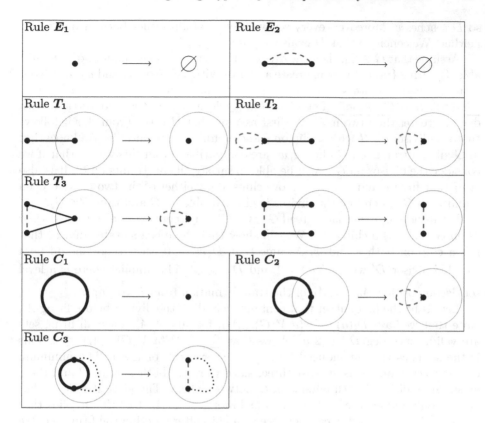

Fig. 3. The reduction rules for divisorial gonality

We denote by $\mathcal{R}$ the set consisting of all the above reduction rules: E_1, E_2, T_1, T_2, T_3, C_1, C_2 and C_3.

In the rest of this paragraph we will present safeness proofs for some of the rules and the more interesting parts of the proof of completeness. Details are found in [9, Sect. 4].

Proposition 1. (Safeness). *The set of rules $\mathcal{R}$ is safe for $\mathcal{G}_2$.*

Proof. We need to proof that all rules are safe, we will show this for rules T_3 and C_3 below, for the other rules, see [9, Lemma 4.6, 4.7, 4.9, 4.10].

Claim 1: Rule T_3 is safe. Let v_1 and v_2 be the vertices with degree one, such that their only constraint is (v_1, v_2) and let u_1 and u_2 be their (possibly equal) neighbors. We first assume that $H \in \mathcal{G}_2$, then there is a suitable divisor on H with one chip on u_1 and another chip on u_2. Consider this divisor on G. Then by firing $V(G) \setminus \{v_1, v_2\}$ we can move a chip to v_1 and v_2. For every vertex $v \in V(G) \setminus \{v_1, v_2\}$ there is a sequence $A_0, A_1, \ldots, A_k \subseteq V(H)$ such that firing this sequence yields a divisor D' with a chip on v. Now add v_i to every set A_j that contains u_i. Firing these sets on G starting from D results in D' on G, so D reaches v. Moreover, every set we fired contains either both v_1 and v_2, or neither. We conclude that D is also suitable on G.

Assume that $G \in \mathcal{G}_2$, then the divisor on G with one chip on v_1 and v_2 is suitable. By firing $\{v_1, v_2\}$ we can create a divisor with a chip on u_1 and u_2 (or two on u_1 if $u_1 = u_2$). It follows that this divisor is suitable when considered on H.

Claim 2: Rule C_3 is safe. Let C be our cycle and v, w the two vertices with degree greater than two in C. We first assume that $H \in \mathcal{G}_2$. From this it follows that the divisor on H with a chip on v and a chip on w is suitable. We know that in G all constraints on C plus (v, w) are compatible. From this we see that if we consider the divisor on G it will be able to satisfy all constraints on C. It is also clear that from v and w we can move chips along either of the two arcs between v and w in C. Therefore the divisor is also suitable on G and thus $G \in \mathcal{G}_2$.

Let us now assume that instead $G \in \mathcal{G}_2$. Clearly there exists a suitable divisor D on G that has a chip on v. We will show that there is a suitable divisor that has a chip on both v and w: Assume that $D(w) = 0$, then there should be a suitable divisor D' with $D'(w) = 1$ and $D \sim_{\mathrm{e}} D'$. This implies there is a level set decomposition $A_0, \ldots, A_k$ of the transformation from D to D'.

Let A_i be the first subset that contains w and D_i the divisor before firing A_i. Note that we have $D_i(a) \geq |E(a, V(G) \setminus A_i)|$ for all $a \in A_i$, since all firing sets are valid. Since $\deg(D_i) = 2$ it follows that $\sum_{a \in A_i} |E(a, V(G) \setminus A_i)| \leq 2$. This is the same as the cut induced by A_i having size two or lower. The minimum cut between v and w is at least three, since they are both part of C and there exists an additional path outside of C between them. Therefore it follows that A_i can only induce a cut of size two or lower if $w \in A_i$. But this implies that $D_i(w) \geq 1$, since a vertex can not receive a chip after entering the firing set. We conclude that $D_i(v) = 1$ and $D_i(w) = 1$.

Also by the fact that the minimum cut between v and w is at least three it follows that a subset firing can only be valid if the subset contains either both v and w or neither (since otherwise the subset would have at least three outgoing edges). It follows we can satisfy the set of constraints including (v, w).

Therefore the divisor D_i gives us a suitable divisor when considered on H. We conclude that $H \in \mathcal{G}_2$. $\qquad \square$

By the previous proposition we now have that membership in $\mathcal{G}_2$ is invariant under the reduction rules in $\mathcal{R}$. For the reduction rules to be useful however we will also need to confirm that any graph can be reduced to the empty graph by a finite sequence of rule applications.

Proposition 2. (Completeness). *The set of rules $\mathcal{R}$ is complete for $\mathcal{G}_2$.*

Proof. Let $G \in \mathcal{G}_2$ be a non-empty graph.

Claim 1: A rule in $\mathcal{R}$ can be applied to G. Assume instead that no rule in $\mathcal{R}$ can be applied to G.

Claim 2: G contains no vertices of degree 1 [9, Lemma 4.14]. It follows that all vertices of G have degree at least 2. Consider the minor H of G created by contracting each path of only degree 2 vertices to an edge. Then any edge in

H was either created by contraction of a path of any number of vertices with degree 2 in G or it already was an edge in G.

If H contains a loop, there is a path of degree 2 vertices in G going from a degree 3 or greater vertex to itself (since G contains no loops), so this path plus the vertex it is attached to forms a cycle with exactly one vertex of degree 3 or greater. Since we cannot apply Rule C_2 to G, it follows that the constraints $\mathcal{C}_C$ are not compatible. This contradicts the following claim:

Claim 3: For every cycle C in G, the constraints $\mathcal{C}_C$ are compatible [9, Lemma 4.15]. Hence H contains no loops.

Now we find a subgraph H' of H with no multiple edges. If H contains no multiple edges, simply let $H' = H$. Otherwise let v and w be two vertices such that there are at least two edges between v and w. Suppose that v and w are still connected to each other after removing two edges e_1, e_2 between them. The removed edges each represent a single edge or a path of degree 2 vertices in G. Thus v, w plus these paths form a cycle C in G with exactly two vertices of degree 3 or greater, where there is also a path between v and w that does not share any edges with C. By Claim 3 we have that the constraints on this cycle are compatible and so we are able to apply Rule C_3 to C. Since we cannot apply any rules to G, it follows that G must be disconnected after removing e_1 and e_2. So any multiple edge in H consists of a double edge, whose removal splits the graph in two connected components. Let H' be the connected component of minimal size over all possible removals of a double edge in H. Note that H' cannot contain any double edge, since this would imply a smaller connected component.

We now have a minor H' of G, which is a simple graph since it has no loops or multiple edges. Also, each vertex of H' has degree at least 3 with at most one exception, namely the vertex that was incident to the two parallel edges that were removed to obtain H'. Since a graph with treewidth at most two has at least two vertices of degree at most two, it follows that $\mathrm{tw}(H') \geq 3$ [12, Lemma 4]. Since treewidth is closed under taking minors we get $\mathrm{tw}(G) \geq 3$. But then by Eq. 1 it follows that $\mathrm{dgon}(G) \geq 3$, creating a contradiction, since $G \in \mathcal{G}_2$. We conclude that our assumption must be wrong and there must be a rule in $\mathcal{R}$ that can be applied to G.

Assume that $G \in \mathcal{G}_2$. By Claim 1 and Proposition 1 we can keep applying rules from $\mathcal{R}$ to G as long as G has not been turned into the empty graph yet. Observe that each rule removes at least one vertex or at least two edges, while never adding more vertices or edges. Since G is finite, rules from $\mathcal{R}$ can only be applied a finite number of times. When no more rules can be applied, it follows that the graph has been reduced to the empty graph. Therefore $\mathcal{R}$ is complete. $\square$

4   Main Algorithm

In this section, we discuss how the reduction rules of Sect. 3 lead to an efficient algorithm that recognize graphs with divisorial gonality 2 or lower.

Theorem 1 (= Theorem A). *There is an algorithm that, given a graph G, decides whether $dgon(G) \leq 2$ in $O(m + n \log n)$ time.*

Proof. We introduce a new rule that shortcuts repeated applications of Rule C_3:

Rule M. *Let u, v be vertices, such that $|E(u, v)| \geq 3$. Remove $2 \left\lfloor \frac{|E(u,v)|-1}{2} \right\rfloor$ edges between u and v and add a constraint (u, v).*

All applications of this rule can be done in $O(m)$ at the start of the algorithm, after which we know that no pair of vertices has more than two edges between them.

Since treewidth is a lower bound on divisorial gonality (Eq. 1), it follows that if $tw(G) > 2$, the algorithm can terminate. Checking whether treewidth is at most 2 can be done in linear time. Hereafter, we assume our graph has treewidth at most 2.

The remainder of the algorithm is of the following form: repeatedly try to apply a safe rule, until none is possible. If no rule is applicable, we can directly decide, as safeness and completeness of our set of rules implies that $dgon(G) \leq 2$, if and only if the resulting graph is empty. We now discuss how this can be done in $O(n \log n)$ time.

As graphs of treewidth k and n vertices have at most kn edges, the underlying simple graph has at most $2n$ edges. There are at most 2 edges between a pair of vertices and no loops, so at most $4n$ edges in total. Note that each rule application decreases the sum of the number of vertices and the number of edges by at least one, so $O(n)$ rules can be applied before we reach the empty graph.

For most rules, standard data structures allow to find applicable rules in amortized constant time. For Rules C_2 and C_3 we employ a technique used in [10]: we use a formulation in monadic second order logic (MSOL), and a data structure, based upon a tree decomposition of G of logarithmic depth and constant width allows to perform queries and updates in $O(\log n)$ time each. (See also [18, 23].)

The main idea is as follows: by [11, Lemma 2.2], we can build in $O(n)$ time a tree decomposition of G of width 8, such that the tree T in the tree decomposition is binary and has $O(\log n)$ depth. We augment the graph by labels that express for vertices and edges whether they are contracted, deleted, or carry a constraint. For each of the Rules C_2 and C_3, we can express the property that these can be applied to the graph obtained after a number of rule applications as a sentence in MSOL on the original graph G augmented with the labeling relations Contracted, Deleted and Carry-a-Constraint. The sentences have free variables that allow to find where in the graph the modification can take place. A modification of Courcelle's algorithm [17] gives that each query and each graph update can be done in time linear in the depth of the tree decomposition, i.e., $O(\log n)$ time. More details are given in [9, Sect. 7].

As the time per application of a safe rule is bounded by $O(\log n)$, and we execute $O(n)$ rule applications, the total time is bounded by $O(n \log n)$.   □

Corollary 1 (= Corollary A). *There is an algorithm that, given a two-edge-connected graph G, decides whether or not G admits an automorphism σ of order two such that the quotient $G/\langle\sigma\rangle$ is a tree, in $O(n\log n + m)$ time.*

Proof. This follows from Theorem 1, since Baker and Norine [5, Theorem 5.12] have shown that a two-edge-connected graph G is hyperelliptic precisely if G admits an automorphism as stated in the theorem.   □

5   Conclusion

In this text, we have focused on *divisorial* gonality, defined by analogy with the theory of divisors on algebraic curves and described in terms of chip-firing games. We gave a quasilinear detection algorithm for dgon ≤ 2. Different flavours of gonality exist, based on analogies with the theory of coverings of algebraic curves; or "stable" versions (in which the graph can be refined, based on ideas from the theory of tropical curves (see [16]). In [9], we give quasilinear time detection algorithms for these variants being two, too.

Finally, we mention some interesting open questions on (divisorial) gonality from the point of view of algorithmic complexity: (a) Can hyperelliptic graphs be recognized in linear time? (b) Which problems become fixed parameter tractable with gonality as parameter? (c) Is there an analogue of Courcelle's theorem for bounded gonality? (d) Is divisorial gonality fixed parameter tractable?

References

1. Arnborg, S., Proskurowski, A.: Characterization and recognition of partial 3-trees. SIAM J. Algebraic Discrete Methods **7**(2), 305–314 (1986)
2. Babai, L.: Graph isomorphism in quasipolynomial time. Preprint arXiv: 1512.03547v2 (2016)
3. Bak, P., Tang, C., Wiesenfeld, K.: Self-organized criticality. Phys. Rev. A **38**(1), 364 (1988)
4. Baker, M.: Specialization of linear systems from curves to graphs. Algebra Number Theory **2**(6), 613–653 (2008). With an appendix by Brian Conrad
5. Baker, M., Norine, S.: Harmonic morphisms and hyperelliptic graphs. Int. Math. Res. Not. IMRN **15**, 2914–2955 (2009)
6. Baker, M., Shokrieh, F.: Chip-firing games, potential theory on graphs, and spanning trees. J. Comb. Theory Ser. A **120**(1), 164–182 (2013)
7. Bienstock, D., Seymour, P.: Monotonicity in graph searching. J. Algorithms **12**, 239–245 (1991)
8. Björner, A., Lovász, L., Shor, P.W.: Chip-firing games on graphs. European J. Combin **12**(4), 283–291 (1991)
9. Bodewes, J.M., Bodlaender, H.L., Cornelissen, G., van der Wegen, M.: Recognizing hyperelliptic graphs in polynomial time. Preprint arXiv: 1706.05670 (2017)
10. Bodlaender, H.L., Drange, P.G., Dregi, M.S., Fomin, F.V., Lokshtanov, D., Pilipczuk, M.: A $c^k n$ 5-approximation algorithm for treewidth. SIAM J. Comput. **45**(2), 317–378 (2016)

11. Bodlaender, H.L., Hagerup, T.: Parallel algorithms with optimal speedup for bounded treewidth. SIAM J. Comput. **27**(6), 1725–1746 (1998)
12. Bodlaender, H.L., Koster, A.M.C.A.: Treewidth computations II. Lower bounds. Inform. and Comput. **209**(7), 1103–1119 (2011)
13. Bosma, W., Cannon, J., Playoust, C.: The Magma algebra system. I. The user language. J. Symbolic Comput. **24**(3–4), 235–265 (1997)
14. Chan, M., Glass, D., Macauley, M., Perkinson, D., Werner, C., Yang, Q.: Sandpiles, spanning trees, and plane duality. SIAM J. Discrete Math. **29**(1), 461–471 (2015)
15. Cohen, H., et al.: Handbook of Elliptic and Hyperelliptic Curve Cryptography, Second Edn. Chapman & Hall/CRC, Boca Raton (2012)
16. Cornelissen, G., Kato, F., Kool, J.: A combinatorial Li-Yau inequality and rational points on curves. Math. Ann. **361**(1–2), 211–258 (2015)
17. Courcelle, B.: The monadic second-order logic of graphs. I: recognizable sets of finite graphs. Inform. and Comput. **85**(1), 12–75 (1990)
18. Courcelle, B., Vanicat, R.: Query efficient implementation of graphs of bounded clique-width. Discrete Appl. Math. **131**(1), 129–150 (2003)
19. Dhar, D.: Self-organized critical state of sandpile automaton models. Phys. Rev. Let. **64**(14), 1613 (1990)
20. van Dobben de Bruyn, J.: Reduced divisors and gonality in finite graphs. Bachelor thesis, Leiden University (2012). https://www.universiteitleiden.nl/binaries/content/assets/science/mi/scripties/bachvandobbendebruyn.pdf
21. van Dobben de Bruyn, J., Gijswijt, D.: Treewidth is a lower bound on graph gonality. Preprint arXiv:1407.7055 (2014)
22. Gijswijt, D.: Computing divisorial gonality is hard. Preprint arXiv:1504.06713 (2015)
23. Hagerup, T.: Dynamic algorithms for graphs of bounded treewidth. Algorithmica **27**(3), 292–315 (2000)
24. Hendrey, K.: Sparse graphs of high gonality. Preprint arXiv:1606.06412 (2016)
25. LaPaugh, A.S.: Recontamination does not help to search a graph. J. ACM **40**(2), 224–245 (1993)
26. Lubiw, A.: Some NP-complete problems similar to graph isomorphism. SIAM J. Comput. **10**(1), 11–21 (1981)
27. Norin, S.: New tools and results in graph minor structure theory. In: Surveys in Combinatorics 2015. London Mathematical Society Lecture Note Series, vol. 424, pp. 221–260. Cambridge University Press (2015)
28. Poonen, B.: Computing rational points on curves. Number theory for the millennium. III (Urbana, IL, 2000), pp. 149–172. A K Peters, Natick (2002)
29. Schicho, J., Schreyer, F.-O., Weimann, M.: Computational aspects of gonal maps and radical parametrization of curves. Appl. Algebra Engrg. Comm. Comput. **24**(5), 313–341 (2013)
30. Tardos, G.: Polynomial bound for a chip firing game on graphs. SIAM J. Discrete Math. **1**(3), 397–398 (1988)

On Directed Feedback Vertex Set Parameterized by Treewidth

Marthe Bonamy[1], Łukasz Kowalik[2], Jesper Nederlof[3], Michał Pilipczuk[2], Arkadiusz Socała[2], and Marcin Wrochna[2(✉)]

[1] CNRS, LaBRI, Talence, France
marthe.bonamy@labri.fr
[2] Institute of Informatics, University of Warsaw, Warsaw, Poland
{kowalik,michal.pilipczuk,as277575,m.wrochna}@mimuw.edu.pl
[3] Eindhoven University of Technology, Eindhoven, Netherlands
j.nederlof@tue.nl

Abstract. We study the DIRECTED FEEDBACK VERTEX SET problem parameterized by the treewidth of the input graph. We prove that unless the Exponential Time Hypothesis fails, the problem cannot be solved in time $2^{o(t \log t)} \cdot n^{\mathcal{O}(1)}$ on general directed graphs, where t is the treewidth of the underlying undirected graph. This is matched by a dynamic programming algorithm with running time $2^{\mathcal{O}(t \log t)} \cdot n^{\mathcal{O}(1)}$. On the other hand, we show that if the input digraph is planar, then the running time can be improved to $2^{\mathcal{O}(t)} \cdot n^{\mathcal{O}(1)}$.

1 Introduction

In the DIRECTED FEEDBACK VERTEX SET (DFVS) problem we are given a digraph G and the goal is to find a smallest *directed feedback vertex set* in it, that is, a subset X of vertices such that $G - X$ is acyclic. The arc-deletion version, DIRECTED FEEDBACK ARC SET (DFAS), differs in that the deletion set X has to consist of edges of G instead of vertices. The parameterized variants of these problems, where we ask about the existence of a solution of size at most k for a given parameter k, are arguably among central problems in the field of parameterized algorithms. Unfortunately, we are still far from a complete understanding of their complexity.

Work supported by the National Science Centre of Poland, grant number 2013/11/D/ST6/03073 (MP, MW). The work of Ł. Kowalik is a part of the project TOTAL that has received funding from the European Research Council (ERC) under the European Union's Horizon 2020 research and innovation programme (grant agreement No 677651). This research is a part of projects that have received funding from the European Research Council (ERC) under the European Union's Horizon 2020 research and innovation programme under grant agreements No 714704 (AS). MP and MW are supported by the Foundation for Polish Science (FNP) via the START stipend programme. JN is supported by NWO Veni grant 639.021.438.

A. Brandstädt et al. (Eds.): WG 2018, LNCS 11159, pp. 65–78, 2018.
https://doi.org/10.1007/978-3-030-00256-5_6

Establishing the fixed-parameter tractability of DFVS was once a major open problem. It has been resolved by Chen et al. [2], who gave an algorithm for both DFVS and DFAS[1] with running time $2^{\mathcal{O}(k \log k)} \cdot n^{\mathcal{O}(1)}$, obtained by combining iterative compression with a smart application of important separators. Very recently, Lokshtanov et al. [15] revisited the algorithm of Chen et al. [2] and improved the running time to $2^{\mathcal{O}(k \log k)} \cdot (n + m)$; that is, the dependence on the size of the graph is reduced to linear, but the dependence on the parameter k is unchanged. Whether the running time can be improved to $2^{\mathcal{O}(k)} \cdot n^{\mathcal{O}(1)}$, or even to $2^{o(k \log k)} \cdot n^{\mathcal{O}(1)}$, remains a challenging open problem [15]. We remark that the question of whether DFVS admits a polynomial kernel on general digraphs remains one of the central open problems in the field of kernelization.

A possible reason for why so little progress has been observed on such an important problem, is that the analysis of cut problems in directed graphs is far more complicated than in undirected graphs, and fewer basic techniques are available. For instance, consider the undirected counterpart of the problem, FEEDBACK VERTEX SET, where the goal is to delete at most k vertices from a given undirected graph in order to obtain a forest. While forests have a very simple combinatorial structure that can be exploited in many ways, acyclic digraphs form a much richer class that cannot be so easily captured. In particular, undirected graphs admitting a feedback vertex set of size k have treewidth at most $k + 1$, and this tree-likeness of positive instances of undirected FVS makes the problem amenable to a variety of techniques related to treewidth; other basic techniques like branching and kernelization are also applicable. Acyclic digraphs may have arbitrarily large treewidth, whereas directed analogues of treewidth offer almost no algorithmic tools useful for the design of FPT algorithms. Therefore, for the study of DFVS and other directed cut problems in the parameterized setting, we are so far left with important separators and a handful of other more involved techniques; cf. [3,4,12,13,18].

In planar digraphs, the complexity of DFAS changes completely. As shown by Lucchesi and Younger [17], it is actually polynomial-time solvable (see also a different presentation by Lovász [16]). More precisely, this is a consequence of the proof of the Lucchesi–Younger theorem [17], which states that in planar digraphs, the minimum size of a directed feedback arc set is equal to the maximum size of a packing of arc-disjoint cycles. The proof is constructive and can be turned into a polynomial-time algorithm that computes a minimum directed feedback arc set together with a maximum cycle packing; see [19] for details.

On the other hand, it is easy to see that DFVS remains NP-hard on planar digraphs, as there is a simple reduction from VERTEX COVER on planar graphs to DIRECTED FEEDBACK VERTEX SET on planar digraphs: just pick an arbitrary ordering of vertices, orient all edges from left to right (giving an acyclic orientation), and replace every edge uv with a directed triangle on u, v, and a fresh vertex. To the best of our knowledge, no algorithm for DFVS with running time

---

[1] In general digraphs, DFVS and DFAS are well-known to be reducible to each other; see [5, Proposition 8.42 and Exercise 8.16]. These reductions, however, do not preserve planarity of the digraph in question.

$2^{o(k \log k)} \cdot n^{\mathcal{O}(1)}$ is known even for planar digraphs, which means that so far we are not able to exploit the planarity constraint in any useful way.

Our Contribution. The goal of this paper is to improve our understanding of DFVS by studying the parametcrization by the treewidth of the input directed graph[2], with a particular focus on the planar setting. We first show that a semi-standard dynamic programming approach yields an algorithm with running time $2^{\mathcal{O}(t \log t)} \cdot n^{\mathcal{O}(1)}$.

Theorem 1. *There is an algorithm that given a digraph G of treewidth t on n vertices, runs in time $2^{\mathcal{O}(t \log t)} \cdot n^{\mathcal{O}(1)}$ and determines the minimum size of a directed feedback vertex set and of a directed feedback arc set in G.*

For the proof of Theorem 1, we define the following dynamic programming table (here for DFVS). For a node x of a tree decomposition of G, let B_x be the associated bag and let G_x be the subgraph induced in G by vertices residing in B_x or below x in the decomposition. Then, for every subset X of B_x and every ordering σ of $B_x \setminus X$, we store the smallest size of a subset Y of $V(G_x) \setminus B_x$ such that $G_x - (X \cup Y)$ is acyclic and admits a topological ordering whose restriction to $B_x \setminus X$ is exactly σ. Dynamic programming algorithm for DFAS is defined similarly. While we believe that this simple formulation of dynamic programming for DFVS and DFAS on a tree decomposition should have been known, we did not find it in the literature and hence we include it in the full version [1].

Our next result states then that the running time of the algorithm of Theorem 1 is tight under the Exponential Time Hypothesis (ETH) (see the Preliminaries section for definitions).

Theorem 2. *Unless ETH fails, there is no algorithm that determines the minimum size of a directed feedback vertex set or of a directed feedback arc set in a given digraph in time $2^{o(t \log t)} \cdot n^{\mathcal{O}(1)}$, where t is the treewidth of the input graph and n is the number of its vertices.*

The proof of Theorem 2 uses the approach of Lokshtanov et al. [14] for proving slightly super-exponential lower bounds for the complexity of parameterized problems. More precisely, we give a parameterized reduction from the $k \times k$ Hitting Set with thin sets problem, for which a lower bound excluding running time $2^{o(k \log k)} \cdot n^{\mathcal{O}(1)}$ under ETH was given in [14]. As an intermediate step, we use problems asking for permutations that satisfy certain constraints; we remark that somewhat similar constraint satisfaction problems, though with different constraints, were previously studied by Kim and Gonçalves [11].

Finally, we move to the setting of planar graphs, where we prove that the running time can be improved to $2^{\mathcal{O}(t)} \cdot n^{\mathcal{O}(1)}$.

Theorem 3. *There is an algorithm that given a planar digraph G of treewidth t on n vertices, runs in time $2^{\mathcal{O}(t)} \cdot n^{\mathcal{O}(1)}$ and determines the minimum size of a directed feedback vertex set in G.*

---

[2] Throughout this paper, the treewidth of a directed graph is defined as the treewidth of its underlying undirected graph.

It is well known that the treewidth of a planar graph on n vertices is bounded by $\mathcal{O}(\sqrt{n})$; see e.g. [7]. This yields the following.

Corollary 1. *There is an algorithm that given a planar digraph G on n vertices, runs in time $2^{\mathcal{O}(\sqrt{n})}$ and determines the minimum size of a directed feedback vertex set in G.*

Note that the algorithm of Corollary 1 is tight under ETH, due to the aforementioned simple reduction from VERTEX COVER to DFVS which preserves planarity. Since VERTEX COVER on planar graphs cannot be solved in time $2^{o(\sqrt{n})}$ under ETH (see [5, Theorem 14.6]), the same lower bound carries over to DFVS on planar digraphs (implying also a tight lower bound of $2^{o(t)} \cdot n^{\mathcal{O}(1)}$ for the parameterization by treewidth on planar digraphs).

The proof of Theorem 3 is perhaps conceptually the most interesting part of this work. The basic idea is to use *sphere-cut decompositions* of plane graphs [6,20]. Namely, as observed by Dorn et al. [6], from the results of Seymour and Thomas [20] it follows that every plane graph admits an optimum-width branch decomposition that respects the plane embedding in the following sense: each subgraph corresponding to a subtree of the decomposition is embedded into a disk so that the interface of the subgraph—vertices adjacent to the remainder of the graph—are embedded on the boundary of the disk. Such a branch decomposition is called a *sphere-cut decomposition*. Since branchwidth is linearly related to treewidth, in the proof of Theorem 3 we may focus on branch decompositions instead of tree decompositions.

As shown by Dorn et al. [6], the topological properties of sphere-cut decomposition can be exploited algorithmically to bound the number of relevant states in dynamic programming. This idea is instantiated in the technique of *Catalan structures* where for some connectivity problems, like HAMILTONIAN CYCLE, the fact that the solution cannot self-intersect in the plane leads to an improvement on the number of states from $2^{\mathcal{O}(b \log b)}$ to $2^{\mathcal{O}(b)}$; here, b is the width of the considered sphere-cut decomposition. However, in the case of DFVS we cannot use Catalan structures directly, since we are not building any connected structure whose plane embedding would impose useful constraints.

Our main contribution here is that nevertheless, an improved upper bound on the number of relevant states can be shown, with a conceptually new reasoning. Consider a directed graph G embedded into a disk Δ and a subset T of its vertices that are placed on the boundary of Δ. Let the *connectivity pattern* induced by G on T be the reachability relation in G restricted to T^2: (s,t) are in the connectivity pattern if and only if in G there is a path from s to t. The crucial combinatorial statement (see Theorem 5) is as follows: the number of different connectivity patterns on T that may be induced by different digraphs G embedded in Δ is bounded by $2^{\mathcal{O}(|T|)}$; note that the naive bound would be $2^{\mathcal{O}(|T|^2)}$. This directly provides the sought upper bound on the number of relevant states in dynamic programming on a sphere-cut decomposition, leading to the proof of Theorem 3. To prove this statement, we show that every realizable connectivity pattern can be encoded using a constant number of simpler relations, each forming a directed outerplanar graph on $|T|$ vertices; the number of different such digraphs is $2^{\mathcal{O}(|T|)}$. In the proof that such an encoding is possible we use the result of Gyárfás that circle graphs are χ-bounded [8,9].

Organization. In Sect. 2 we establish notation and recall known relevant results. Section 3 concerns the main ingredient of the proof of Theorem 3, namely the combinatorial upper bound on the number of different connectivity patterns induced by disk-embedded directed graphs. Section 4 contains the hardness reduction for Theorem 2. Due to space restrictions the proofs of Theorem 3 and 1 and some proofs from Sects. 3 and 4 are deferred to the full version of this paper [1]. In these sections, theorems with deferred proofs are marked with †.

2 Preliminaries

Let $[k] := \{1, 2, \ldots, k\}$, and use standard graph notation, see e.g. [5]. The clique number of graph G is denoted $\omega(G)$, the chromatic number $\chi(G)$.

Chords and Circle Graphs: A *chord* is an unordered pair of distinct points on a circle, called *endpoints* of the chord; one may think of it as a straight line segment between its endpoints. Two chords $\{a, a'\}, \{b, b'\}$ of a circle *cross* if their endpoints are all distinct and a, b, a', b' occur in this order on the circle (clockwise or counter-clockwise). Intuitively this corresponds to the straight line segments aa' and bb' intersecting inside the circle. A *circle graph* is a graph whose vertices correspond to chords of a circle so that two vertices are adjacent if and only if the corresponding chords cross. A *circle graph with directed chords* is a circle graph in which every chord is directed; that is, it is an ordered pair. A directed chord with *tail a* and *head b* is denoted by (a, b). Let T be a finite set of points on a circle and let $R \subseteq T^2$ be a set of chords (directed or undirected). A *crossing* is a pair of crossing chords in R. The circle graph *induced* by R is the one with R as the vertex set where two chords from R are adjacent if they cross.

As introduced by Gyárfás [10], a class $\mathcal{C}$ of graphs closed under induced subgraphs is χ-*bounded* if there exists a function $f : \mathbb{N} \to \mathbb{N}$ such that for every graph $G \in \mathcal{C}$ we have $\chi(G) \le f(\omega(G))$. Gyárfás [8,9] proved the following.

Theorem 4 ([8,9]). *The class of circle graphs is χ-bounded.*

ETH: The Exponential Time Hypothesis (ETH) states that for some $c > 0$, there is no algorithm for 3SAT with running time $\mathcal{O}(2^{cn})$, where n is the number of variables of the input formula. ETH has served as a basic assumption for countless complexity lower bounds of computational problems. We refer to [5, Chap. 14] for a comprehensive overview of applications in parameterized complexity.

3   Connectivity Patterns

In this section we present the main combinatorial result leading to the proof of Theorem 3, which is a reduction of the number of relevant dynamic programming states in the planar setting. This is done by bounding the number of "connectivity patterns" that can be induced by directed graphs embedded in a disk.

Suppose T is a finite set. A *connectivity pattern* on T is any quasi-order on T, that is, a reflexive and transitive relation $P \subseteq T^2$. For a directed graph G and a vertex subset $T \subseteq V(G)$, we define the connectivity pattern *induced by G on T* to be the reachability relation on T in G: (s, t) is in the relation iff there is a path in G from s to t.

The main goal of this section is to prove a result that roughly states the following: for a directed graph G drawn in a closed disk, with T be the vertices lying the boundary of the disk, there are only $2^{O(|T|)}$ different possibilities for the connectivity pattern that G may induce. See Theorem 5 for a formal statement. As mentioned in the introduction, this result will be our main tool for limiting the number of relevant states in dynamic programming for DIRECTED FEEDBACK VERTEX SET on planar graphs. Note that in general directed graphs, the number of different connectivity patterns induced on a vertex subset T may be as large as $2^{\Theta(|T|^2)}$. For instance, any subset of pairs with tail in the first half of T and head in the second half already gives that many possibilities.

The idea of the proof is that such connectivity patterns induced by directed planar graphs embedded in a disk can be generated from simpler relations, which contain enough pairs to infer all the other ones from planarity. This is formalized in the following definition.

Definition 1. *For a set T of points on a circle and a relation $R \subseteq T^2$, define the connectivity pattern on T generated by R, denoted $\mathsf{gen}(R)$, as follows: a pair $(s, t) \in T^2$ is included in $\mathsf{gen}(R)$ if and only if for each partition of the circle into two disjoint arcs X_s, X_t such that $s \in X_s$ and $t \in X_t$, there exist $s' \in X_s$ and $t' \in X_t$ which satisfy $(s', t') \in R$.*

In the above definition, as well as throughout this whole section, arcs on a circle may be open or closed from either side, unless explicitly stated.

It is easy to check that $R \subseteq \mathsf{gen}(R)$ and $\mathsf{gen}(R)$ is indeed reflexive and transitive, for any $R \subseteq T^2$. Hence $\mathsf{gen}(R)$ also contains the reflexive transitive closure of R, but it may be much larger still. Furthermore, one can observe that $\mathsf{gen}(\mathsf{gen}(R)) = \mathsf{gen}(R)$, but we will not use this property. We now show that a connectivity pattern induced by a graph is generated by itself; the goal will be then to find simpler relations generating this pattern.

Lemma 1. *Let G be a planar digraph drawn in a disk Δ, T be a subset of vertices drawn on the boundary of Δ, and P be the connectivity pattern on T induced by G. Then $\mathsf{gen}(P) = P$.*

Proof. Let C be the boundary of Δ; we may assume that C is a circle. Clearly $P \subseteq \mathsf{gen}(P)$. Now assume that $(s, t) \in \mathsf{gen}(P)$, that is, for each partition of C

into two disjoint arcs X_s, X_t such that $s \in X_s$ and $t \in X_t$, there exist $s' \in X_s$ and $t' \in X_t$ which satisfy $(s', t') \in P$. We will show that $(s, t) \in P$.

Assume the contrary, that is, $(s, t) \notin P$. Define $T_s = \{r \in T : (s, r) \in P\}$, see Fig. 1. Let X_t be the largest arc on C that contains t and is disjoint from T_s; this is well-defined since $t \notin T_s$ and $s \in T_s$. Define $X_s = C \setminus X_t$, thus (X_s, X_t) is a partition of C into two disjoint arcs. Since $s \in T_s$, we have $s \notin X_t$ and thus $s \in X_s$. From our assumption that $(s, t) \in \mathsf{gen}(P)$, there exist $s' \in X_s$ and $t' \in X_t$ that satisfy $(s', t') \in P$.

We have two cases: either $s' \in T_s$ or $s' \notin T_s$. If $s' \in T_s$, then $(s, s') \in P$ and consequently $(s, t') \in P$, since P is transitive due to being the reachability relation induced by G. But then $t' \in T_s$ and hence $t' \notin X_t$, a contradiction. Now assume $s' \notin T_s$; in particular $s' \neq s$. Let us move along the circle from s to t such that on the way we meet the point s'. Because the arc X_t was chosen to be the largest possible, between s' and t we meet a point $r \in T_s$. The arc X_t is connected, so between s and r we cannot meet any point from the set X_t, in particular t'. That is, s, s', r, t' appear in this order on the circle (either clockwise or counterclockwise). Since $r \in T_s$, we have $(s, r) \in P$ and $(s', t') \in P$. Therefore, in G there are directed paths from s to r and from s' to t'. These two paths must intersect since they are drawn in a disk, which yields a path in G from s to t'. We conclude that $t' \in T_s$ and hence $t' \notin X_t$, a contradiction. □

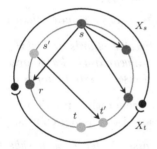

Fig. 1. Proof of Lemma 1: the induced pattern P shown as arrows, points in T_s depicted in green. (Color figure online)

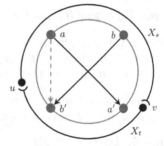

Fig. 2. Proof of Lemma 2.

The next lemma shows that generated connectivity patterns are closed under adding directed chords (a, b') whenever (a, a') and (b, b') cross. This operation (and its inverse) is the only one we will use to simplify the generating relation.

Lemma 2. *Let T be a finite set of points on a circle and let $R \subseteq T^2$. Let $a, b, a', b' \in T$ be distinct points that appear in this order on the circle, such that $(a, a') \in R$ and $(b, b') \in R$. Let $R' = R \cup \{(a, b')\}$. Then $\mathsf{gen}(R) = \mathsf{gen}(R')$.*

Proof. It is enough to prove that for each partition of the circle into two disjoint arcs X_s, X_t, the following two conditions are equivalent:

(1) There exist $s' \in X_s$ and $t' \in X_t$ which satisfy $(s', t') \in R$.
(2) There exist $s' \in X_s$ and $t' \in X_t$ which satisfy $(s', t') \in R'$.

Of course (1) implies (2). Now assume (2). If $(s', t') \in R$ the proof is finished, so suppose $(s', t') = (a, b')$. Let u, v be the ends of the arc X_s, see Fig. 2. We may assume without loss of generality that a, b, a', b' occur clockwise on the circle and are different from u, v; the latter is achieved by moving u, v slightly to points not belonging to T. Let $C_{a,b}$ be the arc of the circle from a (inclusive) to b (exclusive), going clockwise, and define $C_{b,a'}$, $C_{a',b'}$, $C_{b',a}$ analogously; these four arcs form a partition of the circle. Since $a \in X_s$ and $b' \in X_t$, we may assume that $u \in C_{b',a}$ and $v \notin C_{b',a}$. If $v \in C_{a,b}$ or $v \in C_{b,a'}$, then $a \in X_s$ and $a' \in X_t$ satisfy $(a, a') \in R$. Otherwise, if $v \in C_{a',b'}$, then $b \in X_s$ and $b' \in X_t$ satisfy $(b, b') \in R$. In both cases, (1) holds.   □

Next, we prove that the generating relation can be simplified as long as it contains 4 pairwise crossing chords in the right order. The lemma after that shows how to obtain such chords from any set of 7 pairwise crossing chords.

Lemma 3 (†). *Let T be a finite set of points on a circle and let $R \subseteq T^2$. Let $a, b, c, d, x, y, z, u \in T$ be pairwise different points appearing in this order on the circle (clockwise or counterclockwise), such that $(a, x), (b, y), (c, z), (d, u) \in R$. Define*

$$R' = (R \setminus \{(b, y), (c, z)\}) \cup \{(b, z), (c, y)\}.$$

Then $\mathsf{gen}(R') = \mathsf{gen}(R)$ and the number of crossings in R' is smaller than in R.

Lemma 4 (†). *Suppose H is a circle graph with directed chords and $\omega(H) \geq 7$. Then there are distinct points a, b, c, d, x, y, z, u that appear in clockwise order on the circle such that $(a, x), (b, y), (c, z), (d, u)$ are pairwise crossing chords of H.*

Lemmas 3 and 4 allow us to conclude that any generating relation can be iteratively simplified until it contains no set of 7 pairwise crossing chords.

Lemma 5. *Let G be a planar graph drawn in a disk Δ, let T be a subset of vertices of G drawn on the boundary of Δ, and let $P \subseteq T^2$ be the connectivity pattern on T induced by G. Then there exists a relation $R \subseteq T^2$ such that $\mathsf{gen}(R) = P$ and the circle graph induced by R has clique number at most 6.*

Proof. By Lemma 1 there exists a relation $R \subseteq T^2$ (namely $R = P$) such that $\mathsf{gen}(R) = P$. Choose R such that $\mathsf{gen}(R) = P$ and the number of crossings in R is as small as possible. Without loss of generality assume that R does not contain pairs of the form (s, s) for $s \in T$, as such pairs may be removed without changing the generated relation; thus R is a set of directed chords with endpoints in T. Let ω be the clique number of the circle graph induced by R. If $\omega \leq 6$ we are done, so suppose $\omega \geq 7$. By Lemma 4, there are pairwise different points a, b, c, d, x, y, z, u that appear in clockwise order on the circle such that $(a, x), (b, y), (c, z), (d, u) \in R$. Define $R' = (R \setminus \{(b, y), (c, z)\}) \cup \{(b, z), (c, y)\}$. By Lemma 3, $\mathsf{gen}(R') = \mathsf{gen}(R) = P$ and R' has fewer crossings than R, a contradiction.   □

Having obtained a generating relation with no large set of pairwise crossing chords, we will later partition it into a small number of sets of pairwise non-crossing chords using the χ-boundedness of circle graphs (Theorem 4). First, however, we bound the number of such non-crossing sets as follows.

Lemma 6 (†). *Let T be a finite set of points on a circle. Then every set of pairwise non-crossing chords with endpoints in T has at most $2|T| - 3$ chords, and there are $2^{\mathcal{O}(|T|)}$ different such sets.*

We are now ready to prove the main theorem of this section.

Theorem 5. *Let T be a set of n points on the boundary of a closed disk Δ. There exists a family $\mathcal{R}$ of relations $R \subseteq T^2$ such that $|\mathcal{R}| = 2^{\mathcal{O}(n)}$ and the following property is satisfied. For every planar digraph G drawn in Δ such that $T \subseteq V(G)$, the connectivity pattern induced by G on T is generated by some relation in $\mathcal{R}$.*

Proof. Denote by $\mathcal{R}$ the family of all sets of directed chords $R \subseteq T^2$ such that the circle graph induced by R has clique number at most 6. By Lemma 5 this family satisfies the claimed property and it remains to bound its size.

By χ-boundedness of circle graphs (Theorem 4), there exists a number $\chi_{\max}$ such that for $R \in \mathcal{R}$, the chromatic number of the circle graph induced by R is at most $\chi_{\max}$. The chords of any circle graph induced by some $R \in \mathcal{R}$ can thus be partitioned into $\chi_{\max}$ sets (possibly empty) such that no two chords in the same set cross. By Lemma 6, the number of possibilities to choose such a set of undirected, pairwise non-crossing chords is $2^{\mathcal{O}(n)}$, and any such set contains at most $2n - 3$ chords. Hence there are at most 2^{2n-3} possibilities to orient these chords. We conclude that indeed $|\mathcal{R}| \leq (2^{\mathcal{O}(n)} \cdot 2^{2n-3})^{\chi_{\max}} = 2^{\mathcal{O}(n)}$. □

With Theorem 5 in hand, the proof of Theorem 3 boils down to applying standard dynamic programming algorithm on a sphere-cut decomposition of the input graph. Each solved subproblem corresponds to a subgraph H of G embedded in a disk Δ, where each vertex of H that has a neighbor outside of H is embedded on the boundary of Δ; call the set of these vertices B. Then for each partition of B into X and T, and for each connectivity pattern P on T that can be induced by a digraph embedded in Δ, we compute that smallest size of a subset $Y \subseteq V(H) \setminus B$ such that $H - (X \cup Y)$ induces P on T; if there is no such subset, we store $+\infty$. It is straightforward to give recursive equations for this formulation. Moreover, Theorem 3 gives an upper bound of $2^{\mathcal{O}(t)}$ on the number of values computed for each H, where t is the treewidth, implying the running time of $2^{\mathcal{O}(t)} \cdot n^{\mathcal{O}(1)}$. Details, including an overview of sphere-cut decompositions, can be found in the full version [1].

4   Lower Bound

In this section we prove Theorem 2. The hardness reduction happens to work for both problems, producing exactly the same instances. We reduce from a problem shown hard by Lokshtanov et al. [14] (see also [5, Theorem 14.16]):

$k \times k$ HITTING SET WITH THIN SETS
Input: Family $\mathcal{F}$ of subsets of $[k] \times [k]$, each containing at most one element from each row
Question: Is there a set X containing exactly one vertex from each row of $[k] \times [k]$ such that $X \cap F \neq \emptyset$ for each $F \in \mathcal{F}$?

Theorem 6 ([14]). *Unless ETH fails, $k \times k$* HITTING SET WITH THIN SETS *cannot be solved in time $2^{o(k \log k)} \cdot n^{\mathcal{O}(1)}$, where n is the number of input sets.*

We first define an intermediate problem. An *n-permutation d-constraint* is a tuple $(i_1, \ldots, i_d) \in [n]^d$ of d different indices. A permutation $\sigma \colon [n] \to [n]$ *satisfies* such a constraint if $\sigma(i_1) < \sigma(i_2) < \cdots < \sigma(i_d)$. A *k-CNF n-permutation d-formula* is a conjunction of clauses, each of which is a disjunction of at most k n-permutation d-constraints. The *length* of a clause is the number of disjuncts (constraints) in it. Satisfaction of such a formula by a permutation $\sigma \colon [n] \to [n]$ is defined naturally.

We first show hardness for the satisfiability of 3-formulas, with the parameter k denoting both the length of clauses and the number of indices on which the permutation is defined.

Lemma 7. *Unless ETH fails, the satisfiability of a given k-CNF k-permutation 3-formula cannot be decided in time $2^{o(k \log k)} \cdot n^{\mathcal{O}(1)}$, where n is the formula size.*

Proof (sketch†). We only give the construction for the reduction, deferring the proof of its correctness to the appendix. Without loss of generality suppose $k \geq 3$. Let $\mathcal{F}$ be the input instance of $k \times k$ HITTING SET WITH THIN SETS. We construct in polynomial time a k-CNF $(2k+1)$-permutation 3-formula whose satisfiability is equivalent to the input instance $\mathcal{F}$, proving the claim by Theorem 6.

To an initially empty formula ϕ we add the following clauses, each with a single 3-constraint, to ensure that $\{k+1, \ldots, 2k+1\}$ are ordered increasingly by the permutation:

$$(k+1, k+2, k+3), \ (k+2, k+3, k+4), \ \ldots, \ (2k-1, 2k, 2k+1).$$

Then, for each $i \in [k]$ we add a clause with a single 3-constraint $(k+1, i, 2k+1)$. Finally, for each set $F \in \mathcal{F}$, we add the following clause C_F to ϕ: the clause C_F is the disjunction of constraints $(k+j, i, k+j+1)$ over all elements (i,j) of F. Since F contains at most one element of each row, the clause C_F is a disjunction of at most k constraints. □

Next, we show hardness for larger, but structured 2-formulas. For a 3-CNF n-permutation 2-formula ϕ, the *incidence graph* $I(\phi)$ of ϕ is the bipartite graph defined as follows: the vertex set is formed by indices from $[n]$ on one side and clauses of ϕ on the other side, and there is an edge between every clause and each index that occurs in some constraint of the clause. Thus, each clause has degree at most 6 in $I(\phi)$.

Lemma 8. *Unless ETH fails, the satisfiability of a given 3-CNF n-permutation 2-formula with incidence graph of treewidth t cannot be decided in time $2^{o(t \log t)}$. $n^{\mathcal{O}(1)}$. This holds even for formulas in which every clause has length exactly 3 or 1, and has no repeating indices.*

Proof (sketch†). Let ϕ be a k-CNF k-permutation 3-formula. We will construct in polynomial time a 3-CNF n-permutation 2-formula ψ for some $n = \mathcal{O}(k^2)$ such that ψ is satisfiable iff ϕ is and the incidence graph of ψ has treewidth $\mathcal{O}(k)$. The claim then follows by Lemma 7.

The idea is that every 3-constraint (a, b, c) can be thought of as a conjunction $(a, b) \wedge (b, c)$ of two 2-constraints (expressing $\sigma(a) < \sigma(b) \wedge \sigma(b) < \sigma(c)$). Intuitively, we can then transform the obtained 'non-CNF formula' into a 3-CNF in a standard way: a clause $(x \wedge x') \vee (y \wedge y') \vee (z \wedge z') \vee \ldots$ would be replaced by

$$(p_1) \wedge (\neg p_1 \vee x \vee p_2) \wedge (\neg p_2 \vee y \vee p_3) \wedge (\neg p_3 \vee z \vee p_4) \wedge \ldots$$
$$\wedge (\neg p_1 \vee x' \vee p_2) \wedge (\neg p_2 \vee y' \vee p_3) \wedge (\neg p_3 \vee z' \vee p_4) \wedge \ldots \wedge (\neg p_n)$$

where $p_1, p_2, p_3, \ldots, p_n$ are fresh auxiliary variables not appearing anywhere else.

Formally, we will ask for n-permutations with $n := k + (2k + 2)k$; the additional indices are in order to make room for 'auxiliary variables'. We construct ψ as an initially empty conjunction. Each clause C of ϕ is a disjunction $C_1 \vee \cdots \vee C_{k'}$ ($k' \leq k$) of some 3-constraints $C_i = (a_i, b_i, c_i) \in [k]^3$. Let $j_1, j_2, \ldots, j_{2k'+2} \in [n] \setminus [k]$ be some indices that were not yet used in any constructed clause. For each $i \in [k']$, we add the following clauses D_i and D'_i to ψ:

$$D_i = (j_{2i}, j_{2i-1}) \vee (a_i, b_i) \vee (j_{2i+1}, j_{2i+2})$$
$$D'_i = (j_{2i}, j_{2i-1}) \vee (b_i, c_i) \vee (j_{2i+1}, j_{2i+2})$$

We then add two clauses with a single constraint each: $Z = (j_1, j_2)$ and $Z' = (j_{2k'+2}, j_{2k'+1})$. Repeating this for each clause C of ϕ concludes the construction. Let $W(C)$ be the set consisting of clauses and indices used for C: clauses Z, Z', clauses D_i, D'_i for each $i \in [k']$, and indices $j_1, j_2, \ldots, j_{2k'+2}$ as above. Then $[k]$ together with sets $W(C)$ for clauses C of ϕ form a partition of the vertex set of the incidence graph $I(\psi)$ of the constructed formula. Observe that if we remove all the k vertices corresponding to $[k]$, the only remaining edges in $I(\psi)$ have both endpoints within the same $W(C)$ for some clause C of ϕ, and each $W(C)$ has size at most $3k + 4$. This allows to bound the treewidth of $I(\psi)$ by $\mathcal{O}(k)$. Details and the correctness proof of the construction can be found in the appendix. □

We proceed to reducing the satisfiability problem for permutation formulas as described in Lemma 8 to DIRECTED FEEDBACK VERTEX (ARC) SET. Permutations of $[n]$ will be encoded as orderings of a subset of n 'terminal' vertices in the constructed digraph, identified with indices from $[n]$. The digraph will contain gadgets ensuring that a permutation satisfies the original 3-CNF n-permutation

2-formula if and only if the corresponding ordering of terminals can be extended to a topological ordering of the whole digraph, after deleting a prescribed number of vertices (edges). The key element is the or-gadget depicted in Fig. 3, which encodes a clause that is a disjunction of three 2-constraints. Note that this or-gadget has 6 terminal vertices, named x_i, x_i' for $i \in [3]$. The final graph is obtained essentially by taking disjoint copies of the or-gadget and identifying their terminal vertices with terminals.

Lemma 9. *For an ordering $\prec$ of the terminal vertices of the or-gadget, $\prec$ can be extended to a topological ordering of the or-gadget with some 2 vertices (edges) deleted if and only if $x_1 \prec x_1'$ or $x_2 \prec x_2'$ or $x_3 \prec x_3'$. Furthermore, every subgraph of the or-gadget obtained by deleting at most one non-terminal vertex or an edge from it, contains a directed cycle.*

Proof. Given an ordering $\prec$ of the terminal vertices such that $x_1 \prec x_1'$, one can remove e_2 and e_3, or any two vertices incident to them, to create an acyclic subgraph of the or-gadget that admits a topological ordering extending $\prec$. The cases of orderings $\prec$ with $x_2 \prec x_2'$ and with $x_3 \prec x_3'$ are symmetric. Conversely, any removal of two vertices or edges from the or-gadget leaves some directed path $x_i \to x_i'$ ($i \in [3]$) unharmed, implying $x_i \prec x_i'$ in any topological ordering of the obtained subgraph. It is easy to check that two non-terminal vertices or edges of the or-gadget have to be removed to make it acyclic.   □

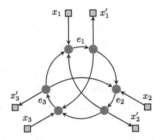

Fig. 3.   The   or-gadget, with   terminal   vertices marked as squares.

Proof (of Theorem 2, sketch†). We give a reduction from the satisfiability problem for 3-CNF n-permutation 2-formulas to DFVS and DFAS. More precisely, on the input of the reduction we are given a 3-CNF n-permutation 2-formula ψ with an incidence graph of treewidth t, where we assume that every clause of ψ has length exactly 3 or 1, and has no repeating indices. We will construct in polynomial time an equivalent instance of (the decision variant of) DFVS (DFAS) of treewidth $\mathcal{O}(t)$. This will prove the claim by Lemma 8.

We construct a digraph G starting from $[n]$ as the vertex set and no edges. For each clause of length 1 in ψ, let $(a, a') \in [n]^2$ be the unique constraint in it. Then we add an edge from a to a' to G. For each clause of length 3 in ψ, let $(a_1, a_1'), (a_2, a_2'), (a_3, a_3') \in [n]^2$ be its constraints. Then we add a new copy of the or-gadget to G, and for each $i \in [3]$ we identify x_i and x_i' with a_i and a_i', respectively. Finally, we set k, the target size of a directed feedback vertex (arc) set, to be twice the number of clauses of length 3 in ψ. The obtained instance (G, k) can be treated both as a DFVS instance and as a DFAS instance.

To bound the treewidth of G, observe that G can be obtained from $I(\psi)$ by replacing each vertex w corresponding to a clause with a copy of the or-gadget (if it represents a clause of length 3), or with just an edge between its original neighbors (if it represents a clause of length 1).   □

5 Concluding Remarks

Our results do not provide any direct insight into the complexity of the classic parameterization: by the target solution size k. We hope, however, that the combinatorial tools we used in the proof of Theorem 3 may be useful for improving the running time for DFVS on planar digraphs, say to running time $2^{\mathcal{O}(k)} \cdot n^{\mathcal{O}(1)}$, or for obtaining a somewhat incomparable running time $n^{\mathcal{O}(\sqrt{k})}$. Observe that there is a large gap between known results in this setting: while the classic reduction from VERTEX COVER on planar graphs gives a lower bound excluding running time $2^{o(\sqrt{k})} \cdot n^{\mathcal{O}(1)}$ under ETH, no faster algorithm than $2^{\mathcal{O}(k \log k)} \cdot (n+m)$ from general digraphs [15] is known.

References

1. Bonamy, M., Kowalik, L., Nederlof, J., Pilipczuk, M., Socała, A., Wrochna, M.: On directed feedback vertex set parameterized by treewidth. arXiv abs/1707.01470 (2017)
2. Chen, J., Liu, Y., Lu, S., O'Sullivan, B., Razgon, I.: A fixed-parameter algorithm for the directed feedback vertex set problem. J. ACM **55**(5), 21:1–21:19 (2008)
3. Chitnis, R.H., Cygan, M., Hajiaghayi, M.T., Marx, D.: Directed subset feedback vertex set is fixed-parameter tractable. ACM Trans. Algorithms **11**(4), 28:1–28:28 (2015)
4. Chitnis, R.H., Hajiaghayi, M., Marx, D.: Fixed-parameter tractability of directed multiway cut parameterized by the size of the cutset. SIAM J. Comput. **42**(4), 1674–1696 (2013)
5. Cygan, M., et al.: Parameterized Algorithms. Springer, Cham (2015). https://doi.org/10.1007/978-3-319-21275-3
6. Dorn, F., Penninkx, E., Bodlaender, H.L., Fomin, F.V.: Efficient exact algorithms on planar graphs: exploiting sphere cut decompositions. Algorithmica **58**(3), 790–810 (2010)
7. Fomin, F.V., Thilikos, D.M.: New upper bounds on the decomposability of planar graphs. J. Graph Theory **51**(1), 53–81 (2006)
8. Gyárfás, A.: On the chromatic number of multiple interval graphs and overlap graphs. Discret. Math. **55**(2), 161–166 (1985)
9. Gyárfás, A.: Corrigendum: on the chromatic number of multiple interval graphs and overlap graphs. Discret. Math. **62**(3), 333 (1986)
10. Gyárfás, A.: Problems from the world surrounding perfect graphs. Applicationes Mathematicae **19**(3–4), 413–441 (1987)
11. Kim, E.J., Gonçalves, D.: On exact algorithms for the permutation CSP. Theor. Comput. Sci. **511**, 109–116 (2013)
12. Kratsch, S., Pilipczuk, M., Pilipczuk, M., Wahlström, M.: Fixed-parameter tractability of Multicut in directed acyclic graphs. SIAM J. Discret. Math. **29**(1), 122–144 (2015)
13. Kratsch, S., Wahlström, M.: Representative sets and irrelevant vertices: new tools for kernelization. In: FOCS 2012, pp. 450–459. IEEE Computer Society (2012)
14. Lokshtanov, D., Marx, D., Saurabh, S.: Slightly superexponential parameterized problems. In: SODA vol. 2011, pp. 760–776 (2011)

15. Lokshtanov, D., Ramanujan, M.S., Saurabh, S.: A linear time parameterized algorithm for directed feedback vertex set. CoRR abs/1609.04347 (2016)
16. Lovász, L.: On two minimax theorems in graph. J. Comb. Theory, Ser. B **21**(2), 96–103 (1976)
17. Lucchesi, C.L., Younger, D.H.: A minimax theorem for directed graphs. J. London Math. Soc **17**, 369–374 (1978)
18. Pilipczuk, M., Wahlström, M.: Directed multicut is $W[1]$-hard, even for four terminal pairs. In: SODA 2016, pp. 1167–1178. SIAM (2016)
19. Schrijver, A.: Combinatorial Optimization - Polyhedra and Efficiency. Springer, Heidelberg (2003)
20. Seymour, P.D., Thomas, R.: Call routing and the ratcatcher. Combinatorica **14**(2), 217–241 (1994)

Optimality Program in Segment
and String Graphs

Édouard Bonnet[1] and Paweł Rzążewski[2(✉)]

[1] ENS Lyon, LIP, Lyon, France
`edouard.bonnet@dauphine.fr`
[2] Faculty of Mathematics and Information Science,
Warsaw University of Technology, Warsaw, Poland
`p.rzazewski@mini.pw.edu.pl`

Abstract. Planar graphs are known to allow subexponential algorithms running in time $2^{O(\sqrt{n})}$ or $2^{O(\sqrt{n}\log n)}$ for most of the paradigmatic problems, while the brute-force time $2^{\Theta(n)}$ is very likely to be asymptotically best on general graphs. Intrigued by an algorithm packing curves in $2^{O(n^{2/3}\log n)}$ by Fox and Pach [SODA'11], we investigate which problems have subexponential algorithms on the intersection graphs of curves (string graphs) or segments (segment intersection graphs) and which problems have no such algorithms under the ETH (Exponential Time Hypothesis). Among our results, we show that, quite surprisingly, 3-COLORING can also be solved in time $2^{O(n^{2/3}\log^{O(1)} n)}$ on string graphs while an algorithm running in time $2^{o(n)}$ for 4-COLORING even on axis-parallel segments (of unbounded length) would disprove the ETH. For 4-COLORING of unit segments, we show a weaker lower bound, excluding a $2^{o(n^{2/3})}$ algorithm (under the ETH). The construction exploits the celebrated Erdős-Szekeres theorem. The subexponential running time also carries over to MIN FEEDBACK VERTEX SET, but not to MIN DOMINATING SET and MIN INDEPENDENT DOMINATING SET.

1 Introduction

Most combinatorial optimization and decision problems admit subexponential algorithms when restricted to planar graphs. More precisely, they can be solved in time $2^{O(\sqrt{n})}$, or $2^{\tilde{O}(\sqrt{n})}$ on planar graphs with n vertices, while under the ETH (Exponential Time Hypothesis, which asserts that 3-SAT cannot be solved in subexponential time [24,25]) they do not admit an algorithm running in time $2^{o(n)}$ on general graphs. The former is due to the facts that planar graphs have treewidth $O(\sqrt{n})$ and that we have efficient algorithms parameterized by the treewidth tw of the graph, namely running in $2^{O(\mathrm{tw})}n^{O(1)}$, or $2^{\tilde{O}(\mathrm{tw})}n^{O(1)}$.

The so-called bidimensionality theory [10,12–14] pushes this square-root phenomenon further by yielding $2^{O(\sqrt{k})}n^{O(1)}$ algorithms where k is the targeted size of a solution (think for example of the problems of finding a maximum independent set or a minimum dominating set of size k). In a nutshell, it exploits a deep

© Springer Nature Switzerland AG 2018
A. Brandstädt et al. (Eds.): WG 2018, LNCS 11159, pp. 79–90, 2018.
https://doi.org/10.1007/978-3-030-00256-5_7

structural result by Robertson, Seymour, and Thomas [38]: planar graphs with treewidth tw have a $\Theta(\text{tw})$-by-$\Theta(\text{tw})$ grid as a minor (i.e., any graph obtained by deleting vertices and edges, and contracting edges). Thus, if the presence of a large grid minor makes the problem trivial (as in, one can always answer yes or always answer no), then one only has to solve efficiently instances with low treewidth; which, as we noted, can often be done. The claimed running time is obtained by defining large grids as $\Theta(\sqrt{k})$-by-$\Theta(\sqrt{k})$, since their absence as minors implies that the treewidth is in $O(\sqrt{k})$. The bidimensionality theory is also used to obtain approximation schemes and linear kernels and could be generalized to graphs with bounded genus and graphs excluding a fixed minor [11].

A natural line of research is to generalize or extend the subexponential (parameterized) algorithms to classes of graphs which do not fall into those categories. For geometric intersection graphs, the situation is much richer than for planar graphs. For instance, Marx and Pilipczuk already observed that packing problems (of the kind of MAX INDEPENDENT SET) are more broadly subject to subexponential algorithms – running typically in $n^{O(\sqrt{k})}$ – than covering problems (of the kind of MIN DOMINATING SET) – for which $n^{O(k)}$ is essentially optimal under the ETH [33, 34].

We briefly survey the existing results concerning subexponential algorithms on geometric intersection graphs. A prominent role is played by intersection graphs of families of *fat objects*, i.e., objects for which the aspect ratio (length divided by width) is bounded. We highlight that fat objects, and in particular disks and squares, often allow fast algorithms and the so-called square-root phenomenon. As we will see, subexponential algorithms are less frequent on intersection graphs of curves and segments but nevertheless present such as exemplified by MAX INDEPENDENT SET, 3-COLORING, and MIN FEEDBACK VERTEX SET.

Subexponential Algorithms on Geometric Intersection Graphs. By a *ply* of a family of geometric objects we denote the maximum number of objects covering a single point. Smith and Wormald show that for any collection of n convex fat objects with ply p there is a balanced separator of size $O(\sqrt{np})$ [42]. This leads to subexponential algorithms when the ply is constant, or in general for problems becoming trivial when the ply is too large, such as K-COLORING. The $2^{\tilde{O}(\sqrt{nk})}$-time algorithm that this win-win provides for coloring n fat objects, say disks, with k colors is shown essentially optimal under the ETH by Biró et al. [5].

A next step may consist of designing FPT (with running time $f(k)n^{O(1)}$) or XP (with running time $n^{f(k)}$) algorithms where the dependency in the parameter is subexponential (for problems of the form "find k vertices such that...."). Using a shifting argument á la Baker [4], Alber and Fiala obtain a $n^{O(\sqrt{k})}$-time algorithm to decide if one can find k disjoint unit disks or squares among n [3]. Marx and Pilipczuk generalize this result to packing k disjoint polygons among n in the same time [33, 34]. Their approach is based on guessing a small separator in the medial axis (i.e., the Voronoi diagram of polygons) of a supposed solution, as suggested by Adamaszek and Wiese and Har-Peled to obtain QPTAS for geometric packing problems [1, 2, 22].

Marx showed that MAX INDEPENDENT SET and MIN DOMINATING SET in the intersection graphs of disks or squares are W[1]-complete, and therefore unlikely to be FPT [32]. Those reductions also show that the $n^{O(\sqrt{k})}$ algorithms [33,34] are essentially optimal under the ETH. Fomin et al. [17] observed that unit disks of bounded degree have treewidth $O(\sqrt{n})$ and used this fact to extend bidimensionality to unit disk graphs for a handful of problems. Recently, a super-set of the previous authors gave $2^{O(\sqrt{k})}n^{O(1)}$-time algorithms for k-FEEDBACK VERTEX SET, k-PATH, k-CYCLE, EXACT k-CYCLE [16].

Non-fat Objects: Segments and Strings. Segment intersection graphs (or segment graphs in short) are the intersection graphs of straight-line segments in the plane. They are called unit segments if all the segments of a representation share the same length. For a fixed integer k, k-DIR is defined as the set of intersection graphs of segments, each parallel to one of fixed k directions. Strings graphs are the intersection graphs of simple curves in the plane. Those curves can be assumed polygonal without loss of generality. The vertices of the polygonal curves in a geometric representation are called *geometric vertices* not to confuse them with the actual vertices of the graph. As shown by Kratochvíl and Matoušek, there are string graphs with n vertices, which require $2^{\Omega(n)}$ geometric vertices in any string representation with polygonal curves [29].

A systematic study of segment graphs and their subclasses was initiated by Kratochvíl and Matoušek [27]. It is interesting to point out that every planar graph is a segment graph, as shown by Chalopin and Gonçalves [9] (this was a long-standing conjecture by Scheinerman [41], see also [21]).

The class of string graphs is very general, as it includes split graphs (i.e., graphs whose vertices can be partitioned into two sets inducing a clique and an independent set), intersection graphs of bodies (i.e., compact shapes with non-empty interior), or incomparability graphs (i.e., graphs whose vertex set is given by the set of elements of a poset, and edges join elements that are incomparable).

Biró et al. showed that even though coloring disks or more generally fat objects with a constant number of colors can be solved in $2^{\tilde{O}(\sqrt{n})}$ [5], 6-coloring axis-parallel segments (2-DIR) in time $2^{o(n)}$ would refute the ETH. This suggests that subexponential algorithms are less frequent on the intersection graphs of non-fat objects such as segments and strings. On the other hand, Fox and Pach presented a subexponential algorithm for MAX INDEPENDENT SET on string graphs [18]. Their approach uses a win-win strategy and is based on the existence of balanced separators in string graphs. Fox, Pach, and Tóth showed that string graphs with m edges have balanced separators of size $O(m^{3/4} \log m)$, and conjectured that there is always a separator of size $O(\sqrt{m})$ [20]. Matoušek showed that string graphs admit a balanced separator of size $O(\sqrt{m} \log m)$ [36]. Finally, recently Lee improved the result of Matoušek, proving the conjecture.

Theorem 1 (Lee [30]). *Every string graph with m edges has a balanced separator of size $O(\sqrt{m})$. Moreover, it can be found in polynomial time, provided that the geometric representation is given.*

Let us point out that this result generalizes the famous planar separator theorem by Lipton and Tarjan [31], as planar graphs are string graphs and the number of edges in a planar graph is linear in the number of vertices. This also shows that Theorem 1 is best possible (up to the constants), as the planar separator theorem is asymptotically tight.

Our Contributions. We show that the subexponential algorithm for MAX INDE-PENDENT SET in string graphs by Fox and Pach [18], running in time $2^{\tilde{O}(n^{2/3})}$, can be extended to 3-COLORING and MIN FEEDBACK VERTEX SET. As in the algorithm of Fox and Pach, the central idea is a win-win: either the graph is rather sparse and the separator of Theorem 1 gives a speed-up, or the graph has a high-degree vertex (used for 3-COLORING) or a large biclique (used for MIN FEEDBACK VERTEX SET) and an efficient branching can be performed. Refining a lower bound of Biró et al. [5], we complement this former result by showing that for any $k \geqslant 4$, k-COLORING cannot be solved in $2^{o(n)}$ even on axis-parallel segments, unless the ETH fails. The reduction relies on having segment lengths with two different orders of magnitude. We therefore ask if unit segments could allow a faster algorithm for k-COLORING for $k \geqslant 4$. Under the ETH, we provide a stronger lower bound than the one for planar graphs (which refutes a running time $2^{o(\sqrt{n})}$) and show that unit segments cannot be k-colored in $2^{o(n^{2/3})}$ for any $k \geqslant 4$. Our construction uses the fact, closely related to the famous Erdős-Szekeres [15] theorem, that any permutation on n totally ordered elements can be partitioned into $O(\sqrt{n})$ monotone subsequences (see Knuth [26, Sect. 5.1.4]).

We then give tight ETH lower bounds for MIN (CONNECTED) DOMINATING SET and MIN INDEPENDENT DOMINATING SET on segment graphs and MAX CLIQUE on string graphs. For that, we design reductions whose number n of produced segments is linear in $N + M$ from satisfiability problems with N variables and M clauses. Indeed, the sparsification lemma of Impagliazzo et al. [23] implies that those satisfiability problems are not solvable in $2^{o(N+M)}$ unless the ETH fails; which enables us to conclude that the problems are not solvable in $2^{o(n)}$ under the ETH, on graphs with n vertices.

Although the NP-hardness of the mentioned problems is known for segment intersection graphs [8,43], getting such linear reductions might be difficult.

For instance, while it is known that planar graphs are a subclass of segment intersection graphs [9], implying the NP-hardness of all the problems of Table 1 except K-COLORING for $k \geqslant 4$ and MAX CLIQUE, this fact does not serve our purpose since they can be solved in time $2^{O(\sqrt{n})}$ on planar graphs. The situation is an interesting intermediate between planar and general graphs. Our objects *can* intersect but we cannot afford crossover gadgets (at least not quadratically many). Certain intersections create unwanted edges, whose importance we have to tame. It is also noteworthy that segment/string graphs cannot be expanders since if they have constant degree, by Theorem 1, they have treewidth $\tilde{O}(\sqrt{n})$. Hence, we are deprived of the *usual hardest instances*.

Table 1. Complexity bounds for classical problems on string and segment graphs. The **upper bounds** work on **string graphs**. The **lower bounds** are designed on **segment graphs**, unless precised otherwise. New results are indicated by the shaded background. By p we denote the number of geometric vertices if a geometric representation is given.

Problem	Upper bound	Lower bound
MAX INDEPENDENT SET	$2^{\tilde{O}(\sqrt{n})}p^{O(1)},\ 2^{\tilde{O}(n^{2/3})}$	$2^{\Omega(\sqrt{n})}$
3-COLORING	$2^{\tilde{O}(n^{2/3})}$	$2^{\Omega(\sqrt{n})}$
k-COLORING for every $k \geqslant 4$	$2^{O(n)}$	$2^{\Omega(n)}$ (even in 2-DIR)
k-COLORING for every $k \geqslant 4$	$2^{O(n)}$	$2^{\Omega(n^{2/3})}$ in unit 3-DIR
MIN FEEDBACK VERTEX SET	$2^{\tilde{O}(n^{2/3})}$	$2^{\Omega(\sqrt{n})}$
MIN (CONNECTED) DOMINATING SET	$2^{O(n)}$	$2^{\Omega(n)}$
MIN INDEPENDENT DOMINATING SET	$2^{O(n)}$	$2^{\Omega(n)}$
MAX CLIQUE	$2^{O(n)}$	$2^{\Omega(n)}$ (on string graphs)

Geometric Representation and Robust Algorithms. In case of graphs with geometric representations, it is important to distinguish between a graph itself (i.e., a pure abstract structure, for which we know that some geometric representation exists), and the representation itself. Note that this is not the case with planar graphs, as finding a plane embedding can be done in linear time [7].

Finding a segment or string representation of a graph was shown to be NP-hard by Kratochvíl [28], and Kratochvíl and Matoušek [27], respectively. However, it was very unclear if the problems are in NP (which is usually the trivial part of an NP-completeness proof). As mentioned above, Kratochvíl and Matoušek [29] showed that some string graphs require a representation of exponential size, which proved that the simple idea of exhaustively guessing the representation cannot work for this problem. Finally, the NP-membership of recognizing string graphs was proven by Schaefer, Sedgwick, and Štefankovič [39].

The story of recognizing segment graphs is even more interesting. On the first sight, the situation seems simpler than for strings, as the number of geometric points in a segment representation is clearly polynomial in n. However, there are segment graphs, whose every segment representation requires points with coordinates doubly exponential in n, i.e., using $2^{\Omega(n)}$ digits (see Kratochvíl and Matoušek [27], and McDiarmid and Müller [37]). Finally, the problem was shown to be complete for the class $\exists\mathbb{R}$ (see Schaefer and Štefankovič [40]), i.e., the class of problems reducible in polynomial time to deciding if a given existential formula over the reals is true. This is a strong evidence that the problem is not in NP. For a nice exposition of the $\exists\mathbb{R}$-completeness proof, see Matoušek [35].

All this shows that a requirement of an explicit geometric representation of an input graph may be a serious drawback of an algorithm. We call an algorithm *robust* if it takes only an abstract structure as an input, and either computes the solution, or concludes (correctly) that the input graph does not belong to the desired class. On the one hand, our algorithms are robust, but work slightly faster if the input is given along with the geometric representation. On the other hand, the lower bounds hold even if the geometric representation is given explicitly.

2   Upper Bounds

Fox and Pach showed that, on string graphs, a maximum independent set can be computed in subexponential time:

Theorem 2 (Fox and Pach [18], Lee [30]). MAX INDEPENDENT SET *can be solved in time* $2^{O(n^{2/3} \log n)}$ *in string graphs with n vertices.*

In their paper, they give a worse running time than the one claimed above, because they used the $O(m^{3/4} \log m)$ separator theorem [20], which has been recently improved to $O(\sqrt{m})$ [30]. The algorithm is a simple win-win argument: if there is a vertex with degree at least $n^{1/3}$, then we branch on including in in the solution or not. Otherwise all degrees are smaller than $n^{1/3}$, and a balanced separator of size $O(\sqrt{m}) = O(n^{2/3})$ can be used for divide-and-conquer.

The result of Fox and Pach was somewhat improved by Marx and Pilipczuk [33,34] based on an approach introduced by Adamaszek, Har-Peled, and Wiese [1]. However, their algorithm requires that the string graph is given with a representation by polygonal curves on a polynomial number of geometric vertices.

Theorem 3 (Marx and Pilipczuk [33]). MAX INDEPENDENT SET *can be solved in time* $2^{O(\sqrt{n} \log n)} p^{O(1)}$ *in string graphs with n vertices, where the strings are given as polygonal curves on a total of p geometric vertices.*

In a nutshell, the idea is to exhaustively guess a small balanced face-separator in the Voronoi diagram of a supposed (although not known) fixed solution, and solve recursively the two subinstances in the inside and outside of this separator.

If this approach does not seem to generalize easily to coloring problems, the algorithm of Fox and Pach can be transported to 3-COLORING with a bit more arguments. The algorithm even works for the more general LIST 3-COLORING (in LIST K-COLORING each vertex v is equipped with a list $L(v) \subseteq [k]$ and we want to find a proper coloring, in which every vertex gets a color from its list).

Theorem 4. LIST 3-COLORING *of a string graph with n vertices can be decided in time* $2^{O(n^{2/3} \log n)}$, *even without geometric representation.*

Proof. Consider an instance (G, L) of LIST 3-COLORING with n vertices. We can assume that each list has two or three elements: if there is a vertex with just one allowed color, we can fix this color and remove it from the list of each neighbor. Let N be the sum of the lengths of the lists; clearly $2n \leqslant N \leqslant 3n$.

First, assume that the maximum degree in G is at most $n^{1/3}$, so the number m of edges is $O(n^{4/3})$. By Theorem 1, G has a balanced separator of size $O(\sqrt{m}) = O(n^{2/3})$. We can find this separator in polynomial time, if the representation is given, or by exhaustive guessing in time $n^{O(n^{2/3})} = 2^{O(n^{2/3} \log n)}$, without a representation. Then we list all colorings of the separator and proceed with a divide-and-conquer approach. The complexity of this step is $2^{O(n^{2/3}\log n)}$.

If there is a vertex v of degree at least $n^{1/3}$, then one among the lists: $\{1,2\},\{1,3\},\{2,3\},\ \{1,2,3\}$ appears on at least $n^{1/3}/4$ of its neighbors. Thus there are two colors (say, 1 and 2) that appear in lists of at least $n^{1/3}/4$ of neighbors of v. Since the list of v has size at least two, one of these colors (say 1) appears on the list of v. We branch into two possibilities: choosing the color 1 for v (then we exclude 1 from the lists of all neighbor of v), and not choosing 1 for v (then we remove 1 from the list of v). The complexity F of this step is given by the recursion $F(N) \leqslant F(N-n^{1/3}/4)+F(N-1) \leqslant F(N-N^{1/3}/(3^{1/3}\cdot 4))+F(N-1)$. This inequality is satisfied by $F(N) = 2^{O(N^{2/3}\log N)} = 2^{O(n^{2/3}\cdot\log n)}$.

Combining these two cases gives the claimed time complexity. Finally, observe that if the input graph is not a string graph, then the exhaustive search for a separator might fail, and then we can report a wrong input instance.   □

In MIN FEEDBACK VERTEX SET problem we ask for the minimum set of vertices, whose removal destroys all cycles. For this problem, there is no obvious branching on a high-degree vertex. Instead, we use the following theorem by Lee.

Theorem 5 (Lee [30]). *There is a constant c such that for any $t \geqslant 1$, $K_{t,t}$-free string graphs on n vertices have fewer than $c \cdot t \log t \cdot n$ edges.*

It is worth mentioning that Fox and Pach [19, Theorem 5] obtained a slightly weaker result with $\log^{O(1)} t$ instead $\log t$. Now we can show the following.

Theorem 6. MIN FEEDBACK VERTEX SET *on string graphs with n vertices can be solved in time $2^{\tilde{O}(n^{2/3})}$.*

Proof. The proof is similar to the proof of Theorem 4, but it involves a slight technical complication. We will solve a more general problem, where the input is a graph G, a set $\mathcal{C}_1$ of constraints of type $\texttt{disconnect}(u,v)$, and another set $\mathcal{C}_2$ of constraints of type $\texttt{stays}(v)$, where u,v are vertices of G. We ask for a minimum feedback vertex set X of G, such that for every constraint $\texttt{disconnect}(u,v)$, the vertices u and v are in different connected components of $G - X$, and for every constraint $\texttt{stays}(v)$ we have $v \notin X$. The algorithm is recursive, the constraints $\mathcal{C}_1$ and $\mathcal{C}_2$ are checked at the leaves of the recursion tree. Clearly, if $\mathcal{C}_1 = \mathcal{C}_2 = \emptyset$, then we just ask for the minimum feedback vertex set.

If G has fewer than $c/3 \cdot n^{4/3} \log n$ edges (where c is a constant from Theorem 5), then by Theorem 1 there is a balanced separator S of size $O(\sqrt{m}) = \tilde{O}(n^{2/3}\log^{1/2} n)$, we can find it in time $n^{O(n^{2/3})} = 2^{O(n^{2/3}\log^{3/2} n)}$ by exhaustive search or in polynomial time, if the geometric representation is given. Let V_1, V_2 be sets such that $V(G) = V_1 \uplus V_2 \uplus S$, there is no V_1-V_2-path in $G-S$, and $V_1, V_2 \leqslant c' \cdot n$ for a constant c'; they exist since S is a balanced separator. We will exhaustively guess the intersection I' of a fixed minimum solution with S, taking into consideration the current constraints $\mathcal{C}_2$ (this represents at most

$2^{|S|} = 2^{\tilde{O}(n^{2/3})}$ possibilities), introduce the new constraints stays(v) for every $v \in S \setminus I'$, and solve the problem in $G_1 := G[V_1 \cup S \setminus I]$ and $G_2 := G[V_2 \cup S \setminus I]$.

However, note that there might be cycles in G that are not contained in $V_1 \cup S$ nor $V_2 \cup S$ and the straightforward approach discussed above would not destroy them. Let us call such cycles *essential*. For each essential cycle C, and for $i = 1, 2$, we call *i-subpath of C* a subpath of C, whose endvertices are in S, and inner vertices are in V_i. To destroy C, we must disconnect the endvertices of some i-subpath of C. We ensure this by introducing appropriate separation constraints. For every partition $\Pi = (S_1, S_2, \ldots, S_k)$ of $S \setminus I$, we run the algorithm recursively in each graph G_i with additional constraints disconnect(u, v) for every $u, v \in S$, such that u and v are in different parts of Π. The number of partitions of $S \setminus I$ is given by the Bell number of $|S \setminus I|$, which is at most $|S|^{|S|} = 2^{|S| \log |S|} = 2^{\tilde{O}(n^{2/3})}$. This gives us a total of $2^{\tilde{O}(n^{2/3})}$ recursive calls at each level of the recursion tree. It is sufficient to only consider connectivity patterns since being connected is a transitive relation: if u and v, and v and w stay connected in G_i, then u and w also stay connected. We combine solutions in G_1 and G_2 which agree on the subset $I \subseteq S$, and such that the essential cycles cannot survive. It is known and relatively easy to see that this happens exactly when the partitions $\Pi^1 = (S_1^1, S_2^1, \ldots, S_{k_1}^1)$ for G_1 and $\Pi^2 = (S_1^2, S_2^2, \ldots, S_{k_2}^2)$ for G_2 are such that for each pair (i, j), $|S_i^1 \cup S_j^2| \leqslant 1$ and the bipartite intersection graph (with an edge between S_i^1 and S_j^2 iff they have non-empty intersection) is a forest. This step has a total running time $2^{\tilde{O}(n^{2/3})}$.

On the other hand, if G has at least $c/3 \cdot n^{4/3} \log n$ edges, then by Theorem 5 it contains $K_{n^{1/3}, n^{1/3}}$ as a subgraph. We can find it exhaustively in time $n^{2n^{1/3}} \cdot n^{O(1)} = 2^{\tilde{O}(n^{1/3})}$. Observe that any feedback vertex set of G must contain all but one vertex of one bipartition class of the biclique. Guessing which vertex is *not* necessarily chosen into the solution gives a branching algorithm, whose complexity is $F(n) \leqslant 2^{\tilde{O}(n^{1/3} \log n)} + 2n^{1/3} F(n - n^{1/3} + 1)$, which is solved by $F(n) = 2^{O(n^{2/3} \log n)}$. We also trim branches which violate a constraint of $\mathcal{C}_2$. Branching does not introduce new constraints. In particular, the vertex which is not added to the solution with the other vertices of its bipartition class might still be included in the solution later. Observe that the branching on a separator and the branching on a biclique are compatible, and can be done in an interleaved fashion. Finally, if the exhaustive search for a separator or a biclique fails, then either we have reached a constant size (and the subproblem can be brute-forced), or we correctly report that the input graph is not a string graph.   □

3   Lower Bounds

Rather surprisingly, the win-win for 3-COLORING ceases to work for k-COLORING if $k \geqslant 4$. First, let us consider the LIST 4-COLORING. Following Kratochvíl and Matoušek [27], by PURE 2-DIR we denote graphs with a 2-DIR representation, in which parallel segments are disjoint. Note that such graphs are bipartite.

Theorem 7 ($\star^1$). LIST 4-COLORING *of a* PURE 2-DIR *graph cannot be solved in time* $2^{o(n)}$, *even if each list has size at most 3, unless the ETH fails.*

The non-list version is obtained similarly to the case for 6-COLORING in [5].

Theorem 8 ($\star$). *For every fixed* $k \geqslant 4$, *the* k-COLORING *problem of a* 2-DIR *graph cannot be solved in time* $2^{o(n)}$, *unless the ETH fails.*

The construction in the proof of Theorem 7 requires segments of length $O(n)$. For unit segments, we show the following weaker lower bound.

Theorem 9 ($\star$). *For* $k \geqslant 4$, LIST k-COLORING *of unit* 2-DIR *graphs or* k-COLORING *of unit* 3-DIR *graphs cannot be solved in time* $2^{o(n^{2/3})}$, *unless the ETH fails.*

Finally, we show that variants of MIN DOMINATING SET on segment graphs and MAX CLIQUE on string graphs are unlikely to have a subexponential algorithm on segment graphs

Theorem 10 ($\star$). MIN (CONNECTED) DOMINATING SET *and* MIN INDEPENDENT DOMINATING SET *cannot be solved in time* $2^{o(n)}$ *on segment graphs with* n *vertices, unless the ETH fails.*

Theorem 11 ($\star$). MAX CLIQUE *cannot be solved in time* $2^{o(n)}$ *on string graphs with* n *vertices, unless the ETH fails.*

4 Perspectives

We have started a generalized optimality program on segment and string graphs for principal graph problems. On the algorithmic side, we extended a subexponential algorithm for MAX INDEPENDENT SET on string graphs [18] to two other problems: 3-COLORING and MIN FEEDBACK VERTEX SET. On the complexity side, we showed that subexponential algorithms are unlikely for, among others, 4-COLORING and variants of MIN DOMINATING SET. It is quite easy to obtain such lower bounds for string graphs. Extending those results to segments requires more ingenuity, and even more so when it comes to unit segments.

A handful of questions remains unsettled. Can we improve the algorithm or give tight ETH lower bounds for the following problems: MAX INDEPENDENT SET without geometric representation, 3-COLORING, and MIN FEEDBACK VERTEX SET on segments/strings? Can we show that MAX CLIQUE does not admit a subexponential algorithm on segment graphs? The mere NP-hardness of MAX CLIQUE on segments [8] answered 21-year-old open question. Hence, it is likely that getting a tight ETH hardness will be difficult. We would also find interesting to have, for a certain problem, an algorithm for segments (resp. unit segments) which beats the ETH lower bound on strings (resp. segments).

Finally, another natural continuation of this work is to determine which fixed-parameter tractable problems can be solved in time $O^*(2^{\tilde{O}(k^{2/3})})$ or $O^*(2^{\tilde{O}(\sqrt{k})})$,

---

[1] Full proofs of theorems marked with ($\star$) can be found in [6].

and which W[1]-hard problems can be solved in time $f(k)n^{O(\sqrt{k})}$ on segments and strings. For instance, MIN VERTEX COVER can be solved in time $O^*(2^{\tilde{O}(k^{2/3})})$ (even in time $O^*(2^{\tilde{O}(\sqrt{k})})$) if a geometric representation is given with $O^*(2^{\tilde{O}(\sqrt{k})})$ intersections) on string graphs due to the linear kernel yielding an equivalent instance on $2k$ vertices and the algorithm for MAX INDEPENDENT SET. The latter problem can be solved in $n^{O(\sqrt{k})}$ in segments or more generally in polygons of polynomial complexity [33], while MIN DOMINATING SET on string graphs cannot be solved in time $f(k)n^{o(k)}$, for any function f, unless the ETH fails.

References

1. Adamaszek, A., Har-Peled, S., Wiese, A.: Approximation schemes for independent set and sparse subsets of polygons. CoRR abs/1703.04758 (2017). http://arxiv.org/abs/1703.04758
2. Adamaszek, A., Wiese, A.: A QPTAS for maximum weight independent set of polygons with polylogarithmically many vertices. In: Proceedings of the Twenty-Fifth Annual ACM-SIAM Symposium on Discrete Algorithms, SODA 2014, Portland, Oregon, USA, 5–7 January 2014, pp. 645–656 (2014). https://doi.org/10.1137/1.9781611973402.49
3. Alber, J., Fiala, J.: Geometric separation and exact solutions for the parameterized independent set problem on disk graphs. J. Algorithms 52(2), 134–151 (2004)
4. Baker, B.S.: Approximation algorithms for NP-complete problems on planar graphs. J. ACM 41(1), 153–180 (1994). https://doi.org/10.1145/174644.174650
5. Biró, C., Bonnet, É., Marx, D., Miltzow, T., Rzążewski, P.: Fine-grained complexity of coloring unit disks and balls. In: 33rd International Symposium on Computational Geometry, SoCG 2017, 4–7 July 2017, Brisbane, Australia, pp. 18:1–18:16 (2017). https://doi.org/10.4230/LIPIcs.SoCG.2017.18
6. Bonnet, É., Rzążewski, P.: Optimality program in segment and string graphs. CoRR abs/1712.08907 (2017). http://arxiv.org/abs/1712.08907
7. Boyer, J.M., Myrvold, W.J.: On the cutting edge: simplified o(n) planarity by edge addition. J. Graph Algorithms Appl. 8(2), 241–273 (2004). http://jgaa.info/accepted/2004/BoyerMyrvold2004.8.3.pdf
8. Cabello, S., Cardinal, J., Langerman, S.: The clique problem in ray intersection graphs. Discret. Comput. Geom. 50(3), 771–783 (2013). https://doi.org/10.1007/s00454-013-9538-5
9. Chalopin, J., Gonçalves, D.: Every planar graph is the intersection graph of segments in the plane: extended abstract. In: Proceedings of the 41st Annual ACM Symposium on Theory of Computing, STOC 2009, Bethesda, MD, USA, 31 May - 2 June 2009, pp. 631–638 (2009). https://doi.org/10.1145/1536414.1536500
10. Demaine, E.D., Fomin, F.V., Hajiaghayi, M.T., Thilikos, D.M.: Bidimensional parameters and local treewidth. SIAM J. Discret. Math. 18(3), 501–511 (2004)
11. Demaine, E.D., Fomin, F.V., Hajiaghayi, M.T., Thilikos, D.M.: Subexponential parameterized algorithms on bounded-genus graphs and H-minor-free graphs. J. ACM 52(6), 866–893 (2005)
12. Demaine, E.D., Hajiaghayi, M.T.: Fast algorithms for hard graph problems: bidimensionality, minors, and local treewidth. In: Proceedings of GD 2014, pp. 517–533 (2004)

13. Demaine, E.D., Hajiaghayi, M.: The bidimensionality theory and its algorithmic applications. Comput. J. **51**(3), 292–302 (2008)
14. Demaine, E.D., Hajiaghayi, M.: Linearity of grid minors in treewidth with applications through bidimensionality. Combinatorica **28**(1), 19–36 (2008)
15. Erdős, P., Szekeres, G.: A Combinatorial Problem in Geometry, pp. 49–56. Birkhäuser Boston, Boston (1987). https://doi.org/10.1007/978-0-8176-4842-8_3
16. Fomin, F.V., Lokshtanov, D., Panolan, F., Saurabh, S., Zehavi, M.: Finding, hitting and packing cycles in subexponential time on unit disk graphs. CoRR abs/1704.07279 (2017). http://arxiv.org/abs/1704.07279
17. Fomin, F.V., Lokshtanov, D., Saurabh, S.: Bidimensionality and geometric graphs. In: Proceedings of SODA 2012, pp. 1563–1575 (2012)
18. Fox, J., Pach, J.: Computing the independence number of intersection graphs. In: Proceedings of the Twenty-Second Annual ACM-SIAM Symposium on Discrete Algorithms, SODA 2011, San Francisco, California, USA, 23–25 January 2011, pp. 1161–1165 (2011). https://doi.org/10.1137/1.9781611973082.87
19. Fox, J., Pach, J.: Applications of a new separator theorem for string graphs. CoRR abs/1302.7228 (2013). http://arxiv.org/abs/1302.7228
20. Fox, J., Pach, J., Tóth, C.D.: A bipartite strengthening of the crossing lemma. J. Comb. Theory, Ser. B **100**(1), 23–35 (2010). https://doi.org/10.1016/j.jctb.2009.03.005
21. Gonçalves, D., Isenmann, L., Pennarun, C.: Planar Graphs as L-intersection or L-contact graphs. In: Czumaj, A. (ed.) Proceedings of the Twenty-Ninth Annual ACM-SIAM Symposium on Discrete Algorithms, SODA 2018, New Orleans, LA, USA, 7–10 January 2018, pp. 172–184. SIAM (2018). https://doi.org/10.1137/1.9781611975031.12
22. Har-Peled, S.: Quasi-polynomial time approximation scheme for sparse subsets of polygons. In: 30th Annual Symposium on Computational Geometry, SOCG 2014, Kyoto, Japan, 08–11 June 2014, p. 120 (2014). https://doi.org/10.1145/2582112.2582157
23. Impagliazzo, R., Paturi, R., Zane, F.: Which problems have strongly exponential complexity? In: Proceedings of FOCS 1998, pp. 653–662, November 1998
24. Impagliazzo, R., Paturi, R.: On the complexity of k-SAT. J. Comput. Syst. Sci. **62**(2), 367–375 (2001). http://www.sciencedirect.com/science/article/pii/S0022000000917276
25. Impagliazzo, R., Paturi, R., Zane, F.: Which problems have strongly exponential complexity? J. Comput. Syst. Sci. **63**(4), 512–530 (2001). https://doi.org/10.1006/jcss.2001.1774
26. Knuth, D.E.: The Art of Computer Programming: Sorting and Searching, vol. III. Addison-Wesley, Boston (1973)
27. Kratochvíl, J., Matoušek, J.: Intersection graphs of segments. J. Comb. Theory, Ser. B **62**(2), 289–315 (1994). http://www.sciencedirect.com/science/article/pii/S0095895684710719
28. Kratochvíl, J.: String graphs. II. Recognizing string graphs is NP-hard. J. Comb. Theory, Ser. B **52**(1), 67–78 (1991). https://doi.org/10.1016/0095-8956(91)90091-W
29. Kratochvíl, J., Matoušek, J.: String graphs requiring exponential representations. J. Comb. Theory, Ser. B **53**(1), 1–4 (1991). https://doi.org/10.1016/0095-8956(91)90050-T
30. Lee, J.R.: Separators in region intersection graphs. CoRR abs/1608.01612 (2016). http://arxiv.org/abs/1608.01612

31. Lipton, R.J., Tarjan, R.E.: Applications of a planar separator theorem. SIAM J. Comput. **9**(3), 615–627 (1980). https://doi.org/10.1137/0209046
32. Marx, D.: Parameterized complexity of independence and domination on geometric graphs. In: Bodlaender, H.L., Langston, M.A. (eds.) IWPEC 2006. LNCS, vol. 4169, pp. 154–165. Springer, Heidelberg (2006). https://doi.org/10.1007/11847250_14
33. Marx, D., Pilipczuk, M.: Optimal parameterized algorithms for planar facility location problems using Voronoi diagrams. In: Bansal, N., Finocchi, I. (eds.) ESA 2015. LNCS, vol. 9294, pp. 865–877. Springer, Heidelberg (2015). https://doi.org/10.1007/978-3-662-48350-3_72
34. Marx, D., Pilipczuk, M.: Optimal parameterized algorithms for planar facility location problems using Voronoi diagrams. CoRR abs/1504.05476 (2015). http://arxiv.org/abs/1504.05476
35. Matoušek, J.: Intersection graphs of segments and ∃ℝ. CoRR abs/1406.2636 (2014). http://arxiv.org/abs/1406.2636
36. Matoušek, J.: Near-optimal separators in string graphs. Comb. Probab. Comput. **23**(1), 135–139 (2014). https://doi.org/10.1017/S0963548313000400
37. McDiarmid, C., Müller, T.: Integer realizations of disk and segment graphs. J. Comb. Theory, Ser. B **103**(1), 114–143 (2013). https://doi.org/10.1016/j.jctb.2012.09.004
38. Robertson, N., Seymour, P.D., Thomas, R.: Quickly excluding a planar graph. J. Comb. Theory, Ser. B **62**(2), 323–348 (1994). https://doi.org/10.1006/jctb.1994.1073
39. Schaefer, M., Sedgwick, E., Štefankovič, D.: Recognizing string graphs in NP. J. Comput. Syst. Sci. **67**(2), 365–380 (2003). https://doi.org/10.1016/S0022-0000(03)00045-X
40. Schaefer, M., Štefankovič, D.: Fixed points, nash equilibria, and the existential theory of the reals. Theory Comput. Syst. **60**(2), 172–193 (2017). https://doi.org/10.1007/s00224-015-9662-0
41. Scheinerman, E.: Intersection classes and multiple intersection parameters of graphs. Ph.D. thesis, Princeton University (1984)
42. Smith, W.D., Wormald, N.C.: Geometric separator theorems and applications. In: Proceedings of FOCS 1998, pp. 232–243. IEEE Computer Society, Washington, DC (1998). http://dl.acm.org/citation.cfm?id=795664.796397
43. Zverovich, I.E., Zverovich, V.E.: An induced subgraph characterization of domination perfect graphs. J. Graph Theory **20**(3), 375–395 (1995). https://doi.org/10.1002/jgt.3190200313

Anagram-Free Chromatic Number Is Not Pathwidth-Bounded

Paz Carmi[1], Vida Dujmović[2], and Pat Morin[3(✉)]

[1] Department of Computer Science, Ben-Gurion University of the Negev,
Beersheba, Israel
[2] School of Computer Science and Electrical Engineering, University of Ottawa,
Ottawa, Canada
[3] School of Computer Science, Carleton University, Ottawa, Canada
morin@scs.carleton.ca

Abstract. The anagram-free chromatic number is a new graph parameter introduced independently by Kamčev, Łuczak, and Sudakov [1] and Wilson and Wood [5]. In this note, we show that there are planar graphs of pathwidth 3 with arbitrarily large anagram-free chromatic number. More specifically, we describe $2n$-vertex planar graphs of pathwidth 3 with anagram-free chromatic number $\Omega(\log n)$. We also describe kn vertex graphs with pathwidth $2k-1$ having anagram-free chromatic number in $\Omega(k \log n)$.

1 Introduction

A string $s = s_1, \ldots, s_{2k}$ is called an *anagram* if $s_1, \ldots, s_k$ is a permutation of $s_{k+1}, \ldots, s_{2k}$. For a graph G, a c-colouring $\varphi : V(G) \to \{1, \ldots, c\}$ is *anagram-free* if, for every odd-length path $v_1, v_2, \ldots, v_{2k}$ in G, the string $\varphi(v_1), \ldots, \varphi(v_{2k})$ is not an anagram. The *anagram-free chromatic number* of G, denoted $\pi_a(G)$, is the smallest value of c for which G has an anagram-free c-colouring.

Answering a long-standing question of Erdős and Brown, Keränen [2] showed that, for any n, the path P_n on n vertices has an anagram-free 4-colouring. A straightforward divide-and-conquer algorithm applied to any n-vertex graph of treewidth[1] at most k yields an anagram-free $O(k \log n)$-colouring. The same

---

This work was partly funded by NSERC and the Ontario Ministry of Research, Innovation and Science. This work is based on work performed while attending the AlgoPARC Workshop on Parallel Algorithms and Data Structures at the University of Hawaii at Manoa, in part supported by the National Science Foundation under Grant No. 1745331.

[1] A *tree decomposition* (T, B) of a graph G consists of a tree T and set $B = \{B_x : x \in V(T)\}$ of subsets of $V(G)$ called *bags* that are indexed by the nodes of T with the following properties:

1. for every edge $uw \in E(G)$, there is at least one bag B_x, $x \in V(T)$ with $u, w \in B_x$; and

© Springer Nature Switzerland AG 2018
A. Brandstädt et al. (Eds.): WG 2018, LNCS 11159, pp. 91–99, 2018.
https://doi.org/10.1007/978-3-030-00256-5_8

divide-and-conquer algorithm, applied to graphs that exclude a fixed minor gives an anagram free $O(\sqrt{n})$-colouring [1].

An interesting variant of this divide-and-conquer algorithm is used by Wilson and Wood [5] to obtain anagram-free $(4k+1)$-colourings of trees of pathwidth[2] k. On the negative side, Kamčev, Łuczak, and Sudakov [1] and Wilson and Wood [5] have shown that there are trees—even binary trees—with arbitrarily large anagram-free chromatic number. These results, and some others, are summarized in Table 1.

Table 1. Bounds on anagram-free chromatic number. Upper bounds apply to all graphs in the class. Lower bounds apply to some graphs in the class.

Graph class	Bounds	Reference
Paths	$\pi_\alpha(G) = 4$	[2, Theorem 1]
Graphs of treewidth k	$\pi_\alpha(G) \in O(k \log n)$	folklore
Graphs excluding a minor of size h	$\pi_\alpha(G) \in O(h^{3/2} n^{1/2})$	[1, Proposition 1.2]
Trees	$\pi_\alpha(G) \in \Omega(\log n / \log \log n)$	[5, Theorem 3]
Trees of pathwidth k	$k \leq \pi_\alpha(G) \leq 4k + 1$	[5, Theorem 5]
Trees of radius r	$r \leq \pi_\alpha(G) \leq r + 1$	[5, Theorem 4]
Binary trees	$\pi_\alpha(G) \in \Omega(\sqrt{\log n / \log \log n})$	[1, Proposition 1.1]
4-regular graphs	$\pi_\alpha(G) \in \Omega(\sqrt{n} / \log n)$	[1, Proposition 3.1]
d-regular graphs	$\pi_\alpha(G) \in \Omega(n)$	[1, Theorem 1.3]
Subdivisions of graphs	$\pi_\alpha(G) \leq 8$	[4, Theorem 6]
Planar graphs	$\pi_\alpha(G) \in O(\sqrt{n})$	[1, Corollary 2.3]
Planar graphs of max. degree 3	$\pi_\alpha(G) \in \Omega(\log n)$	[1, Proposition 2.4] [5, Theorem 1]
Planar graphs of pathwidth 3	$\pi_\alpha(G) \in \Omega(\log n)$	Theorem 1
Graphs of pathwidth $k > 3$	$\pi_\alpha(G) \in \Omega(k \log n)$	Theorem 2

All of the examples of graphs having large anagram-free chromatic number are graphs with large pathwidth [3]. Therefore, an obvious question is whether anagram-free chromatic number is pathwidth-bounded, i.e., can $\pi_\alpha(G)$ be upper bounded by some function of the pathwidth $pw(G)$ of G? Such a result seems plausible, for two reasons:

1. pathwidth is a measure of how path-like a graph is and Keränen showed that paths have anagram-free 4-colourings; and

2. for every vertex $v \in V(G)$, the set $T[v] = \{x \in V(T) : v \in B_x\}$ induces a (connected) subtree of T.

The *width* of a tree decomposition (T, B) is the size $\max\{|B_x| : x \in V(T)\}$ of its largest bag. The *treewidth* of a graph G is the minimum width of any tree decomposition of G.

[2] A *path decomposition* of G is a tree decomposition (P, B) of G where P is a path. The *pathwidth* of G is the minimum width of any path decomposition of G.

2. the result of Wilson and Wood [5] shows that $\pi_a(T) \leq 4\,\mathrm{pw}(T) + 1$ for every tree, T.

The purpose of this note, however, is to show that the result of Wilson and Wood can not be strengthened even to planar graphs of pathwidth 3 and maximum degree 5. (Here and throughout, $\log x = \log_2 x$ denotes the binary logarithm of x.)

Theorem 1. *For every $n \in \mathbb{N}$, there exists a $2n$-vertex planar graph of pathwidth 3 and maximum degree 5 whose anagram-free chromatic number is at least $\log(n + 1)$.*

Theorem 2. *For every $n \in \mathbb{N}$ and every integer $k \geq 3$, there exists a kn-vertex graph of pathwidth $2k - 1$ and maximum degree $3k - 1$ whose anagram-free chromatic number is at least $(k - 2)\log(n/3)$.*

These two results show that the straightforward divide-and-conquer algorithm using separators gives asymptotically worst-case optimal colourings for graphs of pathwidth k and graphs of treewidth k.

2   Proof of Theorem 1

Let $s \in \Sigma^*$ be a string over some alphabet Σ. For each $a \in \Sigma$, we let $n_a(s)$ denote the number of occurences of a in s. We say that s is *even* if $n_a(s)$ is even for each $a \in \Sigma$. The following lemma says that strings with no even substrings must use an alphabet of at least logarithmic size.

Lemma 1. *If $s = s_0, \ldots, s_{2n-1} \in \Sigma^{2n}$ and $|\Sigma| < \log(n + 1)$, then s contains a non-empty even substring $s_{2i}, \ldots, s_{2j-1}$ for some $0 \leq i < j \leq n$.*

Proof. For any string $q \in \Sigma^*$, we define the *parity vector* $P(q) = \langle n_a(q) \bmod 2 : a \in \Sigma \rangle$ and observe that q is even if and only if $P(q) = \langle 0, \ldots, 0 \rangle$. Furthermore, for two strings p and q, the parity vector of their concatenation pq is equal to the xor-sum (i.e., modulo 2 sum) of their parity vectors:

$$P(pq) = P(p) \oplus P(q).$$

Define the strings $t_0, \ldots, t_n$, where t_0 is the empty string and, for each $i \in \{1, \ldots, n\}$, define $t_i = s_0, \ldots, s_{2i-1}$.

Now consider the parity vectors $P(t_0), P(t_1), \ldots, P(t_n)$. Each of these $n + 1$ vectors is a binary string of length $|\Sigma| < \log(n + 1)$ therefore, there must exist two indices $i, j \in \{0, \ldots, n\}$ with $i < j$ such that $P(t_i) = P(t_j)$. However,

$$P(t_j) = P(t_i) \oplus P(s_{2i}, \ldots, s_{2j-1})$$

and since $P(t_i) = P(t_j)$, this implies that $P(s_{2i}, \ldots, s_{2j-1}) = \langle 0, \ldots, 0 \rangle$ and $s_{2i}, \ldots, s_{2j-1}$ is even, as required.   □

The next lemma says that if we split an even string into consecutive pairs, then we can can colour one element of each pair red and the other blue in so that the resulting red and blue multisets are exactly the same.

Lemma 2. *Let* $s = s_0, \ldots, s_{2r-1} \in \Sigma^{2r}$ *be an even string. Then there exists a binary sequence* $v_0, \ldots, v_{r-1}$ *such that the string* $s_v = s_{0+v_0}, s_{2+v_1}, \ldots, s_{2(r-1)+v_{r-1}}$ *has* $n_a(s_v) = n_a(s)/2$ *for all* $a \in \Sigma$.

Proof. Suppose for the sake of contradiction that the lemma is not true, and let s be the shortest counterexample. For each binary vector $v \in \{0,1\}^r$, let $s_{\overline{v}} = s_{0+1-v_0}, s_{2+1-v_1}, \ldots, s_{2(r-1)+1-v_{r-1}}$ be the complement of s_v. Note that, for any $v \in \{0,1\}^r$ and any $a \in \Sigma$, $n_a(s_v)+n_a(s_{\overline{v}}) = n_a(s)$. Therefore, v satisifies the conditions of the lemma if and only if $n_a(s_v) = n_a(s_{\overline{v}})$ for all $a \in \Sigma$. Let $v \in \{0,1\}^r$ be the binary vector that minimizes

$$\sum_{a \in \Sigma} |n_a(s_v) - n_a(s_{\overline{v}})|. \tag{1}$$

Since s is a counterexample to the lemma, (1) is greater than zero.

For each $j \in \{0, \ldots, r-1\}$, let $x_j = s_{2j+v_j}$ and let $y_j = s_{2j+1-v_j}$ so that $s_v = x_0, \ldots, x_{r-1}$ and $s_{\overline{v}} = y_0, \ldots, y_{r-1}$. Since (1) is non-zero, there exists some j_1 such that $n_{x_{j_1}}(s_v) > n_{x_{j_1}}(s_{\overline{v}})$. This means that $n_{y_{j_1}}(s_v) \geq n_{y_{j_1}}(s_{\overline{v}})$, otherwise flipping[3] v_{j_1} would decrease (1) by two. Furthermore, $y_{j_1} \neq x_{j_1}$ since, otherwise, we could remove s_{2j_1} and s_{2j_1+1} from s and obtain a smaller counterexample, since the value of v_j has no effect on (1).

Refer to Fig. 1. Let $a_1 = x_{j_1}$ and for $k = 2, 3, 4 \ldots$, define $a_k = y_{j_{k-1}}$ and define j_k to be any index such that $x_{j_k} = a_k$. Notice that that $n_{a_k}(s_v) \geq n_{a_k}(s_{\overline{v}})$ since, otherwise, flipping $v_{j_1}, \ldots, v_{j_{k-1}}$ would decrease the value of (1). Indeed, flipping $v_{j_1}, \ldots, v_{j_{k-1}}$ decreases $n_{a_1}(s_v)$ by one, increases $n_{a_k}(s_v)$ by one, and does not change $n_a(s_v)$ for any $a \in \Sigma \setminus \{a_1, a_k\}$. This implies that j_k is well-defined since $n_{a_k}(s_v) \geq n_{a_k}(s_{\overline{v}}) \geq 1$.

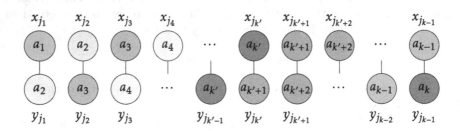

Fig. 1. The proof of Lemma 2.

Since s is finite, there is some minimum value k such that $a_k = a_{k'}$ for some $k' < k$. This defines a sequence of indices $j_{k'}, \ldots, j_{k-1}$ such that

---

[3] Here and throughout, *flipping* a binary variable b means changing its value to $1 - b$.

1. $a_{k'} = x_{j_{k'}} = y_{j_{k-1}} = a_k$;
2. $a_\ell = y_{j_{\ell-1}} = x_{j_\ell}$ for all $\ell \in \{k'+1, \ldots, k-1\}$.

In words, for each $\ell \in \{k', \ldots, k\}$, each occurrence of a_ℓ in s_v is matched with a corresponding occurrence of a_ℓ in $s_{\overline{v}}$. We claim that this contradicts the minimality of s. Indeed, by removing $s_{2j_{k'}}, s_{2j_{k'}+1}, s_{2j_{k'+1}}, s_{2j_{k'+1}+1}, \ldots, s_{2j_{k-1}}, s_{2j_{k-1}+1}$ from s we obtain a smaller counterexample. □

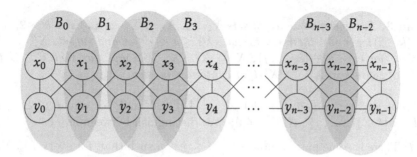

Fig. 2. The graph G in the proof of Theorem 1.

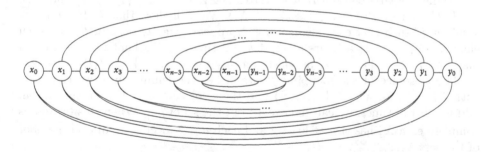

Fig. 3. The graph G in the proof of Theorem 1 is planar and is even a 2-page graph.

Proof of Theorem 1. The graph G used to prove the lower bound has vertex set $V(G) = \{x_0, y_0, \ldots, x_{n-1}, y_{n-1}\}$ and edge set

$$E(G) = \bigcup_{i=0}^{n-2} \{x_i x_{i+1}, x_i y_{i+1}, y_i y_{i+1}, y_i x_{i+1}\} \cup \{x_i y_i : i \in \{0, \ldots, n-1\}\}.$$

The graph G has pathwidth 3 as can be seen from the path decomposition whose bags are $B_0, \ldots, B_{n-2}$ where $B_i = \{x_i, y_i, x_{i+1}, y_{i+1}\}$. See Fig. 2. Although not immediately obvious from Fig. 2, G is also planar—see Fig. 3.

Now, consider some colouring $\varphi : V(G) \to \Sigma$ with $|\Sigma| < \log(n+1)$. Applying Lemma 1 to the string $s = \varphi(x_0), \varphi(y_0), \ldots, \varphi(x_{n-1}), \varphi(y_{n-1})$ we conclude that there is some $i < j$ such that $\varphi(x_i), \varphi(y_i), \ldots, \varphi(x_j), \varphi(y_j)$ is even. By Lemma 2 and the symmetry between each x_i and y_i we can assume that $n_a(\varphi(x_i), \ldots, \varphi(x_j)) = n_a(\varphi(y_i), \ldots, \varphi(y_j))$ for each $a \subset \Sigma$. But then the path $x_i, \ldots, x_j, y_j, y_{j-1}, \ldots, y_i$ has a colour sequence that is an anagram. $\square$

3   Proof of Theorem 2

Lemma 3. *For every sequence of sets $X_1, \ldots, X_n \subseteq \Sigma$, each of size $k > 2$, with $|\Sigma| < (k-2)\log(n/3)$, there exists indices $1 \le i < j \le n$ and subsets $X'_i, \ldots, X'_j$ such that, for each $\ell \in \{i, \ldots, j\}$, $X'_\ell \subseteq X_\ell$, $|X'_\ell| \ge 2$ and, for each $a \in \Sigma$ the number of subsets in $X'_i, \ldots, X'_j$ that contain a is even.*

Proof. For any $1 \le i \le j \le n$, let $\Sigma_{i,j} = \bigcup_{\ell=i}^{j} X_i$ and, for any $I \subset \Sigma_{i,j}$, let $N_{i,j}(I) = \{\ell \in \{i, \ldots, j\} : X_\ell \cap I \neq \emptyset\}$. We distinguish between two cases.

Case 1: There is some pair of indices $1 \le i \le j \le n$ such that, for every $I \subseteq \Sigma_{i,j}$,

$$|N_{i,j}(I)| \ge |I|/(k-2). \tag{2}$$

In this case we will show the existence of the desired sets $X'_i, \ldots, X'_j$. Without loss of generality, assume $i = 1$, $j = n$, and define $N = N_{1,n}$.

Define a bipartite graph H with vertex set $V(H) = \Sigma \cup \{1, \ldots, n\}$ and edge set $E(H) = \{(a,i) : i \in \{1, \ldots, n\}, a \in X_i\}$. We will show that $E(H)$ contains a subset E' such that each element $a \in \Sigma$ appears exactly once in E' and each element of $\{1, \ldots, n\}$ appears at most $k-2$ times in E'. That is, E' defines a mapping $f : \Sigma \to \{1, \ldots, n\}$ in which, for any $i \in \{1, \ldots, n\}$, $|f^{-1}(i)| \le k-2$.

The existence of the mapping f establishes the lemma since we can start with $X'_i = X_i$ for all $i \in \{1, \ldots, n\}$ and then, for each $a \in \Sigma$ that appears an odd number of times, we can remove a from the set $X'_{f(a)}$. When this process is complete each X'_i has size at least 2 and each $a \in \Sigma$ occurs in an even number of the sets $X'_1, \ldots, X'_n$.

All that remains is to prove the existence of the edge set E', which we do using an augmenting paths argument like that used, for example, to prove Hall's Marriage Theorem. Consider an edge set $E' \subseteq E(H)$ that contains exactly one edge incident to each $a \in \Sigma$ and let $f : \Sigma \to \{1, \ldots, n\}$ be the corresponding mapping. Then we define

$$\Phi(E') = \sum_{i=1}^{n} \max\{0, |f^{-1}(i)| - (k-2)\}.$$

Note that the set E' we hope to find has $\Phi(E') = 0$. Now, select some E' that minimizes $\Phi(E')$. If $\Phi(E') = 0$ then we are done, so assume by way of contradiction, that $\Phi(E') > 0$. Thus, there exists some index $i_0 \in \{1, \ldots, n\}$

such that $|f^{-1}(i_0)| \geq k - 1$ and therefore the set $\Sigma_0 = f^{-1}(i_0)$ has size at least $k - 1$. Therefore,

$$|N(\Sigma_0)| \geq \left\lceil \frac{|\Sigma_0|}{k-2} \right\rceil \geq \left\lceil \frac{k-1}{k-2} \right\rceil = 2.$$

In particular, $N(\Sigma_0) \setminus \{i_0\}$ is non-empty. Let $I_0 = \{i_0\}$ and observe that each $i_1 \in N(\Sigma_0) \setminus I_0$ must have $|f^{-1}(i_1)| \geq k - 2$ since, otherwise we could replace the edge (a_1, i_0) with (a_1, i_1) in E' and this would decrease $\Phi(E')$. Let $I_1 = N(\Sigma_0)$ and let $\Sigma_1 = \bigcup_{i_1 \in I_i} f^{-1}(i_1)$. We have just argued that

$$|\Sigma_1| \geq |I_1|(k-2) + 1$$

and therefore,

$$|N(\Sigma_1)| \geq \left\lceil \frac{|\Sigma_1|}{k-2} \right\rceil \geq \left\lceil \frac{|I_1|(k-2)+1}{k-2} \right\rceil \geq |I_1| + 1.$$

But now we can continue this argument, defining $I_j = N(\Sigma_{j-1})$ and $\Sigma_j = \bigcup_{i_j \in I_j} f^{-1}(i_j)$. Again, each $i_j \in I_j \setminus \bigcup_{\ell=1}^{j-1} I_\ell$ must have $|f^{-1}(i_j)| \geq k - 2$, otherwise we can find a path $i_0, a_0, i_1, a_1, \ldots, a_{j-1}i_j$ and replace, in E', the edges $i_0 a_0, \ldots, i_{j-1} a_{j-1}$ with $a_0 i_1, a_1 i_2, \ldots, a_{j-1} i_j$ which would decrease $\Phi(E')$. In this way, we obtain an infinite sequence of subsets $I_0, \ldots, I_\infty \subseteq \{1, \ldots, n\}$ such that $|I_j| > |I_{j-1}|$. This is clearly a contradiction, since each $|I_j|$ is an integer in $\{1, \ldots, n\}$.

Case 2: For every $1 \leq i < j \leq n$, there exists a set $I \subset \Sigma_{i,j}$ such that $|N_{i,j}(I)| < |I|/(k-2)$. In this case, we will show that $|\Sigma| \geq (k-2) \log(n/3)$.

Before jumping into the messy details, we sketch an inductive proof that gives the main intuition for why $|\Sigma| \in \Omega(k \log n)$: There is some set $I_0 \subset \Sigma$ such that $[1, n] \setminus N(I_0)$ consists of $O(|I_0|/k)$ intervals. One such interval contains $\Omega(nk/|I_0|)$ integers $i_0, \ldots, j_0$. By induction on n, $|\Sigma_{i_0, j_0}| = \Omega(k \log(nk/|I_0|))$. But Σ_{i_0, j_0} is disjoint from I_0, so

$$|\Sigma| \geq |I_0| + \Omega(k \log(nk/|I_0|)) = |I_0| + \Omega(k \log n) - O(k \log(|I_0|/k)) = \Omega(k \log n).$$

The messy details occur when $|I| = k - 1$ since then the $|I|$ and $-O(k \log(|I|/k)$ terms are close in magnitude.

Let $n_0 = n$, $i_0 = 1$, $j_0 = n$, $\Sigma_0 = \Sigma$ and let $I_0 \subseteq \Sigma_0$ be such that $|N(I_0)| < |I|/(k-2)$. For each integer ℓ with $n_{\ell-1} \geq 1$, we define

1. i_ℓ and j_ℓ such that $i_{\ell-1} \leq i_\ell < j_\ell \leq j_{\ell-1}$, $\{i_\ell, \ldots, j_\ell\} \cap N_{i_{\ell-1}, j_{\ell-1}}(I_{\ell-1}) = \emptyset$, and $n_\ell = j_\ell - i_\ell + 1$ is maximized.
2. $I_\ell \subset \Sigma_{i_\ell, j_\ell}$ such that $|N_{i_\ell, j_\ell}(I_\ell)| < |I_\ell|/k$;

In words, $N_{i_{\ell-1}, j_{\ell-1}}(I_{\ell-1})$ partitions $i_{\ell-1}, \ldots, j_{\ell-1}$ into intervals and we choose i_ℓ and j_ℓ to be the endpoints of a largest such interval and recurse on that interval using a new set I_ℓ. Letting $y_\ell = |N_{i_\ell, j_\ell}(I_\ell)|$, observe that, for $\ell \geq 1$,

$$n_\ell \geq \frac{n_{\ell-1} - y_{\ell-1}}{y_{\ell-1} + 1} > \frac{n_{\ell-1}}{y_{\ell-1} + 1} - 1.$$

By expanding the preceding equation we can easily show that

$$n_\ell \geq \frac{n}{\prod_{\tau=0}^{\ell-1}(y_\tau + 1)} - 2.$$

Note that $n_{\ell+1}$ is defined until $n_\ell < 1$ so combining this with the preceding equation and taking logs yields

$$\sum_{\tau=0}^{\ell-1}(y_\tau + 1) > \log(n/3) \tag{3}$$

Finally, observe that the sets $I_0, \ldots, I_{\ell-1}$ are disjoint, so

$$|\Sigma| \geq \sum_{\tau=0}^{\ell-1}|I_\tau| > \sum_{\tau=0}^{\ell-1}(k-2)y_\tau. \tag{4}$$

Now, minimizing (4) subject to (3) and using the fact that each $y_\tau \geq 1$ is an integer shows that $|\Sigma| \geq (k-2)\log(n/3)$, as desired. (The minimum is obtained when $\ell = \log(n/3)$ and $y_1 = y_2 = \cdots = y_{\ell-1} = 1$.) □

Proof of Theorem 2. The pathwidth $2k - 1$ graph, G, used in this proof is a natural generalization of the pathwidth 3 graph used in the proof of Theorem 1. The kn vertices of G are partitioned in subsets $V_1, \ldots, V_n$, each size of size k. For each $i \in \{1, \ldots, n\}$, V_i is a clique and, for each $i \in \{1, \ldots, n-1\}$, every vertex in V_i is adjacent to every vertex in V_{i+1}. That this graph has pathwidth $2k - 1$ can be seen from the path decomposition whose bags are $B_1, \ldots, B_{n-1}$ where each $B_i = \{V_i \cup V_{i+1}\}$.

Suppose we have some colouring $\varphi : V(G) \to \Sigma$, with $|\Sigma| < (k-2)\log(n/3)$. Define the sets $X_1, \ldots, X_n$ where $X_i = \{\varphi(v) : v \in V_i\}$. By Lemma 3, we can find indices $i \in \{0, \ldots, n-1\}$ and $r > 0$ and subsets $V_1', \ldots, V_r'$ such that, for each $\ell \in \{1, \ldots, r\}$, $V_\ell' \subseteq V_{i+\ell}$, $|V_\ell'| \geq 2$, and such that each colour $a \in \Sigma$ is used in an even number of $V_1', \ldots, V_r'$.

Next, label the vertices in $V_1', \ldots, V_r'$ red and blue as follows. If $|V_i'|$ is even, then label half its vertices red and half its vertices blue, arbitrarily. Let $Q_1, \ldots, Q_t$ denote the subsequence of $V_1', \ldots, V_r'$ consisting of only sets of odd size (so the vertices in $Q_1, \ldots, Q_t$ are not labelled red or blue yet). Then, for odd values of i, label $\lceil |Q_i|/2 \rceil$ vertices of Q_i red and the remaining blue. For even values of i label $\lfloor |Q_i|/2 \rfloor$ vertices of Q_i red and the remaining blue. Observe that, since $\sum_{i=1}^r |V_i'|$ is even, t is also even, so exactly half the vertices in $\bigcup_{i=1}^r V_i'$ are red and half are blue.

Now, consider the following perfect bichromatic matching of the complete graph whose vertex set is $\bigcup_{i=1}^r V_i'$: In every set V_i' of even size we match each red vertex in V_i' with a blue vertex in V_i'. In each odd size set Q_i, we match $\lfloor |Q_i|/2 \rfloor$ red vertices with blue vertices leaving one vertex v_i unmatched. This leaves t unmatched vertices $u_1, \ldots, u_t$ and these vertices alternate colour between red and blue. To complete the matching, we match u_{2i} with u_{2i-1} for each $i \in \{1, \ldots, t/2\}$.

Now, treat this matching as a long string $s = x_1, y_1, \ldots, x_q, y_q$ where each $x_i = \varphi(v_i)$, each $y_i = \varphi(w_i)$, and each (v_i, w_i) is a matched pair of vertices. By construction, s is an even string of length $2q$ so applying Lemma 2 to s we obtain two sets of vertices $V = \{v_1', \ldots, v_q'\}$ and $W = \{w_1', \ldots, w_q'\}$ such that, for each $a \in \Sigma$, $n_a(\varphi(v_1), \ldots, \varphi(v_q)) = n_a(\varphi(w_1), \ldots, \varphi(w_q))$. Thus, all that remains is to show that G contains a path P whose first half is some permutation of V and whose second half is some permutation of W. But this is obvious, because, for each $i \in \{1, \ldots, r\}$, V_i' contains at least one vertex of V and at least one vertex of W. Thus, the path P first visits all the vertices of $V \cap V_1'$ followed by all the vertices of $V \cap V_2'$, and so on until visiting all the vertices in $V \cap V_r'$. Next, the path returns and visits all the vertices in $W \cap V_r'$, $W \cap V_{r-1}'$, and so on back to $W \cap V_1'$. The existence of the path P shows that no colouring of G with fewer than $(k-2)\log(n/3)$ colours is anagram-free, so $\pi_\alpha(G) \geq (k-2)\log(n/3)$.   □

4   Remarks

We have shown that the anagram-free chromatic number of graphs of pathwidth 3 is unbounded. Graph of pathwidth 1 are caterpillars (a special case of trees) and therefore, by the result of Wilson and Wood [5], have anagram-free chromatic number at most 5. It is still open problem, explicitly stated by Wilson and Wood to determine if graphs of pathwidth 2 have bounded anagram-free chromatic number.

We have show that anagram-free chromatic number is not pathwidth-bounded, even for planar graphs. The graph we use in the proof of Theorem 1 is a 2-page graph; it has a book embedding using two pages. Outerplanar graphs have a book embedding using a single page. Is anagram-free chromatic number pathwidth-bounded for outerplanar graphs? We do not even know if the $2 \times n$ grid has constant anagram-free chromatic number.

References

1. Kamčev, N., Łuczak, T., Sudakov, B.: Anagram-free colorings of graphs. Comb. Probab. Comput. **27**(4), 1–20 (2017). Online first edition published August 2017
2. Keränen, V.: Abelian squares are avoidable on 4 letters. In: Kuich, W. (ed.) ICALP 1992. LNCS, vol. 623, pp. 41–52. Springer, Heidelberg (1992). https://doi.org/10. 1007/3-540-55719-9_62
3. Robertson, N., Seymour, P.D.: Graph minors. I. Excluding a forest. J. Comb. Theor. Ser. B **35**(1), 39–61 (1983)
4. Wilson, T.E., Wood, D.R.: Anagram-free colourings of graph subdivisions (2017)
5. Wilson, T.E., Wood, D.R.: Anagram-free graph colouring (2017)

Tight Lower Bounds for the Number of Inclusion-Minimal st-Cuts

Alessio Conte[1], Roberto Grossi[2], Andrea Marino[2], Romeo Rizzi[3], Takeaki Uno[1], and Luca Versari[2(✉)]

[1] National Institute of Informatics, Tokyo, Japan
{conte,uno}@nii.ac.jp
[2] Università di Pisa, Pisa, Italy
{grossi,marino,luca.versari}@di.unipi.it
[3] Università di Verona, Verona, Italy
rizzi@di.univr.it

Abstract. We study the number of inclusion-minimal cuts in an undirected connected graph G, also called st-cuts, for any two distinct nodes s and t: the st-cuts are in one-to-one correspondence with the partitions $S \cup T$ of the nodes of G such that $S \cap T = \emptyset$, $s \in S$, $t \in T$, and the subgraphs induced by S and T are connected. It is easy to find an exponential upper bound to the number of st-cuts (e.g. if G is a clique) and a constant lower bound. We prove that there is a more interesting lower bound on this number, namely, $\Omega(m)$, for undirected m-edge graphs that are biconnected or triconnected (2- or 3-node-connected). The wheel graphs show that this lower bound is the best possible asymptotically.

1 Introduction

Cuts are among the fundamental notions in graphs. A cut in a graph G represents a bipartition S, T of its node set $V(G)$, and the corresponding cutset is the set of edges in $E(G)$ having one endpoint in S and the other in T. Cutsets have a wide range of applications, such as switching functions, sensitivity analysis of optimization problems, vertex packing, and network reliability [22,23]. Due to the sheer number of cuts, it makes sense to focus on those whose cutsets are minimal under inclusion (i.e. any subset of their edges is not a cutset): these cuts corresponds to those having both induced subgraphs $G[S]$ and $G[T]$ connected.[1]

For any two given distinct nodes s and t in $V(G)$, we consider their *st-cuts* or, equivalently, the bipartitions S, T for which $s \in S$, $t \in T$ and both $G[S]$ and $G[T]$ are connected (a.k.a. bonds). In the following, we refer to just S as a cut, meaning the bipartition S and $T = V(G) \backslash S$ as it is clear from the context.

In this paper, we investigate the number of st-cuts in an undirected connected graph G with $n = |V(G)|$ nodes and $m = |E(G)|$ edges. This is useful to recursively generate st-cuts, as knowing a lower bound on their number can

---

[1] Since G is connected, also $G[S]$ and $G[T]$ are connected, otherwise we could remove at least one edge from the minimal cutset to reconnect $G[S]$ or $G[T]$.

© Springer Nature Switzerland AG 2018
A. Brandstädt et al. (Eds.): WG 2018, LNCS 11159, pp. 100–110, 2018.
https://doi.org/10.1007/978-3-030-00256-5_9

help to better amortize the cost of recursive calls. For general graphs, we can face two kinds of extreme situations. If G is a clique, any choice of $S \subseteq V(G)$ such that $s \in S$ and $t \notin S$ gives rise to an st-cut; this yields 2^{n-2} st-cuts as each such S can be obtained by adding s to any subset of $V(G) \backslash \{s, t\}$. At the other extreme is a single st-cut when G is made up of two cliques connected by an edge $\{s, t\}$, where S is the node set of one of the cliques (see Fig. 1 (left)).

Fig. 1. Left: connected graph with one st-cut. Right: 2-edge-connected graph with two st-cuts. The clouds correspond to cliques.

It is natural to investigate how the number of st-cuts changes if we add some more stringent requirement on the connectivity of G. For 2-edge-connectivity, where G remains connected by any single edge removal, the situation does not change significantly. The graph in Fig. 1 (right) is 2-edge connected and has just two st-cuts. The graph is formed by a triangle where s and t are its nodes, and the remaining node is part of a clique of $n - 2$ nodes: the only possible choices for S are the singleton s or the clique extended with s. On the other hand, choosing both s and t inside the clique (instead of the triangle) would give rise to an exponential number of st-cuts, as discussed before.

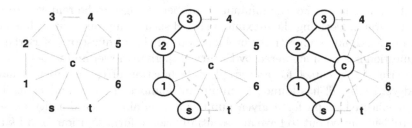

Fig. 2. Left: a wheel graph (triconnected, thus also biconnected). Center and right: st-cuts with respectively $c \notin S$ and $c \in S$.

What if G is biconnected (i.e. 2-node-connected)? For 2-node-connectivity, G remains connected by any single node removal, and the clique is clearly biconnected, thus we still have an exponential number of st-cuts. But this is not the interesting question to pose. What we investigate is the *minimum* number of st-cuts that a biconnected graph can have. For example, the wheel graph in Fig. 2 has $2(n-2)$ st-cuts: the graph is a cycle of $n - 1$ nodes, all connected to a center node, thus $m = 2(n - 1)$. The center node either belongs to S or not: in either cases, we have $n - 2$ ways of choosing the remaining nodes, as they must bipartite the cycle in two sectors of adjacent nodes. One may wonder if it is possible to find a biconnected graph with a constant number of st-cuts: the answer to this question is negative.

Indeed the contribution of this paper is to give a proof that there are $\Omega(m)$ st-cuts in any biconnected graph G for any choice of distinct node s, t. This provides an interesting *gap*, from $\Omega(1)$ to $\Omega(m)$, when we move from either connectivity or 2-edge-connectivity to 2-node-connectivity. The wheel in Fig. 2 shows that the lower bound is tight. It is an open problem to study higher connectivity or other requirements on G: we observe that our lower bound extends to k-node-connected graphs with $k \geq 2$, and matches for $k = 2, 3$ as the wheel is triconnected, but we do not know if the lower bound is tight for $k > 3$.

Related Work. In the literature, the generation of all cuts in undirected graphs has been studied by Abel and Bicker [1], Beltmore and Jensen [3], Tsukiyama et al. [27], Golberg [11]. Others are [15,21,28]. Among these approaches, Tsukiyama et al. is the most efficient as it requires $O(m)$ time per cutset. Algorithms for generating minimum cardinality and minimum weight cuts have been proposed by Ball and Provan [2], Gardner [10], and Picard and Queyranne [20]. In [29], all cuts of G are returned by non-decreasing weights ordering. Other variations include the k-best cuts problem which have been considered in [12]. (For the case of directed graphs see [25].) Enumerating the cutsets between all pairs of nodes reduces to the problem of solving a system of linear equations [19]. The notion of cutset has been generalized to cut conjunctions in [17].

From the above works we see that over the years a lot of listing algorithms have been proposed for generating minimal cutsets. On the other hand, studies about bounds on the number of cutsets have been focused on minimum cutsets without fixing s and t, i.e. the minimum number of edges to be removed to disconnect a graph. Deciding the maximum size of such a minimum cutset has been called the maximum connectivity problem, one of the 14 questions of Berge [4]. This question has been answered by Harary [14], giving lower and upper bounds on the size of this cutset for any graph as a function of the number of nodes and edges. Bixby [6] has found the minimum number of edges and nodes in a k-edge-connected graph for a given number b of minimum cardinality cutsets. This problem turned out to have an essentially closed form solution for all k and b, and for many values of k and b it is possible to build a graph achieving this minimum.

Over the years, this interest towards bounding cutsets have been mainly motivated by the network reliability problem [7]. One of the fundamental results is due to Kruskal [18] and Katona [16] in terms of F_i, which is the number of sets of i edges which do not contain a cutset. Upper bounds for the number of minimum cutsets in terms of the radius, diameter, minimum degree, maximum degree, chordality, girth and other parameters have been given by Chandran et al. [8] for weighted graphs. Harada et al. [13] have provided lower bounds for the number of cutsets of a given size (not necessarily minimal). We remark that all the above works consider set of edges, rather than partition of nodes, which are eventually minimum but in any case never minimal. Hence, we are not aware of previous work on lower bounds for the number of minimal st-cuts as discussed in this paper.

Preliminaries. All the graphs considered in this paper are undirected, connected, and simple (without multiple edges or self-loops). Hence a sequence of nodes cannot induce more than one path or cycle, thus we may refer to paths and cycles simply as sequences of nodes. Two paths are disjoint if they do not share any node (and consequently any edge), and are *internally* disjoint if they share both the first and last nodes in their sequences, but are otherwise disjoint. Two paths that do not share any edge (but may share nodes) are called *edge* disjoint. We call $\kappa(G)$ the *node connectivity* of G, that is the size of the smallest node cut. If $\kappa(G) = k$ we say that G is k-connected (and thus removing $k - 1$ nodes cannot disconnect the graph). By Menger's theorem, we have that for each pair of nodes $x, y \in V(G)$ there are at least $\kappa(G)$ internally disjoint paths between x and y.

An st-numbering for two adjacent nodes s, t is a numbering of the nodes of G such that each node (except t) is adjacent to a node larger than itself, and each node (except s) to one smaller than itself. When G is biconnected, [9] proves that there is an st-numbering for any pair of adjacent nodes s and t, and that it can be found in linear time. Furthermore, we remark that this ordering can be found even if s and t are not adjacent. Indeed, let G' be the graph obtained by adding the edge $\{s, t\}$ to G. G' is still biconnected, and has an st-numbering. Consider the same numbering for G: s still has a neighbor larger than itself in G (any neighbor, since it has the smallest label), and by the same logic t still has a neighbor smaller than itself. All other nodes have the same neighborhood in G as they had in G'. Thus we can remark the following

Observation 1. *There is an st-numbering on a biconnected graph G for any pair of nodes s and t.*

In the rest of the paper, we assume that the nodes are numbered in st-numbering, and thus $x < y$ for any two nodes x and y means that x appears earlier than y in the st-numbering.

2   Number of st-Cuts in a Biconnected Graph

This section illustrates our main result that, for any undirected biconnected graph G and any two distinct nodes s, t, there are at least $\max(n, m-2n) = \Omega(m)$ st-cuts. In order to get this lower bound we will attempt at defining, for each edge, a corresponding cut, ensuring that each such cut is valid. However, we also need these cuts to be distinct from each other.

In the following, we will produce two sets of cuts, corresponding to different kind of edges, which may overlap with each other, but each set will contain distinct elements. One of the sets will contain exactly n distinct cuts, and the other at least $m-2n$: as a result, we obtain that G has at least $\max(n, m-2n) = \Omega(m)$ st-cuts.

One of the main ingredients of our proof will be defining a *backbone* of a graph, which is based on the *st*-numbering of G. Its structure immediately leads to a classification of the edges of G that will be crucial to define the *st*-cuts. Indeed, it helps us to overcome the fact that it is not possible to identify a distinct *st*-cut for each edge in a straightforward way, as some edges are not yielding new *st*-cuts.

For the sake of discussion, we also report some observations on *st*-numbering that can be partially found in previous work [5,24,26].

2.1   Backbone of the Graph

Consider a biconnected graph $G = (V(G), E(G))$, and an *st*-numbering induced by two of its nodes s and t.

Definition 1 (backbone). *The* backbone *is the graph* BB $= (V(G), E(\text{BB}))$ *where* $E(\text{BB}) \subseteq E(G)$ *is defined as* $\{i, j\} \in E(\text{BB})$ *if and only if* j *is either the largest or smallest neighbor of* i *in the st-numbering.*

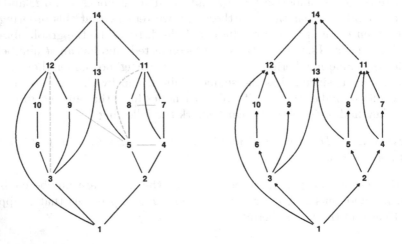

Fig. 3. Left: a biconnected graph G, labeled in *st*-numbering ($s = 1, t = 14$). Backbone edges in bold, shortcut edges dashed, and cross edges in gray. Right: the corresponding graph $\overrightarrow{\text{BB}}$ showing predecessor-successor relationships of G.

In other words, the backbone is obtained by taking for each node just the edge to its largest and its smallest neighbor, as illustrated in Fig. 3 (left). We observe that BB fulfills the following properties.

- All edges having s or t as an extreme are in BB, as s and t are respectively the smallest and largest node.
- BB has at most $2n$ edges, since we take at most 2 edges for each node.

- The st-numbering of G is also an st-numbering for BB, since each node has at least one smaller neighbor (except s) and a larger one (except t).

For the mere purpose of definitions, we consider the oriented version of BB, called bipolar orientation $\overrightarrow{BB}$, where an arc (x, y) belongs to $\overrightarrow{BB}$ iff $\{x, y\} \subseteq E(\text{BB})$ and $x < y$, as illustrated in Fig. 3 (right). Note that s and t are respectively the only source and target in $\overrightarrow{BB}$. For each node $v \in V(G)$, we say that $x \in V(G)$ is a *predecessor* of v iff there is an oriented path from x to v in $\overrightarrow{BB}$, and $y \in V(G)$ is a *successor* of v iff the oriented path in $\overrightarrow{BB}$ is from v to y. Note that x is a predecessor of v iff there is a monotone increasing path from x to v in BB, and y is a successor of v iff there is a monotone decreasing path from y to v in BB. As it can be seen, there can be pairs of nodes such that they are not one predecessor of the other.

In the following, we drop $\overrightarrow{BB}$ and focus on BB alone, keeping the sets of predecessors and successors of each node v, respectively denoted as ANC(v) and DESC(v). Note that $v \notin$ ANC(v) and $v \notin$ DESC(v). Clearly $x \in$ ANC(v) iff $v \in$ DESC(x), and ANC(v) $\cap$ DESC(v) $= \emptyset$. We remark that s and t are respectively a predecessor and a successor of all nodes in $V(G)$ except themselves.

The edges $\{x, y\} \in E(G)$ can be classified in the given BB as three types (see Fig. 3 (left)).

- $\{x, y\}$ is a *backbone* (type B) edge iff $\{x, y\} \in E(\text{BB})$.
- $\{x, y\}$ is a *shortcut* (type S) edge iff $\{x, y\} \in E(G) \backslash E(\text{BB})$ and $x \in$ ANC(y).
- $\{x, y\}$ is a *cross* (type C) edge iff $\{x, y\} \in E(G) \backslash E(\text{BB})$ and $x \notin$ ANC(y).

It is important to remark that each edge of $E(G)$ falls under exactly one of the above types.

Observation 2. *Any edge of G is either a backbone edge, a cross edge, or a shortcut edge.*

2.2 Case Analysis on the st-Cut Types

In the following, we will use the classification of the edges to define st-cuts. In particular, for each edge, depending on its type, we define a corresponding cut. For all these edges, the corresponding st-cuts (S, T) we define are always valid, meaning that both S and $T = V \backslash S$ induce connected subgraphs.

Definition 2 (B - backbone cut). *For each node $v \in V(G) \backslash \{t\}$, its type-B cut is $S = $ ANC(v) $\cup \{v\}$.*

Example (B - backbone cut). On the graph in Fig. 3, $v = 9$ yields the type-B cut $S = \{1, 3, 9\}$ (and $T = V(G) \backslash S$). A visual representation is shown in Fig. 4 (left).

Lemma 3. *Every node except t yields a valid type-B cut, and the type-B cuts are pairwise distinct.*

Proof. Each predecessor of v is on a path from v to s made of predecessors of v, thus $G[S] = G[\text{ANC}(v) \cup \{v\}]$ is connected. $G[T]$ is also connected as any node that is *not* predecessor of v has a path to t made of nodes which are not predecessors of v. Furthermore, note that all nodes in S (except v itself) are predecessors of v, and v is the only node that satisfies this property, thus v can be uniquely deduced from the set S, meaning that two different nodes may not lead to the same type-B cut. □

As a consequence of Lemma 3, we get the following.

Observation 3. *Every biconnected graph G has at least n distinct st-cut for any choice of s and t.*

In order to increase the lower bound in Observation 3, in the following we consider edges $\{x, y\}$ which do *not* belong to the backbone of G. Note that there are at least $m - 2n$ such edges since the backbone has at most $2n$ edges.

Suppose $x < y$ without loss of generality in the rest of the section. Note that y cannot be a predecessor of x, but may or may not be a successor.

Definition 4 (C - cross cut). *For each cross edge $\{x, y\}$, its type-C cut is $S = \text{ANC}(x) \cup \text{ANC}(y) \cup \{x, y\}$.*

Example (C - cross cut). On the graph in Fig. 3, $\{x, y\} = \{5, 9\}$ yields the type-C cut $S = \{1, 2, 3, 5, 9\}$ (and $T = V(G) \backslash S$). A visual representation is shown in Fig. 4 (center).

Lemma 5. *Every cross edge yields a valid type-C cut, and the type-C cuts are pairwise distinct.*

Proof. By the proof of Lemma 3, $G[\text{ANC}(x) \cup \{x\}]$ and $G[\text{ANC}(y) \cup \{y\}]$ are connected. As the subgraphs share the node s, their union $G[\text{ANC}(x) \cup \text{ANC}(y) \cup \{x, y\}] = G[S]$ is connected too. $G[T]$ is connected since it is made of nodes which are not predecessors of x nor y, thus have a path to t made of nodes that are not predecessor of x nor y.

Given the set S, the only two nodes who are not successor of any other node in S are x and y, thus the cut is uniquely identified by the cross edge $\{x, y\}$, meaning a different edge may not yield the same cut. □

Definition 6 (S - shortcut cut). *For each shortcut edge $\{x, y\}$, where $y \in \text{ANC}(x)$, its type-S cut is $S = V(G) \backslash T$, where T is the set of all nodes connected to t in $G[V(G) \backslash (\text{ANC}(y) \cup \{x, y\})]$, including t itself.*

Example (S - shortcut cut). On the graph in Fig. 3, $\{x, y\} = \{3, 12\}$ yields the type-S cut $S = \{1, 3, 6, 10, 12\}$ (and $T = V(G) \backslash S$). A visual representation is shown in Fig. 4 (right).

Lemma 7. *Every shortcut edge yields a valid type-S cut, and the type-S cuts are pairwise distinct.*

Proof. Let $G' = G[V(G)\backslash(\text{ANC}(x) \cup \{x,y\})]$ (recall that $x < y$). In a shortcut cut, T contains all nodes connected to t in G', including t itself. Thus $G[T]$ is connected by definition.

Consider now $S = V(G)\backslash T$. In particular, S will contain $\text{ANC}(x) \cup \{x,y\}$, which by definition of ANC is connected and contains s, plus all the nodes that cannot reach t in G': since these nodes cannot reach t, and G is connected, they must be connected to $\text{ANC}(x) \cup \{x,y\}$ instead, thus $G[S]$ is connected, meaning that S,T is a valid cut. A visual representation of this can be seen in Fig. 4 (right), where $\{x,y\} = \{3,12\}$, 6 and 10 are the nodes that cannot reach t in $G[V(G)\backslash\text{ANC}(3) \cup \{3,12\}]$, while 9 can reach t via the edge $\{9,5\}$.

We now only need to show that any two shortcut edges cannot produce the same cut.

First, all nodes that are *not* predecessors of y have a path to t made of nodes which are not predecessors of y. Since $\text{ANC}(x) \subset \text{ANC}(y)$, these nodes may reach t in G', and thus are in T. This means that $S \subseteq \text{ANC}(y) \cup \{y\}$. Moreover, by the properties of $\text{ANC}()$, we have that y is the only node that satisfies this property.

We will now prove our claim by contradiction. Suppose another edge $\{w,z\}$ yields the same S. Then we have that $S \subseteq \text{ANC}(z) \cup \{z\}$, which implies that $y = z$ because of what we said above. Moreover, we must have that $w \in S$ as both extremes of the shortcut edge go in S when defining a type-s cut.

We can now only consider shortcut edges of the form $\{w,y\}$. Without loss of generality, assume $x < w$ (we cannot have $x = w$ since G is not a multigraph). Note that, by definition, $x \in S$. Furthermore, note that by construction of the backbone, the edge between w and its largest neighbor is in the backbone, thus since $\{w,y\}$ is a shortcut edge (not a backbone edge), w must have a neighbor $v > y$.

As $v > y$ and $w > x$, it follows that $v, w \notin \text{ANC}(x) \cup \{x,y\}$. This means that there is a path from w to t that does not use any node from $\text{ANC}(x) \cup \{x,y\}$, thus $w \in T$. As we supposed $w \in S$, this is a contradiction and the thesis follows. $\square$

Lemma 8. *The sets of type-c cuts and type-s cuts are disjoint.*

Proof. Notice that, as proven in Lemma 7, we have that for each type-s cut there exists a node y that belongs to S and such that $S \subseteq \{y\} \cup \text{ANC}(y)$.

On the other hand, a type-c cut has $S = \{x,y\} \cup \text{ANC}(x) \cup \text{ANC}(y)$, where $\{x,y\}$ is a cross edge. We thus have that $y \notin \text{ANC}(x)$ and $x \notin \text{ANC}(y)$. Moreover, as $x \notin \text{ANC}(z)$ if $z \in \text{ANC}(x)$, any other node of S is not a predecessor of either x or y (or both). This implies that there is no node in S that has the whole S among its predecessors, thus proving that we cannot find the same cut in both the sets of type-s cuts and type-c cuts. $\square$

We finally give the proof of our main result, which is now an immediate consequence of the properties proved so far.

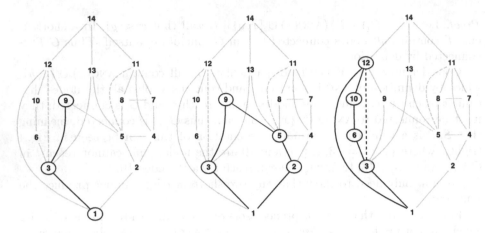

Fig. 4. *st*-cuts corresponding to respectively a type-B cut with $v = 9$ (left), a type-C cut with $\{x, y\} = \{5, 9\}$ (center), and a type-S cut with $\{x, y\} = \{3, 12\}$ (right)

Theorem 9. *For any biconnected graph G and any two distinct nodes s, t, there are at least $\max(n, m - 2n) = \Omega(m)$ st-cuts.*

Proof. Observation 3 proves that we have at least n different cuts. On the other hand, any non-backbone edge gives us either a type-S cut or a type-C cut, and no cut is obtained twice in this way, as proven in Lemma 8. As there are $m - 2n$ non-backbone edges, we have at least $m - 2n$ cuts.   □

2.3   Graphs that Allow for an *st*-Numbering

While we considered biconnected graphs, it can be noted that Theorem 9 holds for any graph admitting an *st*-numbering, as this is sufficient for our proof. This condition is slightly more general than assuming G to be biconnected, and actually corresponds to the biconnected components tree of G being a path, with the components containing s and t in its extremes.

Indeed, as it can be seen in Fig. 5, any node in a biconnected component out of this path (the dashed ones) is separated by both s and t by a single cut node, meaning that there cannot be both a monotone increasing path and a monotone decreasing path from the node to respectively t and s.

On the other hand, it can be easily seen how an *st*-numbering for a path of biconnected component can be computed by combining a suitable *st*-numbering of each of the components between its articulation points.

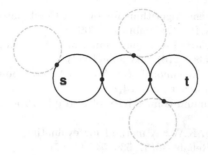

Fig. 5. An example of a graph that does not allow an st-numbering, and of a subgraph (highlighted in black) that does. Circles represent biconnected components.

3 Conclusions and Further Work

In this paper we have proved that there are $\Omega(m)$ st-cuts in any biconnected graph G for any choice of distinct nodes s, t. We have shown that this lower bound is tight for k-node-connected graphs with $k = 2, 3$ as there is a triconnected graph, i.e., the wheel in Fig. 2, matching this lower bound. The natural question which remains open is whether the $\Omega(m)$ bound is tight in k-node-connected graphs for $k > 3$.

Acknowledgements. This work was partially supported by JST CREST, grant number JPMJCR1401, Japan, and MIUR, Italy.

References

1. Abel, U., Bicker, R.: Determination of all minimal cut-sets between a vertex pair in an undirected graph. IEEE Trans. Reliab. **31**(2), 167–171 (1982)
2. Ball, M.O., Provan, J.S.: Calculating bounds on reachability and connectedness in stochastic networks. Networks **13**(2), 253–278 (1983)
3. Bellmore, M., Jensen, P.A.: An implicit enumeration scheme for proper cut generation. Technometrics **12**(4), 775–788 (1970)
4. Berge, C.: La theorie des graphes. Presses Universitaires de France, Paris (1958)
5. Biedl, T.C., Kant, G.: A better heuristic for orthogonal graph drawings. Comput. Geom. **9**(3), 159–180 (1998)
6. Bixby, R.E.: The minimum number of edges and vertices in a graph with edge connectivity n and m n-bonds. Networks **5**(3), 253–298 (1975)
7. Brecht, T.B., Colbourn, C.J.: Lower bounds on two-terminal network reliability. Discrete Appl. Math. **21**(3), 185–198 (1988)
8. Chandran, L.S., Ram, L.S.: On the number of minimum cuts in a graph. SIAM J. Discrete Math. **18**(1), 177–194 (2004)
9. Shimon Even and Robert Endre Tarjan: Computing an st-numbering. Theor. Comput. Sci. **2**(3), 339–344 (1976)
10. Gardner, M.L.: Algorithm to aid in the design of large scale networks. Large Scale Syst. **8**(2), 147–156 (1985)

11. Goldberg, L.A.: Efficient Algorithms for Listing Combinatorial Structures, vol. 5. Cambridge University Press, Cambridge (2009)
12. Hamacher, H.W., Picard, J.-C., Queyranne, M.: On finding the K best cuts in a network. Oper. Res. Lett. **2**(6), 303–305 (1984)
13. Harada, H., Sun, Z., Nagamochi, H.: An exact lower bound on the number of cut-sets in multigraphs. Networks **24**(8), 429–443 (1994)
14. Harary, F.: The maximum connectivity of a graph. Proc. Nat. Acad. Sci. **48**(7), 1142–1146 (1962)
15. Jasmon, G.B., Foong, K.W.: A method for evaluating all the minimal cuts of a graph. IEEE Trans. Reliab. **36**(5), 539–545 (1987)
16. Katona, G.: A theorem for finite sets. In: Theory of Graphs, pp. 187–207 (1968)
17. Khachiyan, L., Boros, E., Borys, K., Elbassioni, K., Gurvich, V., Makino, K.: Generating cut conjunctions in graphs and related problems. Algorithmica **51**(3), 239–263 (2008)
18. Kruskal, J.B.: The number of simplices in a complex. Math. Optim. Tech. **10**, 251–278 (1963)
19. Martelli, A.: A Gaussian elimination algorithm for the enumeration of cut sets in a graph. J. ACM **23**(1), 58–73 (1976)
20. Picard, J.-C., Queyranne, M.: On the structure of all minimum cuts in a network and applications. Math. Program. **22**(1), 121–121 (1982)
21. Prasad, V.C., Sankar, V., Rao, K.S.P.: Generation of vertex and edge cutsets. Microelectron. Reliab. **32**(9), 1291–1310 (1992)
22. Provan, J.S., Ball, M.O.: Computing network reliability in time polynomial in the number of cuts. Oper. Res. **32**(3), 516–526 (1984)
23. Provan, J.S., Shier, D.R.: A paradigm for listing (s, t)-cuts in graphs. Algorithmica **15**(4), 351–372 (1996)
24. Pierre Rosenstiehl and Robert Endre Tarjan: Rectilinear planar layouts and bipolar orientations of planar graphs. Discrete Comput. Geom. **1**, 343–353 (1986)
25. Shier, D.R., Whited, D.E.: Iterative algorithms for generating minimal cutsets in directed graphs. Networks **16**(2), 133–147 (1986)
26. Tamassia, R., Tollis, I.G.: A unified approach a visibility representation of planar graphs. Discrete Comput. Geom. **1**, 321–341 (1986)
27. Tsukiyama, S., Shirakawa, I., Ozaki, H., Ariyoshi, H.: An algorithm to enumerate all cutsets of a graph in linear time per cutset. J. ACM (JACM) **27**(4), 619–632 (1980)
28. Li, Y., Taha, H.A., Landers, T.L.: A recursive approach for enumerating minimal cutsets in a network. IEEE Trans. Reliab. **43**(3), 383–388 (1994)
29. Yeh, L.-P., Wang, B.-F., Hsin-Hao, S.: Efficient algorithms for the problems of enumerating cuts by non-decreasing weights. Algorithmica **56**(3), 297–312 (2010)

Subexponential-Time and FPT Algorithms for Embedded Flat Clustered Planarity

Giordano Da Lozzo[1]([⊠]), David Eppstein[2], Michael T. Goodrich[2], and Siddharth Gupta[2]

[1] Roma Tre University, Rome, Italy
dalozzo@dia.uniroma3.it
[2] University of California, Irvine, USA
{eppstein,goodrich,guptasid}@uci.edu

Abstract. The C-PLANARITY problem asks for a drawing of a *clustered graph*, i.e., a graph whose vertices belong to properly nested clusters, in which each cluster is represented by a simple closed region with no edge-edge crossings, no region-region crossings, and no unnecessary edge-region crossings. We study C-PLANARITY for *embedded flat clustered graphs*, graphs with a fixed combinatorial embedding whose clusters partition the vertex set. Our main result is a subexponential-time algorithm to test C-PLANARITY for these graphs when their face size is bounded. Furthermore, we consider a variation of the notion of *embedded tree decomposition* in which, for each face, including the outer face, there is a bag that contains every vertex of the face. We show that C-PLANARITY is fixed-parameter tractable with the embedded-width of the underlying graph and the number of disconnected clusters as parameters.

1 Introduction

A *clustered graph* (or *c-graph*) is a pair $\mathcal{C}(G, \mathcal{T})$ with *underlying graph* G and *inclusion tree* $\mathcal{T}$, i.e., a rooted tree whose leaves are the vertices of G. Each internal node μ of $\mathcal{T}$ represents a cluster of vertices of G (its leaf descendants) which induces a subgraph $G(\mu)$. A *c-planar drawing* of $\mathcal{C}(G, \mathcal{T})$ (Fig. 1) consists of a drawing of G and of a representation of each cluster μ as a *simple closed region* $R(\mu)$, i.e., a region homeomorphic to a closed disc, such that: (1) Each region $R(\mu)$ contains the drawing of $G(\mu)$. (2)

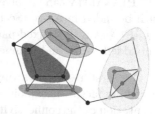

Fig. 1. A c-planar drawing.

For every two clusters $\mu, \nu \in \mathcal{T}$, $R(\nu) \subseteq R(\mu)$ if and only if ν is a descendant of μ in $\mathcal{T}$. (3) No two edges cross. (4) No edge crosses any region boundary more than once. (5) No two region boundaries intersect.

An interesting and challenging line of research in graph drawing concerns the computational complexity of the C-PLANARITY problem, which asks to test the

© Springer Nature Switzerland AG 2018
A. Brandstädt et al. (Eds.): WG 2018, LNCS 11159, pp. 111–124, 2018.
https://doi.org/10.1007/978-3-030-00256-5_10

existence of a c-planar drawing of a c-graph. This problem is notoriously difficult, particularly when (as in Fig. 1) clusters may be disconnected, faces may
have unbounded size, and the cluster hierarchy may have multiple nested levels. No known subexponential-time algorithm solves the (general) C-PLANARITY
problem, and it is unknown whether it is NP-complete. Thus, there is considerable interest in subexponential-time, slice-wise polynomial (for short XP), and
fixed-parameter tractable algorithms for special cases of C-PLANARITY.

C-PLANARITY was introduced by Feng, Cohen, and Eades [18], who solved it
in quadratic time for the *c-connected* case, i.e., when every cluster induces a connected subgraph. Dahlhaus [16] claimed a linear-time algorithm for c-connected
C-PLANARITY (with some details later provided by Cortese *et al.* [13]). Goodrich
et al. [20] gave a cubic-time algorithm for disconnected clusters that satisfy an
"extroverted" property, and Gutwenger *et al.* [21] provided a polynomial-time
algorithm for "almost" c-connected inputs. Cornelsen and Wagner showed poly-
nomiality for *completely connected c-graphs*, i.e., when not only every cluster but
also the complement of each cluster is connected [12]. FPT algorithms have also
been investigated [8,11]. For additional special cases, see, e.g., [2–4].

A c-graph is *flat* when no non-trivial cluster is a subset of another, so T has
only three levels: the root, the clusters, and the leaves. Flat C-PLANARITY can
be solved in polynomial time for embedded c-graphs with at most 5 vertices per
face [17] or at most two vertices of each cluster per face [10], for embedded c-
graphs in which each cluster induces a subgraph with at most two connected components [22], and for c-graphs with two clusters [7] or three clusters [1]. Jelínková
et al. [23] provide efficient algorithms for 3-connected flat c-graphs when each
cluster has at most 3 vertices. Fulek [19] speculates that C-PLANARITY could
be solvable in subexponential time for more general embedded flat c-graphs.

New Results. In this paper, we provide subexponential-time and fixed-
parameter tractable algorithms for broad classes of c-graphs. We show the following results:

◇ C-PLANARITY can be solved in subexponential time for embedded flat c-graphs
with bounded face size (Sect. 3).
◇ C-PLANARITY is fixed-parameter tractable for embedded flat c-graphs with
embedded-width and number of disconnected clusters as parameters (Sect. 4).

Our first result uses divide-and-conquer with a large but subexponential
branching factor. It exploits cycle separators in planar graphs and a concise representation of the connectivity of each cluster in a c-planar drawing. This method
also leads to an XP algorithm for the class of *generalized h-simply-nested graphs*,
which includes the simply-nested graphs with bounded face size (see [15]).

We obtain our second result by expressing c-planarity in extended monadic
second-order logic for embedded flat c-graphs and applying Courcelle's Theorem.
The graphs to which this result applies, with bounded treewidth and bounded
face size, include the *nested triangles graphs*, a standard family of examples that
are hard for many graph drawing tasks, the *dual graphs of bounded-treewidth*

bounded-degree plane graphs, and the *buckytubes*, graphs formed from a planar hexagonal lattice wrapped to form a cylinder of bounded diameter.

We provide full details of omitted and sketched proofs in the full version [15].

2   Definitions and Preliminaries

The graphs considered in this paper are finite, simple, and connected. For standard concepts about planar graphs, their connectivity and embeddings, such as *combinatorial* and *planar embeddings*, *rotation* at a vertex, *faces*, and *embedded (or plane) graphs*, we refer the reader to the full version of the paper [15]. The *length* of a face f is the number of occurrences of edges encountered in a traversal of f. The *maximum face size* of an embedded graph is the length of its largest face.

Tree-Width and Embedded-Width. A *tree decomposition* of a graph G is a tree T whose nodes, called *bags*, are labeled by subsets of vertices of G. For each vertex v the bags containing v must form a nonempty contiguous subtree of T, and for each edge uv at least one bag must contain both u and v. The *width* of the decomposition is one less than the maximum cardinality of any bag, and the *treewidth* $\mathrm{tw}(G)$ of G is the minimum width of any of its tree decompositions.

Recently, Borradaile *et al.* [9] developed a variant of treewidth, specialized for plane graphs, called embedded-width. According to their definitions, a tree decomposition *respects* an embedding of a plane graph G if, for every inner face f of G, at least one bag contains all the vertices of f. They define the *embedded-width* $\mathrm{emw}(G)$ of G to be the minimum width of a tree decomposition that respects the embedding of G. We will use the following result.

Theorem 1 ([9], Theorem 2). *If G is a plane graph where every inner face has length at most ℓ, then* $\mathrm{emw}(G) \leq (\mathrm{tw}(G) + 2) \cdot \ell - 1$.

Borradaile *et al.*do not require the vertices of the outer face to be contained in the same bag. In our applications, we modify this concept so that the tree decomposition also includes a bag containing the outer face, and we denote the minimum width of such a tree decomposition as $\overline{\mathrm{emw}}(G)$. By simply adding the vertices of the outer face to all bags, we have the following.

Lemma 1. *If G is a plane graph whose maximum face size (including the size of the outer face) is ℓ, then* $\overline{\mathrm{emw}}(G) \leq (\mathrm{tw}(G) + 3) \cdot \ell - 1$.

Clustered Planarity. Recall that, in a c-graph $\mathcal{C}(G, \mathcal{T})$, each internal node μ of $\mathcal{T}$ corresponds to the set $V(\mu)$ of vertices of G at leaves of the subtree of $\mathcal{T}$ rooted at μ. Set $V(\mu)$ induces the subgraph $G(\mu)$ of G. We call *clusters* the internal nodes other than the root. A cluster μ is *connected* if $G(\mu)$ is connected and *disconnected* otherwise. C-graph $\mathcal{C}$ is *c-connected* if every cluster is connected.

A c-graph is *c-planar* if it admits a c-planar drawing. Two c-graphs $\mathcal{C}(G, \mathcal{T})$ and $\mathcal{C}'(G', \mathcal{T}')$ are *equivalent* if both are c-planar or neither is. If the root of $\mathcal{T}$

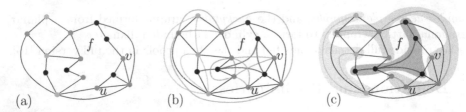

Fig. 2. (a) An embedded flat c-graph $\mathcal{C}(G, \mathcal{T})$. (b) A super c-graph of $\mathcal{C}$ containing all the candidate saturating edges of $\mathcal{C}$ (thick and colored curves); since vertices u and v belong to different components of $X_\mu(f)$ but to the same connected component of $G(\mu)$, edge (u, v) is not a candidate saturating edge. (c) A super c-graph of $\mathcal{C}$ satisfying Condition (iii) of Theorem 2; regions enclosing vertices of each cluster are shaded. (Color figure online)

has leaf children, enclosing each leaf v in a new singleton cluster produces an equivalent c-graph. Therefore, we can safely assume that each vertex belongs to a cluster. A c-graph is *flat* if each leaf-to-root path in $\mathcal{T}$ has exactly three nodes. The clusters of a flat c-graph form a partition of the vertex set.

An *embedded c-graph* $\mathcal{C}(G, \mathcal{T})$ is a c-graph whose underlying graph has a fixed combinatorial embedding. It is *c-planar* if it admits a c-planar drawing that preserves the embedding of G. Since we only deal with embedded flat c-graphs, we will refer to such graphs simply as c-graphs, except in the statements of theorems.

We define the *candidate saturating edges* of a c-graph $\mathcal{C}(G, \mathcal{T})$ as follows. For each face f of G, let $G(f)$ be the closed walk composed of the vertices and edges of f. For each cluster $\mu \in \mathcal{T}$, consider the set $X_\mu(f)$ of connected components of $G(f)$ induced by the vertices of μ and, for each component $\xi \in X_\mu(f)$, assign a vertex of f in ξ as a reference vertex of ξ. We add an edge inside f between the reference vertices of any two components in $X_\mu(f)$ if and only if such vertices belong to different connected components of $G(\mu)$; see Figs. 2a and b. A c-graph obtained from $\mathcal{C}$ by adding to it a subset E^+ of its candidate saturating edges is a *super c-graph* of $\mathcal{C}$. Di Battista and Frati [17] gave the following characterization.

Theorem 2 ([17], **Theorem 1**). *An embedded flat c-graph $\mathcal{C}(G, \mathcal{T})$ is c-planar if and only if: (i) G is planar; (ii) there exists a face f in G such that when f is chosen as the outer face for G no cycle composed of vertices of the same cluster encloses a vertex of a different cluster in its interior; and (iii) there exists a super c-graph $\mathcal{C}'(G', \mathcal{T})$ of $\mathcal{C}$ such that G' is planar and $\mathcal{C}'$ is c-connected (see Fig. 2c).*

Conditions (i) and (ii) of Theorem 2 can be easily verified in linear time. Therefore, we can assume that any c-graph satisfies these conditions. Following [17] we thus view the problem of testing c-planarity as one of testing Condition (iii).

A *cluster-separator* in a c-graph $\mathcal{C}(G, \mathcal{T})$ is a cycle ρ in G for which some cluster $\mu \in \mathcal{T}$ has vertices both in the interior and in the exterior of ρ, but with $V(\mu) \cap V(\rho) = \emptyset$. Condition (iii) immediately yields the following observation.

Observation 1. *A c-graph that has a cluster-separator is not c-planar.*

In the next sections, it will be useful to only consider c-graphs which are at least 2-connected (Sect. 3) and 3-connected (Sect. 4). The next lemma, conveniently stated in a stronger form[1], shows that this is not a loss of generality.

Lemma 2. *Let $\mathcal{C}(G, \mathcal{T})$ be an n-vertex c-graph with maximum face size ℓ. There exists an $O(n)$-time algorithm that constructs an equivalent c-graph $\mathcal{C}^*(G^*, \mathcal{T}^*)$ with $|V(G^*)| = O(n)$ such that: 1. G^* is 3-connected, 2. the maximum face size κ of G^* is $O(\ell)$, and 3. the c-graph $\mathcal{C}^\diamond(G^\diamond, \mathcal{T}^\diamond)$ obtained by augmenting $\mathcal{C}^*(G^*, \mathcal{T}^*)$ with all its candidate saturating edges is such that $\mathrm{tw}(G^\diamond) = O(\overline{\mathrm{emw}}(G))$.*

3   A Subexponential-Time Algorithm for C-Planarity

In this section, we describe a divide-and-conquer algorithm for testing the c-planarity of 2-connected c-graphs exploiting cycle separators in planar graphs.

The "conquer" part of our divide-and-conquer uses the following operation on pairs of c-graphs. Let G_1 and G_2 be plane graphs on overlapping vertex sets such that the outer face of G_1 and an inner face of G_2 are bounded by the same cycle ρ. *Merging G_1 and G_2* constructs a new plane graph G from $G_1 \cup G_2$ as follows. We remove multi-edges (belonging to cycle ρ) and assign each vertex v a rotation whose restriction to the edges of G_2 (of G_1) is the same as the rotation at v in G_2 (in G_1). This is possible as cycle ρ bounds the outer face of G_1 and an inner face of G_2. We say that G is a *merge* of G_1 and G_2. Now consider two c-graphs $\mathcal{C}_1(G_1, \mathcal{T}_1)$ and $\mathcal{C}_2(G_2, \mathcal{T}_2)$ such that (i) $G_1 \cap G_2 = \rho$ is a cycle, (ii) for each vertex $v \in V(\rho)$, vertex v belongs to the same cluster μ both in $\mathcal{T}_1$ and in $\mathcal{T}_2$, and (iii) cycle ρ bounds the outer face of G_1 and an inner face of G_2 (when a choice for their outer faces that satisfies Condition (ii) of Theorem 2 has been made). *Merging $\mathcal{C}_1$ and $\mathcal{C}_2$* is the operation that constructs a c-graph $\mathcal{C}(G, \mathcal{T})$ as follows. Graph G is obtained by merging G_1 and G_2. Tree $\mathcal{T}$ is obtained as follows. Initialize $\mathcal{T}$ to $\mathcal{T}_1$. First, for each cluster $\mu \in \mathcal{T}_2 \cap \mathcal{T}_1$, we add the leaves of μ in $\mathcal{T}_2$ as children of cluster μ in $\mathcal{T}$, removing duplicate leaves. Second, for each cluster $\mu \in \mathcal{T}_2 \backslash \mathcal{T}_1$, we add the subtree of $\mathcal{T}_2$ rooted at μ as a child of the root of $\mathcal{T}$. We say that $\mathcal{C}(G, \mathcal{T})$ is a *merge* of $\mathcal{C}_1(G_1, \mathcal{T}_1)$ and $\mathcal{C}_2(G_2, \mathcal{T}_2)$.

In the "divide" part of the divide-and-conquer, we replace subgraphs of the input by smaller planar components called *cycle-stars* that preserve their c-planarity properties. Let G be a connected plane graph that contains a face whose boundary is a cycle ρ. We say that G is a *cycle-star* if removing all the edges of ρ makes G a forest of stars; refer to Fig. 3c. Also, we say that cycle ρ is *universal* for G and we say that a vertex of G is a *star vertex* of G if it does not belong to ρ. Clearly, the size of G is $O(|\rho|)$.

---

[1] In Sect. 4, we exploit all the properties of the lemma. In Sect. 3, we only exploit the existence of an equivalent 2-connected c-graph with maximum face size $\kappa = O(\ell)$.

For a c-planar c-graph $\mathcal{C}(G, \mathcal{T})$ and a cycle separator ρ, we denote by $\mathcal{C}_\rho^+(G^+, \mathcal{T}^+)$ (by $\mathcal{C}_\rho^-(G^-, \mathcal{T}^-)$) the c-graph obtained from $\mathcal{C}$ by removing all the vertices and the edges of G that lie in the interior of ρ (in the exterior of ρ). Consider a super c-graph $\mathcal{C}'(G', \mathcal{T})$ of $\mathcal{C}$ satisfying Condition (iii) of Theorem 2, which exists since $\mathcal{C}$ is c-planar. We now give a procedure, which will be useful throughout the paper, to construct two special c-planar c-graphs $\mathcal{C}^-(S^-, \mathcal{K}^-)$ and $\mathcal{C}^+(S^+, \mathcal{K}^+)$ associated with $\mathcal{C}'$ whose properties are described in the following lemma.

Lemma 3. *C-graphs $\mathcal{C}^-(S^-, \mathcal{K}^-)$ and $\mathcal{C}^+(S^+, \mathcal{K}^+)$ are such that: 1. graph S^- (S^+) is a cycle-star whose universal cycle is ρ, 2. cycle ρ bounds the outer face of S^- (an inner face of S^+), 3. each star vertex of S^- (S^+) and all its neighbors belong to the same cluster in $\mathcal{K}^-$ ($\mathcal{K}^+$), and 4. the c-graph $\mathcal{C}_{out}$ ($\mathcal{C}_{in}$) obtained by merging $\mathcal{C}^-(S^-, \mathcal{K}^-)$ and $\mathcal{C}_\rho^+(G^+, \mathcal{T}^+)$ (by merging $\mathcal{C}^+(S^+, \mathcal{K}^+)$ and $\mathcal{C}_\rho^-(G^-, \mathcal{T}^-)$) is c-planar.*

We describe how to construct $\mathcal{C}^-(S^-, \mathcal{K}^-)$ from $\mathcal{C}'$; refer to Fig. 3. The construction of $\mathcal{C}^+(S^+, \mathcal{K}^+)$ is symmetric.

First, for each cluster μ such that $V(\mu) \cap V(\rho) = \emptyset$, we remove all the vertices in $V(\mu)$ lying in the interior of ρ together with their incident edges. By Observation 1, the resulting c-graph is still c-planar and c-connected. Also, we remove edges in the interior of ρ whose endpoints belong to different clusters. Clearly, this simplification preserves c-connectedness. We still denote the resulting c-graph as $\mathcal{C}'$.

Second, consider the c-graph H consisting of the vertices and of the edges of $\mathcal{C}'$ lying in the interior and along the boundary of ρ. For each cluster μ and for each connected component c_μ^i of μ in H, we replace all the vertices and edges of c_μ^i lying in the interior of ρ in $\mathcal{C}'$ with a single vertex s_μ^i, assigning it to the same cluster μ and making it adjacent to all the vertices in $V(c_\mu^i) \cap V(\rho)$. Let $\mathcal{C}^*$ be the resulting c-graph. It is easy to see that such a transformation preserves c-connectedness and planarity, therefore $\mathcal{C}^*$ is a c-connected c-planar c-graph. By construction, each vertex $v \in V(\rho)$ is adjacent to a single vertex s_μ^i, where μ is the cluster vertex v belongs to; thus, the vertices and the edges in the interior and along the boundary of ρ in $\mathcal{C}^*$ form c-graph $\mathcal{C}^-(S^-, \mathcal{K}^-)$ whose underlying graph S^- is a cycle-star satisfying Properties (1), (2) and (3) of Lemma 3. Further, since the subgraph of $\mathcal{C}^*$ consisting of the vertices and of the edges lying in the exterior and along the boundary of ρ coincides with $\mathcal{C}_\rho^+(G^+, \mathcal{T}^+)$, we have that $\mathcal{C}^*$ is a c-planar c-connected super c-graph of $\mathcal{C}_{out}$. Thus, by Condition (iii) of Theorem 2, Property (4) of Lemma 3 is also satisfied.

Let $\mathcal{C}_\Delta^-(R^-, \mathcal{J}^-)$ ($\mathcal{C}_\Delta^+(R^+, \mathcal{J}^+)$) be a c-graph obtained by augmenting the c-graph $\mathcal{C}^-(S^-, \mathcal{K}^-)$ ($\mathcal{C}^+(S^+, \mathcal{K}^+)$) of Lemma 3 by introducing new vertices, each belonging to a distinct cluster, and by adding edges only between the vertices in $V(S^-)$ ($V(S^+)$) and the newly introduced vertices in such a way that cycle ρ bounds a face of R^- (R^+) and all the other faces of R^- (R^+) are triangles. From the construction of Lemma 3, we also have the following useful technical remark.

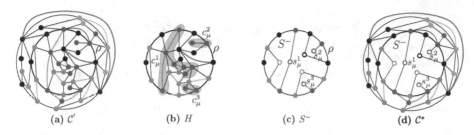

(a) $\mathcal{C}'$　　　　　(b) H　　　　　(c) S^-　　　　　(d) $\mathcal{C}^*$

Fig. 3. (a) Super c-graph $\mathcal{C}'$ of $\mathcal{C}$. (b) Each component of the blue cluster μ in H lies inside a simple closed region. (c) Cycle-star S^- corresponding to H. (d) The c-connected c-planar c-graph $\mathcal{C}^*$ obtained by *replacing* H with S^- in $\mathcal{C}'$. (Color figure online)

Remark 1. *The c-graph obtained by merging $\mathcal{C}_\Delta^-(R^-, \mathcal{J}^-)$ and $\mathcal{C}_\rho^+(G^+, \mathcal{T}^+)$ (by merging $\mathcal{C}_\Delta^+(R^+, \mathcal{J}^+)$ and $\mathcal{C}_\rho^-(G^-, \mathcal{T}^-)$) is c-planar.*

We now describe a divide-and-conquer algorithm based on Lemma 3, called TESTCP, that tests the c-planarity of a 2-connected c-graph $\mathcal{C}(G, \mathcal{T})$ and returns a super c-graph $\mathcal{C}^*$ of $\mathcal{C}$ satisfying Condition (iii) of Theorem 2, if $\mathcal{C}$ is c-planar. See Fig. 4 and [15] for illustrations of the c-graphs constructed by the algorithm.

We first give an intuition on the role of cycle-stars in Algorithm TESTCP.

Let $\mathcal{C}(G, \mathcal{T})$ be a 2-connected c-planar c-graph and let ρ be a cycle separator of G. Consider any c-connected c-planar super c-graph $\mathcal{C}'$ of $\mathcal{C}$. Let I^- be the super c-graph of $\mathcal{C}_\rho^-(G^-, \mathcal{T}^-)$, composed of the vertices of G^- and of the edges in the interior and along the boundary of ρ in $\mathcal{C}'$. By Lemma 3, we can injectively map I^- with a cycle-star S^- whose universal cycle is ρ. This is due to the fact that there exists a *one-to-one correspondence* between the connected components of I^- induced by the vertices of each cluster in $\mathcal{T}^-$ and the star vertices of S^-. Similar considerations hold for the super c-graph I^+ of $\mathcal{C}_\rho^+(G^+, \mathcal{T}^+)$. Although the c-planarity of $\mathcal{C}_\rho^+$ and $\mathcal{C}_\rho^-$ is necessary for the c-planarity of $\mathcal{C}$, it is not a sufficient condition, as the connectivity of clusters inside ρ in I^- (*internal cluster-connectivity*) and the connectivity of clusters outside ρ in I^+ (*external cluster-connectivity*) must also together determine the c-connectedness of $\mathcal{C}'$. The role of cycle-stars S^- and S^+ in the algorithm presented in this section is exactly that of concisely *representing* the internal cluster-connectivity of I^- and the external cluster-connectivity of I^+, respectively, to devise a divide-and-conquer approach to test the c-planarity of $\mathcal{C}$.

Outline of the Algorithm. We overview the main steps of our algorithm below.

– If $n = O(\ell)$, we test c-planarity directly, as a base case for the divide-and-conquer recursion (see [15]). Otherwise, we construct a cycle-separator ρ of G and test whether ρ is a cluster-separator. If so, $\mathcal{C}$ cannot be c-planar (Observation 1), and we halt the search.

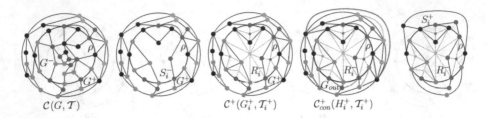

Fig. 4. Illustrations of some of the c-graphs constructed by Algorithm TESTCP.

- We generate all cycle-stars S_i^- with universal cycle ρ. A cycle-star S_i^- represents a potential connection pattern of clusters inside ρ. For each cycle-star S_i^- we apply Procedure OUTERCHECK to test whether this pattern could be augmented by additional connections outside ρ to complete the desired cluster-connectivity. That is, we test whether $\mathcal{C}_\rho^+$ admits a c-connected c-planar super c-graph whose internal cluster-connectivity is represented by S_i^-. To test this, we replace the subgraph G^- of G in $\mathcal{C}$ with an internally-triangulated supergraph R_i^- of S_i^- to obtain a c-graph $\mathcal{C}^+$ and recursively test $\mathcal{C}^+$ for c-planarity. It is important to observe that, the triangulation step prevents $\mathcal{C}^+$ from having saturating edges inside ρ, thus enforcing exactly the same internal-cluster connectivity represented by S_i^- (Remark 1). If $\mathcal{C}^+$ is c-planar, then the procedure returns a c-connected c-planar super c-graph $\mathcal{C}_{con}^+$ of $\mathcal{C}^+$. If no cycle-star passes the test, then $\mathcal{C}$ is not c-planar by Lemma 3. We call all the cycle-stars that passed this test *admissible*.
- We then apply Procedure INNERCHECK to verify whether the internal-cluster connectivity represented by some admissible cycle-star S_i^- can *actually* be realized by a c-connected c-planar super c-graph of $\mathcal{C}$. For each admissible cycle-star S_i^-, the procedure applies the construction of Lemma 3 to obtain a cycle-star S_i^+ representing the external cluster-connectivity of $\mathcal{C}_{con}^+$. Then, it tests whether $\mathcal{C}_\rho^-$ admits a c-connected c-planar super c-graph $\mathcal{C}_{con}^-$ whose external cluster-connectivity is represented by S_i^+. This is done similarly to Procedure OUTERCHECK, by triangulating the exterior of ρ and recursively testing c-planarity of a smaller c-graph. If Procedure INNERCHECK succeeds for any admissible cycle-star S_i^-, then we can merge the subgraphs of $\mathcal{C}_{con}^-$ and of $\mathcal{C}_{con}^+$ induced by the vertices inside and outside ρ, respectively, to obtain a c-connected c-planar super c-graph of $\mathcal{C}$, and we halt the search with a successful outcome. It might be the case that $\mathcal{C}_{con}^-$ has a different internal-cluster connectivity than that represented by S_i^-, but this is not a problem, because the different cluster connectivity (which necessarily corresponds to a different admissible cycle-star) still provides a c-planar drawing of the whole graph.
- If no admissible cycle-star passes Procedure INNERCHECK, $\mathcal{C}$ is not c-planar.

It is crucial in this algorithm that ρ be a cycle-separator. Because it is a cycle, no candidate saturating edges can connect vertices in the interior of ρ to vertices in the exterior of ρ, as such vertices do not share any face. That is, the interaction between G_ρ^- and G_ρ^+ only happens through vertices of ρ. This allows us to split the instance into smaller instances recursively along ρ and model the interaction via cycle-stars (by Lemma 3 and Remark 1) whose universal cycle is ρ.

The complete listing of Algorithm TESTCP is provided in the next page.

Correctness of the Algorithm. We show that, given a 2-connected c-graph $\mathcal{C}(G, \mathcal{T})$, Algorithm TESTCP returns YES, which happens when both procedures OUTERCHECK and INNERCHECK succeed, if and only if $\mathcal{C}(G, \mathcal{T})$ is c-planar.

($\Rightarrow$) Suppose that OUTERCHECK and INNERCHECK succeed for a cycle-star $S_\omega^- \in \mathcal{S}$ constructed at step 2a. We show that $\mathcal{C}$ is c-planar. Consider the c-graph $\mathcal{C}^*$ constructed at step 3(a)v from $\mathcal{C}_{con}^-(H_\omega^-, \mathcal{T}_\omega^-)$ and $\mathcal{C}_{con}^+(H_\omega^+, \mathcal{T}_\omega^+)$. The proof of this direction follows from the next claim about $\mathcal{C}^*$ and from Theorem 2.

Claim 1. *C-graph $\mathcal{C}^*(G^*, \mathcal{T})$ is a c-planar c-connected super c-graph of $\mathcal{C}(G, \mathcal{T})$.*

($\Leftarrow$) Suppose that $\mathcal{C}(G, \mathcal{T})$ is c-planar. We show that Procedure OUTERCHECK and INNERCHECK succeed. Since $\mathcal{C}$ is c-planar, there exists a super c-graph $\mathcal{C}^*(G^*, \mathcal{T})$ of $\mathcal{C}$ such that G^* is planar and $\mathcal{C}^*$ is c-connected, by Theorem 2. By using the construction of Lemma 3 on c-graph $\mathcal{C}^*$, we can obtain a cycle-star S^- whose universal cycle is ρ that represents the connectivity of clusters inside ρ in $\mathcal{C}^*$. The proof of this direction follows from the next claim.

Claim 2. *Procedures OUTERCHECK and INNERCHECK succeed for $S_i^- = S^-$.*

We are finally ready to present the main result of the section.

Theorem 3. *The C-PLANARITY problem can be solved in $2^{O(\sqrt{\ell n} \cdot \log n)}$ time for n-vertex embedded flat c-graphs with maximum face size ℓ.*

Proof Sketch. Given an n-vertex c-graph $\mathcal{C}(G, \mathcal{T})$ with maximum face size ℓ, by Lemma 2, we can replace $\mathcal{C}$ with an equivalent 2-connected c-graph whose vertex set and maximum face size are linear in n and ℓ, respectively, and apply Algorithm TESTCP to such a c-graph to determine whether $\mathcal{C}$ is c-planar.

Since G is 2-connected, we can find a cycle separator of G of size $s(n) = O(\sqrt{\ell n})$ that splits G into two subgraphs each having at most $\frac{2n}{3}$ vertices [24]. By construction, the same bounds hold for G_i^- and G_i^+ with respect to their size. Also, since each cycle-star S_i^- is in one-to-one correspondence with a non-crossing partition of a set containing $s(n)$ elements, the number of cycle-stars satisfying the properties described at step 2a is given by the Catalan number $C_{s(n)} \leq 4^{s(n)}$, and S_i^- has size $g(n)$, which is in $O(s(n))$. Further, the non-recursive running time $f(n)$ of the algorithm is bounded by the time taken by steps 1 and 3(a)i, i.e., $O(n)$ time. Finally, the running time of Algorithm TESTCP is expressed by the following recurrence:

ALGORITHM TESTCP(C-GRAPH $\mathcal{C}(G,\mathcal{T})$)

BASE CASE

If $|V(G)| = O(\ell)$, then we can test C-PLANARITY for $\mathcal{C}(G,\mathcal{T})$ in $O(1)$ time when ℓ is a constant (refer to the full version of the paper [15]), by performing a brute force search to find a subset E' of the candidate saturating edges of $\mathcal{C}$ such that c-graph $\mathcal{C}'(G \cup E', \mathcal{T})$ satisfies Condition (iii) of Theorem 2.

RECURSIVE STEP

1. Select a cycle separator ρ of G. If ρ is a cluster-separator, then **return** NO; otherwise, construct c-graphs $\mathcal{C}_\rho^+(G^+, \mathcal{T}^+)$ and $\mathcal{C}_\rho^-(G^-, \mathcal{T}^-)$ as defined in Lemma 3.

2. OUTERCHECK
   (a) Construct the set $\mathcal{S}$ of all cycle-stars such that, for every $S \in \mathcal{S}$, it holds that (i) ρ is the universal cycle of S, (ii) ρ bounds the outer face of S, and (iii) every star vertex of S is incident only to vertices of ρ belonging to the same cluster.
   (b) For each cycle-star $S_i^- \in \mathcal{S}$:
      i. Construct a c-graph $\mathcal{C}^-(S_i^-, \mathcal{K}_i^-)$ as follows. First, initialize $\mathcal{K}_i^-$ to the subtree of $\mathcal{T}$ whose leaves are the vertices of S_i^-. Then, for each star vertex v of S_i^-, assign v to the cluster $\mu \in \mathcal{K}_i^-$ to which all its neighbors belong.
      ii. Augment $\mathcal{C}^-(S_i^-, \mathcal{K}_i^-)$ to an internally triangulated c-graph $\mathcal{C}_\Delta^-(R_i^-, \mathcal{J}_i^-)$ by introducing new vertices, each belonging to a distinct cluster, and by adding edges only between vertices in $V(S_i^-)$ and the newly introduced vertices (that is, no two non-adjacent vertices in S_i^- are adjacent in R_i^-).
      iii. Merge $\mathcal{C}_\Delta^-(R_i^-, \mathcal{J}_i^-)$ and $\mathcal{C}_\rho^+(G^+, \mathcal{T}^+)$ to obtain a c-graph $\mathcal{C}^+(G_i^+, \mathcal{T}_i^+)$[a].
      iv. Run TESTCP($\mathcal{C}^+(G_i^+, \mathcal{T}_i^+)$) to test whether $\mathcal{C}^+(G_i^+, \mathcal{T}_i^+)$ is c-planar[b].
   (c) If no c-graph $\mathcal{C}^+(G_i^+, \mathcal{T}_i^+)$ is c-planar, then **return** NO; otherwise, initialize $\mathcal{S}'$ as the set of *admissible* cycle-stars, i.e., the cycle-stars in $\mathcal{S}$ whose corresponding c-graph $\mathcal{C}^+(G_i^+, \mathcal{T}_i^+)$ is c-planar.

3. INNERCHECK
   (a) For each admissible cycle-star $S_i^- \in \mathcal{S}'$:
      i. Let $\mathcal{C}_{con}^+(H_i^+, \mathcal{T}_i^+)$ be the c-planar c-connected super c-graph of $\mathcal{C}^+$ returned by TESTCP($\mathcal{C}^+(G_i^+, \mathcal{T}_i^+)$) (step 2(b)iv). Apply the construction of Lemma 3 to c-graph $\mathcal{C}_{con}^+(H_i^+, \mathcal{T}_i^+)$ and cycle ρ to obtain a c-graph $\mathcal{C}^+(S_i^+, \mathcal{K}_i^+)$ satisfying Properties (2) and (3) of the lemma.
      ii. Augment $\mathcal{C}^+(S_i^+, \mathcal{K}_i^+)$ to a c-graph $\mathcal{C}_\Delta^+(R_i^+, \mathcal{J}_i^+)$ by introducing new vertices, each belonging to a distinct cluster, and by adding edges only between the vertices in $V(S_i^+)$ and the newly introduced vertices in such a way that cycle ρ bounds an inner face of R_i^+ and all the other faces of R_i^+ are triangles.
      iii. Merge $\mathcal{C}_\Delta^+(R_i^+, \mathcal{J}_i^+)$ and $\mathcal{C}_\rho^-(G^-, \mathcal{T}^-)$ to obtain a c-graph $\mathcal{C}^-(G_i^-, \mathcal{T}_i^-)$[a].
      iv. Run TESTCP($\mathcal{C}^-(G_i^-, \mathcal{T}_i^-)$) to test whether $\mathcal{C}^-(G_i^-, \mathcal{T}_i^-)$ is c-planar[b].
      v. If TESTCP($\mathcal{C}^-(G_i^-, \mathcal{T}_i^-)$) returns YES, then construct a c-planar c-connected super c-graph $C^*(G^*, \mathcal{T})$ of $\mathcal{C}(G,\mathcal{T})$ as follows. Let $\mathcal{C}_{con}^-(H_i^-, \mathcal{T}_i^-)$ be the c-planar c-connected c-graph returned by TESTCP($\mathcal{C}^-(G_i^-, \mathcal{T}_i^-)$). Remove all the vertices and edges of H_i^- in the exterior of cycle ρ, thus obtaining a new c-graph $\mathcal{C}_{in}(G_{in}, \mathcal{T}_{in})$ in which cycle ρ bounds the outer face. Similarly, remove all the vertices and edges of H_i^+ in the interior of cycle ρ, thus obtaining a new c-graph $\mathcal{C}_{out}(G_{out}, \mathcal{T}_{out})$ in which cycle ρ bounds an inner face. Finally, merge $\mathcal{C}_{in}$ and $\mathcal{C}_{out}$ to obtain c-graph $C^*(G^*, \mathcal{T})$ and **return** YES along with c-graph $C^*(G^*, \mathcal{T})$.

4. **return** NO if no c-graph $\mathcal{C}^-(G_i^-, \mathcal{T}_i^-)$, constructed at step 3(a)iii, is c-planar.

---

[a] The merging operations are well defined as cycle ρ bounds the outer face of R_i^- and an inner face of G^+, as well as an inner face of R_i^+ and the outer face of G^-.

[b] As $\mathcal{C}^+(G_i^+, \mathcal{T}_i^+)$ and $\mathcal{C}^-(G_i^-, \mathcal{T}_i^-)$ are 2-connected, TESTCP can be recursively applied.

$$T(n) = 2C_{s(n)}\left(T\left(\frac{2n}{3} + g(n)\right) + f(n)\right) \tag{1}$$

Since Eq. (1) solves to $T(n) = 2^{O(\sqrt{\ell n} \cdot \log n)}$, the statement follows.   □

4   An MSO$_2$ Formulation for C-Planarity

In this section, we show that the property of a c-graph of admitting a c-planar drawing can be expressed in extended monadic second-order (MSO$_2$) logic. We use this result and the fact that graph properties definable in MSO$_2$ logic can be verified in linear time on graphs of bounded treewidth, by Courcelle's Theorem [14], to build an FPT algorithm for testing the c-planarity of embedded flat c-graphs.

First-order graph logic deals with formulae whose variables represent the vertices and edges of a graph. *Second-order graph logic* also allows quantification over k-ary relations defined on the vertices and edges. *MSO$_2$ logic* only allows quantification over elements and *unary relations*, that is, sets of vertices and edges. Given a graph G and an MSO$_2$ formula ϕ, we say that G *models* ϕ, denoted by $G \models \phi$, if the logic statement expressed by ϕ is satisfied by the vertices, edges, and sets of vertices and edges in G. We will apply Courcelle's theorem not to the underlying graph G of the clustered planarity instance, but to the supergraph $G^\diamond$ of G that includes all the candidate saturating edges of G. This will allow us to quantify over sets of candidate saturating edges, but in exchange we must show that $G^\diamond$, and not just G, has low treewidth (Lemma 2).

Let H be a graph and let $E_1, E_2 \subseteq E(H)$. The following logic predicates can be expressed in MSO$_2$ logic (refer, e.g., to [6] for their detailed formulation):

◇ PLANAR$_H(E_1, E_2) :=$ the subgraph $(V(H), E_1 \cup E_2)$ of H is planar, and
◇ CONN$_H(U, E_1, E_2) :=$ vertices in $U \subseteq V(H)$ are connected by edges in $E_1 \cup E_2$.

Let $\mathcal{C}(G, \mathcal{T})$ be a c-graph and let E^* be the set of all the candidate saturating edges of $\mathcal{C}$. By Property (iii) of Theorem 2, c-graph $\mathcal{C}$ admits a c-planar drawing if and only if there exists a super c-graph $\mathcal{C}'(G', \mathcal{T})$ of $\mathcal{C}$ such that G' is planar and $\mathcal{C}'$ is c-connected. Testing Property (iii) amounts to determining the existence of a set $E^+ \subseteq E^*$ such that (i) the subgraph G' of $G^\diamond$ obtained by adding the edges in E^+ to G is planar and (ii) graph $G'(\mu)$ is connected, for each cluster $\mu \in \mathcal{T}$.

We remark that in an MSO$_2$ formula it is possible to refer to given subsets of vertices or edges of a graph, provided that the elements of such subsets can be distinguished from the elements of other subsets by equipping them with labels from a constant finite set [5]. Therefore, in our formulae we use the unquantified variables V_i to denote the set of vertices belonging to cluster μ_i, for each disconnected cluster $\mu_i \in \mathcal{T}$, E^* to denote the set consisting of all the candidate saturating edges of $\mathcal{C}$, and E_G to denote $E(G)$.

Let c be the number of disconnected clusters in $\mathcal{T}$. We have the formula:

$$\text{C-PLANAR}_{\mathcal{C}(G,\mathcal{T})} \equiv \exists (E^+ \subseteq E^*) \left[\text{PLANAR}_{G^\diamond}(E_G, E^+) \land \bigwedge_{i=1}^{c} \text{CONN}_{G^\diamond}(V_i, E_G, E^+) \right]$$

It is easy to see that formula C-PLANAR$_{\mathcal{C}(G,\mathcal{T})}$ correctly expresses Condition (iii) of Theorem 2 only if G admits a unique combinatorial embedding (up to a flip). In fact, if G has more than one embedding, formula C-PLANAR$_{\mathcal{C}(G,\mathcal{T})}$ might still be satisfiable after a change of the embedding, as formula PLANAR$_{G^\diamond}(E_G, E^+)$ models the planarity of an abstract graph rather than the planarity of a combinatorial embedding. We formalize this fact in the following lemma.

Lemma 4. *Let $\mathcal{C}(G,\mathcal{T})$ be a c-graph such that G has a unique combinatorial embedding and let $\mathcal{C}^\diamond(G^\diamond, \mathcal{T}^\diamond)$ be the c-graph obtained by augmenting $\mathcal{C}$ with all its candidate saturating edges. Then, $\mathcal{C}$ is c-planar iff $G^\diamond \models \text{C-PLANAR}_{\mathcal{C}(G,\mathcal{T})}$.*

Since changes of embedding are not allowed in our context, as we aim at testing the c-planarity of a c-graph given a prescribed embedding, we combine Lemmas 2 and 4, and then invoke Courcelle's Theorem to obtain the following main result.

Theorem 4. *The* C-PLANARITY *problem can be solved in $f(\overline{\text{emw}}, c)O(n)$ time for n-vertex embedded flat c-graphs with c disconnected clusters and whose underlying graph has embedded-width $\overline{\text{emw}}$, where f is a computable function.*

Proof. To test that $\mathcal{C}(G,\mathcal{T})$ admits a c-planar drawing with the given embedding we proceed as follows. First, we apply Lemma 2 to obtain a c-graph $\mathcal{C}^*(G^*, \mathcal{T}^*)$ that is equivalent to $\mathcal{C}$ such that G^* is 3-connected. Note that, the 3-connectivity of G^* implies that it has a unique combinatorial embedding (up to a flip). Then, we construct formula $\phi = \text{C-PLANAR}_{\mathcal{C}^*(G^*, \mathcal{T}^*)}$ and the super c-graph $\mathcal{C}^\diamond(G^\diamond, \mathcal{T}^\diamond)$ of $\mathcal{C}^*$ obtained by augmenting $\mathcal{C}^*$ with all its candidate saturating edges. Finally, we use Courcelle's Theorem to test whether $G^\diamond \models \phi$. The correctness immediately follows from Lemmas 2 and 4.

We now argue about the running time. By Lemma 2, c-graph $\mathcal{C}^*(G^*, \mathcal{T}^*)$ can be constructed in $O(n)$ time. Let κ be the maximum face size of G^*. The number of candidate saturating edges of $\mathcal{C}^*$ is $O(\kappa^2 n)$. By Lemma 2, Hence, we can augment $\mathcal{C}^*(G^*, \mathcal{T}^*)$ to obtain $\mathcal{C}^\diamond(G^\diamond, \mathcal{T}^\diamond)$ in $O(\ell^2 n)$ time.

By Courcelle's theorem [14], it is possible to verify whether $G^\diamond \models \phi$ in $g(tw(G^d iamond), len(\phi))O(|V(G^\diamond)| + |E(G^\diamond)|)$ time, where g is a computable function. By Lemma 2, $|V(G^\diamond)| = |V(G^*)| = O(n)$ and $\text{tw}(G^\diamond) = \overline{\text{emw}}(G)$. Also, by the discussion above, $|E(G^\diamond)| = O(\ell^2 n)$ and, by definition of embedded-width, $\ell = O(\overline{\text{emw}})$; thus, $|E(G^\diamond)| = O(\overline{\text{emw}}^2 n)$. Further, formula ϕ can be constructed in time proportional to its length $len(\phi)$, which is $O(c)$. Therefore, the overall running time can be expressed as $f(\overline{\text{emw}}, c)O(n)$, where f is a computable function. □

5   Conclusions and Open Problems

In this paper, we provide subexponential-time, XP (see [15]), and FPT algorithms to test C-PLANARITY of fairly-broad classes of c-graphs.

Several interesting questions arise from this research: (1) Can our results be generalized from flat to non-flat c-graphs? (2) Is there a fully polynomial-time algorithm to test C-PLANARITY of c-graphs whose underlying graph is a generalized h-simply-nested graph? (3) Are there interesting parameters of the underlying graph such that C-PLANARITY is FPT with respect to a single one of them (e.g., outerplanarity index, maximum face size, notable graph width parameters)? (4) Are there interesting parameters of the c-graph such that C-PLANARITY is FPT with respect to a single one of them (e.g., number of clusters, number of vertices of the same cluster incident to the same face)?

Acknowledgments. Supported in part by MIUR Project "MODE" under PRIN 20157EFM5C, by H2020-MSCA-RISE project 734922 - "CONNECT", by MIUR-DAAD JMP N° 34120, and by NSF grants CCF-1618301 and CCF-1616248. This article also reports on work supported by the U.S. Defense Advanced Research Projects Agency (DARPA) under agreement no. AFRL FA8750-15-2-0092. The views expressed are those of the authors and do not reflect the official policy or position of the Department of Defense or the U.S. Government.

References

1. Akitaya, H.A., Fulek, R., Tóth, C.D.: Recognizing weak embeddings of graphs. In: Czumaj, A. (ed.) SODA 2018, pp. 274–292. SIAM (2018)
2. Angelini, P., Da Lozzo, G.: Clustered planarity with pipes. In: Hong, S. (ed.) ISAAC 2016. LIPIcs, vol. 64, pp. 13:1–13:13. Schloss Dagstuhl - LZI (2016)
3. Angelini, P., Da Lozzo, G., Di Battista, G., Frati, F.: Strip planarity testing for embedded planar graphs. Algorithmica **77**(4), 1022–1059 (2017)
4. Angelini, P., Da Lozzo, G., Di Battista, G., Frati, F., Patrignani, M., Roselli, V.: Relaxing the constraints of clustered planarity. Comput. Geom. **48**(2), 42–75 (2015)
5. Arnborg, S., Lagergren, J., Seese, D.: Easy problems for tree-decomposable graphs. J. Algorithms **12**(2), 308–340 (1991)
6. Bannister, M.J., Eppstein, D.: Crossing minimization for 1-page and 2-page drawings of graphs with bounded treewidth. In: Duncan, C., Symvonis, A. (eds.) GD 2014. LNCS, vol. 8871, pp. 210–221. Springer, Heidelberg (2014). https://doi.org/10.1007/978-3-662-45803-7_18
7. Biedl, T.: Drawing planar partitions III: two constrained embedding problems. Technical report RRR 13–98, Rutcor Research Report (1998)
8. Bläsius, T., Rutter, I.: A new perspective on clustered planarity as a combinatorial embedding problem. Theor. Comput. Sci. **609**, 306–315 (2016)
9. Borradaile, G., Erickson, J., Le, H., Weber, R.: Embedded-width: a variant of treewidth for plane graphs. CoRR abs/1703.07532 (2017). http://arxiv.org/abs/1703.07532
10. Chimani, M., Di Battista, G., Frati, F., Klein, K.: Advances on testing C-planarity of embedded flat clustered graphs. In: Duncan, C., Symvonis, A. (eds.) GD 2014. LNCS, vol. 8871, pp. 416–427. Springer, Heidelberg (2014). https://doi.org/10.1007/978-3-662-45803-7_35

11. Chimani, M., Klein, K.: Shrinking the search space for clustered planarity. In: Didimo, W., Patrignani, M. (eds.) GD 2012. LNCS, vol. 7704, pp. 90–101. Springer, Heidelberg (2013). https://doi.org/10.1007/978-3-642-36763-2_9
12. Cornelsen, S., Wagner, D.: Completely connected clustered graphs. J. Discrete Algorithms 4(2), 313–323 (2006)
13. Cortese, P.F., Di Battista, G., Frati, F., Patrignani, M., Pizzonia, M.: C-planarity of C-connected clustered graphs. JGAA 12(2), 225–262 (2008)
14. Courcelle, B.: The monadic second-order logic of graphs. I. Recognizable sets of finite graphs. Inf. Comput. 85(1), 12–75 (1990)
15. Da Lozzo, G., Eppstein, D., Goodrich, M.T., Gupta, S.: Subexponential-time and FPT algorithms for embedded flat clustered planarity. CoRR abs/1803.05465 (2018). http://arxiv.org/abs/1803.05465
16. Dahlhaus, E.: A linear time algorithm to recognize clustered planar graphs and its parallelization. In: Lucchesi, C.L., Moura, A.V. (eds.) LATIN 1998. LNCS, vol. 1380, pp. 239–248. Springer, Heidelberg (1998). https://doi.org/10.1007/BFb0054325
17. Di Battista, G., Frati, F.: Efficient C-planarity testing for embedded flat clustered graphs with small faces. JGAA 13(3), 349–378 (2009)
18. Feng, Q.-W., Cohen, R.F., Eades, P.: Planarity for clustered graphs. In: Spirakis, P. (ed.) ESA 1995. LNCS, vol. 979, pp. 213–226. Springer, Heidelberg (1995). https://doi.org/10.1007/3-540-60313-1_145
19. Fulek, R.: C-planarity of embedded cyclic c-graphs. In: Hu, Y., Nöllenburg, M. (eds.) GD 2016. LNCS, vol. 9801, pp. 94–106. Springer, Cham (2016). https://doi.org/10.1007/978-3-319-50106-2_8
20. Goodrich, M.T., Lueker, G.S., Sun, J.Z.: C-planarity of extrovert clustered graphs. In: Healy, P., Nikolov, N.S. (eds.) GD 2005. LNCS, vol. 3843, pp. 211–222. Springer, Heidelberg (2006). https://doi.org/10.1007/11618058_20
21. Gutwenger, C., Jünger, M., Leipert, S., Mutzel, P., Percan, M., Weiskircher, R.: Advances in C-planarity testing of clustered graphs. In: Goodrich, M.T., Kobourov, S.G. (eds.) GD 2002. LNCS, vol. 2528, pp. 220–236. Springer, Heidelberg (2002). https://doi.org/10.1007/3-540-36151-0_21
22. Jelínek, V., Jelínková, E., Kratochvíl, J., Lidický, B.: Clustered planarity: embedded clustered graphs with two-component clusters. In: Tollis, I.G., Patrignani, M. (eds.) GD 2008. LNCS, vol. 5417, pp. 121–132. Springer, Heidelberg (2009). https://doi.org/10.1007/978-3-642-00219-9_13
23. Jelínková, E., Kára, J., Kratochvíl, J., Pergel, M., Suchý, O., Vyskocil, T.: Clustered planarity: small clusters in cycles and eulerian graphs. JGAA 13(3), 379–422 (2009)
24. Miller, G.L.: Finding small simple cycle separators for 2-connected planar graphs. J. Comput. Syst. Sci. 32(3), 265–279 (1986)

Computing Small Pivot-Minors

Konrad K. Dabrowski[1]([✉]) [iD], François Dross[2] [iD], Jisu Jeong[3] [iD],
Mamadou Moustapha Kanté[4] [iD], O-joung Kwon[5] [iD], Sang-il Oum[3] [iD],
and Daniël Paulusma[1] [iD]

[1] Department of Computer Science, Durham University, Durham, UK
{konrad.dabrowski,daniel.paulusma}@durham.ac.uk
[2] LIRMM, CNRS, Université de Montpellier, Montpellier, France
francois.dross@lirmm.fr
[3] Department of Mathematical Sciences, KAIST, Daejeon, South Korea
jjisu@kaist.ac.kr, sangil@kaist.edu
[4] Université Clermont Auvergne, LIMOS, CNRS, Aubière, France
mamadou.kante@uca.fr
[5] Department of Mathematics, Incheon National University, Incheon, South Korea
ojoungkwon@gmail.com

Abstract. A graph G contains a graph H as a pivot-minor if H can
be obtained from G by applying a sequence of vertex deletions and edge
pivots. Pivot-minors play an important role in the study of rank-width.
However, so far, pivot-minors have only been studied from a structural
perspective. We initiate a systematic study into their complexity aspects.
We first prove that the PIVOT-MINOR problem, which asks if a given
graph G contains a given graph H as a pivot-minor, is NP-complete.
If H is not part of the input, we denote the problem by H-PIVOT-MINOR.
We give a certifying polynomial-time algorithm for H-PIVOT-MINOR for
every graph H with $|V(H)| \leq 4$ except when $H \in \{K_4, C_3 + P_1, 4P_1\}$,
via a structural characterization of H-pivot-minor-free graphs in terms
of a set $\mathcal{F}_H$ of minimal forbidden induced subgraphs.

1 Introduction

Computing whether a graph H appears as a "pattern" inside some other graph G
is a well-studied problem in the area of structural and algorithmic graph theory.

Kwon is supported by the European Research Council (ERC) under the European
Union's Horizon 2020 research and innovation programme (ERC consolidator grant
DISTRUCT, agreement No. 648527). Dabrowski and Paulusma are supported by
EPSRC (EP/K025090/1) and the Leverhulme Trust (RPG-2016-258). Jeong and
Oum are supported by the National Research Foundation of Korea (NRF) grant
funded by the Korea government (MSIP) (No. NRF-2017R1A2B4005020). Kanté is
supported by French Agency for Research under the GraphEN project (ANR-15-CE-
0009). Underlying research data: source code used to prove Propositions 6 and 10
can be found at [6]. This work made use of the NVIDIA CUDA Research Centre
cluster facility at Durham University. We thank Stephen Bonner, John Brennan,
Ibad Kureshi and Grégoire Payen de La Garanderie for their assistance.

ⓒ Springer Nature Switzerland AG 2018
A. Brandstädt et al. (Eds.): WG 2018, LNCS 11159, pp. 125–138, 2018.
https://doi.org/10.1007/978-3-030-00256-5_11

The definition of a pattern depends on the set of graph operations that we are allowed to use. For instance, if we can obtain H from G via a sequence of vertex deletions, edge deletions and edge contractions, then G contains H as a minor. The MINOR problem is that of testing whether a given graph G contains a given graph H as a minor. This problem is known to be NP-complete even if G and H are trees of small diameter [19]. Hence, it is natural to fix the graph H and let the input consist of only G. This leads to the H-MINOR problem, and a celebrated result of Robertson and Seymour [28] states that the H-MINOR problem can be solved in cubic time for every graph H. If we only allow vertex deletions and edge contractions, then we obtain the H-INDUCED MINOR problem. In contrast, this problem can be NP-complete (see [9] for an example of a "hard" graph H on 68 vertices). Other well-known containment relations include containing a graph H as a contraction, an induced subgraph, a subdivision, or an (induced) topological minor; see, e.g. [3,13,17,18,29] for some complexity results for these relations.

We focus on the pivot-minor containment relation, defined as follows. The *local complementation* at a vertex u in a graph G replaces every edge of the subgraph induced by the neighbours of u by a non-edge, and vice versa. We denote the resulting graph by $G * u$. An *edge pivot* is the operation that takes an edge uv, first applies a local complementation at u, then at v, and then at u again. We denote the resulting graph by $G \wedge uv = G * u * v * u$ and note that $G * u * v * u = G * v * u * v$, so $G \wedge uv = G \wedge vu$. Alternatively, we can define the edge pivot operation as follows. Consider the set S_u of neighbours of u that are non-adjacent to v, the set S_v of neighbours of v that are non-adjacent to u and the set S_{uv} of common neighbours of u and v. Replace every edge between any two vertices in distinct sets from $\{S_u, S_v, S_{uv}\}$ by a non-edge and vice versa. Then delete every edge between u and S_u and add every edge between u and S_v. Similarly, delete every edge between v and S_v and add every edge between v and S_u. See Fig. 1 for an example. A graph G contains a graph H as a *pivot-minor* if G can be modified into (an isomorphic copy of) H by a sequence of vertex deletions and edge pivots.

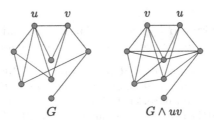

$$G \qquad\qquad G \wedge uv$$

Fig. 1. An example of a graph before and after pivoting an edge.

Pivot-minors were called p-reductions by Bouchet [1] and have been studied from a structural perspective, as they form a very suitable tool for working with rank-width [22,26]. Rank-width is a well-known width parameter (see [25] for a survey) and pivot-minors play a similar role for rank-width as minors do for treewidth. Oum [23] showed that for every positive constant k the class of graphs of rank-width at most k is well-quasi-ordered under the pivot-minor relation. Kwon and Oum [16] proved that every graph of rank-width at most k is a pivot-minor of a graph of treewidth at most $2k$, and that a graph of linear rank-width at most k is a pivot-minor of a graph of path-width at most $k + 1$.

Pivot-minors are closely related to so-called vertex-minors, introduced in the nineties as ℓ-reductions by Bouchet [1]. A graph G contains a graph H as a *vertex-minor* if G can be modified into (an isomorphic copy of) H by a sequence of vertex deletions and local complementations. Hence, if G contains H as a pivot-minor, then G contains H as a vertex-minor (but not necessarily vice versa). Bouchet [1] characterized circle graphs in terms of forbidden vertex-minors and by using this result, Geelen and Oum [12] were able to characterize circle graphs in terms of forbidden pivot-minors. Oum [24] conjectured that for each fixed bipartite circle graph H, every graph G of sufficiently large rank-width contains H as a pivot-minor. This conjecture is known to be true when G is a line graph, a bipartite graph or a circle graph (see [24]).

We study pivot-minors from an *algorithmic* perspective, that is, we consider the following research question:

Can we decide if a graph H is a pivot-minor of a graph G in polynomial time?

If both G and H are part of the input, then we obtain the following problem:

PIVOT-MINOR
    Instance: A pair of graphs G and H.
    Question: Does G have a pivot-minor isomorphic to H?

If H is not part of the input but fixed, then we obtain the H-PIVOT-MINOR problem. Question 7 in [25] asked for the complexity of H-PIVOT-MINOR, which has not been studied so far.

Our Results. We initiate a systematic study into the complexity of computing pivot minors. In Sect. 2 we prove that PIVOT-MINOR is NP-complete. Due to this, it is natural to study the computational complexity of H-PIVOT-MINOR, as proposed in [25]. To get a handle on this problem, we restrict ourselves to small graphs H. For every graph H on at most four vertices except for the complete graph K_4, the edgeless graph $4P_1$ and the triangle plus a vertex $C_3 + P_1$, we give a certifying algorithm that solves H-PIVOT-MINOR in polynomial time.

To explain the idea behind our algorithms, we observe that H-*pivot-minor-free* graphs, that is, graphs that do not contain H as a pivot-minor, are closed under vertex deletion. It is well known and readily seen that a class of graphs is closed under vertex deletion if and only if it can be characterized by a (possibly infinite) set of minimal forbidden induced subgraphs. In Sect. 3, for every graph $H \notin \{K_4, C_3 + P_1, 4P_1\}$ with $|V(H)| \leq 4$ we determine the set $\mathcal{F}_H$ of minimal forbidden induced subgraphs. We then test if the input graph G contains an induced subgraph $F \in \mathcal{F}_H$. If not, then G is H-pivot-minor-free. Otherwise, G contains H as a pivot-minor. As the graph F found by our algorithm contains H

as a pivot-minor, F is a certificate that can be used to verify H-pivot-minor containment in polynomial time: first delete all vertices of G not in F and then apply vertex deletions and edge pivots to obtain H from F. See [20] for a survey on certifying algorithms.

We discuss the graphs $K_4, C_3 + P_1$ and $4P_1$ in Sect. 4. Computer experiments show that $\mathcal{F}_{4P_1}$ contains over 100,000 graphs, so it is likely that $\mathcal{F}_{4P_1}$ is not finite. We prove that $\mathcal{F}_{K_4}$ and $\mathcal{F}_{C_3+P_1}$ each contain infinitely many graphs. In the same section we discuss some further computer experiments and propose a general framework for future research.

2   When H Is Part of the Input

We prove that PIVOT-MINOR is NP-complete. We first introduce some terminology and basic results on matroids, which can be found in [27]. A *matroid* is a pair $M = (E, \mathcal{I})$ of a finite set E, called the *ground set*, and a set $\mathcal{I}$ of subsets of E satisfying the following three properties: (i) $\mathcal{I} \neq \emptyset$; (ii) if $Y \in \mathcal{I}$ and $X \subseteq Y$, then $X \in \mathcal{I}$; and (iii) if $X, Y \in \mathcal{I}$ with $|Y| = |X| + 1$, then there exists an element $y \in Y \setminus X$ such that $X \cup \{y\} \in \mathcal{I}$. A set $X \subseteq E$ is *independent* in $M = (E, \mathcal{I})$ if $X \in \mathcal{I}$, otherwise X is *dependent*. The *rank* of a subset $X \subseteq E$ is the size of a largest independent subset of X. The *rank* of a matroid $M = (E, \mathcal{I})$ is the rank of E. A *base* of a matroid is a maximal independent set. A *circuit* of a matroid is a minimal dependent set. The *dual matroid* M^* of a matroid $M = (E, \mathcal{I})$ is a matroid on E such that X is a base of M^* if and only if $E \setminus X$ is a base in M. For a subset X of E, we define $M \setminus X$ to be the matroid $(E \setminus X, \mathcal{I}')$ such that $\mathcal{I}' = \{X' \subseteq E \setminus X \mid X' \in \mathcal{I}\}$. We define $M/X = (M^* \setminus X)^*$. A matroid N is a *minor* of a matroid M if $N = (M \setminus X)/Y$ for some disjoint sets X and Y. A matroid $M = (E, \mathcal{I})$ is *binary* if there is a matrix over the binary field whose columns are indexed by E such that X is independent in M if and only if the corresponding columns are linearly independent. It is known that the dual matroid of a binary matroid is also binary.

A major example of binary matroids arises from graphs. For a graph $G = (V, E)$, let $\mathcal{I}$ be the set of subsets X of E such that the subgraph (V, X) has no cycles. Then $M(G) = (E, \mathcal{I})$ is a matroid, called the *cycle matroid* of G and such matroids are binary. It is known that circuits of $M(G)$ are precisely the edge sets of cycles of G and if a graph H is a minor of G, then $M(H)$ is a minor of $M(G)$.

If G is connected and has n vertices and m edges, then $M(G)$ has rank $n-1$ because any spanning tree of G has $n-1$ edges, and $(M(G))^*$ has rank $m-n+1$.

For a binary matroid $M = (E, \mathcal{I})$, the *fundamental graph* of M with respect to a base B is the bipartite graph on E with the bipartition $(B, E \setminus B)$ such that $x \in B$, $y \in E \setminus B$ are adjacent if and only if $(B \setminus \{x\}) \cup \{y\}$ is a base of M. Conversely, for a bipartite graph G with a bipartition (A, B), we may define a binary matroid $\mathrm{Bin}(G, A, B)$ on $V(G)$ represented by the $A \times V(G)$ matrix

$$
\begin{array}{cc}
A & B \\
\end{array}
$$
$$
A \begin{pmatrix} I_A & M_{A,B} \end{pmatrix}
$$

over the binary field where I_A is the $A \times A$ identity matrix and $M_{A,B}$ is the $A \times B$ submatrix of the adjacency matrix of G whose (x, y)-entry is 1 if and only if x and y are adjacent. We need the following lemma for our NP-hardness result.

Lemma 1 ([22, Corollary 3.6]). *The following statements hold:*

(i) *Let N, M be binary matroids, and H, G be fundamental graphs of N and M respectively. If N is a minor of M, then H is a pivot-minor of G.*

(ii) *Let G be a bipartite graph with bipartition $A \cup B = V(G)$. If H is a pivot-minor of G, then there is a bipartition $A' \cup B' = V(H)$ such that $\mathrm{Bin}(H, A', B')$ is a minor of $\mathrm{Bin}(G, A, B)$.*

Theorem 1. PIVOT-MINOR *is* NP-*complete.*

Proof. We reduce from the HAMILTON CYCLE problem, which asks if a graph has a Hamilton cycle. This problem is NP-complete even for 3-regular graphs [10]. Let $G = (V, E)$ be a 3-regular graph with n vertices and m edges. We may assume without loss of generality that $n \geq 5$ and that G is connected. As G is 3-regular, $2m = 3n$. Consequently, $(M(G))^*$ has rank $m - n + 1 = \frac{1}{2}n + 1$.

Let T be a spanning tree of G. Let G_T be the fundamental graph of $M(G)$ with respect to $E(T)$, which can be built in polynomial time. We claim G has a Hamilton cycle if and only if the n-vertex star $K_{1,n-1}$ is a pivot-minor of G_T.

For the forward direction, suppose G has a Hamilton cycle C. Then G contains C as a minor and thus $M(G)$ has $M(C)$ as a minor, and so G_T has every fundamental graph of $M(C)$ as a pivot-minor by Lemma 1(i). This proves the forward direction, because every fundamental graph of $M(C)$ is isomorphic to $K_{1,n-1}$.

For the reverse direction, suppose that $K_{1,n-1}$ is a pivot-minor of G_T. Then by Lemma 1(ii), $V(K_{1,n-1})$ has a bipartition (A', B') such that $\mathrm{Bin}(K_{1,n-1}, A', B')$ is a minor of $M(G) = \mathrm{Bin}(G_T, A, B)$ for some partition (A, B) of $V(G_T)$. As $K_{1,n-1}$ is connected, it admits only two possible bipartitions (that is, there is a unique way of partitioning the vertices of $K_{1,n-1}$ into two independent sets and there are two ways to order the sets). So $\mathrm{Bin}(K_{1,n-1}, A', B')$ is either $M(C)$ or its dual $(M(C))^*$, where C is the cycle on n vertices. Therefore $M(C)$ or $(M(C))^*$ is a minor of $M(G)$. Equivalently, $M(C)$ is a minor of $M(G)$ or $(M(G))^*$. Because the rank of $M(C)$ is $n - 1$ and the rank of $(M(G))^*$ is $\frac{1}{2}n + 1 < n - 1$ (as $n \geq 5$) we find that $M(C)$ cannot be a minor of $(M(G))^*$. Thus, $M(C)$ is a minor of $M(G)$ and therefore $M(G)$ has a circuit of length at least n. This implies that G has a cycle of length n. □

3   When H Is Fixed

We give a certifying algorithm for recognizing H-pivot-minor-free graphs for every graph H on at most four vertices except for the cases where $H \in \{K_4, C_3 + P_1, 4P_1\}$ (see Sect. 4 for a further discussion on these three graphs). For each such graph H, we determine the minimal set $\mathcal{F}_H$ such that a graph G contains H

as a pivot-minor if and only if G contains an induced subgraph in $\mathcal{F}_H$. The cases where $H \in \{2P_1 + P_2, 2P_2\}$ are too involved to expect a combinatorial proof, so we rely on a computer-based proof for these cases. Of the remaining cases, the ones where H is not $3P_1$-free are more involved than the others. We therefore consider the $3P_1$-free cases in Sect. 3.1 and the remaining cases in Sect. 3.2.

3.1   When H Is $3P_1$-Free

The graph $\overline{G} = (V, \{uv \mid uv \notin E(G), u \neq v\}$ is the *complement* of a graph G. A *co-component* in a graph G is a maximal set of vertices in G that induces a connected subgraph in $\overline{G}$. The graph $G_1 + G_2 = (V(G_1) \cup V(G_2), E(G_1) \cup E(G_2))$ is the *disjoint union* of two vertex-disjoint graphs G_1 and G_2. Recall that $K_{1,n-1}$ is the star on n vertices. The path and cycle on n vertices are denoted P_n and C_n, respectively; the *length* of a path or cycle is the number of edges it contains. The *paw*, *diamond*, *dart* and *claw* are the graphs $\overline{P_1 + P_3}$, $\overline{2P_1 + P_2}$, $\overline{P_1 + \text{paw}}$ and $K_{1,3}$, respectively (see also Fig. 2). A graph class is *pivot-minor-closed* if it is closed under vertex deletions and edge pivots. A graph G is $(H_1, \ldots, H_p)$-*free* for a set $\mathcal{H} = \{H_1, \ldots, H_p\}$ of graphs if G has no induced subgraph isomorphic to a graph in $\mathcal{H}$. Let $H \notin \{K_4, C_3 + P_1\}$ be a $3P_1$-free graph with $|V(H)| \leq 4$, so $H \in \{P_1, 2P_1, P_1 + P_2, P_2, 2P_2, P_3, P_4, C_3, C_4, \text{paw}, \text{diamond}\}$. The cases $H = P_1$, $H = P_2$ and $H = 2P_1$ are trivial. We now consider the other cases (we omit the proofs of Propositions 1–3).

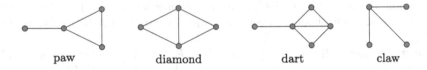

paw   diamond   dart   claw

Fig. 2. Graphs referred to in Sect. 3.

Proposition 1. *For a graph G, P_3 is a pivot-minor of G if and only if P_3 is an induced subgraph of G.*

Proposition 2. *For a graph G, C_3 is a pivot-minor of G if and only if an odd cycle is an induced subgraph of G if and only if G is not bipartite.*

Proposition 3. *The following statements are equivalent for every graph G:*

(i) $P_1 + P_2$ *is a pivot-minor of G.*
(ii) $P_1 + P_2$, C_4 *or the diamond is an induced subgraph of G.*
(iii) G *is neither a complete graph, an edgeless graph nor a star.*

   A graph is a *clique-star* if it consists of pairwise vertex-disjoint cliques K, $L_1, \ldots, L_p$ for some $p \geq 0$, such that every vertex of K is adjacent to every vertex of $L_1 \cup \cdots \cup L_p$ and there is no edge between any two distinct cliques L_i

and L_j. Note that we may assume that $p \neq 1$, as if $p = 1$ then the clique-star is a complete graph, in which case we can set $p = 0$. We need the following lemma (we omit the proof).

Lemma 2. *The class of clique-stars is pivot-minor-closed.*

Proposition 4. *The following statements are equivalent for every graph G.*

(i) P_4 *is a pivot-minor of G.*
(ii) C_4 *is a pivot-minor of G.*
(iii) P_4, C_4 *or the dart is an induced subgraph of G.*
(iv) G *has a component that is not a clique-star.*

Proof. Both the P_4 and C_4 can be obtained from each other by pivoting one edge and so (i) and (ii) are equivalent. Pivoting an edge incident to a vertex of degree 2 and a vertex of degree 3 in the dart yields a bull (see Fig. 3), which contains P_4 as an induced subgraph. Therefore the dart contains P_4 as a pivot-minor, so (iii) implies (i) and (ii). As P_4, C_4 and the dart are not clique-stars, (iii) implies (iv). Lemma 2 implies that the class of graphs all of whose components are clique-stars is pivot-minor-closed, hence (i) and (ii) imply (iv).

It remains to prove that (iv) implies (iii). Suppose that G has a component D that is not a clique-star. Also assume that G is (P_4, C_4)-free. It is well known that the complement of a connected P_4-free graph on at least two vertices is disconnected [4]. Hence we can partition $V(D)$ into two sets A and B, such that every vertex of A is adjacent to every vertex of B. Moreover, as D is not a complete graph, we may assume that B is not a clique. If A is not a clique either, then two non-adjacent vertices of A, together with two non-adjacent vertices of B, form an induced C_4, a contradiction. Hence A is a clique. We may assume that A is chosen to be maximal subject to the condition that every vertex of A is adjacent to every vertex of B and B contains two non-adjacent vertices.

Suppose B induces a connected subgraph. Then, since $G[B]$ is P_4-free, connected, and contains at least two vertices, we can partition B into two non-empty sets B_1 and B_2 such that every vertex of B_1 is adjacent to every vertex of B_2. As B is not a clique, this means that at least one of B_1 and B_2, say B_2, is not a clique. Then, by the same argument as before, B_1 must be a clique. This implies that every vertex of B_1 is adjacent to every other vertex of B_1 and to every vertex of B_2. However, this contradicts the maximality of A, as we could have chosen $A \cup B_1$ instead. Hence B does not induce a connected subgraph of D.

Let $J_1, \ldots, J_r$ be the components of $D[B]$ for some $r \geq 2$. Since D is not a clique-star, one of $J_1, \ldots, J_r$, say J_1, is not complete. If follows that J_1 contains an induced P_3, say on vertices u, v, w. Then u, v, w, together with a vertex of A and a vertex of J_2, induce a dart. $\square$

Proposition 5. *The following statements are equivalent for every graph G.*

(i) *The paw is a pivot-minor of G.*
(ii) *The diamond is a pivot-minor of G.*

(iii) *The paw, the diamond or an odd cycle of length at least* 5 *is an induced subgraph of* G.
(iv) G *has a component that is neither bipartite nor complete.*

Proof. By pivoting one edge, the diamond can be obtained from the paw and so (i) and (ii) are equivalent. Since every odd cycle on at least five vertices contains the paw as a pivot-minor, (iii) implies (i) and (ii). As the classes of complete graphs and bipartite graphs are pivot-minor-closed, (i) and (ii) imply (iv).

To prove that (iv) implies (iii), suppose (iii) is false. Let D be a component of G. We claim that D is bipartite or complete. If not, C_3 is a proper induced subgraph of D. Let K be a maximal clique of D containing the vertices of a C_3. As D is not complete and K is maximal, there is a vertex $u \in V(D) \setminus K$ that has both a neighbour and a non-neighbour in K. Since K is a clique of size at least 3, D contains the paw or diamond as an induced subgraph, a contradiction. □

We proved the next proposition by computer (see [6] for source code).

Proposition 6 (proved by computer). *The set* $\mathcal{F}_{2P_2}$ *has size* 9.

A sequence S of vertex deletions and edge pivots is an H-*pivot-minor-sequence* of a graph G if H can be obtained from G after applying the operations of S.

Theorem 2. *For* $H \in \{P_1, 2P_1, P_2, P_1 + P_2, P_3, C_3, 2P_2, P_4, C_4, paw, diamond\}$, *there is a polynomial-time algorithm for* H-PIVOT-MINOR *that gives an* H-*pivot-minor-sequence (if one exists).*

Proof. The cases when $H \in \{P_1, 2P_1, P_2\}$ are trivial. By Propositions 1, 3, 4 and 6, the set $\mathcal{F}_H$ of minimal obstructions is finite if $H \in \{P_3, P_1 + P_2, P_4, C_4, 2P_2\}$. If $H = C_3$, by Proposition 2, we need to find an odd cycle F, which we do in polynomial time by testing bipartiteness. If $H \in \{paw, diamond\}$, then we use condition (iv) in Proposition 5 to decide if a graph has H as a pivot-minor; this allows us to find a forbidden induced subgraph F efficiently by using the argument in its proof. Then the theorem follows, as in polynomial time we can find the vertex deletions and edge pivots that modify F into H. □

3.2   When H Is Not $3P_1$-Free

We now consider the cases where $H \in \{3P_1, 2P_1 + P_2, P_1 + P_3, claw\}$. The *bull* is the graph obtained from P_5 by adding an edge between the second vertex and the fourth vertex. The graph W_4 is obtained from C_4 by adding one vertex adjacent to all vertices of C_4. The graph BW_3 is the bipartite graph on seven vertices obtained from C_6 by adding one vertex adjacent to three pairwise non-adjacent vertices of the cycle. We will work with the complement of BW_3, denoted by $\overline{BW_3}$. See Fig. 3 for pictures of the bull, W_4 and $\overline{BW_3}$.

We write G/v to denote $(G \wedge zv) - v$ if a vertex v has a neighbour z and $G - v$ if v is isolated. Two graphs are *pivot-equivalent* if they can be obtained from each

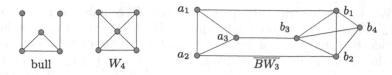

Fig. 3. Forbidden graphs from Sect. 3.2.

other by a sequence of edge pivots. For two distinct neighbours x, y of v, because $(G \wedge xv) - v = (G \wedge yv \wedge xy) - v = (G \wedge yv - v) \wedge xy$, we find that $(G \wedge xv) - v$ is pivot-equivalent to $(G \wedge yv) - v$ and thus the choice of neighbour of v does not change the pivot-equivalence of graphs G/v. We need two lemmas (we omit the proofs). Lemma 4 holds in the context of binary delta-matroids or matrix pivots (see [2,24]) and its proof is inspired by the analogous proof for vertex-minors in [12].

Lemma 3. *Let v, x, y be distinct vertices of a graph G. If $xy \in E(G)$, then $(G \wedge xy) - v$ is pivot-equivalent to $G - v$ and $(G \wedge xy)/v$ is pivot-equivalent to G/v.*

Lemma 4. *If a graph H is a pivot-minor of a graph G and $v \in V(G) \setminus V(H)$, then H is a pivot-minor of $G - v$ or $(G \wedge vw) - v$ for some neighbour w of v in G.*

The proofs for the cases where $H \in \{P_1 + P_3, \text{claw}\}$ rely on the proof for the $H = 3P_1$ case. Our proof for the $H = 3P_1$ case focuses on showing that if a graph G contains $3P_1$ as a pivot-minor, then G contains a graph from $\{3P_1, W_4, \overline{BW_3}\}$ as an induced subgraph. We will do this by induction on $|V(G)|$. Since $3P_1$ is edgeless, we cannot pivot any edge in it. Therefore, the above claim holds if $|V(G)| \leq 3$, and so we may assume that $|V(G)| \geq 4$. If G has a pivot-minor isomorphic to $3P_1$, then by Lemma 4, there is a vertex $v \in V(G)$ such that $G - v$ or G/v contains a pivot-minor isomorphic to $3P_1$ for some neighbour w of v. Clearly, if $G - v$ contains $3P_1$, W_4 or $\overline{BW_3}$ as an induced subgraph then G also contains this graph as an induced subgraph. Therefore, by the induction hypothesis, we may assume that G/v contains $3P_1$ as a pivot-minor. Lemmas 5, 6 and 7, we show that if G/v contains an induced subgraph isomorphic to $3P_1$, W_4 or $\overline{BW_3}$, then G contains an induced subgraph in $\{3P_1, W_4, \overline{BW_3}\}$; these lemmas (we omit the proofs) will form the main steps in our induction.

Lemma 5. *Let vw be an edge of a graph G. If $(G \wedge vw) - v$ contains $3P_1$ as an induced subgraph, then G contains $3P_1$ or W_4 as an induced subgraph.*

Lemma 6. *Let vw be an edge of a graph G. If $G \wedge vw$ contains W_4 as an induced subgraph, then G contains $3P_1$, W_4 or $\overline{BW_3}$ as an induced subgraph.*

Lemma 7. *Let G be a graph containing an edge vw. If $G \wedge vw$ contains $\overline{BW_3}$ as an induced subgraph, then G contains $3P_1$, W_4 or $\overline{BW_3}$ as an induced subgraph.*

Proposition 7. *A graph G contains $3P_1$ as a pivot-minor if and only if G contains a graph from $\{3P_1, W_4, \overline{BW_3}\}$ as an induced subgraph.*

Proof. We first prove the "if" part. Suppose G contains a graph $H \in \{3P_1, W_4, \overline{BW_3}\}$ as an induced subgraph. If $H = W_4$, then by pivoting an edge incident to the vertex of degree 4 we obtain a graph which contains $3P_1$ as an induced subgraph. If $H = \overline{BW_3}$, then let $U_1 = \{a_1, a_2, a_3\}$ and $U_2 = \{b_1, b_2, b_3, b_4\}$ be the two cliques of H and $a_i b_i \in E(H)$ for $i = 1, 2, 3$. By pivoting an edge $a_1 b_1$, we obtain a subgraph induced by $\{a_2, a_3, b_2, b_3, b_4\}$ that is isomorphic to W_4.

Next, we prove the "only if" part. Suppose G contains $3P_1$ as a pivot-minor. We use induction on $|V(G)| = n$ to prove that G contains a graph from $\{3P_1, W_4, \overline{BW_3}\}$ as an induced subgraph. We may assume that $n \geq 4$.

As $n \geq 4 > |V(3P_1)|$, Lemma 4 implies that there is a vertex $v \in V(G)$ such that $G - v$ or $(G \wedge vw) - v$, for some neighbour w of v, contains $3P_1$ as a pivot-minor. If $G - v$ contains $3P_1$ as a pivot-minor, then by the induction hypothesis, $G - v$ contains an induced subgraph in $\{3P_1, W_4, \overline{BW_3}\}$, hence so does G. Now we assume that $(G \wedge vw) - v$, for some neighbour w of v, contains $3P_1$ as a pivot-minor. By the induction hypothesis, $(G \wedge vw) - v$ contains $3P_1$, W_4 or $\overline{BW_3}$ as an induced subgraph. Applying Lemmas 5, 6 and 7, respectively, we find that G contains an induced graph in $\{3P_1, W_4, \overline{BW_3}\}$. $\square$

Proposition 8. *The following statements are equivalent for every graph G.*

(i) $P_1 + P_3$ *is a pivot-minor of G.*
(ii) $P_1 + P_3$, $K_{2,3}$, W_4 *or* $\overline{BW_3}$ *is an induced subgraph of G.*
(iii) G *contains $3P_1$ as a pivot minor and G is not a clique-star.*

Proof. It is easy to verify that $K_{2,3}$, W_4 and $\overline{BW_3}$ contain $P_1 + P_3$ as a pivot-minor. Therefore (ii) implies (i). To prove that (i) implies (iii), suppose that G contains $P_1 + P_3$ as a pivot-minor. Since $3P_1$ is a pivot-minor of $P_1 + P_3$, it follows that G contains $3P_1$ as a pivot-minor. It is easy to verify that all clique-stars are $(P_1 + P_3)$-free. Since the class of clique-stars is pivot-minor-closed by Lemma 2, it follows that all clique-stars are $(P_1 + P_3)$-pivot-minor-free. Hence G is not a clique-star. Therefore (i) implies (iii).

It remains to show that (iii) implies (ii). Suppose (ii) does not hold, that is, G is $(P_1 + P_3, K_{2,3}, W_4, \overline{BW_3})$-free. A graph is $\overline{P_1 + P_3}$-free if and only if every component of it is either complete multipartite or C_3-free [21]. Hence, as G is $(P_1 + P_3)$-free, every co-component of G is either a disjoint union of cliques or $3P_1$-free. If every co-component of G is $3P_1$-free, then since co-components are complete to each other, it follows that G is $3P_1$-free. Then G is $(3P_1, W_4, \overline{BW_3})$-free. Then, by Proposition 7, G is $3P_1$-pivot-minor-free. Assume that G has a co-component D that contains an induced $3P_1$. Then D is a disjoint union of (at least three) cliques. As G is $K_{2,3}$-free, every other co-component of G is $2P_1$-free, in which case it consists of a single vertex. Therefore the vertices in all the other co-components of G form a dominating clique. Hence G is a clique-star. $\square$

For $H = $ claw, we need a lemma (we omit the proof) that allows us to focus on *connected* graphs.

Lemma 8. *A graph G is* (bull, claw, P_5)*-free if and only if every component of G is $3P_1$-free.*

Combining Lemma 8 with Proposition 7, it is easy to prove the following (we omit the proof).

Proposition 9. *A graph G contains the claw as a pivot-minor if and only if G contains a graph from $\{claw, P_5, bull, W_4, \overline{BW_3}\}$ as an induced subgraph.*

We proved the next proposition by computer (see [6] for source code).

Proposition 10 (proved by computer). *The set $\mathcal{F}_{2P_1+P_2}$ has size 19.*

In the same way as for Theorem 2 we use Propositions 7–10 to prove:

Theorem 3. *For $H \in \{3P_1, 2P_1 + P_2, P_1 + P_3, claw\}$, there is a polynomial-time algorithm for H-PIVOT-MINOR that gives an H-pivot-minor-sequence (if one exists).*

4   Future Work

We aim to continue determining the complexity of H-PIVOT-MINOR. We do not know yet if there is a graph H for which H-PIVOT-MINOR is NP-complete. Our current technique for proving polynomial-time solvability is to find the set $\mathcal{F}_H$ of minimal forbidden induced subgraphs or a structural characterization verifiable in polynomial time. Our research led to the following framework for future work.

1. For a graph H, determine if $\mathcal{F}_H$ is finite (or has a polynomial characterization). We have some preliminary results for the remaining graphs H on at most four vertices, namely K_4, $C_3 + P_1$ and $4P_1$. Using a computer, we found that $\mathcal{F}_{4P_1}$ contains over 100,000 graphs even if we only list graphs on at most twelve vertices. As such, it is likely that $\mathcal{F}_{4P_1}$ is not finite. If $H = K_4$ and $H = C_3 + P_1$, then the set $\mathcal{F}_H$ has infinite size. We also started to extend our computer approach to graphs H on more than four vertices, which yielded large finite sets $\mathcal{F}_H$ for certain graphs H. The largest finite set we have found is $\mathcal{F}_{P_2+C_4} = \mathcal{F}_{P_2+P_4}$, which contains 7932 graphs. In addition to $\mathcal{F}_{4P_1}$, we found that $\mathcal{F}_{3P_2}$ also contains over 100,000 graphs, but it is not yet feasible for us to test if the set of minimal forbidden graphs found so far is complete. Besides some further tests by computer, we also need to answer the question of whether $\mathcal{F}_H$ is infinite whenever H contains an induced subgraph H' for which $\mathcal{F}_{H'}$ is infinite.

2. For a graph H, determine if H-pivot-minor-free graphs have bounded rankwidth. If for a fixed graph H, the class of H-pivot-minor-free graphs has rank-width at most k for some constant k, then we can decide in polynomial time if a given graph G contains H as a pivot-minor. We first check in polynomial time [26] if the rank-width $\mathrm{rw}(G)$ of G is at least $k+1$ or at most $3k+1$. If $\mathrm{rw}(G) \geq k+1$, then G has H as a pivot-minor. If $\mathrm{rw}(G) \leq 3k + 1$, then we can decide in cubic time if G has H as a pivot-minor by adapting the approach for vertex-minor testing on graphs of bounded rank-width from [5], namely via expression in monadic second order logic with modulo-2 counting (we refer to a future paper for the details).

3. For a graph H, follow a hybrid approach by combining approaches 1 and 2.
In fact, for a graph H, it suffices to determine a sufficiently precise set $\mathcal{F}' \supseteq \mathcal{F}_H$, after which we can try to prove boundedness of rank-width of the superclass of $\mathcal{F}'$-free graphs using techniques for hereditary graph classes (see e.g. [8, 14, 15]).

4. For a graph H, determine whether the class of H-pivot-minor-free graphs is well-quasi-ordered by the induced subgraph relation.
For every graph H, the set $\mathcal{F}_H$ is an antichain with respect to the induced subgraph relation. Suppose that the class of H-pivot-minor-free graphs is a subclass of a hereditary class $\mathcal{H}$ that is defined by a finite collection of forbidden induced subgraphs such that $\mathcal{H}$ is well-quasi-ordered by the induced subgraph relation. Then all graphs in $\mathcal{F}_H$ are either one of the finitely-many minimal forbidden induced subgraphs for $\mathcal{H}$, or belong to $\mathcal{H}$. Since $\mathcal{H}$ is well-quasi-ordered by the induced subgraph relation and the graphs in $\mathcal{F}_H$ form an antichain, it follows that $\mathcal{F}_H$ is finite. For example, since the graph W_4 contains $3P_1$ as a pivot-minor and the class of $(3P_1, W_4)$-free graphs is well-quasi-ordered by the induced subgraph relation [7], it follows that $\mathcal{F}_{3P_1}$ is finite. Thus, even without finding the precise graphs in $\mathcal{F}_H$, it may be possible to establish that the class of H-pivot-minor-free graphs is well-quasi-ordered by the induced subgraph relation, and so conclude that the H-PIVOT MINOR problem is polynomial-time solvable by finiteness of $\mathcal{F}_H$.

We note that approaches 2 and 3 do not yield certifying algorithms, while approach 4 only gives a non-constructive proof that such an algorithm exists. Besides the above, a proof for the Minor Recognition conjecture [11] for binary matroids would also yield a technique to obtain complexity results for pivot-minors. In particular, if this conjecture is true, then for every graph H the H-PIVOT-MINOR problem is polynomial-time solvable for bipartite graphs. This follows from Lemma 1, which implies that a connected bipartite graph H is a pivot-minor of a bipartite graph G if and only if for binary matroids M and N that have G and H as fundamental graphs, respectively, N or the dual of N is a minor of M (if H is not connected, then we try all possible ways of making duals per component of H).

Finally, it would be interesting to perform a similar complexity study with respect to vertex-minors, starting by taking both G and H as part of the input.

References

1. Bouchet, A.: Circle graph obstructions. J. Comb. Theory, Ser. B **60**(1), 107–144 (1994)
2. Bouchet, A., Duchamp, A.: Representability of Δ-matroids over GF(2). Linear Algebra Appl. **146**, 67–78 (1991)
3. Brouwer, A.E., Veldman, H.J.: Contractibility and NP-completeness. J. Graph Theory **11**(1), 71–79 (1987)
4. Corneil, D.G., Lerchs, H., Stewart, L.: Complement reducible graphs. Discrete Appl. Math. **3**(3), 163–174 (1981)
5. Courcelle, B., Oum, S.: Vertex-minors, monadic second-order logic, and a conjecture by Seese. J. Comb. Theory, Ser. B **97**(1), 91–126 (2007)
6. Dabrowski, K.K.: Computing small pivot-minors [computer software]. Durham University (2018). https://doi.org/10.15128/r1t722h881p
7. Dabrowski, K.K., Lozin, V.V., Paulusma, D.: Clique-width and well-quasi-ordering of triangle-free graph classes. In: Bodlaender, H.L., Woeginger, G.J. (eds.) WG 2017. LNCS, vol. 10520, pp. 220–233. Springer, Cham (2017). https://doi.org/10.1007/978-3-319-68705-6_17
8. Dabrowski, K.K., Paulusma, D.: Clique-width of graph classes defined by two forbidden induced subgraphs. Comput. J. **59**(5), 650–666 (2016)
9. Fellows, M.R., Kratochvíl, J., Middendorf, M., Pfeiffer, F.: The complexity of induced minors and related problems. Algorithmica **13**(3), 266–282 (1995)
10. Garey, M.R., Johnson, D.S., Tarjan, R.E.: The planar Hamiltonian circuit problem is NP-complete. SIAM J. Comput. **5**(4), 704–714 (1976)
11. Geelen, J., Gerards, B., Whittle, G.: Towards a structure theory for matrices and matroids. In: Proceedings of ICM 2006, vol. III, no. 827–842 (2006)
12. Geelen, J., Oum, S.: Circle graph obstructions under pivoting. J. Graph Theory **61**(1), 1–11 (2009)
13. Grohe, M., Kawarabayashi, K.-i., Marx, D., Wollan, P.: Finding topological subgraphs is fixed-parameter tractable. In: Proceedings of STOC 2011, pp. 479–488 (2011)
14. Gurski, F.: The behavior of clique-width under graph operations and graph transformations. Theory Comput. Syst. **60**(2), 346–376 (2017)
15. Kamiński, M., Lozin, V.V., Milanič, M.: Recent developments on graphs of bounded clique-width. Discrete Appl. Math. **157**(12), 2747–2761 (2009)
16. Kwon, O., Oum, S.: Graphs of small rank-width are pivot-minors of graphs of small tree-width. Discrete Appl. Math. **168**, 108–118 (2014)
17. Lévêque, B., Lin, D.Y., Maffray, F., Trotignon, N.: Detecting induced subgraphs. Discrete Appl. Math. **157**(17), 3540–3551 (2009)
18. Lévêque, B., Maffray, F., Trotignon, N.: On graphs with no induced subdivision of K_4. J. Comb. Theory Ser. B **102**(4), 924–947 (2012)
19. Matoušek, J., Thomas, R.: On the complexity of finding iso- and other morphisms for partial k-trees. Discrete Math. **108**(1–3), 343–364 (1992)
20. McConnell, R.M., Mehlhorn, K., Näher, S., Schweitzer, P.: Certifying algorithms. Comput. Sci. Rev. **5**(2), 119–161 (2011)
21. Olariu, S.: Paw-free graphs. Inf. Process. Lett. **28**(1), 53–54 (1988)
22. Oum, S.: Rank-width and vertex-minors. J. Comb. Theory Ser. B **95**(1), 79–100 (2005)
23. Oum, S.: Rank-width and well-quasi-ordering. SIAM J. Discrete Math. **22**(2), 666–682 (2008)

24. Oum, S.: Excluding a bipartite circle graph from line graphs. J. Graph Theory **60**(3), 183–203 (2009)
25. Oum, S.: Rank-width: algorithmic and structural results. Discrete Appl. Math. **231**, 15–24 (2017)
26. Oum, S., Seymour, P.D.: Approximating clique-width and branch-width. J. Comb. Theory Ser. B **96**(4), 514–528 (2006)
27. Oxley, J.G.: Matroid Theory. Oxford Graduate Texts in Mathematics, vol. 21, Second edn. Oxford University Press, Oxford (2011)
28. Robertson, N., Seymour, P.D.: Graph minors. XIII. The disjoint paths problem. J. Comb. Theory Ser. B **63**(1), 65–110 (1995)
29. van't Hof, P., Kamiński, M., Paulusma, D., Szeider, S., Thilikos, D.M.: On graph contractions and induced minors. Discrete Appl. Math. **160**(6), 799–809 (2012)

Saving Probe Bits by Cube Domination

Peter Damaschke$^{(\boxtimes)}$

Department of Computer Science and Engineering, Chalmers University,
41296 Göteborg, Sweden
ptr@chalmers.se

Abstract. We consider the problem of storing a single element from an m-element set as a binary string of optimal length, and comparing any queried string to the stored string without reading all bits. This is the one-element version of the problem of membership testing in the bit probe model, and solutions can serve as building blocks of general membership testers. Our principal contribution is the equivalence of saving probe bits with some generalized notion of domination in hypercubes. This domination variant requires that every vertex outside the dominating set belongs to a sub-hypercube, of fixed dimension, in which all other vertices belong to in the dominating set. This fixed dimension equals the number of saved probe bits. We give specific constructions showing that up to three probe bits can be ignored when m is far enough from the next larger power of 2. The main technical idea is to use low-dimensional (grid) relaxations of the problem. The design of optimal schemes remains an open problem, however one has to notice that even usual domination in hypercubes is far from being completely understood.

Keywords: Bit probe model · Dominating set · Hypercube

1 Introduction

1.1 The Setting

Consider a fixed set U of m elements. We wish to perform two actions:

(i) *Store a single element of U in memory.*
(ii) *For any $u \in U$ (given to us by some external questioner) answer the question whether the stored element equals u.*
    Memory may also be empty, and in this case the answer to (ii) should always be negative. An action being similar to (ii) is simply:
(iii) *Return the stored element.*

One can obviously use (iii) to do (ii), but for a negative answer to (ii) it may not be necessary to identify the stored element.

We suppose that memory consists of some number s of bits. Clearly, the smallest possible memory size for the above task is $s = \lceil \log(m+1) \rceil$, where

© Springer Nature Switzerland AG 2018
A. Brandstädt et al. (Eds.): WG 2018, LNCS 11159, pp. 139–151, 2018.
https://doi.org/10.1007/978-3-030-00256-5_12

$\log := \log_2$. Then every element of U can be represented by a unique string of s bits that may be stored. Furthermore it is trivial to recover the stored element by reading all s bits. However it may be sufficient to read a smaller number $t < s$ of bits.

1.2   Contributions

Only at first glance it might appear counterintuitive that not all bits need to be read. However, it suffices to probe a subset of bits that distinguishes the queried element from all other potentially stored elements, and in fact, $t < s$ is possible to achieve if m is far enough from the next larger power of 2. We will see that designing schemes that utilize this idea is equivalent to domination in punctured hypercubes. In a nutshell: Binary strings form a hypercube, but not all vertices need to be used to encode elements, and we can decide on the unused ones. When checking an element, we can stop reading more bits as soon as the already known bits lead to a unique encoding. The vertices described by the read bits form a sub-hypercube. This naturally defines a generalized domination problem in hypercubes that we call cube domination. Since already usual domination in hypercubes is difficult, we relax the problem to low-dimensional grids in order to be able to construct specific solutions. We can save up to 3 probe bits for certain large ranges of m. The main open question is how to prove lower bounds on t, or equivalently, upper bounds on m. Therefore our numerical results are likely to be improved by further research.

Other complexity models and other ways of implementing the same data structure (other than by binary strings) are beyond the scope of this paper.

1.3   Background

A more general task than the one considered here is to store a set of up to n elements and to answer set membership queries, i.e., tell whether a queried element is in the stored set or not. This fundamental data structure problem has been studied thoroughly [4,5,15,16].

The bit-probe model considers the number s of memory bits and the number t of bits that must be read (probed) in order to answer a query. In that model, memory access is assumed to be the expensive part, while the external computations needed to decide and decode the probes are a secondary concern. In general, so called succinct data structures are concerned with the trade-off between memory space and probes for answering queries or executing other data structure operations. There are more recent results on set membership queries in the bit-probe model in different ranges of the parameters [8,9,13].

Membership testers for small numbers n of elements can be relevant in various ways. For memory management reasons, a large membership tester may be split into many small blocks, where some hash function determines the blocks where elements are stored. The question of optimizing blocks for fixed small n was raised in [6]. In the present paper we consider only the case $n = 1$. However, testers for single elements can be building blocks of any membership tester that first finds the location in memory where the queried element would be stored, and then checks whether the stored candidate actually equals the queried element. Storing and testing single elements is also a relevant task in its own right, e.g., for authentication and access control of software agents. There, the stored element may be a secret key.

As mentioned, there is a trade-off between s and t. We focus on the smallest possible memory size s per element and ask how many probe bits can be saved nevertheless. In big data applications, memory space can be a bottleneck or an expensive resource.

The immediate relevance of a few saved probe bits might not be large, however for practical s they can make up some percent of the memory bits, and queries may be frequent, such that certain applications may become slightly faster. However the more interesting aspect is an understanding of the combinatorial nature of the problem, and the question which savings are possible at all. Domination is a classic notion in graph theory, and connections between properties of binary codes and the combinatorics of hypercubes are common.

In our notion of cube domination, every vertex not being in the dominating set must belong to some sub-hypercube of fixed dimension, in which all other vertices do belong to the dominating set. Some other, remotely similar generalizations of domination have been considered in the literature. One example are k-dominating sets, where every vertex outside the set must have at least k neighbors in the set. The concept of k-tuple dominating sets is similar. We refer to [1, 7, 11] for several approximation results. In our cube dominating sets, the "dominating" vertices are not all adjacent to the "dominated" ones, instead, they must form some low-diameter subgraph of a fixed shape that includes the dominated vertex. We define our notion only for hypercubes and their subgraphs. Similar concepts of subgraph domination in general graphs might also be interesting, should some motivation for them arise.

2 Preliminaries

To avoid the invention of private terminology we use the established notation for membership testers. Informally, an (m, n, s, t)-scheme is a data structure that can store, within s bits, up to n elements from a universe of m elements and test whether a queried element is among the stored ones, by reading at most t bits. In our case we always have $n = 1$, and hence the "data structure" shall answer question (ii) from the Introduction.

Upon a query, some *strategy* decides on the bits to read and eventually outputs an answer. An *adaptive* strategy decides sequentially which bit to read next, depending on the query and on the bits already seen.

Definition 1. *An $(m, 1, s, t)$-scheme consists of:*

- *a partitioning of the set $\{0, 1\}^s$ of binary strings of length s into a set E of m element strings, a set D of dummy strings, and a set $\{o\}$ containing only one special string (without loss of generality the zero string),*

- an injective function $\varphi : U \longrightarrow E$ that encodes the elements of a fixed set U with $|U| = m$ as element strings,
- a deterministic adaptive strategy that takes a string $e \in E \cup \{o\}$ and a queried element $u \in U$, reads at most t bits of e, and outputs "yes" if $e = \varphi(u)$, and "no" otherwise.

We may informally identify an element $u \in U$ with its element string $\varphi(u)$. The string o represents the empty set, that is, the state when no element is stored. For our worst-case results, adaptivity is not really needed (see also the later remark after Proposition 1). When checking an element, we expect specific bits at specific positions, hence we can also read these bits simultaneously. However, adaptivity can make a difference for the efficiency of negative tests.

Definition 2. *Given an $(m, 1, s, t)$-scheme, we define the following concepts and expressions, where x_i denotes the i-th symbol of a string x.*

- *A ternary string is a string x from $\{0, 1, *\}^s$. We call $*$ the wildcard and let $|x|$ denote the number of non-wildcard symbols.*
- *For $x, y \in \{0, 1, *\}^s$ we define $x \subset y$ and call x a substring of y, if $\forall i : x_i = y_i \lor x_i = *$.*
- *A ternary string $p \subset \varphi(u)$ is a probe for $u \in U$, if $p \subset e$ is false for all $e \in (E \cup \{o\}) \setminus \{u\}$.*
- *A probe p for u is a minimal probe if no ternary string q with $q \subset p$, $q \neq p$ is still a probe for u.*
- *To read a probe means to look up the symbols of the probe which are not wildcards.*

Now we obtain a more combinatorial characterization of $(m, 1, s, t)$-schemes and get rid of the consideration of a strategy. Remember that the zero string o represents the empty set and does not need a probe.

Proposition 1. *An $(m, 1, s, t)$-scheme is uniquely determined by φ, E, and some minimal probe $p(u) \subset \varphi(u)$ for every $u \in U$. Moreover, every $p(u)$ contains at least one bit 1. For every $u \in U$, the test whether the stored string is $\varphi(u)$ is done by reading $p(u)$ entirely. Consequently, $t = \max_{u \in U} |p(u)|$.*

Proof. In order to decide the presence of a queried element u, some minimal probe w of u must be read completely (both in the positive and negative case). The reason is that, by minimality, no deviating bit may be found, until the entire probe is read. We give the argument more formally: Let $w' \subset w$ with $|w'| = |w| - 1$. Since w' is not a probe of u, there exists another element $v \in U$ with $w' \subset \varphi(v)$. That is, $\varphi(u)$ and $\varphi(v)$ agree on w' and differ in the bit omitted from w. Without reading this bit, too, a tester cannot tell apart u and v. In particular, the tester cannot safely confirm or rule out the presence of u.

Finally observe that every minimal probe of an element must also contain some 1 in order to distinguish it from the zero string. □

While Definition 1 allowed adaptive reading, the above proof shows that the bits of a minimal probe can be read in any order, even nonadaptively, since we must read them all. In the adaptive setting, reading the probe bits in random order reveals a deviating bit in expected time at most $t/2$, hence negative answers can be given faster, but not positive ones.

Finally, we define $L := \lfloor \log m \rfloor$. Note that $s = L + 1$ is the optimal space that can be used by any $(m, 1, s, t)$-scheme.

3   Cube Domination in Hypercubes

The following graph-theoretic definitions are quite common.

The (Hamming) *weight* of a binary string x is the number of 1s in that string: $|\{i : x_i = 1\}|$. The (Hamming) *distance* between two binary strings x and y of equal lengths is the number of different bit positions: $|\{i : x_i \neq y_i\}|$. The s-*dimensional hypercube* Q_s is the graph with vertex set $\{0, 1\}^s$ where two vertices are adjacent if and only if their Hamming distance is 1. The *sub-hypercube* of Q_s described by the ternary string $x \in \{0, 1, *\}^s$ is the subgraph induced by all vertices $y \in \{0, 1\}^s$ with $x \subset y$. The *punctured s-dimensional hypercube* Q_s^o is Q_s after removal of one vertex and its indicent edges.

Note that we do not strictly distinguish between elements u, their strings $\varphi(u)$, and their corresponding vertices in the hypercube Q_s; this should not cause confusion. Therefore we may also speak of *element vertices* and *dummy vertices* in Q_s, corresponding to the element and dummy strings, respectively, as introduced in Definition 1.

A subset D of vertices in a graph is called *dominating* if every vertex outside D has a neighbor in D. The *domination number* $\gamma(G)$ of a graph G is the size of a minimum dominating set. We say that a vertex is *dominated* by itself and by its neighbors.

In hypercubes and their subgraphs we will now generalize the notion of domination. As far as we know, this concept is novel. Note that the special case $\ell = 1$ is usual domination in hypercubes.

Definition 3. *In a hypercube Q_s or in an induced subgraph of Q_S, we call a set $D \subseteq \{0, 1\}^s$ an ℓ-cube dominating set if for every vertex $u \notin D$ there exists an ℓ-dimensional sub-hypercube consisting of u and $2^\ell - 1$ vertices from D. We say that u is ℓ-cube dominated by these vertices from D. We call a vertex $v \in D$ redundant if every $u \notin D$ is already ℓ-cube dominated by some $2^\ell - 1$ vertices from $D \setminus \{v\}$.*

Informally, a redundant vertex is not needed to ℓ-cube dominate any other vertex. The connection of cube domination to our bit-probe number problem is given by the following statement that has a straightforward proof.

Proposition 2. *An $(m, 1, L+1, L+1-\ell)$-scheme exists if and only if the vertex set of the punctured $(L + 1)$-dimensional hypercube can be divided into a set E of m element vertices and a set D of $2^{L+1} - 1 - m$ dummy vertices, such that D is an ℓ-cube dominating set. Specifically, a probe for any element $u \in U$ can be chosen as the ternary string of an ℓ-dimensional sub-hypercube which ℓ-cube dominates u.*

The final lemma in this section is also simple but will be a useful tool for concrete constructions of schemes. Techniques for the extension of schemes to larger sizes are also known for general membership testers.

Lemma 1. *Let D be an ℓ-cube dominating set in the full $(L+1)$-dimensional hypercube Q_{L+1} with $|D| = 2^{L+1} - m$, hence with exactly m vertices outside D. Furthermore, let f be any positive integer. Then we have:*

(i) *There exists an $(m-1, 1, L+1, L+1-\ell)$-scheme. If D has some redundant vertex, then there exists an $(m, 1, L+1, L+1-\ell)$-scheme.*

(ii) *There exists a $(2^f m - 1, 1, L+1+f, L+1+f-\ell)$-scheme. If D has some redundant vertex, then there exists a $(2^f m, 1, L+1+f, L+1+f-\ell)$-scheme.*

Proof. (i) We remove some vertex not being in D, to get a punctured hypercube with dummy vertices in D and $m-1$ element vertices. Since no vertex of D has been removed, D is still ℓ-cube dominating, and Proposition 2 yields the first assertion. If some vertex in D is redundant, we remove this vertex instead, and retain m element vertices.

(ii) We append to every string f further bits, in all 2^f possible ways. If the original string was an element (dummy) string, then we let all its 2^f extensions be element (dummy) strings as well. In other words, we glue together 2^f copies of the given hypercube equipped with the given set D. Every element vertex is still ℓ-cube dominated within its own copy, and redundant vertices stay redundant. Now both assertions follow as above.   □

4   Ignoring One Bit

Now we use the equivalence to cube domination to obtain concrete $(m, 1, s, t)$-schemes with $s = L+1$ and $t < s$, that is, schemes that use optimal space but need not read all bits. First we deal with $t = L$, that is, $\ell = 1$ and usual dominatiom, according to Proposition 2.

We have to state that the knowledge of exact domination numbers of (full) hypercubes is rather fragmentary. Only very few domination numbers are known [2, 3, 10, 12, 14, 17]:

Theorem 1. *The domination number $\gamma(Q_{L+1})$ of the full $(L+1)$-dimensional hypercube is:*

- *7, 12, 16, 32, 62, for $L = 4, 5, 6, 7, 8$, respectively,*
- *2^{L+1-k} for $2^k - 2 \leq L \leq 2^k - 1$,*
- *at most 2^{L-2} for every $L > 6$.*

Let $G - v$ denote the graph G after removal of vertex v and its incident edges. A very simple fact is:

Proposition 3. *For every graph G we have $\gamma(G) - 1 \leq \gamma(G - v) \leq \gamma(G)$.*

Specifically we get the following results to start with.

Proposition 4. *There exists a $(5, 1, 3, 2)$-scheme, but no $(6, 1, 3, 2)$-scheme.*

Proof. We have $\gamma(Q_3^\circ) = 2$: One example of a minimum dominating set is given by $D = \{100, 011\}$, which yields a $(5, 1, 3, 2)$-scheme that uses the element strings $010, 001, 110, 101, 111$ and the corresponding probes $01*, 0*1, 1*0, 10*, *11$. Due to Proposition 2, a $(6, 1, 3, 2)$-scheme would require a single dominating vertex, which can however dominate only 4 of the 7 vertices. $\square$

Proposition 5. *There exists an $(11, 1, 4, 3)$-scheme, but no $(12, 1, 4, 3)$-scheme.*

Proof. Similar but slightly more complex; omitted due to space limitations. $\square$

Similarly, the known domination numbers from Theorem 1 yield a $(24, 1, 5, 4)$-scheme, a $(51, 1, 6, 5)$-scheme, a $(111, 1, 7, 6)$-scheme, and so on, but in these cases we could not figure out whether m can be improved by 1 (which might be possible due to Proposition 3). However, a more rewarding question than closing these tiny gaps is the bit probe number for general L:

Theorem 2. *For every $L \geq 6$ and every $m < 1.75 \cdot 2^L$, there exists an $(m, 1, L + 1, L)$-scheme.*

Proof. By Theorem 1, the $(L + 1)$-dimensional hypercube, and thus also the punctured one, has a dominating set of size 2^{L-2}. Due to Proposition 2, the result now follows for $m = 2^{L+1} - 1 - 2^{L-2}$ and for all smaller numbers m. $\square$

An open problem is whether the factor 1.75 can be further increased for larger numbers L. It might even tend to 2 for $L \to \infty$; we do not know a non-trivial bound. According to Theorem 1, this problem is equivalent to the notoriously difficult domination numbers of hypercubes. Therefore we do not investigate this problem further and look into the opposite direction instead: We show that even more probe bits can be saved when the ratios $m/2^L$ are somewhat smaller.

5   Ignoring Two Bits

5.1   Approach

In this section we will construct $(m, 1, L + 1, L - 1)$-schemes. First we give some intuition. According to Proposition 2 applied to $\ell = 2$, every element string must belong to a *quadrangle* (2-dimensional sub-hypercube) together with three dummy vertices. On the other hand, since $s = L + 1$, the majority of vertices must be element vertices. To satisfy these seemingly conflicting requirements, any $(m, 1, L + 1, L - 1)$-scheme needs conglomerates of such quadrangles that share many dummy vertices but include even more different element vertices. In this situation, our key observation is that a star of dummy vertices, consisting of one "central" vertex and $k \leq L+1$ of its neighbors, can indeed form quadrangles with up to $\binom{k}{2}$ element vertices at distance 2 from the central vertex. Since $\binom{k}{2} > k+1$ for $k \geq 4$, the element vertices form the majority. However, the challenge is to pave the entire hypercube by such stars, thereby maximizing m.

Next, a practical approach to reduce the search space for this puzzle, possibly at the price of getting only suboptimal m, is to map the hypercube to a low-dimensional grid and treat all vertices with the same image equally, i.e., make them either dummy or element vertices. In other words, we relax our problem to some weighted problem in homomorphic images of hypercubes. (Remember that even the usual domination problem in hypercubes is not well understood, hence suboptimal yet "positive" results for $\ell = 2$ should be valuable.) Below we will develop the technical details of our constructions. First we define several special objects. Recall that $s = L + 1$.

Definition 4. *Let π be a partitioning of the ordered set of bit positions $\{1, \ldots, s\}$ into g segments of $s_1, \ldots, s_g$ positions, where $\forall i : s_i > 0$ and $\sum_{i=1}^{g} s_i = s$. Given π, we set up a g-dimensional grid of points with integer coordinate vectors $(x_1, \ldots, x_g)$, where $\forall i : 0 \leq x_i \leq s_i$. We also partition the $\prod_{i=1}^{g}(s_i + 1)$ grid points into a set $\tilde{D}$ of dummy points and a set $\tilde{E}$ of element points.*

Finally, we define an $(m, 1, s, t)$-scheme from π, $\tilde{D}$, $\tilde{E}$ as follows. We map every vertex v of the punctured s-dimensional hypercube (binary string v of s bits) to the grid point $(x_1, \ldots, x_g)$, where x_i is the number of $1s$ in the i-th segment of the string v. We let D and E be the set of strings mapped onto $\tilde{D}$ and $\tilde{E}$, respectively.

Recall from Proposition 1 that an $(m, 1, s, t)$-scheme is characterized by the set of the element strings and their minimal probes. In Definition 4 we haven't yet specified the probes (and t), but we will do this by cube domination, using Proposition 2. Due to Lemma 1 we consider full hypercubes and redundant vertices in D. For $\ell = 2$ we will now derive a condition on the dummy points that is sufficient to meke D a 2-cube dominating set.

For this purrpose, we can alternatively view a grid as a graph where two vertices (grid points) are adjacent if and only if their Euclidean distance is 1. We call a *straight path* a path of grid points where all coordinates but one are equal (figuratively, a path without bends). A *quadrangle* in the grid is a subgraph of four grid points that form a cycle.

Lemma 2. *For any scheme defined by a grid as specified in Definition 4, the following statements are equivalent:*

- *Every element point in the grid is an end point of a straight path with three points, where the two others are dummy points, or belongs to a quadrangle with three dummy points.*
- *The set of dummy vertices is 2-cube dominating in Q_s.*

The two cases are illustrated here, with dummy points and element points displayed as * and e, respectively:

$$\boxed{*\ *\ e} \qquad \boxed{\begin{array}{ll} * & * \\ * & e \end{array}}$$

Proof. Consider any quadrangle in the hypercube. Its four strings have the form $u0v0w$, $u0v1w$, $u1v0w$, $u1v1w$, where u, v, w are substrings that may be empty. Remember the notion of segment from Definition 4.

If the two changing bits belong to the same segment, then the four elements are mapped to grid points whose coordinates except one are constant, and the changing coordinate has the values i, $i + 1$, $i + 1$, $i + 2$, for some integer i. If the two changing bits belong to different segments, then the four elements are mapped to grid points whose coordinates except two are constant, and the changing coordinates have the values (i, j), $(i + 1, j)$, $(i, j + 1)$, $(i + 1, j + 1)$, for some integers i and j.

Suppose that the set of dummy vertices is 2-cube dominating. Then it follows: Every element vertex is in some quadrangle with three vertices, and the image of this quadrangle is either a straight path with three points or a quadrangle in the grid, as seen above. In the former case, the element vertex is an end point of the path, since the inner point represents two vertices which cannot be one element and one dummy vertex.

This shows one direction of the equivalence. For proving the opposite direction, consider any element vertex and its image point p in the grid.

Let p be the end of a straight path with dummy points q and r. The strings of vertices mapped to the path differ in only one segment σ. Let the number of 1s in p, q, r in σ be i, $i+1$, $i+2$, respectively. We choose any two positions of bits in σ that are 0 in the considered string mapped to p. Such positions exist since r represents all strings with $i + 2$ bits 1 in σ, in particular, the length of σ is at least $i + 2$. We change any one of these 0s into 1 to obtain two strings mapped to q, and we change both 0s into 1 to obtain a string mapped to r. These three strings are dummy strings and together 2-cube dominate our considered element string. The case where the strings in p, q, r have $i + 2$, $i + 1$, i, respectively, bits 1s in σ is symmetric. There we change two 1s into 0s instead.

It should be evident that the reasoning for quadrangles in the grid is similar, except that only one bit is changed in each of two segments. □

5.2   Results

We are ready to derive specific results on $(m, 1, L + 1, L - 1)$-schemes, using the above tools plus further heuristics (see below).

First we mention that no $(8, 1, 4, 2)$-scheme exists, hence no $(9, 1, 4, 2)$-scheme etc. This can be shown by tedious exhaustive case examinations, but we have not found a concise proof.

However, for sizes $L \geq 4$ we do obtain $(m, 1, L + 1, L - 1)$-schemes. Note that we want to maximize m for a given L, and that any $(m, 1, L + 1, L - 1)$-scheme trivially implies $(m', 1, L + 1, L - 1)$-schemes for all $m' < m$.

The plan is to place the dummy points preferably close to the borders and corners, where the element numbers are small. But we must also observe the condition from Lemma 2 and "reach" all grid points. We first try to dominate the heavy central points cheaply.

As a convenient notation, we display grids by indicating the numbers of strings mapped to each grid point. From Definition 4 it is obvious that they are certain binomial coefficients or products thereof. We mark each dummy point by an asterisk (*).

Proposition 6. *There exists a* $(19, 1, 5, 3)$*-scheme.*

Proof. Apply Lemma 1 (i), Lemma 2, and the following 1-dimensional grid.

$$\boxed{*1 \; *5 \; 10 \; 10 \; *5 \; *1}$$

Note that we get only $m = 19$, due to lack of a redundant dummy vertex. $\square$

Corollary 1. *There exist a* $(39, 1, 6, 4)$*-scheme, a* $(79, 1, 7, 5)$*-scheme, a* $(159, 1, 8, 6)$*-scheme, and a* $(319, 1, 9, 7)$*-scheme.*

Proof. Apply Lemma 1 (ii) to the scheme from Proposition 6. $\square$

The next item would be a $(639, 1, 10, 8)$-scheme, but this can be improved:

Proposition 7. *There exists a* $(723, 1, 10, 8)$*-scheme.*

Proof. Apply Lemma 1 (i), Lemma 2, and the following 2-dimensional grid.

$$\begin{array}{|cccccc|}
\hline
1 & *5 & *10 & *10 & 5 & 1 \\
5 & 25 & *50 & 50 & 25 & *5 \\
*10 & 50 & 100 & 100 & *50 & *10 \\
*10 & *50 & 100 & 100 & 50 & *10 \\
*5 & 25 & 50 & *50 & 25 & 5 \\
1 & 5 & *10 & *10 & *5 & 1 \\
\hline
\end{array}$$

Indeed, all element points satisfy the condition in Lemma 2. $\square$

Next we might apply Lemma 1 (ii) once more and obtain a $(1447, 1, 11, 9)$-scheme. But we can raise m somewhat. Note that a new pattern of dummy points in the grid emerges:

Proposition 8. *There exists a* $(1464, 1, 11, 9)$*-scheme.*

Proof. Apply Lemma 1 (i), Lemma 2, and the following 2-dimensional grid.

$$\begin{array}{|cccccc|}
\hline
*1 & 6 & *15 & *20 & *15 & 6 & *1 \\
5 & 30 & 75 & *100 & 75 & 30 & 5 \\
*10 & *60 & 150 & 200 & 150 & *60 & *10 \\
*10 & *60 & 150 & 200 & 150 & *60 & *10 \\
5 & 30 & 75 & *100 & 75 & 30 & 5 \\
*1 & 6 & *15 & *20 & *15 & 6 & *1 \\
\hline
\end{array}$$

Note as a minor detail that the dummy vertices at the corners are redundant, so we can remove one of them. $\square$

Finally we get a general result for saving two probe bits. It follows instantly from Lemma 1, Proposition 8, and $1464/1024 > 1.42$.

Theorem 3. *For every $L \geq 10$ and every $m < 1.42 \cdot 2^L$, there exists an $(m, 1, L + 1, L - 1)$-scheme.*

Proof. This follows now instantly from Lemma 1 and Proposition 8, noticing that $1464/1024 > 1.42$. □

Once more, the factor 1.42 is probably not the last word and may be raised further. We remark that we have not found better values of m via grids of dimension $g > 2$. An explanation might be that the quadrangles themselves are 2-dimensional, such that higher-dimensional grids might not give better opportunities to glue them.

5.3   Ignoring Even More Bits

Next we can show that even some $(m, 1, L + 1, L - 2)$-schemes exist. Similarly as in case $\ell = 2$, the condition for a scheme described by a 2-dimensional grid is that every element point is the end of a straight path of four points (a (1×4)-subgrid) or the corner of some (2×3)-subgrid, where all other points in the respective subgrid are dummy points. The proof is analogous to Lemma 2.

Theorem 4. *For every $L \geq 7$ and every $m < 1.09 \cdot 2^L$, there exists an $(m, 1, L + 1, L - 2)$-scheme.*

Proof. The following 1-dimensional grid describes a $(69, 1, 7, 4)$-scheme (where we remove one of the $35 + 35$ element vertices in the punctured hypercube).

$$\boxed{*1 \ *7 \ *21 \ 35 \ 35 \ *21 \ *7 \ *1}$$

Note that every element point is the end of a path with three dummy points. Since $70/64 > 1.09$, Lemma 1 implies the assertion for every larger L. □

Again, it could be interesting to raise the factor 1.09. We conclude with a

Conjecture. There exist no $(m, 1, L + 1, L - 3)$-schemes.

We were led to this conjecture by reasoning on the growth of binomial coefficients, and from failed attempts to construct such schemes from grids despite much experimentation. But actually it is open whether $(m, 1, L + 1, L + 1 - \ell)$-schemes are possible even for arbitrarily large ℓ. A weaker conjecture is that no such schemes exist from some fixed ℓ on.

6   Conclusions

We have constructed several non-obvious schemes for storing one out of m elements, that use optimal space and save some probe bits when querying the stored element. They work for large ranges of m. Although being based on fairly general ideas (cube domination in hypercubes, low-dimensional relaxation, and doubling), the specific results have been found in an ad-hoc way. It could be nice to find good general patterns of dummy points.

The impact of extra memory bits ($s = L + 2$, $s = L + 3$, etc.) was beyond our scope. Of course, larger probe bit savings would not be surprising there, but it remains the (difficult) question of exactly quantifying them. Some other open questions were already mentioned. Finally, these schemes for single elements could be used inside membership testers for subsets of up to n elements.

Acknowledgment. Special thanks go to the anonymous reviewer who pointed out additional references around Theorem 1.

References

1. Argiroffo, G.R., Leoni, V., Torres, P.: On the complexity of k-domination and k-tuple domination in graphs. Info. Proc. Lett. **115**, 556–561 (2015)
2. Arumugam, S., Kala, R.: Domination parameters of hypercubes. J. Indian Math. Soc. **65**, 31–38 (1998)
3. Azarija, J., Henning, M.A., Klavzar, S.: (Total) domination in prisms. Electron. J. Combin. **24**, paper 1.19 (2017)
4. Buhrman, H., Miltersen, P.B., Radhakrishnan, J., Venkatesh, S.: Are bitvectors optimal? SIAM J. Comput. **31**, 1723–1744 (2002)
5. Carter, L., Floyd, R.W., Gill, J., Markowsky, G., Wegman, M.N.: Exact and approximate membership testers. In: Lipton, R.J., et al. (eds.) STOC 1978, pp. 59–65. ACM (1978)
6. Fan, B., Andersen, D.G., Kaminsky, M., Mitzenmacher, M.: Cuckoo filter: practically better than bloom. In: Seneviratne, A., et al. (eds.) CoNEXT 2014, pp. 75–88. ACM (2014)
7. Förster, K.T.: Approximating fault-tolerant domination in general graphs. In: Nebel, M.E., Szpankowski, W. (eds.) ANALCO 2013, pp. 25–32. SIAM (2013)
8. Garg, M., Radhakrishnan, J.: Set membership with a few bit probes. In: Indyk, P. (ed.) SODA 2015, pp. 776–784. ACM-SIAM (2015)
9. Garg, M., Radhakrishnan, J.: Set membership with non-adaptive bit probes. In: Vollmer, H., Vallée, B. (eds.) STACS 2017. LIPIcs, vol. 66, paper 38, Dagstuhl (2017)
10. Harary, F., Livingston, M.: Independent domination in hypercubes. Appl. Math. Lett. **6**, 27–28 (1993)
11. Klasing, R., Laforest, C.: Hardness results and approximation algorithms of k-tuple domination in graphs. Info. Proc. Lett. **89**, 75–83 (2004)
12. Klavzar, S., Ma, M.: The domination number of exchanged hypercubes. Info. Proc. Lett. **114**, 159–162 (2014)
13. Lewenstein, M., Munro, J.I., Nicholson, P.K., Raman, V.: Improved explicit data structures in the Bitprobe model. In: Schulz, A.S., Wagner, D. (eds.) ESA 2014. LNCS, vol. 8737, pp. 630–641. Springer, Heidelberg (2014). https://doi.org/10.1007/978-3-662-44777-2_52
14. Östergård, P.R.J., Blass, U.: On the size of optimal binary codes of length 9 and covering radius 1. IEEE Trans. Inform. Theory **47**, 2556–2557 (2001)

15. Pagh, R.: On the cell probe complexity of membership and perfect hashing. In: Vitter, J.S., Spirakis, P.G., Yannakakis, M. (eds.) STOC 2001, pp. 425–432. ACM (2001)

16. Raman, R., Raman, V., Satti, S.R.: Succinct indexable dictionaries with applications encoding k-ary trees, prefix sums and multisets. ACM Trans. Algor. **3**, paper 43 (2007)

17. van Wee, G.J.M.: Improved sphere bounds on the covering radius of codes. IEEE Trans. Inform. Theory **34**, 237–245 (1988)

Graph Amalgamation Under Logical Constraints

Mateus de Oliveira Oliveira$^{(\boxtimes)}$

University of Bergen, Postboks 7803, 5020 Bergen, Norway
`mateus.oliveira@uib.no`

Abstract. We say that a graph G is an H-amalgamation of graphs G_1 and G_2 if G can be obtained by gluing G_1 and G_2 along isomorphic copies of H. In the AMALGAMATION RECOGNITION problem we are given connected graphs H, G_1, G_2, G and the goal is to determine whether G is an H-amalgamation of G_1 and G_2. Our main result states that AMALGAMATION RECOGNITION can be solved in time $2^{O(\Delta \cdot t)} \cdot n^{O(t)}$ where n, t, Δ are the number of vertices, the treewidth and the maximum degree of G respectively.

We generalize the techniques used in our algorithm for H-amalgamation recognition to the setting in which some of the graphs H, G_1, G_2, G are not given explicit at the input but are instead required to satisfy some topological property expressible in the counting monadic second order logic of graphs (CMSO logic). In this way, when restricting ourselves to graphs of constant treewidth and degree our approach generalizes certain algorithmic decomposition theorems from structural graph theory from the context of clique-sums to the context in which the interface graph H is given at the input.

Keywords: Graph amalgamation · CMSO logic · Logical constraints

1 Introduction

Amalgamation of graphs and related structures have been studied for at least four decades and have been used as a crucial tool in many branches of graph theory and combinatorics [14,15,18,25,28]. In particular, already in the special case where the interface graph H is a clique, the notion of H-amalgamation has played a major role in the development of structural graph theory [17,19,26], in the study of the chromatic number of graphs [17] and in algorithmics [10].

Definition 1.1 (H-Amalgamation). *A graph G is an H-amalgamation of graphs G_1 and G_2, if there exist injective morphisms[1] $\mu : H \to G$, $\{\mu_i : H \to G_i\}_{i \in \{1,2\}}$ and $\{\eta_i : G_i \to G\}_{i \in \{1,2\}}$ such that $G = \eta_1(G_1) \cup \eta_2(G_2)$ and $\mu = \eta_i \circ \mu_i$ for each $i \in \{1,2\}$.*

---

[1] Graph morphisms will be properly defined in Sect. 2.

© Springer Nature Switzerland AG 2018
A. Brandstädt et al. (Eds.): WG 2018, LNCS 11159, pp. 152–163, 2018.
https://doi.org/10.1007/978-3-030-00256-5_13

Intuitively, G is an H-amalgamation of G_1 and G_2 if G can be obtained by identifying G_1 and G_2 along isomorphic copies of H. In this sense, we may think of H as an interface between G_1 and G_2. Since G_1 and G_2 may have many distinct subgraphs isomorphic to H, the injective morphisms μ_1 and μ_2 are used to specify which of these subgraphs will be used as an interface between G_1 and G_2. The morphisms μ, η_1, η_2 together with the conditions $G = \eta_1(G_1) \cup \eta_2(G_2)$, $\mu = \eta_1 \circ \mu_1$ and $\mu = \eta_2 \circ \mu_2$ are used to formalize the intuition that the graph G is obtained by identifying the image of μ_1 in G_1 with the image of μ_2 in G_2.

Perhaps the most natural question that arises when considering the notion of H-amalgamation of graphs is the question of recognizing whether a given graph G is an H-amalgamation of graphs G_1, G_2. Below, we write $|G|$ to denote the number of vertices of G.

AMALGAMATION RECOGNITION
INPUT: Connected graphs H, G_1, G_2 and G where $|H| \leq |G_1|, |G_2| \leq |G|$.
OUTPUT: YES if G is an H-amalgamation of G_1 and G_2 and NO otherwise.

Our main result (Theorem 3.8) states that AMALGAMATION RECOGNITION can be solved in time $2^{O(\Delta \cdot t)} \cdot n^{O(t)}$ where n, t, Δ, are the number of vertices, the treewidth and the maximum degree of the graph G respectively. We note that AMALGAMATION RECOGNITION trivially generalizes SUBGRAPH ISOMORPHISM (by setting $G = G_1$ and $H = G_2$), and therefore our algorithm also shows that this latter problem can be solved in time $2^{O(\Delta \cdot t)} \cdot n^{O(t)}$. This gives an alternative to the celebrated algorithm of Matoušek and Thomas [24] which works in time $f(\Delta') \cdot n^{O(t)}$ for some fast growing function f in the maximum degree Δ' of the interface graph H. Under the assumption that $P \neq NP$, there is no algorithm that solves SUBGRAPH ISOMORPHISM in time $f(t) \cdot n^{O(1)}$. This is because SUBGRAPH ISOMORPHISM is already NP complete for $t = 4$ [23]. Additionally, under the assumption that $W[1] \neq FPT$, there is no algorithm for SUBGRAPH ISOMORPHISM that runs in time $f_1(t) \cdot |G|^{f_2(t')}$ where f_1, f_2 are computable functions, t is the treewidth of G and t' is the treewidth of H [23].

Building on our main result we devise algorithms for variants of the AMALGAMATION RECOGNITION problem where either the target graph G, or the interface graph H is not explicitly given at the input, but is instead required to satisfy some property expressible in the counting monadic second-order logic of graphs (CMSO logic) [4–6,8]. This logic is expressive enough to define several interesting graph properties, such as planarity, k-outerplanarity, embeddability on a surface of genus k, k-connectivity, k-colorability, Hamiltonicity [4]. Additionally, on graphs of constant treewidth[2], this logic can be used to define several graph polynomials of interest in topological combinatorics, such as the Tutte's polynomial, the Farrel polynomial, the Jones and Kauffman polynomials for knot diagrams, etc. [20–22].

---

[2] Definability of these polynomials require access to a total ordering of the edge set of the graph. On graphs of constant treewidth such orderings are by themselves CMSO definable.

We note that the problem of decomposing a graph G into a clique-sum of a graph G_1 satisfying a CMSO property Π_1 and a graph G_2 satisfying a CMSO property Π_2 has found many important applications in the field of structural graph theory [11,16,17,27]. One of our results (Theorem 4.4) can be regarded as a generalization of this type of problem to the context in which the interface graph is given at the input.

2 Preliminaries

For each $n \in \mathbb{N}$, we let $[n] = \{1, ..., n\}$. As a degenerate case, we let $[0] = \emptyset$.

Graphs: A *graph* is a triple $G = (V_G, E_G, \mathrm{Inc}_G)$ where V_G is a set of vertices, E_G is a set of edges, and $\mathrm{Inc}_G \subseteq E_G \times V_G$ is an incidence relation. For each $e \in E_G$ we let $endpts(e) = \{v \mid \mathrm{Inc}_G(e, v)\}$ be the set of endpoints of e, and we assume that $|endpts(e)|$ is either 0 or 2. We note that our graphs are allowed to have multiple edges, but no loops. We say that a graph H is a subgraph of G if $V_H \subseteq V_G$, $E_H \subseteq E_G$ and $\mathrm{Inc}_H = \mathrm{Inc}_G \cap E_H \times V_H$. Alternatively, we say that G is a supergraph of H. The degree of a vertex $v \in V_G$ is the number $d(v)$ of edges incident with v. We let $\Delta(G)$ denote the maximum degree of a vertex of G.

A *path* in a graph G is a sequence $v_1 e_1 v_2 ... e_{n-1} v_n$ where $v_i \in V_G$ for $i \in [n]$, $e_i \in E_G$ for $i \in [n-1]$, $v_i \neq v_j$ for $i \neq j$, and $\{v_i, v_{i+1}\} = endpts(e_i)$ for each $i \in [n-1]$. We say that G is *connected* if for every two vertices $v, v' \in V_G$ there is a path whose first vertex is v and whose last vertex is v'.

Let G and H be graphs. A morphism from H to G is a pair of functions $\mu = (\dot{\mu}: V_H \to V_G, \overline{\mu} : E_H \to E_G)$ such that for every $e \in E_H$, if $endpts(e) = \{v, v'\}$ then $endpts(\overline{\mu}(e)) = \{\dot{\mu}(v), \dot{\mu}(v')\}$. We say that μ is injective if both $\dot{\mu}$ and $\overline{\mu}$ are injective. As an abuse of notation we may write $\mu : H \to G$ to denote a morphism from H to G. Additionally, we may write $\mu(H)$ to denote the subgraph of G formed by vertices $\dot{\mu}(V_H)$ and edges $\overline{\mu}(E_H)$.

Terms. Let $[r]^*$ denote the set of all strings over $[r]$ and let λ denote the empty string. A subset $N \subseteq [r]^*$ is *prefix closed* if for every $p \in [r]^*$ and every $j \in [r]$, $pj \in N$ implies that $p \in N$. We note that the empty string λ is an element of any prefix closed subset of $[r]^*$. We say that $N \subseteq [r]^*$ is *well numbered* if for every $p \in [r]^*$ and every $j \in [r]$, the presence of pj in N implies that $p1, ..., p(j-1)$ also belong to N. We say that a subset $N \subseteq [r]^*$ is *tree-like* if N is both prefix-closed and well-numbered.

Let Σ be a finite set of symbols. An *r-ary term* over Σ is a function $\tau : N \to \Sigma$ whose domain N is a tree-like subset of $[r]^*$. We may write $Pos(\tau)$ to denote the domain of τ. We say that τ is a *binary term* if τ is a 2-ary term. We use $Ter(\Sigma)$ to denote the set of all terms over Σ. If $\tau_1, \tau_2, ..., \tau_k$ are terms in $Ter(\Sigma)$ and $a \in \Sigma$, then we let $\tau = a(\tau_1, ..., \tau_k)$ be the term where $\tau[\lambda] = a$ and for each $jp \in Pos(\tau)$, $\tau[jp] = \tau_j[p]$.

Tree Automata: Let Σ be a finite set of symbols. A *bottom-up tree automaton* over Σ is a tuple $\mathcal{A} = (Q, \Sigma, F, \mathfrak{R})$ where Q is a set of states, $F \subseteq Q$ a set of final states, and $\mathfrak{R}$ is a set of transitions of the form $(\mathfrak{q}_1, ..., \mathfrak{q}_r, a, \mathfrak{q})$ with $a \in \Sigma$,

$0 \leq r \leq 2$, and $\mathfrak{q}_1, ..., \mathfrak{q}_r, \mathfrak{q} \in Q$. The size of $\mathcal{A}$, which is defined as $|\mathcal{A}| = |Q| + |\mathfrak{R}|$, measures the number of states in Q plus the number of transitions in $\mathfrak{R}$. The set $\mathcal{L}(\mathcal{A}, \mathfrak{q}, i)$ of all terms reaching a state $\mathfrak{q} \in Q$ in depth at most i is inductively defined as follows: If $(a, \mathfrak{q})$ is a transition in $\mathfrak{R}$, then a reaches state $\mathfrak{q}$ in depth 1. If $(\mathfrak{q}_1, ..., \mathfrak{q}_k, a, \mathfrak{q})$ is a transition in $\mathfrak{R}$, and $\tau_1, ..., \tau_k$ are terms in $Ter(\Sigma)$ such that τ_j reaches state $\mathfrak{q}_j$ in depth at most i for each $j \in [k]$, then the term $a(\tau_1, ..., \tau_k)$ reaches $\mathfrak{q}$ in depth $i+1$. The language accepted by $\mathcal{A}$, denoted by $\mathcal{L}(\mathcal{A})$, is defined as the set of terms in $Ter(\Sigma)$ that reach some state $\mathfrak{q} \in F$ in any finite depth.

Let $\pi : \Sigma \to \Sigma'$ be a map between finite sets of symbols Σ and Σ'. Such mapping can be homomorphically extended to a mapping $\pi : Ter(\Sigma) \to Ter(\Sigma')$ between terms by setting $\pi(\tau)[p] = \pi(\tau[p])$ for each position $p \in Pos(\tau)$. Additionally, π can be further extended to a set of terms $\mathcal{L} \subseteq Ter(\Sigma)$ by setting $\pi(\mathcal{L}) = \{\pi(\tau) \mid \tau \in Ter(\Sigma)\}$. Below we state some well known closure and decidability properties for tree automata.

Lemma 2.1 (Properties of Tree Automata [3]). *Let Σ and Σ' be finite sets of symbols. Let $\mathcal{A}_1$ and $\mathcal{A}_2$ be tree automata over Σ, and $\pi : \Sigma \to \Sigma'$ be a mapping.*

1. *One can construct in time $O(|\Sigma|)$ a tree automaton $\mathcal{A}(\Sigma)$ accepting the language $Ter(\Sigma)$.*
2. *One can construct in time $O(|\mathcal{A}_1| \cdot |\mathcal{A}_2|)$ a tree automaton $\mathcal{A}_1 \cap \mathcal{A}_2$ such that $\mathcal{L}(\mathcal{A}_1 \cap \mathcal{A}_2) = \mathcal{L}(\mathcal{A}_1) \cap \mathcal{L}(\mathcal{A}_2)$.*
3. *One can determine whether $\mathcal{L}(\mathcal{A}_1) \neq \emptyset$ in time $|\mathcal{A}_1|^{O(1)}$.*
4. *One can construct in time $O(|\mathcal{A}_1|)$ a tree automaton $\pi(\mathcal{A}_1)$ such that $\mathcal{L}(\pi(\mathcal{A}_1)) = \pi(\mathcal{L}(\mathcal{A}_1))$.*

2.1 Concrete Tree Decompositions

In this section we define the notion of t-concrete tree decomposition of a graph following closely the exposition in [9]. Intuitively, a t-concrete tree decomposition may be regarded as a way of representing a graph together with one of its tree decompositions. Such a representation is convenient because it allows one to represent families of graphs of constant treewidth via tree automata that accept concrete decompositions. We note that similar ideas are widespread in texts dealing with recognizable properties of graphs [1,2,7,12,13].

A t-concrete bag is a pair $\mathbf{B} = (B, b)$ where $B \subseteq [t]$, and $b \subseteq B$ with $b = \emptyset$ or $|b| = 2$. We note that B is allowed to be empty. We let $\mathcal{B}(t)$ be the set of all t-concrete bags[3]. Note that $|\mathcal{B}(t)| \leq t^2 \cdot 2^t$. We regard the set $\mathcal{B}(t)$ as a finite

---

[3] We note that in texts dealing with similar notions of decomposition, it is customary to define a bag of width t as a graph with at most t vertices together with a function that labels the vertices of these graphs with numbers from $\{1, ..., t\}$. Our notion of t-concrete bag, on the other hand, may be regarded as a representation of a graph with at most t vertices injectively labeled with numbers from $\{1, ...t\}$ and at most *one* edge. Within this point of view, the representation used here is a syntactic restriction of the former. On the other hand, any decomposition which uses bags with arbitrary graphs of size t can be converted into a t-concrete decomposition, by expanding each bag into a sequence of t^2 concrete bags.

alphabet which will be used to construct terms representing tree decompositions of graphs.

A t-concrete tree decomposition is a *binary* term $\tau \in Ter(\mathcal{B}(t))$. We let $\tau[p] = (\tau[p].B, \tau[p].b)$ be the t-concrete bag at position p of τ. For each $s \in [t]$, we say that a subset $P \subseteq Pos(\tau)$ is s-*maximal* if the following conditions are satisfied.

1. P is connected in $Pos(\tau)$.
2. $s \in \tau[p].B$ for every $p \in P$.
3. If P' is a connected subset of $Pos(\tau)$ and $s \in \tau[p].B$ for every $s \in P'$, then $P' \subseteq P$.

Note that if P and P' are s-maximal then either $P = P'$, or $P \cap P' = \emptyset$. Additionally, $p \in Pos(\tau)$ and each $s \in \tau[p].B$, there exists a unique subset $P \subseteq Pos(\tau)$ such that P is s-maximal and $p \in P$. We denote this unique set by $P(p, s)$.

Definition 2.2. *Let $\tau \in Ter(\mathcal{B}(t))$. The graph $\mathcal{G}(\tau)$ associated with τ is defined as follows.*

1. $V_{\mathcal{G}(\tau)} = \{v_{s,P} \mid s \in [t], P \subseteq Pos(\tau), P \text{ is } s\text{-maximal}\}$.
2. $E_{\mathcal{G}(\tau)} = \{e_p \mid p \in Pos(\tau), b \neq \emptyset\}$.
3. $Inc_{\mathcal{G}(\tau)} = \{(e_p, v_{s,P(p,s)}) \mid e_p \in E_{\mathcal{G}(\tau)}, s \in \tau[p].b\}$.

The following observation is immediate, using the fact that if a graph has treewidth t, then it has a rooted tree decomposition in which each node has at most two children (see for instance [12]).

Observation 1. *A graph G has treewidth t if and only if there exists some $(t+1)$-concrete tree decomposition $\tau \in Ter(\mathcal{B}(t+1))$ such that $\mathcal{G}(\tau)$ is isomorphic to G.*

3 An Algorithm for Amalgamation Recognition

In this section we will show that the problem of determining whether a connected graph G is an H-amalgamation of connected graphs G_1 and G_2 can be solved in time $2^{O(t \cdot \Delta)} \cdot n^{O(t)}$ where t is the treewidth of G and Δ is the maximum degree of G. To this end we will introduce some machinery for the manipulation of tree automata accepting families of t-concrete decompositions.

We say that a t-concrete bag (B, b) is a *sub-bag* of a t-concrete bag (B', b') if $B \subseteq B'$ and $b \subseteq b'$.

Definition 3.1. *We say that a t-concrete tree decomposition $\tau \in Ter(\mathcal{B}(t))$ is a sub-decomposition of a t-concrete tree decomposition $\tau' \in Ter(\mathcal{B}(t))$ if the following conditions are satisfied.*

S1. *$Pos(\tau) = Pos(\tau')$.*
S2. *For each $p \in Pos(\tau)$, $\tau[p]$ is a sub-bag of $\tau'[p]$.*

S3.   *For each $p, pj \in Pos(\tau)$, and for each $s \in [t]$, if $s \in \tau'[p].B$ and $s \in \tau'[pj].B$, then $s \notin \tau[p].B$ if and only if $s \notin \tau[pj].B$.*

We write $\tau \sqsubseteq \tau'$ to denote that τ is a sub-decomposition of τ'. Intuitively, if a t-concrete tree decomposition τ represents a graph G then sub-decompositions of τ represent subgraphs of G. The following lemma states that sub-decompositions of τ are in one to one correspondence with subgraphs of $\mathcal{G}(\tau)$.

Lemma 3.2 ([9]).

1. *Let $\tau, \tau' \in Ter(\mathcal{B}(t))$. If τ is a sub-decomposition of τ' then $\mathcal{G}(\tau)$ is a subgraph of $\mathcal{G}(\tau')$.*
2. *Let $\tau' \in Ter(\mathcal{B}(t))$ and let G be a subgraph of $\mathcal{G}(\tau')$. Then there exists a unique $\tau \in Ter(\mathcal{B}(t))$ such that τ is a sub-decomposition of τ' and $\mathcal{G}(\tau) = G$.*

Let $\mathcal{B}(t)$ be the set of t-concrete bags. We let $\mathcal{B}(t)^{\otimes 2}$ be the set of ordered pairs of bags in $\mathcal{B}(t)$. If τ_1 and τ_2 are t-concrete decompositions with $Pos(\tau_1) = Pos(\tau_2)$, then the *tensor product* of τ_1 with τ_2 is the term $\tau_1 \otimes \tau_2 \in Ter(\mathcal{B}(t)^{\otimes 2})$ defined as follows.

1. $Pos(\tau_1 \otimes \tau_2) = Pos(\tau_1)$.
2. For each $p \in Pos(\tau_1)$, $(\tau_1 \otimes \tau_2)[p] = (\tau_1[p], \tau_2[p])$.

Intuitively $\tau_1 \otimes \tau_2$ is obtained by placing τ_1 side by side with τ_2. The condition that $Pos(\tau_1) = Pos(\tau_2)$ guarantees that both terms have the same tree structure, and therefore the definition of tensor product is well defined. If $\mathcal{L}_1$ and $\mathcal{L}_2$ are sets of terms in $Ter(\mathcal{B}(t))$, then we let $\mathcal{L}_1 \otimes \mathcal{L}_2 = \{\tau_1 \otimes \tau_2 \mid Pos(\tau_1) = Pos(\tau_2)\}$ be the set of tensor products between terms in $\mathcal{L}_1$ and terms in $\mathcal{L}_2$. The following proposition states that tensor products of languages represented by tree automata can be constructed efficiently.

Proposition 3.3. *Let $\mathcal{A}_1$ and $\mathcal{A}_2$ be tree automata over $\mathcal{B}(t)$. Then one can construct in time $O(|\mathcal{A}_1| \cdot |\mathcal{A}_2|)$ a tree automaton $\mathcal{A}_1 \otimes \mathcal{A}_2$ such that $\mathcal{L}(\mathcal{A}_1 \otimes \mathcal{A}_2) = \mathcal{L}(\mathcal{A}_1) \otimes \mathcal{L}(\mathcal{A}_2)$.*

We note that the notions of tensor product of bags, terms and tree automata can be extended straightforwardly to any arbitrary number of factors. In particular, we let $\mathcal{B}(t)^{\otimes 4}$ be the set of 4-tuples of elements from $\mathcal{B}(t)$, $Ter(\mathcal{B}(t)^{\otimes 4})$ be the set of terms over the $\mathcal{B}(t)^{\otimes 4}$, and for tree automata $\mathcal{A}_1, \mathcal{A}_2, \mathcal{A}_3, \mathcal{A}_4$ over $Ter(\mathcal{B}(t))$, we let $\mathcal{A}_1 \otimes \mathcal{A}_2 \otimes \mathcal{A}_3 \otimes \mathcal{A}_4$ be the tree automaton accepting precisely those terms $\tau_1 \otimes \tau_2 \otimes \tau_3 \otimes \tau_4 \in Ter(\mathcal{B}(t)^{\otimes 4})$ such that $\tau_i \in \mathcal{L}(\mathcal{A}_i)$ for $i \in \{1, 2, 3, 4\}$.

Next we will provide a local characterization of tuples of graphs (H, G_1, G_2, G) satisfying the property that G is an H-amalgamation of G_1 and G_2. This local characterization is given in terms of the notion of *interface sequence*.

Definition 3.4 (Interface Sequence). *We say that a sequence $\tau, \tau_1, \tau_2, \tau'$ of t-concrete decompositions is a t-concrete interface sequence if $\tau \sqsubseteq \tau_i$ and $\tau_i \sqsubseteq \tau'$ for each $i \in \{1, 2\}$, and $\tau'[p] = \tau_1[p].B \cup \tau_2[p].B$ and $\tau'[p] = \tau_1[p].b \cup \tau_2[p].b$ for each $p \in Pos(\tau')$.*

Intuitively, $\tau, \tau_1, \tau_2, \tau'$ is an interface sequence if for each $i \in \{1,2\}$, τ can be embedded bag-wise into τ_i, τ_i can be embedded bag-wise into τ', and τ' is the bag-wise union of τ_1 and τ_2.

Lemma 3.5. *Let H, G and G_1, G_2 be graphs of treewidth at most t. Then G is an H-amalgamation of G_1 and G_2 if and only if there exists a $(t+1)$-concrete interface sequence $\tau, \tau_1, \tau_2, \tau'$ satisfying the following properties.*

1. $\mathcal{G}(\tau) \simeq H$.
2. $\mathcal{G}(\tau') \simeq G$.
3. $\mathcal{G}(\tau_i) \simeq G_i$ for each $i \in \{1,2\}$.

Proof. Let G be an H-amalgamation of G_1, G_2. Then there exist injective morphisms $\mu : H \to G$, $\{\mu_i : H \to G_i\}_{i \in \{1,2\}}$ and $\{\eta_i : G_i \to G\}_{i \in \{1,2\}}$ such that $G = \eta_1(G_1) \cup \eta_2(G_2)$, and $\mu = \eta_i \circ \mu_i$ for each $i \in \{1,2\}$. Let $G'_i = \eta_i(G_i)$ be the image of η_i in G_i, and let $H' = \mu(H)$ be the image of μ in G. Then H' and G'_i are subgraphs of G such that $G = G'_1 \cup G'_2$ and $H' = G'_1 \cap G'_2$. Since G has treewidth at most t, there is a $(t+1)$-concrete tree decomposition $\tau' \in Ter(\mathcal{B}(t+1))$ such that $G \simeq \mathcal{G}(\tau')$.

We may assume that $G = \mathcal{G}(\tau')$, that H, G_1, G_2 are subgraphs of G and that H is a subgraph of G_1, G_2, since otherwise we could simply rename the vertices of G, G_1, G_2 and H appropriately. Since G_1, G_2 and H are subgraphs of G, Lemma 3.2 implies that there exist unique sub-decompositions τ, τ_1 and τ_2 such that $\mathcal{G}(\tau) = H$, $\mathcal{G}(\tau_1) = G_1$ and $\mathcal{G}(\tau_2) = G_2$. Additionally, since $G_1 \cup G_2 = G$, we have that for each position $p \in Pos(\tau')$, $\tau'[p].B = \tau_1[p].B \cup \tau_2[p].B$ and $\tau'[p].b = \tau_1[p].b \cup \tau_2[p].b$. Since H is also a subgraph of G_1 and of G_2, again by Lemma 3.2, there exists a sub-decomposition $\tilde{\tau}_1$ of τ_1 such that $\mathcal{G}(\tilde{\tau}_1) = H$ and a sub-decomposition $\tilde{\tau}_2$ of τ_2 such that $\mathcal{G}(\tilde{\tau}_2) = H$. But since the sub-decomposition relation is transitive, $\tilde{\tau}_1$ and $\tilde{\tau}_2$ are also sub-decompositions of τ'. Finally, since τ, $\tilde{\tau}_1$ and $\tilde{\tau}_2$ are sub-decompositions of τ' such that $\mathcal{G}(\tau) = \mathcal{G}(\tilde{\tau}_1) = \mathcal{G}(\tilde{\tau}_2) = H$, by uniqueness, we have that $\tau = \tilde{\tau}_1 = \tilde{\tau}_2$. This shows that the sequence $\tau, \tau_1, \tau_2, \tau'$ is an interface sequence.

For the converse, let $\tau, \tau_1, \tau_2, \tau'$ be an interface sequence satisfying Conditions 3.5 to 3.5. Then for each $i \in \{1,2\}$, $\mathcal{G}(\tau)$ is a subgraph of $\mathcal{G}(\tau_i)$, and $\mathcal{G}(\tau_i)$ is a subgraph of $\mathcal{G}(\tau')$. Let $\mu : \mathcal{G}(\tau) \to \mathcal{G}(\tau')$ be the inclusion map from $\mathcal{G}(\tau)$ to $\mathcal{G}(\tau')$, and for each $i \in \{1,2\}$, let $\mu_i : \mathcal{G}(\tau) \to \mathcal{G}(\tau_i)$ be the inclusion map from $\mathcal{G}(\tau)$ to $\mathcal{G}(\tau_i)$, and $\eta_i : \mathcal{G}(\tau_i) \to \mathcal{G}(\tau')$ be the inclusion map from $\mathcal{G}(\tau_i)$ to $\mathcal{G}(\tau')$. Then these morphisms are injective, and $\mu = \eta_i \circ \mu_i$ for each $i \in \{1,2\}$. Additionally, the condition that $\tau'[p].B = \tau_1[p].B \cup \tau_2[p].B$ and $\tau'[p].b = \tau_1[p].b \cup \tau_2[p].b$ implies that the graph $\mathcal{G}(\tau')$ is the union of the graphs $\mathcal{G}(\tau_1)$ and $\mathcal{G}(\tau_2)$. Therefore, $\mathcal{G}(\tau')$ is an $\mathcal{G}(\tau)$-amalgamation of $\mathcal{G}(\tau_1)$ and $\mathcal{G}(\tau_2)$. $\square$

The next theorem states that for each $t \in \mathbb{N}$ one can construct a tree automaton $\mathcal{I}(t)$ over $\mathcal{B}(t)^{\otimes 4}$ that accepts precisely those terms in in $Ter(\mathcal{B}(t)^{\otimes 4})$ that correspond to t-concrete interface sequences.

Theorem 3.6 (All Interface Sequences). *For each $t \in \mathbb{N}$ one can construct in time $2^{O(t)}$ a tree automaton $\mathcal{I}(t)$ over $\mathcal{B}(t)^{\otimes 4}$ that accepts a term $\tau \otimes \tau_1 \otimes \tau_2 \otimes \tau' \in Ter(\mathcal{B}(t)^{\otimes 4})$ if and only if $\tau, \tau_1, \tau_2, \tau'$ is a t-concrete interface sequence.*

The next theorem states that given a connected graph G of maximum degree Δ, and a positive integer t, one can construct in time $2^{O(\Delta \cdot t)} \cdot |V_G|^{O(t)}$ a tree automaton $\mathcal{A}(G, t)$ over $\mathcal{B}(t)$ that accepts precisely those t-concrete tree decompositions of G.

Theorem 3.7 ([9])**.** *Let G be a connected graph of treewidth t and maximum degree Δ. Then one can construct in time $2^{O(\Delta \cdot t)} \cdot |V_G|^{O(t)}$ a tree automaton $\mathcal{A}(G, t)$ over $\mathcal{B}(t)$ such that for each $\tau \in Ter(\mathcal{B}(t))$, $\tau \in \mathcal{L}(\mathcal{A}(G, t))$ if and only if τ is a concrete tree decomposition of G.*

The following theorem is the main result of this section.

Theorem 3.8 (Amalgamation Recognition). *Given connected graphs H, G_1, G_2 and G, each with at most n vertices, treewidth at most t, and maximum degree at most Δ, one can determine whether G is an H-amalgamation of G_1, G_2 in time $2^{O(\Delta \cdot t)} \cdot n^{O(t)}$.*

Proof. By Theorem 3.6, one can construct in time $2^{O(t)}$ a tree automaton $\mathcal{I}(t+1)$ that accepts a term $\tau \otimes \tau_1 \otimes \tau_2 \otimes \tau' \in Ter(\mathcal{B}(t))$ if and only if $\tau, \tau_1, \tau_2, \tau'$ is a $(t+1)$-concrete interface sequence.

Consider the following tree automaton over $\mathcal{B}(t+1)^{\otimes 4}$.

$$\mathcal{A}(H, G_1, G_2, G, t+1) = \mathcal{A}(H, t+1) \otimes \mathcal{A}(G_1, t+1) \otimes \mathcal{A}(G_2, t+1) \otimes \mathcal{A}(G, t+1).$$

Then $\mathcal{A}(H, G_1, G_2, G, t+1)$ accepts a term $\tau \otimes \tau_1 \otimes \tau_2 \otimes \tau' \in Ter(\mathcal{B}(t+1)^{\otimes 4})$ if and only if $\mathcal{G}(\tau) \simeq H$, $\mathcal{G}(\tau_i) \simeq G_i$ for each $i \in \{1, 2\}$, and $\mathcal{G}(\tau') \simeq G'$. Additionally, this automaton can be constructed in time $2^{O(\Delta \cdot t)} \cdot n^{O(t)}$ by a combination of Theorem 3.7 and Proposition 3.3.

Now consider the following tree automaton over $\mathcal{B}(t+1)^{\otimes 4}$.

$$\mathcal{A} = \mathcal{A}(H, G_1, G_2, G, t+1) \cap \mathcal{I}(t+1).$$

Then $\mathcal{A}$ accepts a $(t+1)$-concrete decomposition $\tau \otimes \tau_1 \otimes \tau_2 \otimes \tau'$ if and only if $\tau, \tau_1, \tau_2, \tau'$ is an interface sequence , $\mathcal{G}(\tau) \simeq H$, $\mathcal{G}(\tau_i) \simeq G_i$ for each $i \in \{1, 2\}$ and $\mathcal{G}(\tau') \simeq G$. By Lemma 3.5, this happens if and only if G is an H-amalgamation of G_1, G_2.

Therefore, in order to determine whether G is an H-amalgamation of G_1, G_2 it is enough to determine whether the language accepted by $\mathcal{A}$ is non-empty. Since $|\mathcal{I}(t+1)| = 2^{O(t)}$ and $|\mathcal{A}(H, G_1, G_2, t+1)| = 2^{O(\Delta \cdot t)} \cdot n^{O(t)}$, by Lemmas 2.1.2 and 2.1.3, we can determine whether $\mathcal{L}(\mathcal{A})$ is non-empty in time $2^{O(\Delta \cdot t)} \cdot n^{O(t)}$. $\square$

4   Amalgamating Graphs Under CMSO Constraints

The counting monadic second-order logic of graphs (CMSO) extends first order logic by allowing quantifications over sets of vertices and over sets of edges. This logic can be used to define several natural properties of graphs such as planarity, Hamiltonicity, r-colorability, etc.

In this section we consider variants of the amalgamation recognition problem where the interface graph H, the factor graphs G_1 and G_2 or the final graph G is not explicitly given at the input but it is instead required to satisfy some constraint specified in CMSO logic. Before proceeding we briefly recall the syntax of CMSO logic. We refer to the monograph [7] for an extensive study of the links between CMSO logic and graphs of bounded treewidth.

CMSO Logic: The counting monadic second-order logic of graphs, here denoted by CMSO, extends first order logic by allowing quantifications over sets of vertices and edges, and by introducing the notion of modular counting predicates. More precisely, the syntax of CMSO logic includes the logical connectives $\vee, \wedge, \neg, \Leftrightarrow, \Rightarrow$, variables for vertices, edges, sets of vertices and sets of edges, the quantifiers $\exists, \forall$ that can be applied to these variables, and the following atomic predicates:

1. $x \in X$ where x is a vertex variable and X a vertex-set variable;
2. $y \in Y$ where y is an edge variable and Y an edge-set variable;
3. $\mathrm{Inc}(x, y)$ where x is a vertex variable, y is an edge variable, and the interpretation is that the edge x is incident with the edge y.
4. $card_{s,r}(Z)$ where $0 \leq s < r$, $r \geq 2$, Z is a vertex-set or edge-set variable, and the interpretation is that $|Z| = s \pmod{r}$;
5. equality of variables representing vertices, edges, sets of vertices and sets of edges.

A CMSO *sentence* is a CMSO formula without free variables. If φ is a CMSO sentence, then we write $G \models \varphi$ to indicate that G satisfies φ.

The next theorem may be regarded as a variant of Courcelle's theorem [7].

Theorem 4.1 (Courcelle's Theorem). *There exists a computable function* $f : \mathbb{N} \times \mathbb{N} \to \mathbb{N}$ *and an algorithm* $\mathfrak{U}$ *which takes a CMSO sentence* φ *and an integer t as input and constructs in time $f(|\varphi|, t)$ a tree automaton $\mathcal{A}(\varphi, t)$ accepting the following tree language.*

$$\mathcal{L}(\mathcal{A}(\varphi, t)) = \{\tau \in Ter(\mathcal{B}(t)) \mid \mathcal{G}(\tau) \models \varphi\}. \tag{1}$$

4.1   Amalgamation Recognition

Let φ be a CMSO sentence and G_1, G_2, G be connected graphs. We say that G is a φ-amalgamation of graphs G_1, G_2 if there exists a graph H satisfying φ, such that G is an H-amalgamation of G_1, G_2. For instance, if φ_{pl} is a CMSO sentence defining planar graphs [7], then G is a φ_{pl}-amalgamation of G_1, G_2 if G is an H-amalgamation of G_1 and G_2 for some planar graph H.

φ-AMALGAMATION RECOGNITION: Let φ be a CMSO sentence. Given connected graphs G_1, G_2 and G determine whether G is a φ-amalgamation of G_1, G_2.

Theorem 4.2 (φ-AMALGAMATION RECOGNITION). *There is a computable function $f : \mathbb{N} \times \mathbb{N} \to \mathbb{N}$ and an algorithm $\mathfrak{U}$ that takes as input a CMSO-sentence φ and connected graphs G_1, G_2, G of treewidth at most t and maximum degree at most Δ, and determines in time $f(|\varphi|, t) \cdot |G|^{O(t)}$ whether G is a φ-amalgamation of G_1 and G_2.*

4.2 Amalgamability

Let G_1, G_2 and H be connected graphs, $t \in \mathbb{N}$ and φ be a CMSO sentence. We say that G_1 and G_2 are (φ, t)-amalgable along H if there exists a graph G of treewidth at most t such that G satisfies φ and G is an H-amalgamation of G_1, G_2. For instance, if φ_{pl} is the MSO sentence defining planar graphs, then G_1, G_2 are (φ_{pl}, t)-amalgable along H if there is a way of gluing G_1, G_2 to H in such a way that the resulting graph is planar and has treewidth at most t.

φ-AMALGAMABILITY: Let φ be a CMSO sentence. Given connected graphs H, G_1, G_2 and a positive integer t, determine whether G_1, G_2 are (φ, t)-amalgable along H.

Theorem 4.3 (φ-AMALGAMABILITY). *There is a computable function $f : \mathbb{N} \to \mathbb{N}$ and an algorithm $\mathfrak{U}$ that takes as input a CMSO-sentence φ, and connected graphs H, G_1, G_2 of treewidth at most t and maximum degree at most Δ, and determines in time $f(|\varphi|, t) \cdot 2^{O(\Delta \cdot t)} \cdot (|G_1| + |G_2|)^{O(t)}$ whether G_1 and G_2 are φ-amalgable along H.*

4.3 (φ_1, φ_2)-Factors

The problem of decomposing a graph G into a clique sum of a graph G_1 satisfying a CMSO property Π_1 and a graph G_2 satisfying a CMSO property Π_2 has found many important applications in the field of structural graph theory [11,16,17,27]. In this section we generalize such type of problems to the context in which the interface graph H is given at the input. More precisely, we consider the problem of factorizing G into an H amalgamation of graphs G_1 and G_2 where G_1 satisfies a given CMSO-sentence φ_1 and G_2 satisfies a given CMSO-sentence φ_2. One of many natural questions that fit in this framework is the following[4]: Given connected graphs G and H, does there exist a planar graph G_1 and a Hamiltonian graph G_2 such that G is an H-sum of G_1 and G_2? More formally, we consider the following problem.

(φ_1, φ_2)-FACTORS: Let φ_1 and φ_2 be CMSO formulas and let G and H be connected graphs. Do there exist graphs G_1 and G_2 such that $G_1 \models \varphi_1$, $G_2 \models \varphi_2$ and G is an H-amalgamation of G_1 and G_2?

Theorem 4.4 (φ_1, φ_2-FACTORS). *There is a computable function $f : \mathbb{N} \times \mathbb{N} \times \mathbb{N} \to \mathbb{N}$ and an algorithm $\mathfrak{U}$ that takes as input connected graphs G, H of treewidth at most t and maximum degree at most Δ, and CMSO-sentences φ_1, and φ_2 and determines in time $f(\varphi_1, \varphi_2, t) \cdot 2^{O(\Delta \cdot t)} \cdot |G|^{O(t)}$ whether there exist graphs G_1 and G_2 such that $G_1 \models \varphi_1$, $G_2 \models \varphi_2$ and G is an H-amalgamation of G_1 and G_2.*

---

[4] Both Planarity and Hamiltonicity are CMSO-definable (note that our definition of CMSO logic allows edge-set quantifications).

5   Conclusion

In this work we introduced an algorithmic framework to deal with graph amalgamation problems parameterized by the treewidth and maximum degree of the involved graphs. In particular we have shown that the problem of deciding whether a connected graph G is an H-amalgamation of connected graphs G_1 and G_2 can be solved in time $2^{O(\Delta \cdot t)} \cdot |G|^{O(t)}$ where Δ and t are the maximum degree and the treewidth of G respectively.

We have also considered variants of amalgamation problems where the host graph G, the interface graph H or the factor graphs G_1 and G_2 are not given at the input but are instead required to satisfy certain CMSO property. In general we have shown that such problems can be solved in time $f(\Delta, t) \cdot n^{O(t)}$ where n is the size of the largest considered graph. We believe that these problems may serve as an useful tool in the study of decomposition of graphs.

Acknowledgements. The author thanks Michael Fellows for many valuable comments. This work was supported by the Bergen Research Foundation.

References

1. Adler, I., Grohe, M., Kreutzer, S.: Computing excluded minors. In: Proceedings of SODA 2008, pp. 641–650. SIAM (2008)
2. Bojańczyk, M., Pilipczuk, M.: Definability equals recognizability for graphs of bounded treewidth. In Procedings of LICS 2016, pp. 407–416. ACM (2016)
3. Comon, H., et al.: Tree Automata Techniques and Applications (2007)
4. Courcelle, B.: The monadic second-order logic of graphs. I. Recognizable sets of finite graphs. Inf. Comput. **85**(1), 12–75 (1990)
5. Courcelle, B.: The monadic second-order logic of graphs xii: planar graphs and planar maps. Theor. Comput. Sci. **237**(1), 1–32 (2000)
6. Courcelle, B.: The monadic second-order logic of graphs xiii: graph drawings with edge crossings. Theor. Comput. Sci. **244**(1–2), 63–94 (2000)
7. Courcelle, B., Engelfriet, J.: Graph Structure and Monadic Second-Order Logic: A Language-Theoretic Approach, vol. 138. Cambridge University Press, Cambridge (2012)
8. Courcelle, B., Oum, S.: Vertex-minors, monadic second-order logic, and a conjecture by Seese. J. Combina. Theory, Ser. B **97**(1), 91–126 (2007)
9. de Oliveira Oliveira, M.: On supergraphs satisfying CMSO properties. In: Proceedings of the 26th Annual Conference on Computer Science Logic (CSL 2017). LIPIcs, vol. 82, pp. 33:1–33:15 (2017)
10. Demaine, E.D., Fomin, F.V., Hajiaghayi, M., Thilikos, D.M.: Subexponential parameterized algorithms on bounded-genus graphs and h-minor-free graphs. J. ACM (JACM) **52**(6), 866–893 (2005)
11. Diestel, R.: A separation property of planar triangulations. J. Graph Theory **11**(1), 43–52 (1987)
12. Elberfeld, M.: Context-free graph properties via definable decompositions. In: Proceedings of the 25th Conference on Computer Science Logic (CSL 2016). LIPIcs, vol. 62, pp. 17:1–17:16 (2016)

13. Flum, J., Frick, M., Grohe, M.: Query evaluation via tree-decompositions. J. ACM (JACM) **49**(6), 716–752 (2002)
14. Gross, J.L.: Genus distribution of graph amalgamations: self-pasting at root-vertices. Australas. J. Combin. **49**, 19–38 (2011)
15. Hilton, A.J., Johnson, M., Rodger, C.A., Wantland, E.B.: Amalgamations of connected k-factorizations. J. Combin. Theory, Ser. B **88**(2), 267–279 (2003)
16. Johnson, C.R., McKee, T.A.: Structural conditions for cycle completable graphs. Discret. Math. **159**(1–3), 155–160 (1996)
17. Kriz, I., Thomas, R.: Clique-sums, tree-decompositions and compactness. Discret. Math. **81**(2), 177–185 (1990)
18. Leach, C.D., Rodger, C.: Hamilton decompositions of complete multipartite graphs with any 2-factor leave. J. Graph Theory **44**(3), 208–214 (2003)
19. Lovász, L.: Graph minor theory. Bull. Am. Math. Soc. **43**(1), 75–86 (2006)
20. Makowsky, J.A.: Coloured tutte polynomials and Kauffman brackets for graphs of bounded tree width. Discret. Appl. Math. **145**(2), 276–290 (2005)
21. Makowsky, J.A., Marino, J.P.: Farrell polynomials on graphs of bounded tree width. Adv. Appl. Math. **30**(1–2), 160–176 (2003)
22. Makowsky, J.A., Rotics, U., Averbouch, I., Godlin, B.: Computing graph polynomials on graphs of bounded clique-width. In: Fomin, F.V. (ed.) WG 2006. LNCS, vol. 4271, pp. 191–204. Springer, Heidelberg (2006). https://doi.org/10.1007/11917496_18
23. Marx, D., Pilipczuk, M.: Everything you always wanted to know about the parameterized complexity of Subgraph Isomorphism (but were afraid to ask). In: Proceedings of the 31st International Symposium on Theoretical Aspects of Computer Science (STACS 2014), pp. 542 (2014)
24. Matoušek, J., Thomas, R.: On the complexity of finding ISO-and other morphisms for partial k-trees. Discret. Math. **108**(1), 343–364 (1992)
25. Nešetřil, J.: Amalgamation of graphs and its applications. Ann. New York Acad. Sci. **319**(1), 415–428 (1979)
26. Robertson, N., Seymour, P.D.: Graph minors. XVI. Excluding a non-planar graph. J. Combin. Theory, Ser. B **89**(1), 43–76 (2003)
27. Seymour, P.D., Weaver, R.: A generalization of chordal graphs. J. Graph Theory **8**(2), 241–251 (1984)
28. Yang, Y., Chen, Y.: The thickness of amalgamations and cartesian product of graphs. Discuss. Math. Graph Theory **37**(3), 561–572 (2017)

∀∃ℝ-Completeness and Area-Universality

Michael Gene Dobbins[1], Linda Kleist[2(✉)], Tillmann Miltzow[3],
and Paweł Rzążewski[4]

[1] Binghamton University, Binghamton, NY, USA
mdobbins@binghamton.edu
[2] Technische Universität Berlin, Berlin, Germany
kleist@math.tu-berlin.de
[3] Université libre de Bruxelles, Brussels, Belgium
t.miltzow@gmail.com
[4] Faculty of Mathematics and Information Science,
Warsaw University of Technology, Warsaw, Poland
p.rzazewski@mini.pw.edu.pl

Abstract. In the study of geometric problems, the complexity class $\exists\mathbb{R}$ plays a crucial role since it exhibits a deep connection between purely geometric problems and real algebra. Sometimes $\exists\mathbb{R}$ is referred to as the "real analogue" to the class NP. While NP is a class of computational problems that deals with existentially quantified *boolean* variables, $\exists\mathbb{R}$ deals with existentially quantified *real* variables.

In analogy to Π_2^p and Σ_2^p in the famous polynomial hierarchy, we study the complexity classes $\forall\exists\mathbb{R}$ and $\exists\forall\mathbb{R}$ with *real* variables. Our main interest is focused on the AREA UNIVERSALITY problem, where we are given a plane graph G, and ask if for each assignment of areas to the inner faces of G there is an area-realizing straight-line drawing of G. We conjecture that the problem AREA UNIVERSALITY is $\forall\exists\mathbb{R}$-complete and support this conjecture by a series of partial results, where we prove $\exists\mathbb{R}$- and $\forall\exists\mathbb{R}$-completeness of variants of AREA UNIVERSALITY. To do so, we also introduce first tools to study $\forall\exists\mathbb{R}$. Finally, we present geometric problems as candidates for $\forall\exists\mathbb{R}$-complete problems. These problems have connections to the concepts of imprecision, robustness, and extendability.

1 Introduction

In this paper we investigate problems related to face areas in straight-line drawings of planar graphs. We consider two crossing-free drawings of a planar graph to be *equivalent* if they have the same outer face and rotation system, i.e., for each vertex the cyclic ordering of the incident edges coincides. Recall that a plane

---

A video presenting this paper is available at https://youtu.be/OQkACiNS66o. Proofs omitted due to space constraints can be found in the full version of the manuscript [6]

T. Miltzow—Partially supported by the ERC grant PARAMTIGHT: "Parameterized complexity and the search for tight complexity results", no. 280152.

ⓒ Springer Nature Switzerland AG 2018
A. Brandstädt et al. (Eds.): WG 2018, LNCS 11159, pp. 164–175, 2018.
https://doi.org/10.1007/978-3-030-00256-5_14

graph is a planar graph together with a crossing-free drawing, and the faces of a plane graph are determined by its rotation system. Let G be a plane graph and let F be the set of inner faces of G. A *face area assignment* is a function $\mathcal{A}: F \to \mathbb{R}_0^+$. We say that G' is an $\mathcal{A}$-*realizing drawing*, if G' is an equivalent straight-line drawing of G in which the area of each $f \in F$ is exactly $\mathcal{A}(f)$. If $\mathcal{A}$ has an area-realizing drawing, we say that $\mathcal{A}$ is *realizable*. A plane graph G is *area-universal* if every face area assignment is realizable. Since we only consider crossing-free straight-line drawings, we simply call them drawings from now on.

Since area-universality seems to be a strong property, it is somewhat surprising that many graphs indeed are easily seen to be area-universal. It is straightforward to observe that *stacked triangulations*, also known as *planar 3-trees* or *Apollonian networks*, are area-universal. A stacked triangulation T is defined recursively by subdividing a triangle t of a stacked triangulation T' into three smaller triangles. An area assignment of T can be realized by first realizing T' so that t has the total area of the three smaller triangles, and then subdividing t accordingly. Moreover, it is easy to see that if a graph is area-universal, then each of its subgraphs is also area-universal. These two observations together imply that partial planar 3-trees are area-universal [3]. In 1992, Thomassen [17] proved that plane cubic graphs are area-universal. More recently, Kleist [9] showed that all 1-subdivisions of plane graphs are area-universal. In other words, every area assignment of every plane graph could be realized if we allowed each edge to have at most one bend instead of only allowing straight-line drawings.

For a long time, the only graph known not to be area-universal was the octahedron graph (or graphs containing the octahedron), which was proven by Ringel [15] in 1990. Kleist [9] introduced the first non-trivial infinite family of non-area-universal graphs. In particular, she showed that all Eulerian triangulations and the icosahedron graph are not area-universal. This implies that high connectivity of a graph does not imply area-universality. Moreover, area-universality is not a minor-closed property, as the grid is area-universal [8], but the octahedron graph is not area-universal, although it is a minor of the grid.

In this paper we are interested in the computational problem of deciding if a given plane graph is area-universal; which we denote by AREA UNIVERSALITY.

AREA UNIVERSALITY

Input: A connected plane graph, given by the closed walks around the faces with one specified *outer* walk.
Question: Is every face area assignment realizable?

When investigating natural geometric problems, one often discovers that an instance of such a problem can be described by a system of polynomial equations and inequalities Φ so that real-valued variable assignments that satisfy Φ correspond to solutions of the original geometric problem. EXISTENTIAL THEORY OF THE REALS (ETR) is a computational problem that takes a first-order formula containing only existential quantifiers: $\exists X_1, X_2, \ldots, X_n: \Phi$, where Φ has symbols $0, 1, +, *, =, <, \wedge, \neg, (,), X_1, \ldots, X_n$ as an input and asks whether it is true or not over the reals. The complexity class $\exists\mathbb{R}$ consists of all problems that are many-one reducible to ETR by a Turing machine in at most a polynomial number of steps. Surprisingly many natural geometric problems appear to be ∃ℝ-complete, i.e., ETR is also reducible to these problems in ∃ℝ in the above sense. A prominent example is the stretchability of a pseudoline arrangement (see [12,13,16]). A *pseudoline arrangement* in the plane is a set of unbounded Jordan curves where every pair of curves intersects in exactly one crossing point. A pseudoline arrangement is *stretchable* if there exists an arrangement of straight lines with the same face structure. STRETCHABILITY is a computational problem which asks whether a given pseudoline arrangement is stretchable. Here the input is the order type of a pseudoline arrangement, which is a rank 3 chirotope. Since STRETCHABILITY is ∃ℝ-complete, there is little hope to find a simple algorithm for STRETCHABILITY, since simple algorithms to decide ETR are not known despite tremendous work in real algebraic geometry. The ∃ℝ-completeness of STRETCHABILITY reflects the deep algebraic connections between line arrangements and real algebra. For instance, the smallest non-strechable pseudoline arrangement, depicted in Fig. 1, is based on Pappus's Hexagon Theorem [10], dating back to the 4th century. It considers two different lines with three points each, the points are denoted by A, B, C and X, Y, Z, see Fig. 1. If the lines $\overline{AY}, \overline{BZ}, \overline{CX}$ intersect the lines $\overline{BX}, \overline{CY}, \overline{AZ}$, respectively, then the three points of intersection are collinear. Although the statement is intrinsically geometric, most known proofs are algebraic, see [14].

Fig. 1. Pappus's Hexagon Theorem and non-stretchable pseudoline arrangement.

Geometric problems that are ∃ℝ-complete usually ask for the existence of certain objects, satisfying some semialgebraic properties. However, the nature of AREA UNIVERSALITY seems to be different. We therefore define the new complexity class ∀∃ℝ as the set of all problems that reduce in polynomial time to UNIVERSAL EXISTENTIAL THEORY OF THE REALS (UETR).

UNIVERSAL EXISTENTIAL THEORY OF THE REALS (UETR)
Input: A formula over the reals $(\forall\, Y_1, Y_2, \ldots, Y_m)\,(\exists\, X_1, X_2, \ldots, X_n): \Phi$, where Φ is a quantifier-free with variables $Y_1, Y_2, \ldots, Y_m, X_1, X_2, \ldots, X_n$.
Question: Is the input formula true?

The class ∃∀ℝ is defined analogously. Clearly ∃ℝ is contained in ∀∃ℝ. It is easy to observe and well-known that NP is contained in ∃ℝ. Highly nontrivial is the containment of ∀∃ℝ in PSPACE, which follows from a more general result that deciding first-order formulae over the reals with bounded number of quantifier blocks is in PSPACE (see [2]). For all we know, all these complexity classes could collapse, as we do not know whether NP and PSPACE constitute two different or the same complexity class, see Fig. 2. However, $\exists\mathbb{R} \neq \forall\exists\mathbb{R}$ can be believed with similar confidence as $NP \neq \Pi_2^p$. In addition, it is known that the algebraic expressibility of $\forall\exists\mathbb{R}$-formulae is larger than $\exists\mathbb{R}$-formulae, see [5].

It is worth mentioning that Blum et al. [4] also introduce a hierarchy of complexity classes analogous to the complexity class NP, but over the reals (this generalizes to other rings). Their canonical model of computation is the so-called Blum-Shub-Smale machine (BSS). The main difference between these approaches is that BSS accepts real numbers as input, while the classes ($\exists\mathbb{R}$, $\forall\exists\mathbb{R}$, $\exists\forall\mathbb{R}$) work with ordinary Turing machines, accepting only strings over finite alphabets.

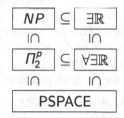

Fig. 2. Relation of complexity classes.

Our Results

It is straightforward to show that AREA UNIVERSALITY belongs to $\forall\exists\mathbb{R}$:

Proposition 1. AREA UNIVERSALITY *is in* $\forall\exists\mathbb{R}$.

The idea is to use a block of universal quantifiers to describe the face area assignment and the block of existential quantifiers to describe the placement of the vertices of the drawing of G. We believe that a stronger statement holds.

Conjecture 1. AREA UNIVERSALITY is $\forall\exists\mathbb{R}$-complete.

While this conjecture, if true, would show that AREA UNIVERSALITY is a really difficult problem in an algebraic and combinatorial sense, it would also give the first known natural geometric problem that is complete for $\forall\exists\mathbb{R}$.

As a first step towards proving our conjecture, we consider three variants of AREA UNIVERSALITY, each approaching the conjecture from a different direction. In Sect. 2 we introduce restricted variants of ETR and UETR which are still complete and may be useful to show hardness for other problems.

Note that two variants that we consider have the spirit of extending a partial drawing with some extra constraints. This problem was shown recently to be $\exists\mathbb{R}$-complete [11]. This work is the *first* to show $\exists\mathbb{R}$-hardness for a problem of drawing a planar graph in the plane.

As a starting point we drop the planarity restriction. For a plane graph G with vertex set V, the *face hypergraph* of G has vertex set V, and its edges correspond to sets of vertices forming the faces (see e.g. [7]). Observe that the face hypergraph of a plane triangulation is 3-uniform, i.e. each hyperedge has 3 vertices. It is clear that AREA UNIVERSALITY can be equivalently formulated in the language of face hypergraphs. This relation motivates the following *partial assignment* (PA) version of the problem.

AREA UNIVERSALITY FOR TRIPLES PA
Input: A set V of vertices, a collection of vertex-triples $F \subseteq \binom{V}{3}$, and a partial area assignment $\mathcal{A}' \colon F' \to \mathbb{R}_0^+$ for some $F' \subseteq F$.
Question: Is it true that for every $\mathcal{A} \colon F \to \mathbb{R}_0^+$, such that $\mathcal{A}(f) = \mathcal{A}'(f)$ for all $f \in F'$, there exists a placement of V in the plane, such that the area for each $f \in F$ is $\mathcal{A}(f)$?

Theorem 1. AREA UNIVERSALITY FOR TRIPLES PA *is* $\forall \exists \mathbb{R}$-*complete*.

For the proof of Theorem 1 we use gadgets similar to the *von Staudt constructions* used to show the $\exists \mathbb{R}$-hardness of order-types, see [12].

Our second result concerns a variant, where we investigate the complexity of realizing a specific area assignment. PRESCRIBED AREA denotes the following problem: Given a plane graph G with an area assignment $\mathcal{A}$, does there exist a crossing-free drawing of G that realizes $\mathcal{A}$? We study a *partial extension* (PE) version of PRESCRIBED AREA, where some vertex positions are fixed and we seek for an area-realizing placement of the remaining vertices.

PRESCRIBED AREA PE
Input: Plane graph $G = (V, E)$, a face area assignment $\mathcal{A}$, fixed positions for $V' \subseteq V$.
Question: Does there exist an $\mathcal{A}$-realizing drawing of G respecting the positions of all $v \in V'$?

We show the following hardness result for PRESCRIBED AREA PE.

Theorem 2. PRESCRIBED AREA PE *is* $\exists \mathbb{R}$-*complete*.

The next two results consider the corresponding question for simplicial complexes in three dimensions. Recall that an *abstract simplicial complex* is a family Σ of non-empty finite sets over a ground set $V = \bigcup \Sigma$, which is closed under taking non-empty subsets. We say Σ is *pure* when the inclusion-wise maximal sets of Σ all have the same number of elements. We say Σ is *realizable* when there is a simplicial complex $\mathcal{S}$ in $\mathbb{R}^3$ that has a vertex for each element of V and a simplex corresponding to each set in Σ.

A *crossing-free drawing* of Σ is a mapping of every $i \in V$ to a point $p_i \in \mathbb{R}^3$, such that the following holds. For any pair of sets $\sigma_1, \sigma_2 \in \Sigma$ there is a separating hyperplane $h = \{x \in \mathbb{R}^3 : \langle a, x \rangle = b\}$ such that $\langle a, p_i \rangle \leq b$ for all $i \in \sigma_1$ and $\langle a, p_i \rangle \geq b$ for all $i \in \sigma_2$. A *volume assignment* for Σ is a non-negative-valued function on the collection T of all 4-element sets in Σ, and a crossing-free drawing of Σ *realizes* a volume assignment $\mathcal{V} : T \to \mathbb{R}_0^+$ when for each $\tau \in T$, the convex hull of the points $\{p_i : i \in \tau\}$ has volume $\mathcal{V}(\tau)$. The analogous questions are:

PRESCRIBED VOLUME
Input: A pure abstract simplicial complex Σ realizable in $\mathbb{R}^3$, and a volume assignment $\mathcal{V}$.
Question: Is there a crossing-free drawing of Σ that realizes $\mathcal{V}$?

VOLUME UNIVERSALITY PA
Input: A pure abstract simplicial complex Σ realizable in $\mathbb{R}^3$, and a partial volume assignment $\mathcal{V}' : T' \to \mathbb{R}_0^+$ for some of the 4-element sets $T' \subseteq T$ of Σ.
Question: Is it true that for every $\mathcal{V} : T \to \mathbb{R}_0^+$, such that $\mathcal{V}(\tau) = \mathcal{V}'(\tau)$ for all $\tau \in T'$, there exists a crossing-free drawing of Σ that realizes $\mathcal{V}$?

Proposition 2. VOLUME UNIVERSALITY PA *is in* ∀∃ℝ.

Note that 3-dimensional simplicial complexes are the analogue of planar triangulations. Indeed, PRESCRIBED AREA for triangulations reduces to PRESCRIBED VOLUME in the following sense:

Proposition 3. *There is a polynomial time algorithm that takes as input any plane triangulation G with positive area assignment $\mathcal{A}$ and outputs a simplicial complex $\mathcal{S}$ with volume assignment $\mathcal{V}$ such that $\mathcal{A}$ is realizable for G if and only if $\mathcal{V}$ is realizable for $\mathcal{S}$.*

Moreover, the analogues of PRESCRIBED AREA and AREA UNIVERSALITY are hard. The two versions read as follows:

Theorem 3. PRESCRIBED VOLUME *is* ∃ℝ-*complete*.

Theorem 4. VOLUME UNIVERSALITY PA *is* ∀∃ℝ-*complete*.

2   Toolbox: Hard Variants of ETR and UETR

In this section we introduce some restricted variants of ETR and UETR which enable us to show hardness. Recently, Abrahamsen et al. showed that the following problem is also ∃ℝ-complete [1].

ETRINV
Input: A formula over the reals of the form $(\exists\, X_1, X_2, \dots, X_n) : \Phi$, where Φ is a conjunction of constraints of the following form: $X = 1$ (introducing a constant 1), $X + Y = Z$ (addition), $X \cdot Y = 1$ (inversion), with $X, Y, Z \in \{X_1, \dots, X_n\}$. Additionally, Φ is either unsatisfiable or has a solution where each variable is within $[1/2, 2]$.
Question: Is the input formula true?

In order to define an even more restricted variant of ETRINV, we need one more definition. Consider a formula Φ of the form $\Phi = \Phi_1 \wedge \Phi_2 \wedge \dots \wedge \Phi_m$, where each Φ_i is a quantifier-free formula of the first-order theory of the reals with variables $X_1, X_2, \dots, X_n$, which uses arithmetic operators and comparisons $(=, <, \leq)$ but no logic symbols. The *incidence graph* of Φ is the bipartite graph with vertex set $\{X_1, X_2, \dots, X_n\} \cup \{\Phi_1, \Phi_2, \dots, \Phi_m\}$ that has an edge $X_i\Phi_j$ if and only if the variable X_i appears in the subformula Φ_j. By PLANAR-ETRINV we denote the variant of ETRINV where the incidence graph of Φ is planar and Φ is either unsatisfiable or has a solution with all variables within $(0, 5)$.

Theorem 5. PLANAR-ETRINV *is* ∃ℝ-*complete*.

Proof. Let $(\exists\, X_1, X_2, \dots, X_n) : \Phi$ be an instance of ETRINV and let G be some embedding of $G(\Phi)$ in $\mathbb{R}^2$. Suppose that G is not crossing-free and consider a pair of crossing edges. Let X and Y denote the variables corresponding to (one endpoint of) these edges. We introduce three new existential variables X', Y', Z

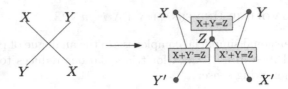

Fig. 3. The crossing gadget.

and three constraints: $X + Y = Z$; $X + Y' = Z$; $X' + Y = Z$. Observe that these constraints ensure that $X = X'$ and $Y = Y'$. Moreover, the embedding of G can be modified so that the new incidence graph has strictly fewer crossings (see Fig. 3): the considered crossing is removed and no new crossing is introduced. We repeat this procedure until the incidence graph of the obtained formula is planar. Finally, note that $0 < 1 \leq Z = X + Y \leq 4 < 5$ whenever $1/2 \leq X, Y \leq 2$. Note that the number of new variables and constraints is at most $O(|\Phi|^4)$, since each constraint in ETRINV has at most three variables.   □

Now we introduce a restricted variant of UETR.

CONSTRAINED-UETR
Input: A formula $(\forall\, Y_1, \ldots, Y_m \in \mathbb{R}^+)\,(\exists\, X_1, \ldots, X_n \in \mathbb{R}^+) : \Phi(X, Y)$ over the reals, where Φ is a conjunction of constraints of the form: $X = 1$ (introducing constant 1), $X + Y = Z$ (addition), $X \cdot Y = Z$ (multiplication), with $X, Y, Z \in \{X_1, \ldots, X_n, Y_1, \ldots, Y_m\}$.
Question: Is the input formula true?

CONSTRAINED-UETR can be seen as a variant of $\forall\exists\mathbb{R}$ that is simplified in a way analogous to a $\exists\mathbb{R}$-complete variant of ETR called INEQ [12,16]. Similarly, we will show that CONSTRAINED-UETR is $\forall\exists\mathbb{R}$-complete.

Theorem 6. CONSTRAINED-UETR *is* $\forall\exists\mathbb{R}$-*complete.*

3 Hardness of AREA UNIVERSALITY FOR TRIPLES PA

Here we prove Theorem 1.

Theorem 1. AREA UNIVERSALITY FOR TRIPLES PA *is* $\forall\exists\mathbb{R}$-*complete.*

Proof. For the containment, it is easy to express the area of a triangle by a polynomial equation: Denoting the coordinates of a vertex v_i by (x_i, y_i), the signed area $A(v_1, v_2, v_3)$ of a counter-clockwise triangle $v_1 v_2 v_3$ can be computed by

$$2 \cdot A(v_1, v_2, v_3) = \det \begin{pmatrix} x_1 & x_2 & x_3 \\ y_1 & y_2 & y_3 \\ 1 & 1 & 1 \end{pmatrix} =: \mathrm{Det}(v_1, v_2, v_3).$$

Thus, we take a conjunction of equations of the above form for each triple.

For the hardness, we reduce from CONSTRAINED-UETR. For every instance Ψ of CONSTRAINED-UETR, we give a set of points V and unordered triples T, along with a partial area assignment $\mathcal{A}'$. Let Ψ be a formula of the form:
$$\Psi = (\forall Y_1, \ldots, Y_m \in \mathbb{R}^+)(\exists X_1, \ldots, X_n \in \mathbb{R}^+) \colon \Phi(Y_1, \ldots, Y_m, X_1, \ldots, X_n).$$
Recall that Φ is a conjunction of constraints of the form $X = 1$, $X + Y = Z$, and $X \cdot Y = Z$. First, we show how to express Φ. Our gadgets are similar to the ones for showing $\exists \mathbb{R}$-hardness of ORDER TYPE (see [12]). All variables are represented by points on one line; which we denote by ℓ for the rest of the proof. First, we enforce points to be on ℓ. Afterwards we construct gadgets for mimicking addition and multiplication.

Introduce three points p_0, p_1, and r and define $\mathcal{A}'(p_0, p_1, r) := 1$. The positive area ensures that the points are not collinear and pairwise different. Without loss of generality we assume that $\|p_0 p_1\| = 1$ and interpret p_0 as 0 and p_1 as 1. Denoting a line through two points a and b by $\ell_{a,b}$, we set $\ell := \ell_{p_0, p_1}$. To force a point x on ℓ, we set $\mathcal{A}'(x, p_0, p_1) := 0$. This introduces no other constraints on the position of x. Each variable X is represented by a point x on ℓ. Additionally, since all variables are non-zero, we introduce a triangle forcing x to be different from p_0. In general, we can ensure that two points x_1 and x_2 are distinct, by introducing a point q and adding a triangle (x_1, x_2, q) with $\mathcal{A}'(x_1, x_2, q) := 1$. The absolute value of X is defined by $\|p_0 x\|$; if x and p_1 lie on the same side of p_0, then the value of X is positive, otherwise it is negative. Here, we allow negative values, but later we force the original variables to be positive.

Now, we describe the addition gadget for a constraint $X + Y = Z$. Let x, y, z be the points encoding the values of X, Y, Z, respectively. Recall that $x, y, z \in \ell$ and $x, y, z \neq p_0$. We introduce a point q_1 and prescribe the areas $\mathcal{A}'(p_0, x, q_1) = \mathcal{A}'(y, z, q_1) = 1$, see on the left of Fig. 4. Since the two triangles have the same height, it holds that $\|yz\| = \|p_0 x\|$. Thus, the value of Z is either $X + Y$ or $X - Y$. Analogously, we introduce a point q_2 and define $\mathcal{A}'(p_0, y, q_2) = \mathcal{A}'(x, z, q_2) = 1$, implying that Z is either $Y + X$ or $Y - X$. Therefore either $Z = X + Y$ (the intended solution) or $Z = X - Y = Y - X$. The latter case implies $X = Y$ and thus $Z = 0$. This contradicts the fact that $z \neq p_0$.

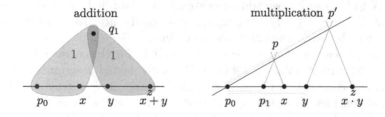

Fig. 4. Gadgets for addition and multiplication.

For the multiplication gadget, we show how to enforce on four pairwise different points p, p', s, s' that $\ell_{p,p'}$ is parallel to $\ell_{s,s'}$, without introducing additional constraints on any of the four points. We insert two new points h_1 on line $\ell_{p,p'}$ and

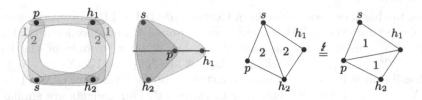

Fig. 5. Forcing a trapezoid.

h_2 on line $\ell_{s,s'}$ by defining $\mathcal{A}'(p,p',h_1) = \mathcal{A}'(s,s',h_2) = 0$. We aim for a trapezoid with points p, h_1, s, h_2 such that ph_1 is parallel to sh_2. For this, we prescribe the areas $\mathcal{A}'(p,h_1,s) = \mathcal{A}'(p,h_1,h_2) = 1$ and $\mathcal{A}'(s,h_2,p) = \mathcal{A}'(s,h_2,h_1) = 2$, see Fig. 5. Indeed, s and h_2 must lie on the same side of the line ℓ_{p,h_1}: Assume by contradiction that ℓ_{p,h_1} separates s and h_2. If p, h_1 are on the same side of ℓ_{s,h_2} then the triangle (s,h_2,p) is contained in or contains the triangle (s,h_2,h_1), see the middle of Fig. 5. However this contradicts the fact that both triangles have the same area and $p \neq h_1$. Consequently, ℓ_{s,h_2} separates p and h_1 and the quadrangle psh_1h_2 can be partitioned by either diagonal sh_2 or ph_1. Thus, $2 = \mathcal{A}(p,h_1,s) + \mathcal{A}(p,h_1,h_2) = \mathcal{A}(s,h_2,h_1) + \mathcal{A}(s,h_2,p) = 4$, which is again a contradiction. Thus s, h_2 lie on the same side of ℓ_{p,h_1}. By the prescribed area, s and h_2 have the same distance to ℓ_{p,h_1}. Hence, the segments ph_1 and sh_2 and the lines $\ell_{p,p'}$ and $\ell_{s,s'}$ are parallel and no further constraints are imposed p, p', s, s'. To construct a multiplication gadget for the constraint $X \cdot Y = Z$, let x, y, z ($\neq p_0$) be the points encoding the values of X, Y, Z, respectively. We introduce two points p, p' not on ℓ, but collinearity with p_0 is forced by $\mathcal{A}'(p_0, p, p') := 0$. By the parallel-line construction we force that $\ell_{p_1, p}$ with $\ell_{y, p'}$ and $\ell_{x, p}$ with $\ell_{z, p'}$ are parallel, see Fig. 4. By the intercept theorem, the following ratios coincide (also for negative variables): $|p_0 p|/|p_0 p'| = |p_0 p_1|/|p_0 y| = |p_0 x|/|p_0 z|$. By definition of x, y, z, we obtain $1/Y = X/Z$, and hence $X \cdot Y = Z$. Recall that $p_1 = 1$. For every universally quantified Y_i, let y_i be the point encoding its value with $y_i \in \ell$ and $y_i \neq p_0$. We introduce a triple $f_i = (p_0, r, y_i)$, whose area is universally quantified. Recall that r is a point with $\mathcal{A}'(p_0, p_1, r) = 1$. To enforce each original variable X to be positive, we add an existentially quantified variable S_X and the constraint $X = S_X \cdot S_X$ where S_X may or may not be positive. This finishes the reduction which clearly runs in polynomial time.

It remains to argue that Ψ is true if and only if, for the constructed instance of AREA UNIVERSALITY FOR TRIPLES PA with partial assignment $\mathcal{A}'$, every assignment $\mathcal{A}$ consistent with $\mathcal{A}'$ is realizable. Suppose Ψ is true, let $\mathcal{A}'$ be as above, and consider an assignment $\mathcal{A}$ that is consistent with $\mathcal{A}'$. Let $V(Y_i)$ be the area assigned to the triple f_i, and let $V(X_i)$ be the value of the variable X_i in some satisfying assignment for Φ. Let $y_1, \ldots, y_m, x_1, \ldots, x_n$ be points positioned on a line at distances from a point p_0 corresponding to these values. Since addition and multiplication relations specified by Φ hold, the corresponding gadgets can be realized, so $\mathcal{A}$ is realizable. Suppose that every assignment $\mathcal{A}$ that is consistent with $\mathcal{A}'$ is realizable and consider values $V(Y_1), \ldots, V(Y_m) \in \mathbb{R}^+$ of the universally quantified variables of Ψ. Then there is a realization of $\mathcal{A}$ where $p_0 \neq p_1$ and each f_i has area $V(Y_i)$. So $V(X_i) = \|x_i - p_0\| / \|p_1 - p_0\|$ is a satisfying assignment for Φ, thus Ψ is true.   □

4 Hardness of PRESCRIBED AREA PE

Here we sketch a proof of Theorem 2 by reducing from PLANAR-ETRINV.

Theorem 2. PRESCRIBED AREA PE *is* $\exists\mathbb{R}$-*complete.*

Proof (Sketch). Let $\Psi = \exists X_1 \ldots X_n : \Phi(X_1, \ldots, X_n)$ be an instance of PLANAR-ETRINV. Recall that we can assume that if Ψ is a positive instance, then it has a solution in which the values of variables are in the interval $(0, 5)$. We construct a plane graph $G_\Psi = (V, E)$, a face area assignment $\mathcal{A}$ of inner faces of G_Ψ, and fixed positions of a subset of vertices, such that G_Ψ has a realizing drawing respecting the position of pre-drawn vertices if and only if Φ is satisfiable by real values from the interval $(0, 5)$. Consider the incidence graph of Φ and fix an orthogonal plane drawing on an integer grid, see Fig. 6 for an example.

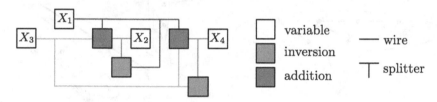

Fig. 6. Incidence graph of $(X_1 + X_2 = X_3) \wedge (X_1 X_2 = 1) \wedge (X_1 + X_4 = X_3) \wedge (X_4 X_3 = 1)$.

To represent each part of G_Ψ, we design several gadgets: *variable gadgets* to represent the variables, as well as *inversion* and *addition gadgets* to realize the constraints. Moreover, we construct *wires* and *splitters* in order to copy and transport information. For an illustration consider Fig. 7. Some vertices in our gadgets have prescribed positions; we call them *fixed*. The remaining vertices are *flexible*. Most flexible vertices lie on a specific segments where the distance to one end of the segment encodes the value of the variable.   □

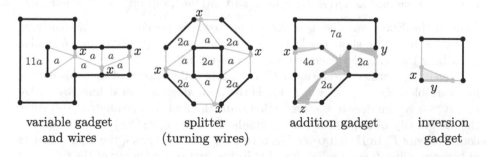

| variable gadget and wires | splitter (turning wires) | addition gadget | inversion gadget |

Fig. 7. Gadgets for Theorem 2; black vertices are fixed, gray vertices are flexible.

5   Volume-Universality

In order to show Theorem 3, we reduce from ETRINV and for Theorem 4 from CONSTRAINED-UETR. We construct a simplicial complex $\mathcal{S} = (V, F)$ and a volume assignment $\mathcal{V}$, such that $\mathcal{S}$ has a $\mathcal{V}$-realization iff Φ is satisfiable. Our essential building block is the *coplanar gadget*. It forces several triangles of *equal area* to lie in a common plane, see Fig. 8 (left). These triangles will be free to one half-space and thus accessible for our further construction. Indeed, all but one vertices lie in the same plane. We use the coplanar gadget to force a set of points representing the values of the variables to lie on a common line ℓ. In order to do so, we take two coplanar gadgets and enforce that their base planes E, E_ℓ are not parallel, see Fig. 8 (right). This allows us to mimic addition and inversion on a line as before. It turns out that in three dimensions we can guarantee crossing-free simplices.

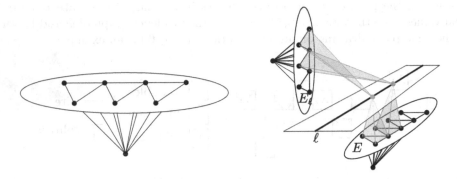

Fig. 8. Two gadgets for PRESCRIBED VOLUME and VOLUME UNIVERSALITY PA.

6   Potential Complete Problems

To motivate the research on $\forall \exists \mathbb{R}$ and $\exists \forall \mathbb{R}$, we present some candidates of problems that might be complete for these classes. A very natural one was suggested by Marcus Schaefer. It is the well-known problem of determining the Hausdorff distance of two semi-algebraic sets: For two sets $A, B \subseteq \mathbb{R}^d$, the Hausdorff distance d_H is defined as $d_H(A, B) = \max\{\sup_{a \in A} \inf_{b \in B} \|ab\|, \sup_{b \in B} \inf_{a \in A} \|ab\|\}$, where $\|ab\|$ denotes the Euclidean distance. As a step towards proving the $\forall \exists \mathbb{R}$-hardness of this problem, we show hardness for a variant where quantifier-free formulas describing the semi-algebraic sets are part of the input.

Given a quantifier-free formula Γ of the first-order theory of the reals with n free variables, $S_\Gamma := \{x \in \mathbb{R}^n : \Gamma(x)\}$ is the semi-algebraic set defined by Γ. By $\pi_k : \mathbb{R}^n \to \mathbb{R}^k$ we denote the projection onto the first k coordinates. Note that the complexity of a quantifier-free formula Γ' defining $\pi_k(S_\Gamma)$ may exceed the complexity of Γ. In HAUSDORFF DISTANCE OF PROJECTIONS, the input consists of two quantifier-free formulas Φ and Ψ in the first-order theory of the reals and $k \in \mathbb{N}$. The question is whether $d_H(\pi_k(S_\Phi), \pi_k(S_\Psi)) = 0$.

Lemma 1. HAUSDORFF DISTANCE OF PROJECTIONS *is ∀∃ℝ-complete.*

Several other candidates of ∀∃ℝ- and ∃∀ℝ-complete problems are related to the notion of *imprecision*, where we assume that our input is only a rough approximation of the 'real' input. Nevertheless, we seek a universal solution that is valid in any case, i.e., for every possible realization of the imprecise data. As an example, consider UNIVERSAL GUARD SET, a variant of the ART GALLERY PROBLEM [1]. For a set of unit disks specifying the imprecise placement of polygon vertices, we ask for a minimum set of guards (points), that can guard every polygon formed by points from the unit disks.

References

1. Abrahamsen, M., Adamaszek, A., Miltzow, T.: The art gallery problem is ∃ℝ-complete. In: Proceedings of the 50th Annual ACM SIGACT Symposium on Theory of Computing. ACM (2018)
2. Basu, S., Pollack, R., Roy, M.-F.: Algorithms in Real Algebraic Geometry. Springer, Heidelberg (2006). https://doi.org/10.1007/3-540-33099-2
3. Biedl, T.C., Velázquez, L.E.R.: Drawing planar 3-trees with given face areas. Comput. Geom. **46**(3), 276–285 (2013)
4. Blum, L., Cucker, F., Shub, M., Smale, S.: Complexity and Real Computation. Springer, New York (2012). https://doi.org/10.1007/978-1-4612-0701-6
5. Davenport, J.H., Heintz, J.: Real quantifier elimination is doubly exponential. J. Symb. Comput. **5**(1), 29–35 (1988)
6. Dobbins, M.G., Kleist, L., Miltzow, T., Rzążewski, P.: ∀∃ℝ-completeness and area-universality. CoRR, abs/1712.05142 (2017)
7. Dvořák, Z., Král', D., Škrekovski, R.: Coloring face hypergraphs on surfaces. Eur. J. Comb. **26**(1), 95–110 (2005)
8. Evans, W.S., et al.: Table cartogram. Comput. Geom. **68**, 174–185 (2018)
9. Kleist, L.: Drawing planar graphs with prescribed face areas. J. Comput. Geom. **9**(1), 290–311 (2018)
10. Levi, F.: Die Teilung der projektiven Ebene durch Gerade oder Pseudogerade. Ber. Math.-Phys. Kl. Sächs. Akad. Wiss **78**, 256–267 (1926)
11. Lubiw, A., Miltzow, T., Mondal, D.: The complexity of drawing a graph in a polygonal region. CoRR, abs/1802.06699 (2018). Accepted at Graph Drawing 2018 (GD 2018)
12. Matoušek, J.: Intersection graphs of segments and ∃ℝ. CoRR, abs/1406.2636 (2014)
13. Mnev, N.E.: The universality theorems on the classification problem of configuration varieties and convex polytopes varieties. In: Viro, O.Y., Vershik, A.M. (eds.) Topology and Geometry—Rohlin Seminar. LNM, vol. 1346, pp. 527–543. Springer, Heidelberg (1988). https://doi.org/10.1007/BFb0082792
14. Richter-Gebert, J.: Perspectives on Projective Geometry: A Guided Tour Through Real and Complex Geometry. Springer, Heidelberg (2011). https://doi.org/10.1007/978-3-642-17286-1
15. Ringel, G.: Equiareal graphs. In: Contemporary Methods in Graph Theory, pp. 503–505 (1990)
16. Schaefer, M., Štefankovič, D.: Fixed points, Nash equilibria, and the existential theory of the reals. Theory Comput. Syst. **60**(2), 172–193 (2017)
17. Thomassen, C.: Plane cubic graphs with prescribed face areas. Comb. Probab. Comput. **1**, 371–381 (1992)

Optimal General Matchings

Szymon Dudycz$^{(\boxtimes)}$ and Katarzyna Paluch

Institute of Computer Science, University of Wrocław, Wrocław, Poland
szymon.dudycz@gmail.com, abraka@cs.uni.wroc.pl

Abstract. Given a graph $G = (V, E)$ and for each vertex $v \in V$ a subset $B(v)$ of the set $\{0, 1, \ldots, d_G(v)\}$, where $d_G(v)$ denotes the degree of vertex v in the graph G, a B-matching of G is any set $F \subseteq E$ such that $d_F(v) \in B(v)$ for each vertex v, where $d_F(v)$ denotes the number of edges of F incident to v. The general matching problem asks the existence of a B-matching in a given graph. A set $B(v)$ is said to have a *gap of length p* if there exists a number $k \in B(v)$ such that $k + 1, \ldots, k + p \notin B(v)$ and $k + p + 1 \in B(v)$. Without any restrictions the general matching problem is NP-complete. However, if no set $B(v)$ contains a gap of length greater than 1, then the problem can be solved in polynomial time and Cornuéjols [5] presented an algorithm for finding a B-matching, if it exists. In this paper we consider a version of the general matching problem, in which we are interested in finding a B-matching having a maximum (or minimum) number of edges.

We present the first polynomial time algorithm for the maximum/minimum B-matching for the case when no set $B(v)$ contains a gap of length greater than 1. This also yields the first pseudopolynomial algorithm for the weighted version of the problem, in which each edge of the graph is assigned a weight and the goal is to compute a minimum or maximum weight B-matching.

1 Introduction

Given a graph $G = (V, E)$ and for each vertex $v \in V$ a subset $B(v)$ of the set $\{0, 1, \ldots, d_G(v)\}$, where $d_G(v)$ denotes the degree of vertex v in the graph G, a B-matching of G is any set $F \subseteq E$ such that $d_F(v) \in B(v)$ for each vertex v, where $d_F(v)$ denotes the number of edges of F incident to v. The general matching problem asks the existence of a B-matching in a given graph. A set $B(v)$ is said to have a *gap of length p* if there exists a natural number $k \in B(v)$ such that $k + 1, \ldots, k + p \notin B(v)$ and $k + p + 1 \in B(v)$. Without any restrictions the general matching problem is NP-complete [14]. However, for the case when no set $B(v)$ contains a gap of length greater than 1, Lovász [14] developed a structural description and Cornuéjols [5] presented a polynomial time algorithm for finding a B-matching, if it exists. It is then one of the strongest generalizations of matchings which is polynomially solvable, unless $\mathcal{P} = \mathcal{NP}$. In

Partly supported by Polish National Science Center grant UMO-2013/11/B/ST6/01748.

A. Brandstädt et al. (Eds.): WG 2018, LNCS 11159, pp. 176–189, 2018.
https://doi.org/10.1007/978-3-030-00256-5_15

the maximum/minimum cardinality variant the goal is to find a B-matching having a maximum/minimum number of edges. In the weighted version of the problem a weight function $w : E \rightarrow \mathbb{R}$ is given and the aim is to find a B-matching that maximizes or minimizes the sum of the weights of the edges.

Previous Work. If $B(v) = \{0, 1\}$ for each vertex v, then a B-matching is in fact a *matching*, i.e., a set of vertex-disjoint edges. A *perfect matching* is a B-matching such that $B(v) = 1$ for each vertex v. Given a function $b : V \rightarrow \mathbb{N}$, a b-matching is any set $F \subseteq E$ such that $d_F(v) \leq b(v)$ for each vertex v and a perfect b-matching or a b-factor is any set $F \subseteq E$ such that $d_F(v) = b(v)$ for each vertex v. If in addition to a function b we are also given a function $a : V \rightarrow \mathbb{N}$, then an (a, b)-matching is any set $F \subseteq E$ such that $a(v) \leq d_F(v) \leq b(v)$ for each vertex v.

All these special cases of the general matching problem are well-solved, both in unweighted and weighted versions. For instance, for the maximum weight b-matching there exist algorithms with the following running times: $\mathcal{O}(n^2 B)$ by Pulleyblank [19], $\mathcal{O}(n^2 m \log B)$ by Marsh [16], $\mathcal{O}(m^2 \log n \log B)$ by Gabow [7], $\mathcal{O}(n^2 m + n \log B(m + n \log n))$ and $\mathcal{O}(n^2 \log n(m + n \log n))$ by Anstee [1], and $\tilde{\mathcal{O}}(W\phi^\omega)$ by Gabow and Sankowski [8], where $n = |V|$, $m = |E|$, $B = \max b(v)$, $\phi = \sum b(v)$, W is a maximum weight of edge and n^ω is the time required to multiply two $n \times n$ matrices. For a good survey on these problems see [22].

In the *antifactor* problem for each vertex v we have $|\{0, 1, \ldots, d_G(v)\} \backslash B(v)| = 1$, meaning that for each vertex there is exactly one degree excluded from the set $B(v)$. Graphs that have an antifactor have been characterized by Lovász in [13].

For the more general case when no set $B(v)$ contains a gap of length greater than 1, Cornuéjols [5] in 1988 presented two solutions to the problem of finding such B-matching, if it exists. One uses a reduction to the edge-and-triangle partitioning problem, in which we are given a graph $G = (V, E)$ and a set T of triangles (cycles of length 3) of G and are to decide if the set of vertices V can be partitioned into sets of cardinality of 2 and 3 so that each set of cardinality 2 is an edge of E and each set of cardinality 3 is a triangle of T. The other is based on an augmenting path approach applied in the modified graph $G' = (V \cup V', E')$ in which each edge e of G is split with two new vertices into three edges. For each new vertex v' the set $B(v')$ is defined to be $\{1\}$ and we start from the set $F \subseteq E'$ such that all requirements regarding vertices of G are satisfied, i.e., $d_F(v) \in B(v)$ for each vertex $v \in V$ and for each vertex $v' \in V'$ it is $d_F(v') \leq 1$. Next we aim to gradually augment F so that it also satisfies the requirements regarding new vertices V' and $d_F(v') = 1$ for each $v' \in V'$. In either case, the computed B-matching is not guaranteed to be of maximum or minimum cardinality. A good characterization of graphs that have a B-matching [23] was provided in 1993 by Sebö [23].

General matchings in bipartite graphs were also studied in terms of their parameterized complexity. Gutin et al. showed that for graphs $G = (U \uplus V, E)$, such that $|B(u)| = 1$ for every $u \in U$, there exists a fixed-parameter tractable algorithm parametrized by the size of V [9].

For the optimization variant of the general matching with no gap greater than 1 Carr and Parekh provided a linear relaxation which is $\frac{1}{2}$-integral [4].

A B-matching is said to be *uniform* if each $B(v)$ is either an interval, i.e., has the form $\{a(v), a(v)+1, \ldots, b(v)\}$ for some nonnegative integers $a(v) \leq b(v)$ or an interval intersected with either even or odd numbers, i.e., has the form $\{a(v), a(v) + 2, \ldots, b(v)\}$ for two nonnegative integers $a(v) \leq b(v)$ such that $b(v) - a(v)$ is even. A maximum/minimum weight uniform B-matching problem was shown to be solvable in polynomial time by Szabó [25]. In the solution to the weighted uniform B-matching Szabó uses the following result of Pap [18]. Let $\mathcal{F}$ be an arbitrary set of odd length cycles of graph G, where a single vertex is considered a cycle of length 1. A *perfect $\mathcal{F}$-matching* is any set of cycles and edges of G such that each vertex belongs to exactly one edge or cycle from $\mathcal{F}$. Pap gave a polynomial time algorithm which minimizes a linear function over the convex hull of perfect $\mathcal{F}$-matchings.

Our Results. We give the first polynomial time algorithm for the maximum/minimum B-matching and $B(v)$ for the case when no set contains a gap of length greater than 1. Our solution yields also the first pseudopolynomial algorithm for the maximum/minimum weight B-matching for the case when no set $B(v)$ contains a gap of length greater than 1.

We provide a structural result for both cardinality and weighted variants, which states that given two B-matchings M and N, their symmetric difference $M \oplus N = (M \setminus N) \cup (N \setminus M)$ can be decomposed into a set of *canonical paths*, a notion which we define precisely later and which plays an analogous role as that of an *alternating path* in the context of standard matchings. A path P is alternating with respect to a matching M if its edges alternate between edges of M and edges not belonging to M. Roughly speaking, a canonical path (with respect to a given B-matching M) consists of a meta-path, that is a sequence of alternating paths, and possibly some number of meta-cycles attached to the endpoints of this meta-path. A meta-cycle is a sequence of alternating paths such that the beginning of the first alternating path coincides with the end of the last alternating path in the sequence. After the application of a canonical path $\mathcal{P}$ to a B-matching M we obtain another B-matching $M' = M \oplus \mathcal{P}$ such that only the parities of the degrees in M and M' of the endpoints of $\mathcal{P}$ are different.

Equipped with this structural result we show how finding a maximum/minimum B-matching can be reduced to a series of computations of a maximum/minimum weight uniform B-matching. In fact we prove that in order to verify if a given B-matching M has maximum/minimum weight it suffices to check if there exists a uniform B-matching of so called *neighbouring type* to M, whose weight is greater/smaller than that of M.

Additionally, we show a very simple reduction of a weighted uniform B-matching to a weighted (a, b)-matching, which yields a more efficient and simpler algorithm than the one by Szabó.

A remaining open problem is whether there exists a polynomial time algorithm for a maximum weight B-matching for the case when no set contains a gap of length greater than 1. It is also possible that our algorithm runs in polynomial time.

Motivation. Matchings, b-matchings and factors are basic combinatorial notions that lie at the foundation of combinatorial optimization. The general matching problem restricted to gaps of at most 1 is one of the strongest generalizations of matching, that was not proven $\mathcal{NP}$-hard. As such it is of theoretical importance to find a polynomial time algorithm for a maximum/minimum cardinality/weight B-matching with gaps at most 1 or in the case of a maximum weight B-matching, to prove that it is $\mathcal{NP}$-hard.

As for practical applications, the general matching is related to the extended global cardinality constraint problem (EGCC). Given a set of variables X, a set of values D, a domain for each variable $D(x) \subseteq D$ and a cardinality set $K(d)$ for each $d \in D$, the goal is to find a valuation of variables, such that the number of variables with value d belongs to $K(d)$. Algorithms for general matchings were used to solve some restricted variants of EGCC [9,21], while the general EGCC is $\mathcal{NP}$-hard. The EGCC problem was used for, among others, staff scheduling in healthcare [3], optical network design [24] or car sequencing [20]. For empirical survey on EGCC see [17].

Related Work. In the deficiency problems the task consists in finding a matching that is as close as possible to given sets $B(v)$. Hell and Kirkpatrick [10] gave an algorithm for finding a minimum deficiency (a,b)-matching among all $(0,b)$-matchings, where the deficiency is measured as the sum of differences $a(v) - d(v)$ over all vertices whose degree is not between $a(v)$ and $b(v)$. They also proved that for another measure of deficiency, namely number of vertices whose degree is outside $(a(v), b(v))$, the problem is NP-hard.

Another related problem consists in decomposing a graph into (a,b)-matchings - a graph that can be decomposed into (a,b)-matchings is called (a,b)-factorable. In [12] Kano gave a sufficient condition for a graph to be $(2a, 2b)$-factorable. Cai [15] generalized this result to $(2a - 1, 2b)$, $(2a, 2b + 1)$ and $(2a - 1, 2b + 1)$-factorable graphs. Hilton and Wojciechowski showed another sufficient condition for an $(r, r + 1)$-factorization of graphs [11].

(a,b)-matchings were also studied in the stable framework - Biró et al. proved that checking whether a stable (a,b)-matching exists is NP-hard [2].

Organization. In Sect. 2 we present a simple reduction for a uniform B-matching. In Sect. 3 we introduce the notion of a canonical paths, followed by the proof of the main theorem of our paper. The proof of a key technical lemma is omitted and available in full version [6]. In Sect. 4 we present an algorithm for a maximum B-matching and a maximum weight B-matching.

2   Uniform B-matching

In this section we show a reduction of a uniform B-matching to an (a,b)-matching.

Suppose an instance of a uniform B-matching involves a graph $G = (V, E)$ and for each vertex $v \in V$ a subset $B(v)$ of the set $\{0, 1, \ldots, d_G(v)\}$. We construct a graph $G' = (V, E \cup E')$ and functions $a, b : V \to \mathbb{N}$ as follows.

If for a vertex v, the set $B(v)$ is an interval $\{c(v), c(v) + 1, \ldots, d(v)\}$ for some nonnegative integers $c(v) \leq d(v)$, then we set $a(v) = c(v)$ and $b(v) = d(v)$. If for a vertex v the set $B(v)$ has the form $\{c(v), c(v) + 2, \ldots, d(v)\}$, i.e., $B(v)$ contains all odd numbers between $c(v)$ and $d(v)$, and $c(v)$ and $d(v)$ are also odd, or $B(v)$ contains all even numbers between $c(v)$ and $d(v)$, and $c(v)$ and $d(v)$ are even, then we add $\frac{d(v)-c(v)}{2}$ loops incident to v and set $a(v) = b(v) = d(v)$. Each loop has weight 0. Apart from this each edge $e \in E$ has the same weight in G and G'. Thus E' consists of some number of loops that are added to each vertex v such that $B(v)$ is not an interval.

Theorem 1. *There is a one-to-one correspondence between uniform B-matchings of G and (a, b)-matchings of G'. A maximum weight (a, b)-matching of G' yields a maximum weight B-matching of G.*

After this reduction the number of edges may increase by at most n^2 and the number of vertices remains the same. To solve the (a, b)-matching we can use an algorithm by Gabow [7]. Its running time on a (multi-)graph with n vertices and m edges is $\sum_{v \in V} b(v) \min(m \log n, n^2)$, which we bound by n^4. As the number of vertices does not change in the reduction, a uniform B-matching can also be found in time $\mathcal{O}(n^4)$.

3   Structure of General B-matchings

In this section we will consider the weighted version of the problem - for the maximum cardinality variant it is enough to set all weights to 1.

Let us first recall and generalise some notions and facts from matching theory. In the case of matchings, it is often convenient to consider the symmetric difference of two matchings. Given two matchings M and N the symmetric difference of M and N, denoted as $M \oplus N$, is equal to $(M \setminus N) \cup (N \setminus M)$. The symmetric difference $M \oplus N$ of two matchings M and N can be decomposed into a set of edge-disjoint alternating paths and alternating cycles, where a path or cycle is said to be *alternating* if its edges belong alternately to M and N. We extend the definition of an alternating path and cycle to the context of B-matchings.

Definition 1. *Let M be any B-matching of G. A sequence of edges $P = ((v_1, v_2), (v_2, v_3), \ldots, (v_{2k-1}, v_{2k}), (v_{2k}, v_1))$ is said to be an alternating cycle (with respect to M) if*

- *for every i such that $1 \leq i \leq k$ the edge (v_{2i-1}, v_{2i}) belongs to M,*
- *$(v_{2k}, v_1) \notin M$ and for every i such that $1 \leq i \leq k - 1$, $(v_{2i}, v_{2i+1}) \notin M$,*
- *each edge of G occurs in P at most once,*
- *vertices $v_1, \ldots, v_{2k}$ are not necessarily distinct.*

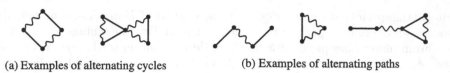

(a) Examples of alternating cycles   (b) Examples of alternating paths

Fig. 1. Alternating cycles and paths.

An alternating path *(with respect to M)* is a sequence of edges $P = ((v_1, v_2), (v_2, v_3), \ldots, (v_k, v_{k+1}))$ *such that*

- *for every i such that* $1 \leq i \leq k - 1$ *exactly one of the edges* $(v_i, v_{i+1}), (v_{i+1}, v_{i+2})$ *belongs to M,*
- *each edge of G occurs in P at most once,*
- *vertices* $v_1, \ldots, v_k$ *are not necessarily distinct,*
- *if* $v_1 = v_{k+1}$, *then either both edges* (v_1, v_2) *and* (v_k, v_1) *are in M, or both are not in M.*

Vertices v_1 and v_{k+1} are called the endpoints of P and edges $(v_1, v_2), (v_k, v_{k+1})$ the ending edges of P.

Examples of alternating paths and cycles are shown in Fig. 1. Throughout the paper we will draw matched edges using wavy lines, and unmatched edges using straight lines.

The decomposition of the symmetric difference of two B-matchings into alternating paths and cycles is not unique. Nevertheless we are interested in *maximal* decompositions, i.e., such that the concatenation of any two alternating paths from the decomposition does not result in a new alternating path or cycle.

By *applying* an alternating path or cycle P to a B-matching M we mean the operation, whose result is $M \oplus P$. We can notice that given any alternating cycle P with respect to a B-matching M, the set $M' = M \oplus P$ is also a B-matching, because $d_{M'}(v) = d_M(v)$ for each vertex v. However, it is not true that for every alternating path P with respect to a B-matching M, $M' = M \oplus P$ is a also B-matching. If v_1, v_2 are the endpoints of P, then $d_{M'}(v_1) \neq d_M(v_1)$ and $d_{M'}(v_2) \neq d_M(v_2)$, so it may happen that $d_{M'}(v_1) \notin B(v_1)$ or $d_{M'}(v_2) \notin B(v_2)$.

We observe the following.

Fact 1. *Given two B-matchings M and N. Let D_- and D_+ denote the sets, respectively, $\{v \in V : d_N(v) < d_M(v)\}$ and $\{v \in V : d_N(v) > d_M(v)\}$ and let D denote $D_- \cup D_+$. Then any maximal decomposition of $M \oplus N$ has the property that each endpoint of an alternating path from the decomposition belongs to D. Also, every ending edge of an alternating path P incident to a vertex v in D_- such that v is an endpoint of P, belongs to M and similarly, every ending edge of an alternating path P incident to a vertex v in D_+ such that v is an endpoint of P, belongs to N.*

Since the application of an alternating path to a B-matching does not necessarily lead to a new B-matching, we need to introduce some generalisation of an alternating path that can be applied in the context of B-matchings in a similar way as an alternating path in the context of (standard) matchings.

From alternating paths of a maximal decomposition of the symmetric difference of two B-matchings M and N we build *meta-paths* and *meta-cycles*. Let $P(u,v)$ denote an alternating path with the endpoints u and v (note that $u, v \in D$). A meta-cycle $\mathcal{C}$ (w.r.t. M) is a sequence of alternating paths of the form $(P(v_1, v_2), P(v_2, v_3), \ldots, P(v_k, v_1))$ such that vertices $v_1, \ldots, v_k$ are pairwise distinct. Analogously, a meta-path $\mathcal{P}(v_1, v_{k+1})$ (w.r.t. M) is a sequence of alternating paths of the form $(P(v_1, v_2), P(v_2, v_3), \ldots, P(v_k, v_{k+1}))$ such that vertices $v_1, \ldots, v_{k+1}$ are pairwise distinct. Let us note that a meta-cycle may consist of one alternating path of the form $P(v, v)$.

For a vertex v and $k \in B(v)$ let $u_k(v)$ be a maximum element of $B(v)$, such that $B(v) \cap [k, u_k(v)]$ does not contain an element of different parity than k. Because $B(v)$ has a gap of length at most 1 we obtain that $B(v) \cap [k, u_k(v)] = \{k, k+2, k+4, \ldots, u_k(v)\}$. Also, either $u_k(v) + 1 \in B(v)$ or $u_k(v)$ is a maximum element of $B(v)$, as otherwise we could increase $u_k(v)$. Similarly let us define $l_k(v)$ to be a minimum element of $B(v)$, such that $B(v) \cap [l_k(v), k]$ does not contain an element of different parity than k.

We define $B_k(v)$ to be

$$B_k(v) := B(v) \cap [l_k(v), u_k(v)] = \{l_k(v), l_k(v) + 2, \ldots, k, \ldots, u_k(v)\}$$

Note that $\{B_k(v)\}_{k \in B(v)}$ is a partition of the set $B(v)$. For a B-matching M we also define $B_M(v) = B_{d_M(v)}(v)$.

Given a B-matching M we say that a B-matching N is **of the same uniform type** as M if for every vertex v it holds that $d_N(v) \in B_M(v)$.

A B-matching N is said to be **of neighbouring type** to a B-matching M if there exists a set W consisting of at most two vertices such that $\forall w \in W : d_N(w) \notin B_M(w)$ and $\forall v \notin W : d_N(v) \in B_M(v)$ and:

- $|W| = 0$, or
- $|W| = 2$ and for $w \in W$ $B_M(w)$ and $B_N(w)$ are adjacent, that is $\max(B_M(w)) + 1 = \min B_N(w)$ or $\max(B_N(w)) + 1 = \min B_M(w)$, or
- $|W| = 1$ and for $w \in W$ there exists k, such that $B_k(w)$ is adjacent to both $B_M(w)$ and $B_N(w)$.

This means that we allow two vertices to have degree outside of $B_M(v)$, but we place limits on how much they can deviate from that set.

We are now ready to give a definition of a **canonical path** - a notion that is going to prove crucial in further analysis and which plays an analogous role as an alternating path in the context of matchings.

Definition 2. *A canonical path $\mathcal{S}(v_1, v_k)$ (with respect to a B-matching M) in a graph G consists of some number of meta-cycles $\mathcal{C}_1, \mathcal{C}_2, \ldots, \mathcal{C}_p$ incident to a vertex v_1, some number of meta-cycles $\mathcal{C}'_1, \mathcal{C}'_2, \ldots, \mathcal{C}'_q$ incident to v_k and in the case $v_1 \neq v_k$ - of a meta-path $\mathcal{P}(v_1, v_k)$ such that the application of all meta-cycles $\mathcal{C}_1, \mathcal{C}_2, \ldots, \mathcal{C}_p, \mathcal{C}'_1, \mathcal{C}'_2, \ldots, \mathcal{C}'_q$ and the meta-path $\mathcal{P}(v_1, v_k)$ to M results in a B-matching of neighbouring type to M.*

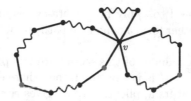

(a) $W = \varnothing$ and $M \oplus N$ is an alternating cycle.

(b) $B_v = \{0, 1, 3, 5, 6\}$. Then $W = \{v\}$ and $M \oplus N$ is a canonical path with one endpoint.

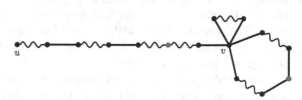

(c) If $B_v = \{0, 2\}$ then $W = \varnothing$. If $B_v = \{0, 1, 2\}$ then $W = \{v\}$. In both cases it is a canonical path with one endpoint.

(d) $B_u = \{0, 1\}$ and $B_v = \{0, 1, 3, 5\}$. Then $W = \{u, v\}$ and $M \oplus N$ is a canonical path with two endpoints.

Fig. 2. Examples of matchings of neighbouring types. Solid edges belong to matching M and wavy edges belong to matching N. For every red vertex w $B_w = \{1\}$ and for every blue vertex z $B_z = \{0, 2\}$. (Color figure online)

Two variants of canonical path - with one endpoint or two endpoints - correspond to different cases in the definition of neighbouring type. Namely, if $v_1 \neq v_k$, then set $W = \{v_1, v_k\}$. Otherwise $v_1 = v_k$ and $W = \{v_1\}$ or $W = \emptyset$. The examples of these cases are presented on Fig. 2.

We will often refer to the weight of a canonical path - that is the effect its application has on a B-matching M. More precisely, for a canonical path $\mathcal{S}$ $w_M(\mathcal{S}) = w(M \oplus \mathcal{S}) - w(M) = \sum_{e \in \mathcal{S} \setminus M} w(e) - \sum_{e \in \mathcal{S} \cap M} w(e)$. Observe that for two edge-disjoint canonical paths $\mathcal{S}_1$ and $\mathcal{S}_2$ we have that $w_M(\mathcal{S}_1) = w_{M \oplus \mathcal{S}_2}(\mathcal{S}_1)$. We will usually write $w(\mathcal{S})$ instead of $w_M(\mathcal{S})$ when the choice of M is clear. Also, when constructing new canonical paths, we will use the notion of a *fine vertex* - we say that a vertex v is *fine in* $\mathcal{S}$ if the number of edges incident to v in $M \oplus \mathcal{S}$ belongs to $B(v)$ and *wrong* otherwise. We say that an endpoint of $\mathcal{S}$ is fine (wrong) if it is fine (wrong) in $\mathcal{S}$. We will say that a path (or cycle) is *positive* if its weight is positive.

In our algorithm we want to subsequently find and apply positive weight canonical paths until a B-matching is optimal. Let us start by showing that it is necessary to consider canonical paths, that is that it may happen that a B-matching is not optimal, but there is no meta-path or meta-cycle augmenting it (i.e. increasing its size). Consider an unweighted graph in Fig. 2d. Then we cannot apply any of the meta-cycles incident to v, because the degree of v would be 2. On the other hand applying the meta-path decreases the size of the B-matching.

Hence we need to apply both meta-cycles and the meta-path at the same time (which together form a canonical path) to obtain a feasible B-matching of greater size.

In the remainder of this section we will prove Theorem 2, which states that if a B-matching M is not optimal, then there exists a canonical path improving it, i.e., such that its application to M gives rise to a B-matching of greater weight. The outline of the proof is as follows. First, in Lemma 1 we prove that any B-matching can be transformed into an optimal one by a sequence of canonical paths. As an optimal B-matching has greater weight, at least one of those paths has positive weight. Next, in Lemma 2 we prove that we can change the order of the canonical paths in such a way that positive weight paths occur earlier in the sequence. The section finishes with the proof of Theorem 2, in which we apply a key technical Lemma 2 to show that we may assume that already the first path in the sequence has positive weight.

We will need a more restricted version of a canonical path. In the example above we have seen that we cannot consider only minimal (with respect to inclusion) canonical paths. Therefore, we introduce another notion, similar to a minimal canonical path but taking into account the weight of a path.

Definition 3. *We say that $\mathcal{S}$ is a basic (canonical) path if it is a canonical path and for no proper subset $\mathcal{S}' \subsetneq \mathcal{S}$, $\mathcal{S}'$ is a canonical path such that $w(\mathcal{S}') \geq w(\mathcal{S})$ or $w(\mathcal{S}') > 0$.*

Observation 1. *Let M be a B-matching. If there exists a canonical path $\mathcal{S}$ w.r.t. M, then there exists a basic canonical path $\mathcal{S}' \subseteq \mathcal{S}$ w.r.t M.*

Lemma 1. *Let M, N be two B-matchings. Then there exists a sequence $\mathcal{S}_1, \mathcal{S}_2, \ldots, \mathcal{S}_k$ and a set of alternating cycles $C_1, C_2, \ldots, C_l$ that satisfy the following.*

1. *Let M_0 denote $M \oplus \bigcup_{i=1}^{l} C_i$. For each i such that $0 < i \leq k$ $\mathcal{S}_i$ is a basic canonical path with respect to M_{i-1} and $M_i = M_{i-1} \oplus \mathcal{S}_i$. Also, $M_k = N$.*
2. *$M \oplus N = \bigcup_{i=1}^{k} \mathcal{S}_i \cup \bigcup_{i=1}^{l} C_i$, where every two elements of the set. $\{\mathcal{S}_1, \ldots, \mathcal{S}_k, C_1, \ldots, C_l\}$ are edge-disjoint.*

Proof. Let us consider some fixed maximal decomposition of $M \oplus N$. Let $C_1, C_2, \ldots, C_l$ denote all alternating cycles of this decomposition. By M_0 we denote $M \oplus \bigcup_{i=1}^{l} C_i$.

If $d_M(v) = d_N(v)$ for every vertex v, then $M \oplus N$ consists solely of alternating cycles $C_1, C_2, \ldots C_l$ and $M_0 = N$ and we are done.

The maximal decomposition of $M_0 \oplus N$ consists only of alternating paths. The *distance* of two B-matchings M and N, denoted as $dist(M, N)$, is defined as

$$dist(M, N) = \sum_{v \in V} |d_N(v) - d_M(v)|$$

In the distance of two B-matchings it is enough to consider the vertices belonging to D, i.e., $dist(M, N) = \sum_{v \in D} |d_N(v) - d_M(v)|$.

Let M_0 and N be two matchings such that the set D corresponding to them is not empty, i.e. there exists a vertex v such that $d_{M_0}(v) \neq d_N(v)$ and hence $dist(M_0, N) > 0$. We show how to construct some canonical path S with respect to M_0 such that the B-matching $M_1 = M_0 \oplus S$ satisfies: $D(M_1, N) \subseteq D(M_0, N), D_-(M_1, N) \subseteq D_-(M_0, N), D_+(M_1, N) \subseteq D_+(M_0, N)$ and $dist(M_1, N) \prec dist(M_0, N)$.

We start by setting S to be any alternating path P that belongs to a maximal decomposition of $M_0 \oplus N$. P may have two different endpoints or one endpoint. If P is not a canonical path, then it means that after its application for at least one of its endpoints v_1 or v_2 it holds that $d_{M_0 \oplus P}(v_i) \notin B(v_i)$, where $i \in \{1, 2\}$. We can notice that apart from this P satisfies all the other conditions of a canonical path. We are going to gradually extend S so that we obtain a canonical path. At each stage of the construction the candidate S for a canonical path has all the properties of a canonical path except for the fact that for one or two of its endpoints it holds that $d_{M_0 \oplus S}(v_i) \notin B(v_i)$, where $i \in \{1, 2\}$.

If S has two endpoints, then both endpoints have degree one. If v_i is not fine in S it means that either $B(v_i)$ contains $d_{M_0}(v_i)$ and $d_{M_0}(v_i) + 2$, but does not contain $d_{M_0}(v_i) + 1$ ($v_i \in D_+$), or $B(v_i)$ contains $d_{M_0}(v_i)$ and $d_{M_0}(v_i) - 2$, but does not contain $d_{M_0}(v_i) - 1$ ($v_i \in D_-$). Then if we add another alternating path starting at v_i, it will cease to be an endpoint of S and its degree will belong to $B_{M_0}(v_i)$. This will be true at each step of our construction - a vertex v that is not an endpoint satisfies $d_{M_0 \oplus S}(v) \in B_{M_0}(v)$. Another invariant that will be maintained during the construction is the following: if there are two endpoints of S their degrees will be odd in S, and if the two endpoints join into one (thus $v_1 = v_2$), then their degree is even in S.

Assume then that we have some candidate path with one endpoint v_1 or two endpoints v_1, v_2, which is not a canonical path, so $d_{M_0 \oplus S}(v_1) \notin B(v_1)$. Since N is a B-matching, there exists an alternating path P' in the maximal decomposition of $(M_0 \oplus S) \oplus N$ with one endpoint v_1. This path has the property that either P and S both diminish the number of edges incident to v_1, or they both increase the number of edges incident to M_0, or our alternating paths would not be maximal. After adding P to S the following things may happen:

1. P has two different endpoints v_1, v_3. Then vertex v_1 is fine in $S \cup P$. If v_3 is not an endpoint of any alternating path belonging to S, then v_3 is a new endpoint of $S \cup P$ and either (i) v_3 is fine in $S \cup P$ and we have decreased the number of wrong endpoints by one or (ii) v_3 is wrong in $M \oplus (S \cup P)$ and the number of wrong endpoints of $S \cup P$ is the same as the number of wrong endpoints of S and we continue the process treating $S \cup P$ as the new candidate for a canonical path. If v_3 is an endpoint of some alternating path belonging to S, then we have created a new meta-cycle C incident to v_3. If v_3 is fine in $S \cup C$, then we decreased the number of wrong endpoints. If v_3 is fine in C then C is a canonical path with respect to M_0. Otherwise it means that $d_{M_0}(v_3) + 2 \notin B(v_3)$ (if $v_3 \in D_+(M_0, N)$) or $d_{M_0}(v_3) - 2 \notin B(v_3)$ (if $v_3 \in D_-(M_0, N)$), so v_3 must be the other endpoint of S. In this case we have only one wrong endpoint left, v_3, and we continue extending S from v_3.

Note that now that two endpoints have joined in v_3, we seemingly have only one endpoint. However, after the addition of an alternating path with two endpoints v_3 and v', $\mathcal{S}$ will have two endpoints - v_3 and v', where v_3 is fine.

2. P has one endpoint v_1. If v_1 is fine in $\mathcal{S} \cup \mathcal{P}$, then we have decreased the number of wrong endpoints of a candidate for a canonical path. Otherwise if P is a canonical path we are done. The only case left is when v_1 is not fine but $d_{M_0}(v_1) + 2 \notin B(v_1)$ (if $v_3 \in D_+(M_0, N)$) or $d_{M_0}(v_3) - 2 \notin B(v_3)$ (if $v_3 \in D_-(M_0, N)$). This may only happen if both endpoints of $\mathcal{S}$ are the same vertex and then we continue extending $\mathcal{S}$ with only one wrong endpoint left.

That way we have constructed a canonical path $\mathcal{S}$ w.r.t. M. By Observation 1 it means that there exists a basic canonical path $\mathcal{S}'$. We can continue finding canonical paths in the same way, this time in $M_0 \oplus \mathcal{S}' \oplus N$. Each such basic canonical path decreases the distance between M and N, which means that way we can decompose $M_0 \oplus N$ into a finite number of basic canonical paths.

Now we are ready to state the key technical lemma.

Lemma 2. *Let M and N be two B-matchings, such that $w(M) < w(N)$. Let $\mathcal{Q}$ be a basic canonical path w.r.t. M contained in $M \oplus N$ and $\mathcal{R}$ a basic canonical path w.r.t. $M \oplus \mathcal{Q}$ and N such that $w(\mathcal{Q}) \leq 0$ and $w(\mathcal{R}) > 0$. Then there exists a canonical path $\mathcal{T}$ w.r.t. M such that $w(\mathcal{T}) > w(\mathcal{Q})$.*

We defer the proof of this lemma to the full version of this paper [6] and let us focus on its consequences.

Theorem 2. *If there exists a B-matching of greater weight than M, then there exists a B-matching of greater weight than M that is of the same uniform type as M or that is of neighbouring type to M.*

Proof. Suppose that there does not exist a B-matching M' of the same uniform type as M and with greater weight than M but there exists a B-matching N having greater weight than M.

By Lemma 1 we know that there exists a sequence of basic canonical paths $\mathcal{S}_1, \mathcal{S}_2, \ldots, \mathcal{S}_k$ and a set of alternating cycles $C_1, C_2, \ldots, C_l$ such that $M \oplus N = \bigcup_{i=1}^{k} \mathcal{S}_i \cup \bigcup_{i=1}^{l} C_i$. The weight of N satisfies $w(N) = w(M) + \sum_{i=1}^{l} w(C_i) + \sum_{i=1}^{k} w(\mathcal{S}_i)$. Since $w(N) > w(M)$ there exists some alternating cycle C_i among the cycles $C_1, \ldots, C_l$ with positive weight or there exists some canonical path $\mathcal{S}_i$ among the canonical paths $\mathcal{S}_1, \mathcal{S}_2, \ldots, \mathcal{S}_k$ with positive weight.

We may, however, observe that if some alternating cycle C_i has positive weight, then $M \oplus C_i$ is of the same uniform type as M and has greater weight than M, which finishes the proof. Assume then, that all alternating cycles have nonpositive weight. As alternating cycles do not change the degree of any vertex, we may apply them after canonical paths. Therefore, let $N' = M \oplus \bigcup_{i=1}^{k} \mathcal{S}_i$ and note that it is also a B-matching, as $\forall v \, d'_N(v) = d_N(v)$. Its weight, however, is not smaller than the weight of N, because we omitted negative weight alternating cycles. Therefore, we can assume that the decomposition of $M \oplus N$ does not contain any alternating cycles.

By Lemma 1 there exists some sequence of basic canonical paths that forms a decomposition of $M \oplus N$, but it is not necessarily unique. From all such sequences let us choose that one, in which S_1 has maximum weight. Let $M_0 = M$ and $M_1 = M \oplus S_1$. For each $i > 1$, S_i is a basic canonical path with respect to M_{i-1} of maximum weight and $M_i = M_{i-1} \oplus S_i$.

Note that when choosing S_i of maximum weight, we will always be able to complete the sequence of canonical paths, because M_i is a B-matching and thus we can apply Lemma 1.

Some basic canonical path S_i must of course have positive weight. Let i be the smallest such index. We will show, that in the chosen decomposition $i = 1$. Assume then, that $i > 1$.

It means that S_i has positive weight and $w(S_{i-1}) \leq 0$. Then, by Lemma 2 and Observation 1, there exists a basic canonical path S'_{i-1} with respect to M_{i-2} such that $w(S'_{i-1}) > w(S_{i-1})$, which contradicts the properties of our decomposition, because instead of adding S_{i-1}, we would choose S'_{i-1}.

4 Algorithm for Computing a Maximum Cardinality B-matching

In this section we will show the algorithmic consequences of Theorem 2, namely we will present a polynomial time algorithm for a maximum cardinality B-matching.

First, let us assume that we have some B-matching M. We want to be able to either verify that it is maximum or find a B-matching of greater cardinality. According to Theorem 2, M is not maximum if and only if there exists a larger B-matching M' such that at most two vertices' degrees are not in $B_M(v)$. Therefore, we can consider all possible sets of at most two vertices, whose degrees would not be restricted to $B_M(v)$. As resulting B-matching would be of neighboring type, we can limit their degrees to adjacent uniform interval of degrees. For the rest of vertices we allow them to have any degree in $B_M(v)$. Therefore, we obtain 4 instances of a uniform B-matching, hence we use Theorem 1 to solve it.

This approach requires solving $\mathcal{O}(n^2)$ instances of a maximum weight uniform B-matching problem and each of them takes $\mathcal{O}(n^4)$ time.

In order to find a maximum cardinality B-matching we start by running Cornuejols' algorithm, which finds any B-matching or verifies that the graph does not admit a B-matching. Then we subsequently augment this matching until it is maximum. The size of a maximum matching can be bounded by the number of edges in the graph, thus the total complexity is $\mathcal{O}(mn^6)$.

This algorithm can be also used for finding a maximum weight B-matching, however, since the value maximum weight B-matchings can be bounded only by mW, where $W = \max |w(e)|$, the algorithm becomes pseudopolynomial.

Algorithm Max B-Matching

1. Using Cornuejols' algorithm find some B-matching M.
2. **while** there exists a B-matching M' of neighbouring type to M with cardinality greater than that of M **do:**
   $$M \leftarrow M'$$
3. Output M.

References

1. Anstee, R.P.: A polynomial algorithm for b-matchings: an alternative approach. Inf. Process. Lett. **24**(3), 153–157 (1987)
2. Biró, P., Fleiner, T., Irving, R.W., Manlove, D.F.: The college admissions problem with lower and common quotas. Theor. Comput. Sci. **411**(34), 3136–3153 (2010)
3. Bourdais, S., Galinier, P., Pesant, G.: HIBISCUS: a constraint programming application to staff scheduling in health care. In: Rossi, F. (ed.) CP 2003. LNCS, vol. 2833, pp. 153–167. Springer, Heidelberg (2003). https://doi.org/10.1007/978-3-540-45193-8_11
4. Carr, R., Parekh, O.: A 12-integral relaxation for the a-matching problem. Oper. Res. Lett. **34**(4), 445–450 (2006)
5. Cornuéjols, G.: General factors of graphs. J. Comb. Theory, Ser. B **45**(2), 185–198 (1988)
6. Dudycz, S., Paluch, K.E.: Optimal general matchings. CoRR, abs/1706.07418 (2017)
7. Gabow, H.N.: An efficient reduction technique for degree-constrained subgraph and bidirected network flow problems. In: Proceedings of the Fifteenth Annual ACM Symposium on Theory of Computing, STOC 1983, pp. 448–456. ACM, New York (1983)
8. Gabow, H.N., Sankowski, P.: Algebraic algorithms for b-matching, shortest undirected paths, and f-factors. In: 54th Annual IEEE Symposium on Foundations of Computer Science, FOCS 2013, Berkeley, CA, USA, 26–29 October 2013, pp. 137–146 (2013)
9. Gutin, G., Kim, E.J., Soleimanfallah, A., Szeider, S., Yeo, A.: Parameterized complexity results for general factors in bipartite graphs with an application to constraint programming. Algorithmica **64**(1), 112–125 (2012)
10. Hell, P., Kirkpatrick, D.G.: Algorithms for degree constrained graph factors of minimum deficiency. J. Algorithms **14**(1), 115–138 (1993)
11. Hilton, A.J.W., Wojciechowski, J.: Semiregular factorization of simple graphs. AKCE Int. J. Graphs Comb. **2**(1), 57–62 (2005)
12. Kano, M.: [a, b]-Factorization of a graph. J. Graph Theory **9**(1), 129–146 (1985)
13. Lovász, L.: Antifactors of graphs. Periodica Mathematica Hungarica **4**(2), 121–123 (1973)
14. Lovász, L.: The factorization of graphs. ii. Acta Mathematica Hungarica **23**(1–2), 223–246 (1972)

15. Cai, M.-C.: [a, b]-Factorizations of graphs. J. Graph Theory **15**(3), 283–301 (1991)
16. Marsh III, A.B.: Matching algorithms. Ph.D. thesis, The Johns Hopkins University (1979)
17. Nightingale, P.: The extended global cardinality constraint: an empirical survey. Artif. Intell. **175**(2), 586–614 (2011)
18. Pap, G.: A TDI description of restricted 2-matching polytopes. In: Bienstock, D., Nemhauser, G. (eds.) IPCO 2004. LNCS, vol. 3064, pp. 139–151. Springer, Heidelberg (2004). https://doi.org/10.1007/978-3-540-25960-2_11
19. Pulleyblank, W.R.: Faces of matching polyhedra. Ph.D. thesis, Department of Combinatorics and Optimization, University of Waterloo (1973)
20. Régin, J.-C., Puget, J.-F.: A filtering algorithm for global sequencing constraints. In: Smolka, G. (ed.) CP 1997. LNCS, vol. 1330, pp. 32–46. Springer, Heidelberg (1997). https://doi.org/10.1007/BFb0017428
21. Samer, M., Szeider, S.: Tractable cases of the extended global cardinality constraint. In: Proceedings of the Fourteenth Symposium on Computing: The Australasian Theory, CATS 2008, Darlinghurst, Australia, vol. 77, pp. 67–74. Australian Computer Society Inc. (2008)
22. Schrijver, A.: Combinatorial Optimization: Polyhedra and Efficiency, vol. 24. Springer, Heidelberg (2002)
23. Sebö, A.: General antifactors of graphs. J. Comb. Theory, Ser. B **58**(2), 174–184 (1993)
24. Simonis, H.: A hybrid constraint model for the routing and wavelength assignment problem. In: Gent, I.P. (ed.) CP 2009. LNCS, vol. 5732, pp. 104–118. Springer, Heidelberg (2009). https://doi.org/10.1007/978-3-642-04244-7_11
25. Szabó, J.: Good characterizations for some degree constrained subgraphs. J. Comb. Theory, Ser. B **99**(2), 436–446 (2009)

Quasimonotone Graphs

Martin Dyer and Haiko Müller[✉]

School of Computing, University of Leeds, Leeds LS2 9JT, UK
{M.E.Dyer,H.Muller}@leeds.ac.uk

Abstract. For any class $\mathcal{C}$ of bipartite graphs, we define quasi-$\mathcal{C}$ to be the class of all graphs G such that every bipartition of G belongs to $\mathcal{C}$. This definition is motivated by a generalisation of the switch Markov chain on perfect matchings from bipartite graphs to nonbipartite graphs. The monotone graphs, also known as bipartite permutation graphs and proper interval bigraphs, are such a class of bipartite graphs. We investigate the structure of quasi-monotone graphs and hence construct a polynomial time recognition algorithm for graphs in this class.

1 Introduction

In [5] (with Jerrum) and [6] we considered the *switch* Markov chain on perfect matchings in bipartite and nonbipartite graphs. This chain repeatedly replaces two matching edges with two non-matching edges involving the same four vertices. We considered the ergodicity and mixing properties of the chain.

In particular, we proved in [5] that the chain is *rapidly mixing* (*i.e.* converges in polynomial time) on the class of *monotone* graphs. This class of bipartite graphs was defined by Diaconis, Graham and Holmes in [4], motivated by statistical applications of perfect matchings. The biadjacency matrices of graphs in the class have a "staircase" structure. Diaconis *et al.* conjectured the rapid mixing property shown in [5]. We also showed in [5] that this class is, in fact, identical to the known class of *bipartite permutation* graphs [14], which is itself known to be identical to the class of *proper interval bigraphs* [9].

In extending the work of [5] to nonbipartite graphs in [6], we showed that the rapid mixing proof for monotone graphs extends easily to a class of graphs which includes, beside the monotone graphs themselves, all *proper*, or *unit*, *interval graphs* [1]. In this class the bipartite graph given by the cut between any bipartition of the vertices of the graph must be a monotone graph. We called these graphs *quasimonotone*.

In fact, "quasi-" is an operator on bipartite graph classes, and can be applied more generally. In this view, quasimonotone graphs are quasi-monotone graphs, as formally defined in Sect. 2, and discussed in Sect. 2.1, below. For any class of bipartite graphs that is recognisable in polynomial time, the definition of its quasi-class implies membership in co-$\mathbb{NP}$. Thus an immediate question is whether we can recognise the quasi-class in polynomial time. The main contribution of this paper is a polynomial time recognition algorithm for quasimonotone graphs.

© Springer Nature Switzerland AG 2018
A. Brandstädt et al. (Eds.): WG 2018, LNCS 11159, pp. 190–202, 2018.
https://doi.org/10.1007/978-3-030-00256-5_16

1.1   Definitions and Notation

If $G = (V, E)$ is a graph and $U \subseteq V$, then $G[U]$ is the subgraph induced by U. Often we do not distinguish between the set U and the subgraph it induces. So a cycle in G is either a subgraph or the set of its vertices. Similarly, we will write $G - H$ when G is isomorphic to H. A subgraph of G is a *cycle* in G if it is connected and 2-regular. The *length* or *size* of a cycle is the number of its edges (or vertices). A *chord* of a cycle (U, F) in G is an edge in $U^{(2)} \cap E \setminus F$. A chord in a cycle of even length is *odd* if the distance between its endpoints on the cycle is odd. That is, an *odd chord* splits an even cycle into two cycles of even length. An *even chord* splits an even cycle into two cycles of odd length.

A *hole* in a graph is a chordless cycle of length at least five. A cycle of length three is a *triangle*, and a cycle of length four a *quadrangle*. A hole is *odd* if it has an odd number of vertices, otherwise *even*. Let HoleFree be the class of graphs without a hole, and EvenHoleFree the class of graphs without even holes. A *long* hole is an odd hole of size at least 7.

For a graph $G = (V, E)$, $L \subseteq V$ and $R = V \setminus L$ the graph $G[L{:}R]$ is the bipartite graph with *bipartition* L, R, and edge set the cut $L{:}R = \{xy \in E : x \in L, y \in R\}$. We refer to $G[L{:}R]$ as a *bipartition* of G.

The *distance* $\mathrm{dist}(u, v)$ between two vertices u and v is the length of a shortest $(u, \ldots, v)$ path in G. For vertices x and y in a subgraph H of G we denote their distance in H by $\mathrm{dist}_H(x, y)$. If $v \in V$, $\mathrm{dist}(v, H)$ is the smallest distance $\mathrm{dist}(v, w)$ from v to any vertex $w \in H$. The maximum distance between two vertices in G is the *diameter* of G. The *neighbourhood* of a vertex v is $N(v)$.

1.2   Structure of the Paper

In 2 we discuss quasi-classes and give examples in Sect. 2.1. Sections 3 to 6 show that quasimonotone graphs can be recognised in polynomial time. In Sect. 3.1 we prove some properties of quasimonotone graphs, using their characterisation by forbidden induced subgraphs. The anticipated recognition algorithm first looks for flaws (defined in Sect. 3.1) and then branches into different procedures depending on the length of a short hole (defined in Sect. 3.3) in the input graph. The remaining forbidden subgraphs are preholes, also defined in Sect. 3.1.

Sections 4 and 5 deal with graphs containing a long hole. We start with lemmas showing that the long hole enforces an annular structure in the absence of flaws. The structure is determined by *splitting*, described in Sect. 5.1. Possible preholes must wind round this annulus once or twice. We complete the process by checking for preholes, using a procedure given in Sect. 5.2. The remaining cases where no long hole exists are considered in Sect. 6. Finally Sect. 7 concludes the paper.

A more detailed version (also with more examples) is available, see [7].

2   Quasi-classes and Pre-graphs

A *hereditary* class of graphs is closed under induced subgraphs. Let BIPARTITE denote the class of bipartite graphs, and let $C \subseteq$ BIPARTITE. Then we will say that the graph G is *quasi-C* if $G[L{:}R] \in C$ for all bipartitions L, R of V.

Lemma 1. *If $C \subseteq$ BIPARTITE is a hereditary class that is closed under disjoint union then $C =$ BIPARTITE $\cap$ quasi-C.*

Proof. First let $G = (L \cup R, E)$ be any bipartite graph that does not belong to C. Since $G = G[L{:}R]$ the graph G does not belong to quasi-C. Hence $C \supseteq$ BIPARTITE $\cap$ quasi-C.

Next we show $C \subseteq$ BIPARTITE $\cap$ quasi-C. Let $G = (X \cup Y, E)$ be a graph in C and let $L{:}R$ be a bipartition of $X \cup Y$. Now $G[L{:}R]$ is the disjoint union of $G_1 = G[(X \cap L) \cup (Y \cap R)]$ and $G_2 = G[(X \cap R) \cup (Y \cap L)]$. The graphs G_1 and G_2 belong to C since the class is hereditary, and hence $G[L{:}R]$ is in C because C is closed under disjoint union. Thus $G \in$ quasi-C.   □

A hereditary graph class can equally well be characterised by a set $\mathcal{F}$ of forbidden subgraphs. The set $\mathcal{F}$ is minimal if no graph in $\mathcal{F}$ contains any other as an induced subgraph. For a bipartite graph H, a graph $G = (V, E)$ is a *pre-H* if there is a bipartition L, R of V such that $G[L{:}R] = H$. In this case H is a spanning subgraph of G. Clearly any bipartite graph H is itself a pre-H.

Lemma 2. *If $C \subseteq$ BIPARTITE is characterised by a set $\mathcal{F}$ of forbidden induced subgraphs, let pre-$\mathcal{F} = \{$pre-$H \mid H \in \mathcal{F}\}$. Then quasi-$C$ is characterised by the set of forbidden induced subgraphs pre-$\mathcal{F}$.*

Proof. Suppose $G = (V, E)$ contains $H' = (V', E')$, a pre-H for some $H \in \mathcal{F}$. Then V' has a bipartition L', R' such that $H'[L'{:}R'] = H$. Extending L', R' to a bipartition L, R of V, $G[L{:}R]$ contains H. Then $G[L{:}R] \notin C$, so $G \notin$ quasi-C. Conversely, if $G \in$ quasi-C, every $G[L{:}R] \in C$, so no $G[L{:}R]$ contains H, for any $H \in \mathcal{F}$. Thus G contains no pre-H, for any $H \in \mathcal{F}$, that is, no $H' \in$ pre-$\mathcal{F}$.   □

2.1   Examples

The class quasi-BIPARTITE is clearly the set of all graphs.

If C is the class of complete bipartite graphs, it is easy to see that quasi-C is the class of complete graphs. Note however, that this class is not closed under disjoint union. Now, if C becomes the class of graphs for which every component is complete bipartite, then quasi-C is the class of graphs without P_4, paw or diamond. These three graphs are the pre-P_4's, see Fig. 1.

If C_d is the class of bipartite graphs with degree at most d, for a fixed integer $d > 0$, then quasi-C_d is the class of all graphs with degree at most d. The unique forbidden subgraph for C_d is clearly the star $K_{1,d+1}$. Therefore, the class quasi-C_d is characterised by forbidding pre-$K_{1,d+1}$'s, a set with size $O(d^2)$. Hence quasi-C_d can be recognised in polynomial time, for fixed d.

Fig. 1. The pre-P_4's: the path P_4, the paw and the diamond

A less obvious example is for the class $\mathcal{C}$ of *linear forests*, which are disjoint unions of paths. Its quasi-class contains all graphs with connected components that are either a path or an odd cycle.

CHORDALBIPARTITE is the class of hole-free bipartite graphs. ODDCHORDAL is the class of graphs in which every even cycle of length at least six has an odd chord. We show in [6] that quasi-CHORDALBIPARTITE = ODDCHORDAL. The complexity of the recognition problem for the class ODDCHORDAL is open, even though CHORDALBIPARTITE can be recognised in almost linear time [12].

3   The Structure of Quasimonotone Graphs

3.1   Flaws and Preholes

A bipartite graph is *monotone* if and only if the rows and columns of its biadjacency matrix can be permuted such that the ones appear consecutively and the boundaries of these intervals are monotonic functions of the row or column index. That is, all the ones are in a staircase-shaped region in the biadjacency matrix. A bipartite graph is monotone if and only if it does not contain a hole, tripod, stirrer or armchair as induced subgraph, see Fig. 2 and [11] Lemma 1.46 on page 52 or [2] Proposition 6.2.1 on page 93. Monotone graphs are also called *bipartite permutation graphs* [14] and *proper interval bigraphs* [9].

Fig. 2. The tripod, the stirrer and the armchair.

Let MONOTONE denote the class of monotone graphs, then the QUASIMONO-TONE will denote the class quasi-MONOTONE. Two example graphs are shown in Fig. 3. Let FLAW be the class containing all pre-tripods, pre-stirrers and pre-armchairs. We will say that any graph in FLAW is a *flaw*. A *flawless* graph G will be one which contains no flaw as an induced subgraph. Since all flaws have seven vertices, we can test in $O(n^7)$ time whether an input graph G on n vertices is flawless. Let FLAWLESS denote the class of flawless graphs, and let QUASIMONOTONE be the class of quasimonotone graphs.

Let $P = (p_1, p_2, \ldots, p_\ell)$ be a path or even cycle in G. The *alternating bipartition* L, R of P assigns $L = \{p_1, p_3, \ldots\}$ and $R = \{p_2, p_4, \ldots\}$. The path P is

Fig. 3. Two quasimonotone graphs

prechordless if it is an induced path $G[L{:}R]$. Similarly, let $C = (p_1, p_2, \ldots, p_\ell)$ be an even cycle in G. Then C is a prehole if it is a hole in $G[L{:}R]$. Thus C must be an even cycle, and all chords must run between L and L or R and R in an alternating bipartition L, R of C. This is equivalent to requiring that C has no odd chord. The alternating partition is inconsistent for an odd cycle, so an odd cycle C cannot be a prehole.

3.2   Properties of Flawless Graphs

Lemma 3. *Let $G \in$ FLAWLESS. Let $P = (p_1, p_2, p_3, p_4, p_5, p_6, p_7)$ be a prechordless path in G, $(p_2, p_3, p_4, p_5, p_6)$ be a hole in G, or $(p_1, p_2, p_3, p_4, p_5, p_6)$ be a prehole in G. If $v \notin P$ is such that $\mathrm{dist}(v, P) = \mathrm{dist}(v, p_4)$, then $\mathrm{dist}(v, p_4) = 1$.*

Lemma 4. *Every hole or prehole in a connected flawless graph is dominating.*

Proof. Let C be an odd hole or prehole in the connected flawless graph G. We show $\mathrm{dist}(v, C) \leq 1$ for every vertex v of G. If $v \in C$, this is obvious. Otherwise, let w be a vertex such that $\mathrm{dist}(v, C) = \mathrm{dist}(v, w)$. Consider the subpath $P = (p_1, p_2, \ldots, p_7)$ of C such that $w = p_4$, where this path wraps around C if $|C| < 7$. Since C is a hole or a prehole, P is prechordless. The result then follows from Lemma 3. $\square$

If C is an odd hole we will call $n(C) = \{v \in V : \mathrm{dist}(v, C) \leq 1\}$, the neighbourhood of C. If G is connected then $G = N(C)$ for any odd hole $C \subseteq G$.

Lemma 5. *Suppose $G \in$ FLAWLESS $\cap$ EVENHOLEFREE, and that C is an odd hole in G, of length at least seven. Then every vertex $v \in V$ has at most three neighbours in C. If there are two neighbours, w, x, then $\mathrm{dist}_C(w, x) = 2$. If there are three neighbours, w, x, y, then $\mathrm{dist}_C(w, x) = \mathrm{dist}_C(x, y) = 2$. If C is a short odd hole (see Sect. 3.3) in G, then v has at most two neighbours on C.*

Lemma 6. *Let C be a prehole in $G \in$ FLAWLESS. Then every vertex $v \in C$ has at most five neighbours in C. Two of these are via edges of C, so v is incident to at most three chords. If there are two chords, vw, vx, then $\mathrm{dist}_C(w, x) = 2$. If there are three chords, vw, vx, vy, then $\mathrm{dist}_C(w, x) = \mathrm{dist}_C(x, y) = 2$.*

Proof. Otherwise, v must have at least four chords. These must be even chords to c_0, c_2, c_4, c_6, where $P = (c_0, c_1, \ldots, c_6, c_7)$ is a subpath of C, since C is a prehole and G has no even holes. We now move v from L to R. The only new edges which appear in $G[L{:}R]$ are those adjacent to v. But now $c_0, v, c_3, c_4, c_5, c_6, c_7$ induce an armchair in $G[L{:}R]$, contradicting $G \in$ FLAWLESS, see Fig. 4.   □

Fig. 4. An armchair

The degree bound of Lemma 6 is tight, see Fig. 5.

Fig. 5. A prehole with a vertex of degree 5

Lemma 7. *Let C be an odd hole in $G \in$ FLAWLESS such that $v \notin C$ and $x \notin C$ are adjacent. Then vertices $w, y \in C$ exist such that (v, x, y, w) is a quadrangle.*

3.3   Determining a Short Odd Hole

We can test whether G contains a hole in time $O(|E|^2)$, using the algorithm of [13]. Moreover, the algorithm returns a hole if one exists. If the hole is even, we can conclude $G \notin$ QUASIMONOTONE. If $G \in$ FLAWLESS, we will show that it has a well-defined structure.

Lemma 8. *If C is an odd cycle in a graph G, there is a triangle or an odd hole C' in G.*

Proof. The claim is clearly true if $|C| \leq 3$. Otherwise, assume by induction that it is true for all cycles shorter than C. If C is not already a hole, it has a chord that divides it into a smaller odd cycle C_1, and an even cycle C_1'. The lemma now follows by induction on C_1.   □

The proof of Lemma 8 can easily be turned into an efficient algorithm to find C'. An odd hole C is *short* if $\text{dist}(v, w) = \text{dist}_C(v, w)$ for all pairs $v, w \in C$.

Lemma 9. *If G is a triangle-free graph containing an odd hole C, then G contains a short odd hole.*

Fig. 6. Short odd holes of unequal size in a quasimonotone graph.

Note that the proof of Lemma 9 gives an efficient algorithm for finding a short odd hole H, given any odd hole C. Clearly the shortest hole in G is a short hole, but the converse need not be true in general, even for quasimonotone graphs.

Corollary 1. *If G has a short odd hole C, diam$(G) \geq$ diam$(C) = (|C| - 1)/2$.*

If C is a prehole, $G' = G[C]$, and $L{:}R$ is the alternating bipartition of C, then $G'[L{:}R]$ contains no edge other than those of C. A *minimal* prehole C is such that $G[C]$ contains no prehole with fewer than $|C|$ vertices.

4   Flawless Graphs Containing a Long Hole

4.1   Triangles

Lemma 10. *Let G be a quasimonotone graph containing an odd hole C of size at least 7. Then G contains no triangle that has a vertex in C (Fig. 7).*

Fig. 7. In a quasimonotone graph a 5-hole and a triangle can share a vertex.

Lemma 11. *Let G be a quasimonotone graph containing an odd hole C of size at least 7. Then G contains no triangle which is vertex-disjoint from C.*

4.2   Long Odd Holes

Lemma 12. *Let C, C' be odd holes in a quasimonotone graph G such that $C' \cap C \neq \varnothing$, and $|C|, |C'| \geq 7$. Let $G' = G[(C' \cup C) \setminus (C' \cap C)]$, Then G' has no odd hole or prehole.*

Corollary 2. *Let C, C' be odd holes in a quasimonotone graph G, such that $C' \cap C \neq \varnothing$. Let $G' = G[(C' \cup C) \setminus (C' \cap C)]$. Then G' is a monotone graph.*

Note that the holes C, C' in Corollary 2 can have different size. See Fig. 6, where G' is a *ladder* (see [5]) with two pendant edges. However, if we have vertex-disjoint odd holes they cannot have different lengths.

A *prism* is the graph given by joining corresponding vertices in two cycles of the same length. It is an *n-prism* if the cycles have length n [10].

Lemma 13. *Let G be a quasimonotone graph containing an odd hole C. Then G contains no vertex-disjoint hole C' with $|C'| \neq |C|$. Moreover, if $|C| \geq 7$, any two vertex-disjoint holes with $|C'| = |C|$ induce a prism in G.*

5   Preholes in Flawless Graphs

Lemma 14. *If $G \in$ FLAWLESS and has an odd hole of size $\ell \geq 7$, any minimal prehole C in G is either an even hole or (a) two odd holes intersecting in an edge or (b) two disjoint odd holes connected by a quadrangle. See Fig. 8.*

Thus, if G contains an odd hole of size at least 7, minimal preholes have only two types, case (a) and case (b). From Lemma 13, case (b) are crossover preholes (Fig. 9).

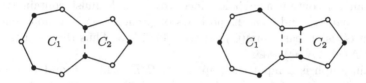

Fig. 8. Preholes with odd holes C_1, C_2, cases (a) left and (b) right.

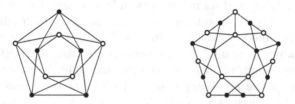

Fig. 9. Flawless crossover preholes.

So let us consider the case (a) preholes. We will call these *Möbius preholes*, since we will show that such a prehole must be a *Möbius ladder* [8,10].

Lemma 15. *Every Möbius prehole in a flawless graph is a Möbius ladder (Fig. 10).*

Fig. 10. Two different drawings of a Möbius ladder.

5.1   Splitting

Let G be a flawless graph with a hole C of length $|C| \geq 6$. If $|C|$ is even, we conclude $G \notin$ QUASIMONOTONE, so $|C| \geq 7$ is odd. Thus G does not contain a triangle, from Lemmas 10 and 11. We will assume that this has been tested. We will now show that G must have the annular structure referred to in Sect. 1.2, rather like a monotone graph with its ends identified.

Now suppose G has a short odd hole C with $C \geq 7$, determined by the procedure of Lemma 9. Thus, by Corollary 1, $\mathrm{diam}(G) \geq \frac{1}{2}(|C|-1) \geq 3$. Choose any $v \in C$, and consider the graph $G_v = G[V \setminus N[v]]$. Then G_v contains no holes, since any hole H in G_v must be a hole in G. But any hole H in G either contains v, or has a vertex w adjacent to v, by Lemma 4. Since $v, w \notin G_v$, $H \not\subseteq G_v$. Neither can G_v contain a prehole, since any prehole must contain two holes. Thus G_v is flawless and contains no holes or preholes, so is a monotone graph. Now $\mathrm{diam}(G)$ is at least $\mathrm{diam}(C) = (|C|-1)/2 \geq 3$. Thus there exists a $w \in C$ such that $N(v) \cap N(w) = \varnothing$.

A chain graph is a bipartite graph $(L \cup R, E)$ where L and R are linearly ordered by inclusion of neighbourhoods. Its biadjacency matrix has the form indicated in Fig. 11, see [5] for details. In the monotone representation, it is an easy observation that the graph has a decomposition into chain graphs, as indicated in Fig. 12, where L is partitioned in $D_1, D_3, \ldots$ and R into $D_2, D_4, \ldots$. Brandstädt and Lozin showed in [3] that such a partition exists. For vertices v and w as above, $N(w)$ and its neighbours induce a monotone subgraph N_w of G, as indicated in Fig. 12. The vertex set of N_w is $\{x \in L \cup R : \mathrm{dist}(w, x) \leq 2\}$. Clearly N_w is the union of two chain graphs C_w, C'_w, with C_w lying in the rows below and including w, and C'_w in the rows above. Using the algorithm of [14], the monotone representation of G_v determines this split. Then we can construct a representation of the adjacency matrix $A(G)$ of G as indicated in the first diagram in Fig. 13, where $D_2 = N(w)$, $C_1 = C_w$ (transposed), and $C_7 = C'_w$. The chain graphs $C_2, \ldots, C_6$ are a decomposition of the monotone graph G_w. Note that the ordering of the chain graphs in the decomposition is circular, and the second diagram in Fig. 13 gives an equivalent representation to the first, where C_1 (transposed) is moved from the first to the last position.

Lemma 16. *A flawless graph G which has an odd hole of size at least 7 is quasimonotone if and only if it has such a decomposition and does not contain a prehole. If there are k chain graphs in the decomposition, then k is odd, and the shortest hole in G has k vertices.*

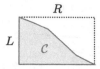

Fig. 11. Chain graph structure

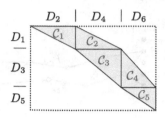
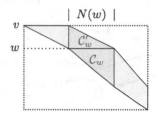

Fig. 12. Decomposition of a monotone graph/neighbourhood of w in G_v

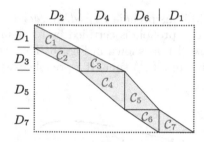
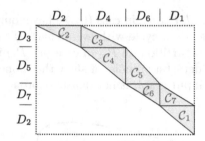

Fig. 13. Decomposition of $A(G)$ for a quasimonotone graph G

5.2 Recognising Preholes

Let $G = (V, E)$ be a flawless graph with a hole of size $\ell \geq 7$. Lemma 16 can determine whether or not G is quasimonotone provided it does not contain a prehole. We now consider recognition of a prehole in such a graph.

We use the partition of V from Sect. 5.1 into independent sets $D_1, D_2, \ldots, D_\ell$, where $D_{\ell+1} \equiv D_1$. All edges in E run between D_i and D_{i+1} ($i \in [\ell]$). Let $G_i = G[D_i \cup D_{i+1}]$, with edge set E_i, and let $\overline{G}_i = (V, E \setminus E_i)$. Note that G_i is a chain graph and $\overline{G}_i$ is a monotone graph. Thus $\overline{G}_i$ is bipartition, with bipartition $L{:}R$, say, with $D_i, D_{i+1} \in L$.

We search for possible crossovers in G_i. These are pairs $a, b \in D_{i+1}, c, d \in D_i$, such that $ac, ad, bc, bd \in E$. We list all such quadruples a, b, c, d, $O(n^4)$ in total, see Fig. 14. Given any quadruple, we attempt to determine vertex disjoint paths P_{ac}, P_{bd} in $\overline{G}_i$ between a, c and b, d or between a, d and b, c. See Fig. 15, cases (a) and (b). We can do this in $O(n|E|) = O(n^3)$ time by network flow. Both paths are even length, since G_i is bipartite and $a, b, c, d \in L$.

If these paths do not exist, we discard this quadruple and consider the next in the list. If these paths do exist, in case (a) we have found a crossover prehole

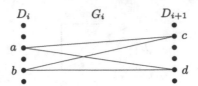

Fig. 14. Possible crossover

Fig. 15. Vertex-disjoint paths, case (a) left, (b) right

P_{ac}, ad, P_{bd}, bc, in case (b) we have found a Möbius prehole P_{ad}, bd, P_{bc}, ac. This is clearly a cycle with even length. That it is a prehole is certified by reversing the bipartition on P_{ac} in case (a), P_{ad} in case (b), as shown in Fig. 16. Thus we can detect a prehole, or show that none exists, in $O(n^7)$ time. If a prehole exists the input graph is not quasimonotone.

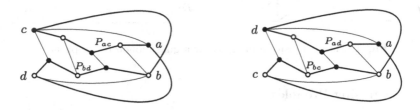

Fig. 16. Preholes, left with crossover (a), right of Möbius type (b).

6   Flawless Graphs Without Long Holes

6.1   Minimal Preholes in Hole-Free Graphs

Let C be any minimal prehole in a flawless hole-free graph G. A triangle in $G[C]$ will be called an *interior* triangle of C if it has no edge in common with C, a *crossing* triangle if it has one edge in common with C, and a *cap* of C if it has two edges in common with C.

Lemma 17. *If C is a minimal prehole in a flawless graph with $|C| > 12$, then $G[C]$ has no interior or crossing triangles, and C is determined by two edge-disjoint caps.*

Let T_1, T_2 be caps of C, such that $v_i \in T_i$ is adjacent to two edges of C ($i = 1, 2$). Then there are two edge-disjoint $(v_1, \ldots, v_2)$ paths P_1, P_2 in C (Fig. 17).

Fig. 17. A prehole and its Hamilton subgraph

Lemma 18. *Let C, with $|C| > 12$, be a minimal prehole in a flawless hole-free graph determined by v_1, v_2, and let $C' = C \setminus \{v_1, v_2\}$. Then $G[C']$ is a Hamilton monotone graph, and all chords of C' connect P_1 to P_2.*

Proof. Clearly $G[C']$ is Hamilton, since $G[C]$ is Hamilton. Now C' cannot be a prehole, since it is strictly smaller than C. So $G[C']$ cannot contain a triangle, by Lemma 17. It cannot contain a larger odd cycle, since then it would contain a triangle, by the argument of Lemma 17. Therefore, $G[C']$ is bipartite and, since $G \in \text{HoleFree}$, contains no hole. So, since $G \in \text{Flawless}$, $G[C']$ is a monotone graph. Suppose uv is an edge of $G[C']$ with $u, v \in P_1$. Then, since $G[C]$ has only even chords, the even chord uv and the segment of P_1 between u and v forms an odd cycle, giving a contradiction. □

Thus any minimal prehole C comprises a Hamilton monotone graph $G[C']$, to which we add two caps T_1, T_2. We may also add edges from v_1 and v_2 to C', as long as they are even chords in C.

Lemma 19. *Let C be a minimal prehole with a cap at $v \in \{v_1, v_2\}$. Then there are at most two chords from v, and both must be connected to either P_1 or P_2.*

Let $T_1 = \{v_1, u_1, w_1\}$, $T_2 = \{v_2, u_2, w_2\}$ be any two edge-disjoint triangles in a flawless graph G. Let M be the component of $G \setminus \{v_1, v_2\}$ containing $u_1 w_1$, $u_2 w_2$, if such a component exists. If M does not exist then v_1, v_2 clearly do not determine a prehole.

Lemma 20. *$C = (v_1, u_1, \ldots, u_2, v_2, w_2, \ldots, w_1, v_1)$ determines a minimal prehole if and only if M is a monotone graph containing two vertex-disjoint paths between u_1, u_2 and v_1, v_2.*

6.2 Preholes Containing 5-Holes and Triangles

It remains to consider preholes in graphs which contain 5-holes. Preholes determined by two triangles will be dealt with as in Sect. 6.1.

Lemma 21. *Let C be a minimal prehole in a flawless graph G which contains no odd hole of size greater than five. If C connects a 5-hole and a triangle, or if C connects two 5-holes, then $|C| \leq 12$.*

7   Conclusion and Discussion

In [6] we considered the problem of ergodicity and rapid mixing of the switch chain in hereditary graph classes. We gave a complete answer to the ergodicity question, and showed rapid mixing for the new class of quasimonotone graphs. This led us to introduce a new "quasi-" operator on bipartite graph classes, which is of independent interest. Quasimonotone graphs are a particular case of this construction. Another interesting class is the class of odd-chordal graphs, which are the quasi-chordal bipartite graphs. This is close to the largest class for which the switch chain is ergodic.

A more straightforward approach to recognising quasimonotone graphs would be provided by a polynomial time recognition algorithm for odd-chordal graphs. This is equivalent to the detection of preholes in a graph. We have considered this question, but we leave it as an open problem. The only evidence we can provide is that it is NP-complete to determine if a graph is a prehole, which may be a harder question. Nonetheless, the NP-completeness proof suggests that an efficient algorithm for recognising odd-chordal graphs may be elusive.

References

1. Bogart, K.P., West, D.B.: A short proof that 'proper = unit'. Discret. Math. **201**(1), 21–23 (1999)
2. Brandstädt, A., Le, V.B., Spinrad, J.P.: Graphclasses: A Survey. Society for Industrial and Applied Mathematics, Philadelphia (1999)
3. Brandstädt, A., Lozin, V.V.: On the linear structure and clique-width of bipartite permutation graphs. Ars Combinatoria **67**, 273–281 (2003)
4. Diaconis, P., Graham, R., Holmes, S.P.: Statistical problems involving permutations with restricted positions. In: State of the Art in Probability and Statistics. Lecture Notes-Monograph Series, vol. 36, pp. 195–222. Institute of Mathematical Statistics (2001)
5. Dyer, M., Jerrum, M., Müller, H.: On the switch Markov chain for perfect matchings. J. ACM **64**(2), 12 (2017)
6. Dyer, M., Müller, H.: Counting perfect matchings and the switch chain. CoRR abs/1705.05790 (2017)
7. Dyer, M., Müller, H.: Quasimonotone graphs. CoRR abs/1801.06494 (2018)
8. Guy, R.K., Harary, F.: On the Möbius ladders. Can. Math. Bull. **10**, 493–496 (1967)
9. Hell, P., Huang, J.: Interval bigraphs and circular arc graphs. J. Graph Theory **46**, 313–327 (2004)
10. Hladnik, M., Marušič, D., Pisanski, T.: Cyclic Haar graphs. Discret. Mat. **244**(1), 137–152 (2002)
11. Köhler, E.: Graphs without asteroidal triples. Ph.D. thesis, TU Berlin (1999)
12. Lubiw, A.: Doubly lexical orderings of matrices. SIAM J. Comput. **16**(5), 854–879 (1987)
13. Nikolopoulos, S.D., Palios, L.: Detecting holes and antiholes in graphs. Algorithmica **47**(2), 119–138 (2007)
14. Spinrad, J., Brandstädt, A., Stewart, L.: Bipartite permutation graphs. Discret. Appl. Math. **18**(3), 279–292 (1987)

Equiangular Polygon Contact Representations

Stefan Felsner[1], Hendrik Schrezenmaier[1]([✉]), and Raphael Steiner[2]

[1] Institut für Mathematik, Technische Universität Berlin, Berlin, Germany
{felsner,schrezen}@math.tu-berlin.de
[2] Fachgebiet Mathematik, FernUniversität in Hagen, Hagen, Germany
steiner.raphael@gmx.de

Abstract. Planar graphs are known to have contact representations of various types. The most prominent example is Koebe's 'kissing coins theorem'. Its rediscovery by Thurston lead to effective versions of the Riemann Mapping Theorem and motivated Schramm's Monster Packing Theorem. Monster Packing implies the existence of contact representations of planar triangulations where each vertex v is represented by a homothetic copy of some smooth strictly-convex prototype P_v.

With this work we aim at computable approximations of Schramm representations. For fixed K approximate P_v by an equiangular K-gon Q_v with horizontal basis. From Schramm's work it follows that the given triangulation also has a contact representation with homothetic copies of these K-gons. Our approach starts by guessing a *K-contact-structure*, i.e., the combinatorial structure of a contact representation. From the combinatorial data, we build a system of linear equations whose variables correspond to lengths of boundary segments of the K-gons. If the system has a non-negative solution, this yields the intended contact representation. If the solution of the system contains negative variables, these can be used as sign-posts indicating how to change the K-contact-structure for another try.

In the case $K = 3$ the K-contact-structures are Schnyder woods, in the case $K = 4$ they are transversal structures. As in these cases, for $K \geq 5$ the K-contact-structures of a fixed graph are in bijection to certain integral flows, and can be viewed as elements of a distributive lattice.

The procedure has been implemented, it computes the solution with few iterations. The experiments involved graphs with up to one hundred vertices.

1 Introduction

Representations of graphs by contacts of geometric objects are actively studied in graph theory and geometry. An early result in this direction is Koebe's

A full version of the paper is available at http://page.math.tu-berlin.de/~felsner/Paper/kgons.pdf.

S. Felsner and H. Schrezenmaier—Partially supported by DFG grant FE-340/11-1.

A. Brandstädt et al. (Eds.): WG 2018, LNCS 11159, pp. 203–215, 2018.
https://doi.org/10.1007/978-3-030-00256-5_17

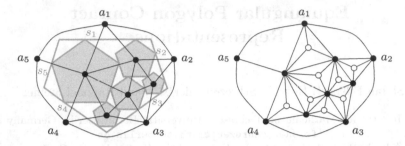

Fig. 1. Left: An equiangular pentagon contact representation of the graph G shown in black where each inner vertex is represented by a regular pentagon. Right: The stack extension G^* of G.

Circle Packing Theorem from 1936. It states that every planar graph can be represented as the contact system of a set of interiourly disjoint disks. Koebe arrived at this result in the context of conformal mapping of 'contact domains'. Unaware of Koebe's work, Thurston reproved the Circle Packing Theorem and connected it to the Riemann Mapping Theorem. This line of research resulted in discretizations of conformal mappings and has strong impact in the area of discrete differential geometry. We refer to [1,19] for further details on those connections.

A very strong generalization of Koebe's theorem is Schramm's Convex Packing Theorem from 1990 [14]. The theorem states that if each vertex v of a planar triangulation G has a prescribed convex prototype P_v, then there is a contact representation of G where each vertex is represented by a (possibly degenerate) homothet of its prototype. When the prototypes have a smooth boundary there are no degeneracies. With this work we aim at efficiently computable approximations of Schramm representations. The idea is to approximate the prototypes P_v with simpler shapes; we use equiangular K-gons. Clearly, a sequence of approximating contact representations with K-gons, one for each positive integer K and each of them confined to the unit square, will contain a subsequence converging to a representation with the prototypes P_v.

Contact representations of graphs with polygons have also been studied widely. Triangle contact representations have been investigated by De Fraysseix et al. [5]. They observed that Schnyder woods can be considered as combinatorial encodings of triangle contact representations of triangulations and that any Schnyder wood can be used to construct a corresponding triangle contact system. Gonçalves et al. [11] observed that Schramm's Convex Packing Theorem can be used to prove the existence of contact representations with homothetic triangles for all 4-connected triangulations. A more combinatorial approach to this result, which aims at computing the representation as the solution of a system of linear equations, which are based on a Schnyder wood, was described by Felsner [7]. On the basis of this approach, Schrezenmaier [17] reproved the existence of homothetic triangle contact representations.

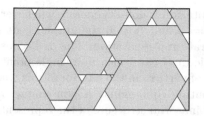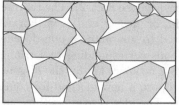

Fig. 2. Parts of equiangular 6-gon and 7-gon contact representations of the same graph.

Representations of graphs with side contacts of rectangles have applications in architecture and VLSI design. For links into the extensive literature we recommend [3,8]. Representations of graphs using squares or, more precisely, graphs as a tool to model packings of squares already appear in classical work of Brooks et al. [2] from 1940. Schramm [15] proved that every 5-connected inner triangulation of a 4-cycle admits a square contact representation. Again there is a combinatorial approach to this result which aims at computing the representation as the solution of a system of linear equations, see Felsner [8]. In this context *transversal structures* play the role of Schnyder woods. As in the case of homothetic triangles, this approach is based on an iterative procedure, however, a proof that the iteration terminates is still missing. On the basis of the approach, Schrezenmaier [16] reproved Schramm's Squaring Theorem.

Before stating our results, we introduce some precise terminology. A *K-gon contact system* S is a finite system of convex K-gons in the plane such that the interiors of any two K-gons are disjoint. If all K-gons of S are equiangular K-gons (i.e., all interior angles are $\frac{K-2}{K}\pi$) with a horizontal segment at the bottom, we call S an *equiangular K-gon contact representation*. The contact system has an *exceptional touching* if there is a point where two corners of K-gons meet. The *contact graph* $\mathcal{G}(S)$ of S is the graph that has a vertex for every K-gon and an edge for every contact of two K-gons in S. Note that $\mathcal{G}(S)$ inherits a crossing-free embedding from S. For a given plane graph G and a K-gon contact system S with $\mathcal{G}(S) = G$ we say that S is a *K-gon contact representation* of G.

We will only consider the case that G is an *inner triangulation of a K-cycle*, i.e., the outer face of G is a K-cycle with vertices $a_1, \ldots, a_K$ in clockwise order, all inner faces are triangles, there are no loops or multiple edges, and there are no additional edges between the outer vertices. Our interest lies in regular K-gon contact representations of G with the additional property that $a_1, \ldots, a_K$ are represented by line segments $s_1, \ldots, s_K$ which together form an equiangular K-gon. The line segment s_1 is always horizontal and at the top, and $s_1, \ldots, s_K$ is the clockwise order of the segments of the K-gon. Figure 1 (left) shows an example for $K = 5$. Figure 2 shows contact systems of 6-gons and 7-gons, respectively.

Let G be an inner triangulation of a K-cycle and for each inner vertex v of G let P_v be a prescribed equiangular K-gon. From Schramm's Convex Packing Theorem it follows that G has a representation as contact graph of homothets of the prototypes (see Sect. 2). The representation is non-degenerate whenever $K \geq$ 5 and odd, or $K \geq 8$ and even. For $K = 3$ and $K = 6$ the graph needs to be 4-connected to guarantee a non-degenerate representation. This is because the three K-gons corresponding to a triangle in G can touch in a single point such that there is no space left for the K-gons of vertices in the interior of this triangle.

We propose a new method for computing equiangular K-gon contact representations. The idea is to guess the combinatorial structure of the representation of G, i.e., for each edge uv of G guess whether the contact involves a corner of P_u or a corner of P_v and also guess which corner of the respective prototype is involved. The guess is encoded in a K-*contact-structure*. The K-contact-structure leads to a system of linear equations whose variables correspond to lengths of boundary segments of the K-gons. The system is non-singular. If it has a non-negative solution, the values of the variables determine the geometry of a K-gon contact representation. If the solution of the system contains negative values, then it is possible to locally modify the K-contact-structure in the neighborhood of negative variables. The modified K-contact-structure corresponds to a new system of equations which has a new solution. This yields an iterative procedure which *hopefully* stops with a positive solution, i.e., with a K-gon contact representation.

We could not prove that the above iterative procedure stops. However the algorithm has been implemented and was used for extensive experiments (Sect. 7). These have always been successful. Similar algorithms for the computation of contact representations by homothetic triangles or squares have been described by Felsner [7,8]. These have also been implemented and successfully tested, c.f. Rucker [13] and Piccetti [12], respectively. We therefore conjecture that the proposed algorithm for computing equiangular K-gon contact representations always terminates with a solution.

In Sect. 3 we introduce K-contact-structures of the graph G. These are certain weighted orientations of a supergraph of G. In Sect. 4 we enhance K-contact-structures with a K-coloring of the edges. The color classes are directed forests that resemble the trees of a Schnyder wood. In Sect. 5 we show that there is a distributive lattice on the set of K-contact-structures of a fixed graph G and describe the combinatorial change in K-contact-structures that form a cover pair. In Sect. 6 we discuss the system of linear equations and prove that it is non-singular. Section 7 describes the iteration which is proposed as a heuristic for computing equiangular K-gon contact representations.

In this paper we focus on odd $K \geq 5$. The case $K = 3$ is well-studied and the case $K \geq 6$ and even will be added in a later version of this paper. The case $K = 5$ was first studied in the bachelor thesis of Steiner [18] (a coauthor in this paper) and further elaborated by the present team of authors [10].

2   The Existence of Equiangular K-gon Contact Representations

In this section let G be an inner triangulation of a K-cycle and let V_{inner} be the set of inner vertices of G. Further, for each $v \in V_{\text{inner}}$, let P_v be an equiangular

K-gon with a horizontal segment at the bottom. We call P_v the *prototype* of v. A *homothetic* copy of a prototype P_v is a set in the plane that can be obtained from P_v by scaling and translation.

Theorem 1. *For odd $K \geq 5$ there is an equiangular K-gon contact representation of G in which each $v \in V_{\text{inner}}$ is represented by a homothetic copy of P_v.*

This theorem is an immediate consequence of the Convex Packing Theorem by Schramm [14] which guarantees a contact representation of G with homothets of the given prototypes if we also allow the inner vertices to be represented by a single point, i.e., a homothetic copy of the prototype with scaling factor 0. The interesting point is that for odd $K \geq 5$ this cannot happen because the interior angles of the equiangular K-gons are too large (combined with the fixed alignment of the K-gons) to allow more than two equiangular K-gons to meet at a given point. Similar proofs have been given for the case $K = 3$ in [11] and $K = 5$ in [10,18].

3   The Combinatorial Structure of Equiangular Polygon Contact Representations

For the entire section let G be an inner triangulation of a K-cycle, $K \geq 3$ odd. We call an inner face of G a *strictly inner face* if it is only incident to inner edges. We denote the set of inner edges of a planar graph H by $E_{\text{inner}}(H)$. For the directed graphs used later in this section we denote the sets of incoming and outgoing edges of a vertex v by $E_{\text{in}}(v)$ and $E_{\text{out}}(v)$, respectively.

Definition 1. *The* stack extension *$G^\star$ of G is the extension of G that contains an extra vertex in every strictly inner face. These new vertices are connected to all three vertices of the respective face. We call the new vertices* stack vertices *and the vertices of G* normal vertices. *Analogously, we call the new edges* stack edges *and the edges of G* normal edges. *See Fig. 1 (right) for an example.*

Definition 2. *A K-contact-structure on G is an orientation and weighting $w : E_{\text{inner}}(G^\star) \to \mathbb{N}$ of the inner edges of $G^\star$ such that*

(P1) $w(e) = 1$ for each normal edge e,
(P2) each stack edge is oriented towards its incident stack vertex,
(P3) the out-flow of each normal vertex u is K, i.e., $\sum_{e \in E_{\text{out}}(u)} w(e) = K$,
(P4) the in-flow of each stack vertex v is $\frac{K-3}{2}$, i.e., $\sum_{e \in E_{\text{in}}(v)} w(e) = \frac{K-3}{2}$.

Definition 3. *Let $\mathcal{A}$ be a K-contact-structure on G. Then we can associate with $\mathcal{A}$ a modified version of $G^\star$ where each inner edge e is replaced by $w(e)$ parallel edges (if $w(e) = 0$, the edge e is deleted) and all edges are oriented as in $\mathcal{A}$. We denote this graph by $G^\star_+(\mathcal{A})$.*

The following theorem shows the key correspondence between K-contact-structures and equiangular K-gon contact representations.

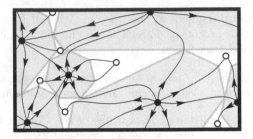

Fig. 3. A contact representation of equiangular 7-gons and the graph $G_+^*(\mathcal{A})$ for its induced 7-contact-structure $\mathcal{A}$.

Theorem 2. *Let S be an equiangular K-gon contact representation of the graph $G = \mathcal{G}(S)$. Then S induces a K-contact-structure on G (see Fig. 3 for an illustration).*

First we consider the case that S has no exceptional touchings. Then the construction of the induced K-contact-structure of S is as follows: Let e be an inner normal edge of G^*. Then e corresponds to the contact of a corner of a K-gon A and a segment of a K-gon B in S. We orient the edge e from the vertex corresponding to A to the vertex corresponding to B and set $w(e) = 1$. Now let $e = uv$ be a stack edge with normal vertex u and stack vertex v. Then u corresponds to a K-gon A of S and v to an area F in S which is enclosed by A and two more K-gons or outer segments s_i. Note that F is a pseudotriangle, i.e., a polygon with exactly three convex corners and arbitrarily many concave corners. We define $w(e)$ to be the number of concave corners of F which are also corners of A, and orient e from u to v.

Properties (P1) and (P2) are fulfilled directly by construction. Property (P3) corresponds to the fact that each K-gon has exactly K corners, and property (P4) corresponds to the fact that each pseudotriangle has exactly $\frac{K-3}{2}$ concave corners.

In the case that S has exceptional touchings, each exceptional touching of two K-gon corners can be interpreted in two ways as a corner-segment contact with infinitesimal distance to the other corner. We choose one of these interpretations and proceed as before. Hence, the K-contact-structure induced by an equiangular K-gon contact representation with exceptional touchings is not unique.

Theorem 3. *Let G be an inner triangulation of a K-cycle. Then there exists a K-contact-structure on G.*

Theorem 3 immediately follows from Theorems 1 and 2. Since we aim for a theory independent from the Monster Packing Theorem by Schramm, we give another elementary proof of Theorem 3. The idea of the proof is the following: We replace each stack edge of G^* by $\frac{K-3}{2}$ parallel edges. Then we show that there exists an orientation of this graph such that each normal vertex has out-degree K and each stack vertex has in-degree $\frac{K-3}{2}$. Such orientations with prescribed

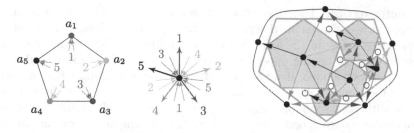

Fig. 4. Left: The local conditions of a K-proper coloring in the case $K = 5$. Right: An example for the K-proper coloring of an induced K-contact-structure. (Color figure online)

vertex degrees have been studied in [6] under the name of α-*orientations*. There are sufficient conditions related to Hall's matching criterion for the existence of α-orientations. In our case the conditions are fulfilled. The existence of the appropriate orientation implies the existence of a K-contact-structure.

4   Coloring K-contact-structures

In this section let G be an inner triangulation of a K-cycle, let $\mathcal{A}$ be a K-contact-structure on G and let $G^\star_+ := G^\star_+(\mathcal{A})$. In the following, the set of colors $1, \ldots, K$ is to be understood as representatives modulo K, i.e., colors c and $c + zK$ are the same for any $z \in \mathbb{Z}$.

Definition 4. *A K-proper coloring of $G^\star_+$ is a coloring of the inner edges of $G^\star_+$ in the colors $1, \ldots, K$ such that (see Fig. 4 (left) for an illustration)*

(C1) for $i = 1, \ldots, K$ all edges incident to the outer vertex a_i have color i,
(C2) each normal vertex has exactly one outgoing edge in each color and the clockwise order of the colors is $1, \ldots, K$,
(C3) incoming edges of a normal vertex, which are located between the outgoing edges of colors c and $c + 1$, have color $c - \frac{K-1}{2}$.

An equiangular K-gon contact representation $\mathcal{S}$ induces a K-contact-structure together with a K-proper coloring. To see this, recall the construction below Theorem 2. Each inner edge of $G^\star_+$ corresponds to a corner of a K-gon of $\mathcal{S}$. We color the corners of each K-gon of $\mathcal{S}$ in the colors $1, \ldots, K$ in clockwise order, starting with color 1 at the corner at the top. Then each inner edge of $G^\star_+$ gets the color of the corner it corresponds to. Figure 4 (right) shows an example.

The following theorem shows that this coloring is a property of the K-contact-structure itself and independent of an inducing contact representation.

Theorem 4. *The graph $G^\star_+$ has a unique K-proper coloring.*

The idea of the construction of the colors is as follows: We start with an inner edge e of $G^\star_+$ and follow a properly defined path that at some point reaches one of the outer vertices. Then the color of this outer vertex will be the color of e. This approach is similar to the proof of the bijection of Schnyder Woods and 3-orientations in [4]. In the definition of these paths, we aim at continuing with the outgoing edge on the opposite side of a vertex. This is motivated by the following geometric idea: If we are already given an equiangular K-gon contact representation, such paths keep a constant slope and therefore run into an outer segment with corresponding slope. If we run into a stack vertex, there is no unique opposite edge. Therefore the path of e is not unique, but we can associate a unique outer vertex with e by showing that all properly defined paths starting with e end in the same outer vertex.

5   The Distributive Lattice of K-contact-structures

Let G be an inner triangulation of a K-cycle. The following definitions give us a formalism how to change a K-contact-structure of G to obtain a new one.

Definition 5. *Let $\mathcal{A}$ be a K-contact-structure of G. We call a multiset E of oriented edges of $G^\star$ flippable in $\mathcal{A}$ if (i) $\mathrm{indeg}_E(v) = \mathrm{outdeg}_E(v)$ for each vertex v; (ii) each normal edge is contained at most once in E and only in the orientation of $\mathcal{A}$; (iii) each stack edge $e = uv$ with stack vertex v is contained at most $w_{\mathcal{A}}(e)$ times in E in the orientation from u to v (no restriction for the other direction).*

Definition 6. *Let $\mathcal{A}$ be a K-contact-structure of G and let E be a flippable set of edges in $\mathcal{A}$. Then we can perform a flip on $\mathcal{A}$ and obtain a new K-contact-structure $\mathcal{A}'$ by changing the orientation of all normal edges in E, and by setting $w_{\mathcal{A}'}(e) := w_{\mathcal{A}}(e) - a + b$ for each stack edge $e = uv$ with normal vertex u and stack vertex v if e is contained a times in E oriented from u to v and b times oriented from v to u.*

It can easily be seen that a flip indeed yields a new K-contact-structure. We can even reach every K-contact-structure $\mathcal{A}'$ from $\mathcal{A}$ by flipping a suitable flippable set of edges.

These flipping operations already show the close relation between K-contact-structures and integral flows on $G^\star$. We now want to formalize this relation and thereby obtain the structure of a distributive lattice on the set of K-contact-structures of G. In particular, K-contact-structures can be equivalently modeled as flows $f : E_{\mathrm{inner}}(\overrightarrow{G^\star}) \to \mathbb{Z}$ on a fixed orientation $\overrightarrow{G^\star}$ of $G^\star$ where each stack edge is oriented towards the incident stack vertex and each normal edge obtains an arbitrary fixed orientation. In such a flow the *excess* of a vertex v is defined as $\omega(v) := \sum_{e \in E_{\mathrm{in}}(v)} f(e) - \sum_{e \in E_{\mathrm{out}}(v)} f(e)$.

Definition 7. *A flow $f : E_{\mathrm{inner}}(\overrightarrow{G^\star}) \to \mathbb{Z}$ is called a K-contact-flow if*

(i) $f(e) \in \{0,1\}$ for each normal edge e;
(ii) $f(e') \in \{0,\ldots,\frac{K-3}{2}\}$ for each stack edge e';

(iii) $\omega(u) = \text{indeg}(u) - K$ *for each normal vertex u;*
(iv) $\omega(v) = \frac{K-3}{2}$ *for each stack vertex v.*

For each normal edge e we set $c_l(e) := 0$ and $c_u(e) := 1$. For each stack edge e' we set $c_l(e') := 0$ and $c_u(e') := \frac{K-3}{2}$. Then the first two conditions can be formulated as $c_l(e'') \leq f(e'') \leq c_u(e'')$ for each edge e''. The set of integral flows $\mathcal{F}(H, \omega, c_l, c_u)$ of a directed planar graph H fulfilling such constraints (bounds c_l, c_u on the flow values and prescribed excesses ω) has been studied in [9].

The following describes a bijection between the set of K-contact-structures and the set of K-contact-flows of G. Let $\mathcal{A}$ be a K-contact-structure on G. If a normal edge e has the same orientation in $\overrightarrow{G^\star}$ and in $\mathcal{A}$, we set $f(e) = 1$, otherwise $f(e) = 0$. For a stack edge e' we set $f(e') = w_{\mathcal{A}}(e')$.

It has been shown in [9] that the set $\mathcal{F}(H, \omega, c_l, c_u)$ carries the structure of a distributive lattice. For a flow $f \in \mathcal{F}(H, \omega, c_l, c_u)$ let the *residual graph*, denoted by H_f, be the following reorientation of H: An edge vw of H is oriented from v to w in H_f if $f(vw) > c_l(vw)$ and from w to v if $f(vw) < c_u(vw)$. Note that in H_f an edge can have no, one, or two orientations. If we decrease the flow f by one on a subgraph H'_f of H_f with the property $\text{indeg}_{H'_f}(v) = \text{outdeg}_{H'_f}(v)$ for each vertex v, we obtain a new flow $f' \in \mathcal{F}(H, \omega, c_l, c_u)$. This operation corresponds to a flip in the K-contact-structure.

Definition 8. *A* chordal path *of a simple cycle C is a directed path consisting of edges inside C (referring to a plane embedding) whose first and last vertex are vertices of C. These two vertices are allowed to coincide.*

A simple cycle C is an essential cycle *if there is a flow f such that C is a directed cycle in H_f and has no chordal path in H_f.*

Theorem 5 ([9]). *The following relation on the set $\mathcal{F}(H, \omega, c_l, c_u)$ of flows of a planar graph H is the cover relation of a distributive lattice: A flow f' covers a flow f if and only if f' can be obtained from f by subtracting one unit of flow on a counterclockwise oriented essential cycle in H_f.*

Now we can apply this to the set of K-contact-flows of G.

Theorem 6. *The set of all K-contact-structures of G carries the structure of a distributive lattice. In this lattice a K-contact-structure $\mathcal{A}'$ covers a K-contact-structure $\mathcal{A}$ if there is a flippable counterclockwise oriented facial cycle in $G^\star$ such that $\mathcal{A}'$ can be obtained from $\mathcal{A}$ by flipping this cycle.*

6   System of Linear Equations

Let G be an inner triangulation of a K-cycle and let $\mathcal{A}$ be a K-contact-structure of G. Let $G^\star_+ := G^\star_+(\mathcal{A})$. We will propose a system of linear equations that allows us to compute an equiangular K-gon contact representation of G with induced K-contact-structure $\mathcal{A}$ if such a representation exists. If such a representation does not exist, the solution of the system will have negative variables.

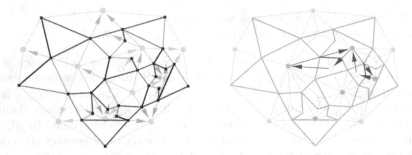

Fig. 5. Left: The skeleton graph corresponding to the K-contact-structure of Fig. 4. Right: The signs of the variables in the solution, green for non-negative and red for negative, and the corresponding sign-separating edges in blue. (Color figure online)

We start by describing how to obtain the skeleton graph of the contact representation. The skeleton graph G_{skel} is the medial graph of G_+^* without the edges corresponding to the outer face of G_+^*. We color the edges of G_{skel} according to the following rules: If the edge corresponds to an angle of an inner normal vertex and this angle lies between the outgoing edges of colors c and $c+1$ in the K-proper coloring of G_+^*, it gets the color $c - \frac{K-1}{2}$. If the edge corresponds to an angle of the outer normal vertex a_i, it gets the color i. See Fig. 5 (left) for an example.

The colors of the edges of G_{skel} correspond to their required slopes in the following way: Let B be an equiangular K-gon with a horizontal side at the top and its sides colored in the colors $1, \ldots, K$ in clockwise order, starting at the top. Then a crossing-free straight-line drawing of G_{skel} is an equiangular K-gon contact representation of G with induced K-contact-structure $\mathcal{A}$ if and only if each edge e has the same slope as the side of B that has the same color as e.

The purpose of the system of linear equations is to find edge lengths for the edges of G_{skel} in such a drawing with the additional property that the K-gons are homothets of the given prototypes. Therefore we have a variable x_e for each edge e of G_{skel} representing its length, and a variable x_v for each inner vertex v of G representing the scaling factor of its prototype P_v. We have equations which ensure that the scaling factor x_v of each normal vertex fits together with the edge lengths x_e of the K-gon corresponding to v. For $i = 1, \ldots, K$ let $\ell_i(P_v)$ be the length of the ith segment of P_v, starting with the horizontal segment and then proceeding in clockwise direction. Further let $E_i(v)$ be the edges of color i in G_{skel} corresponding to angles of v. Then the sum of the lengths of the edges in $E_i(v)$ has to be equal to $x_v \ell_i(P_v)$, the scaled segment length of the prototype: $\sum_{e \in E_i(v)} x_e - \ell_i(P_v) x_v = 0$. Further we have 2 equations for each inner face of G ensuring that the edges of G_{skel} corresponding to this face form a closed curve (these are the pseudotriangles). Finally, we add one more equation to our system stating that the lengths of the edges building the line segment corresponding to the outer vertex a_1 of G sum up to 1: $\sum_{e \in E_1(a_1)} x_e = 1$. This equation is the only inhomogeneous equation and will ensure that the solution of the system is unique. We denote the entire system by $A_{\mathcal{A}}\mathbf{x} = \mathbf{e}_1$ where $A_{\mathcal{A}}$ is a matrix depending on the K-contact-structure $\mathcal{A}$ and $\mathbf{e}_1 = (1, 0, 0, \ldots, 0)$.

Theorem 7. *The system $A_{\mathcal{A}}\mathbf{x} = \mathbf{e}_1$ has a unique solution. This solution is non-negative if and only if the K-contact-structure $\mathcal{A}$ is induced by an equiangular K-gon contact representation of G with the given prototypes.*

One direction of the latter part of the statement is trivial because the edge lengths of a contact representation yield a non-negative solution of the system. For the other direction we show that we can construct a contact representation from the edge lengths given by a non-negative solution by gluing together the K-gons and pseudotriangles resulting from this solution.

7   A Heuristic

In this section we propose a heuristic to compute an equiangular K-gon contact representation of a given triangulation G of a K-cycle. The basic idea of our heuristic is to start with an arbitrary K-contact-structure $\mathcal{A}$ of G and to solve the system $A_{\mathcal{A}}\mathbf{x} = \mathbf{e}_1$. If the solution is non-negative, we can construct the contact representation from the edge lengths given by the solution and are done. Otherwise, we can use the negative variables of the solution as sign-posts indicating how to change the K-contact-structure for another try. The goal is to find a flippable set of edges in G^* (see Sect. 5) that separates the edges of G_{skel} with negative solution value from these with non-negative solution value.

Definition 9. *We call these three types of oriented edges $e = (v, w)$ in G^* sign-separating edges (see Fig. 6): (A) v, w are normal vertices, the abstract K-gons of both vertices have a sign-change at the contact, the two involved abstract pseudotriangles do not have a sign-change at the contact; (B) v is a normal vertex, w is a stack vertex, and there is a sign-change at the corner corresponding to e; (C) v is a stack vertex, w is a normal vertex, the abstract pseudotriangle corresponding to v has a sign-change in a convex corner, the abstract K-gon corresponding to w has a sign-change at the same point, but not a corner.*

Let E_{+-} be the set of all sign-separating edges. It might be that there is a normal vertex v and a stack vertex u such that $(u, v), (v, u) \in E_{+-}$. In this case let w be the normal vertex corresponding to the abstract K-gon touching the abstract K-gon of v in the contact point where (u, v) and (v, u) have their assigned sign-changes. Then we change E_{+-} in the following way: We remove (v, u) from E_{+-} and add (v, w) and (w, u) instead. We call this a *repairing step*. It guarantees that a flip of the edges in E_{+-} changes the K-contact-structure.

Theorem 8. *After performing all possible repairing steps, the set E_{+-} is flippable in $\mathcal{A}$ and the corresponding flip leads to a K-contact-structure $\mathcal{A}' \neq \mathcal{A}$.*

We could not prove that iterating to flip the edges in E_{+-} can guarantee any kind of progress. Therefore a proof is still missing that this heuristic always

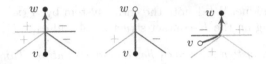

Fig. 6. The sign-separating edges of types (A), (B) and (C).

terminates with a solution. However, the heuristic has been subject to extensive experiments. We tested the heuristic with a total of 1000 random graphs with up to 100 inner vertices and up to 23 outer vertices. The heuristic terminated for each graph after few seconds.[1] Therefore we have the following conjecture.

Conjecture 1. The heuristic described in this section terminates with a solution for all K, for every graph G which is an inner triangulation of a K-cycle, and for every K-contact-structure of G to start the heuristic.

The experiments also suggest that the number of iterations is polynomial or even linear (sub-linear in the average case). Since the equation system has linear size and systems of linear equations can be solved exactly in polynomial time, this would imply that the heuristic has polynomial running time.

Acknowledgements. We want to thank Manfred Scheucher for supporting us with the implementation of the heuristic.

References

1. Bowers, P.L.: Circle packing: a personal reminiscence. In: Pitici, M. (ed.) The Best Writing on Mathematics 2010, pp. 330–345. Princeton University Press, Princeton (2010)
2. Brooks, R.L., Smith, C., Stone, A.H., Tutte, W.T.: The dissection of rectangles into squares. Duke Math. J. **7**(1), 312–340 (1940)
3. Buchsbaum, A.L., Gansner, E.R., Procopiuc, C.M., Venkatasubramanian, S.: Rectangular layouts and contact graphs. ACM Trans. Algorithms **4**, 28 (2008). Article no 8
4. de Fraysseix, H., de Mendez, P.O.: On topological aspects of orientations. Discret. Math. **229**(1), 57–72 (2001)
5. de Fraysseix, H., de Mendez, P.O., Rosenstiehl, P.: On triangle contact graphs. Comb. Probab. Comput. **3**, 233–246 (1994)
6. Felsner, S.: Lattice structures from planar graphs. Electron. J. Comb. **11**(1), R15 (2004)
7. Felsner, S.: Triangle contact representations. In: Midsummer Combinatorial Workshop, Praha (2009)
8. Felsner, S.: Rectangle and square representations of planar graphs. In: Pach, J. (ed.) Thirty Essays on Geometric Graph Theory, pp. 213–248. Springer, New York (2013). https://doi.org/10.1007/978-1-4614-0110-0_12

---

[1] Visualizations of some examples can be found at https://www3.math.tu-berlin.de/diskremath/research/kgon-representations/index.html.

9. Felsner, S., Knauer, K.: ULD-lattices and Δ-bonds. Comb. Probab. Comput. **18**(5), 707–724 (2009)
10. Felsner, S., Schrezenmaier, H., Steiner, R.: Pentagon contact representations. In: Proceedings of the Eurocomb, pp. 421–427 (2017)
11. Gonçalves, D., Lévêque, B., Pinlou, A.: Triangle contact representations and duality. In: Proceedings of the Graph Drawing, pp. 262 273 (2011)
12. Picchetti, T.: Finding a square dual of a graph (2011)
13. Rucker, J.: Kontaktdarstellungen von planaren Graphen. Diplomarbeit, Technische Universität Berlin (2011)
14. Schramm, O.: Combinatorically prescribed packings and applications to conformal and quasiconformal maps. Modified version of Ph.D. thesis from 1990. arXiv.org/0709.0710v1
15. Schramm, O.: Square tilings with prescribed combinatorics. Isr. J. Math. **84**(1–2), 97–118 (1993)
16. Schrezenmaier, H.: Zur Berechnung von Kontaktdarstellungen. Masterarbeit, Technische Universität Berlin (2016)
17. Schrezenmaier, H.: Homothetic triangle contact representations. In: Bodlaender, H.L., Woeginger, G.J. (eds.) WG 2017. LNCS, vol. 10520, pp. 425–437. Springer, Cham (2017). https://doi.org/10.1007/978-3-319-68705-6_32
18. Steiner, R.: Existenz und Konstruktion von Dreieckszerlegungen triangulierter Graphen und Schnyder woods. Bachelorarbeit, FernUniversität in Hagen (2016)
19. Stephenson, K.: Introduction to Circle Packing. The Theory of Discrete Analytic Functions. Cambridge University Press, Cambridge (2005)

Temporal Graph Classes: A View Through Temporal Separators

Till Fluschnik, Hendrik Molter, Rolf Niedermeier, and Philipp Zschoche[✉]

Institut für Softwaretechnik und Theoretische Informatik, TU Berlin,
Berlin, Germany
{till.fluschnik,h.molter,rolf.niedermeier,zschoche}@tu-berlin.de

Abstract. We investigate the computational complexity of separating two distinct vertices s and z by vertex deletion in a temporal graph. In a temporal graph, the vertex set is fixed but the edges have (discrete) time labels. Since the corresponding TEMPORAL (s, z)-SEPARATION problem is NP-hard, it is natural to investigate whether relevant special cases exist that are computationally tractable. To this end, we study restrictions of the underlying (static) graph—there we observe polynomial-time solvability in the case of bounded treewidth—as well as restrictions concerning the "temporal evolution" along the time steps. Systematically studying partially novel concepts in this direction, we identify sharp borders between tractable and intractable cases.

1 Introduction

Reachability, connectivity, and robustness in networks depend often on time. For instance, in public transport or human contact networks, available connections or contacts are time-dependent. To model such time-dependent aspects, one turns from static graphs to temporal graphs. Formally, an undirected *temporal graph* $G = (V, E, \tau)$ is an ordered triple consisting of a set V of vertices, a set $E \subseteq \binom{V}{2} \times \{1, \dots, \tau\}$ of *time-edges*, and a maximal time label $\tau \in \mathbb{N}$. We study the problem of finding a small set of vertices in a temporal graph whose removal disconnects two designated terminals: a classic, polynomial-time solvable problem in (static) graph theory.

TEMPORAL (s, z)-SEPARATION

Input: A temporal graph $G = (V, E, \tau)$, two distinct vertices $s, z \in V$, and $k \in \mathbb{N}$.

Question: Does G admit a temporal (s, z)-separator of size at most k?

Due to the space constraints, missing details and proofs (marked with ⋆) are deferred to a long version [9] of this paper, see https://arxiv.org/abs/1803.00882.

T. Fluschnik—Supported by the DFG, project DAMM (NI 369/13).

H. Molter—Supported by the DFG, project MATE (NI 369/17).

A. Brandstädt et al. (Eds.): WG 2018, LNCS 11159, pp. 216–227, 2018.
https://doi.org/10.1007/978-3-030-00256-5_18

Herein, a vertex set $S \subseteq V \setminus \{s, z\}$ is a *temporal (s, z)-separator* for a given temporal graph $G = (V, E, \tau)$ with $s, z \in V$ if there is no temporal (s, z)-path in $G - S := (V \setminus S, \{(\{v, w\}, t) \in E \mid v, w \in V \setminus S\}, \tau)$. A *temporal (s, z)-path* of length ℓ in G is a sequence $P = ((\{v_0, v_1\}, t_1), (\{v_1, v_2\}, t_2), \ldots, (\{v_{\ell-1}, v_\ell\}, t_\ell))$ of time-edges in E, where $s = v_0$, $z = v_\ell$, $v_i \neq v_j$ for all $i, j \in \{0, \ldots, \ell\}$ with $i \neq j$, and $t_i \leq t_{i+1}$ for all $i \in \{1, \ldots, \ell - 1\}$. TEMPORAL (s, z)-SEPARATION is NP-hard [11]. In this work, we study TEMPORAL (s, z)-SEPARATION on restricted classes of temporal graphs with the goal to identify computationally tractable cases.

So far, in the literature one basically finds two different directions concerning the definition of temporal graph classes. One direction is defining temporal graph classes through the underlying graph (that is, essentially, the graph obtained by forgetting about the time labels of the edges) [1,7,17]. Herein, one restricts the input temporal graph to have its underlying graph being contained in some specific graph class. The other direction considers properties expressible through temporal aspects [5,8,13,15]. Such properties are, for instance, each layer being a subgraph of its succeeding layer, or the temporal graph being periodic, that is, having a subsequence of layers which is repeated in the same order for some periods. In this work, we study TEMPORAL (s, z)-SEPARATION on temporal graph classes from both directions.

Our Contributions. We show that TEMPORAL (s, z)-SEPARATION remains NP-complete on many restricted temporal graph classes.

- TEMPORAL (s, z)-SEPARATION remains NP-complete on temporal graphs whose underlying graph falls into a class of graphs containing complete-but-one graphs (that is, complete graphs where exactly one edge is missing) or line graphs. However, if the underlying graph is of bounded treewidth, TEMPORAL (s, z)-SEPARATION becomes polynomial-time solvable (see Fig. 1 for an overview).
- TEMPORAL (s, z)-SEPARATION remains NP-complete on temporal graphs where each layer contains only one edge (Corollary 1). In contrast, if we require each layer to be a unit interval graph with respect to the same global vertex ordering, then TEMPORAL (s, z)-SEPARATION becomes solvable in polynomial time (Theorem 1).
- Regarding temporal graph classes defined through temporal aspects, TEMPORAL (s, z)-SEPARATION becomes solvable in polynomial time on single-peaked temporal graphs, on graphs where all layers are identical (1-periodic or 0-steady), or when the number of periods is at least the number of vertices. In all other considered cases TEMPORAL (s, z)-SEPARATION remains NP-complete (see Table 1 in Sect. 4 for an overview).

Related Work. Kempe et al. [11] proved TEMPORAL (s, z)-SEPARATION to be NP-complete. Zschoche et al. [17] proved that TEMPORAL (s, z)-SEPARATION remains NP-complete on temporal graphs with bipartite or planar underlying graphs. Moreover, TEMPORAL (s, z)-SEPARATION is W[1]-hard when parameterized by the solution size k [17].

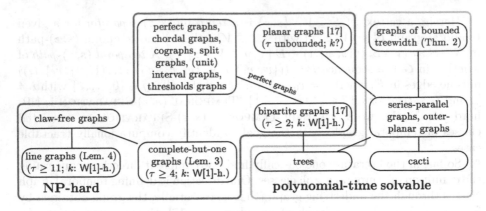

Fig. 1. Computational complexity of TEMPORAL (s, z)-SEPARATION for some graph classes of the underlying graph. An edge between two classes indicates containment of the lower in the upper class. For the classes of line, complete-but-one, bipartite, and planar graphs, we provide for which values of the maximum time label τ NP-hardness is proven as well as the parameterized complexity of TEMPORAL (s, z)-SEPARATION when parameterized by the solution size k. We point out that in the case of planar graphs, neither a bound on τ nor the parameterized complexity regarding k is known.

Casteigts et al. [5] defined twelve different classes of temporal graphs and showed a corresponding inclusion diagram. Among these classes, they define temporal graph classes with recurrence or periodicity of edges. On a slightly different notion of the latter class, Flocchini et al. [8] studied the problem of exploring a temporal graph. Kuhn et al. [13] studied the problem of token dissemination on temporal graphs where in each time-interval of length T, at least the edges of an arbitrary spanning tree appear.

The class of temporal graphs with underlying graphs of bounded treewidth are considered in the context of temporal graph exploration [7] and single-source temporal connectivity [1]. Erlebach et al. [7] studied the problem of temporal graph exploration on temporal graphs with underlying graphs being planar and of bounded vertex degree. They also introduced the class of temporal graphs with regularly present edges, where the absence of each edge in consecutive time steps is lower- and upper-bounded by two values. Michail and Spirakis [15] studied a temporal version of the TRAVELING SALESPERSON PROBLEM on temporal graphs with bounded dynamic diameter, where the dynamic diameter is the smallest number d such that every vertex can reach any other vertex at any time in at most d time steps.

2   Preliminaries

As a convention, $\mathbb{N}$ denotes the natural numbers without zero.

Static Graphs. We use basic notations from (static) graph theory. Let $G = (V, E)$ be an *undirected, simple graph*. We use $V(G)$ and $E(G)$ to denote the set of vertices and the set of edges of G, respectively. We denote by $G - V'$ $:= (V \setminus V', \{\{v, w\} \in E \mid v, w \in V \setminus V'\})$ the graph G without the vertices in $V' \subseteq V$. For $V' \subseteq V$, $G[V'] := G - (V \setminus V')$ denotes the *induced subgraph* of G on the vertices V'. A *path* of length ℓ is sequence of edges $P = (\{v_1, v_2\}, \{v_2, v_3\}, \ldots, \{v_\ell, v_{\ell+1}\})$ where $v_i \neq v_j$ for all $i, j \in \{1, \ldots, \ell - 1\}$ with $i \neq j$. We set $V(P) = \{v_1, v_2, \ldots, v_{\ell+1}\}$. Path P is an (s, z)-*path* if $s = v_1$ and $z = v_{\ell+1}$. A set $S \subseteq V$ of vertices is an (s, z)-*separator* in G if there is no (s, z)-path in $G - S$.

Temporal Graphs. Let $\boldsymbol{G} = (V, \boldsymbol{E}, \tau)$ be a temporal graph. The graph $G_i(\boldsymbol{G}) = (V, E_i(\boldsymbol{G}))$ is called *layer* i of the temporal graph $\boldsymbol{G} = (V, \boldsymbol{E}, \tau)$ if and only if $\{v, w\} \in E_i(\boldsymbol{G}) \Leftrightarrow (\{v, w\}, i) \in \boldsymbol{E}$. The *underlying graph* $G_\downarrow(\boldsymbol{G})$ of a temporal graph $\boldsymbol{G} = (V, \boldsymbol{E}, \tau)$ is defined as $G_\downarrow(\boldsymbol{G}) := (V, E_\downarrow(\boldsymbol{G}))$, where $E_\downarrow(\boldsymbol{G}) = \{e \mid \exists t : (e, t) \in \boldsymbol{E}\}$. (We drop $\boldsymbol{G}$ in the notations if it is clear from the context.) For $X \subseteq V$ we define the *induced temporal subgraph* of $\boldsymbol{G}$ by X by $\boldsymbol{G}[X] := (X, \{(\{v, w\}, t) \in \boldsymbol{E} \mid v, w \in X\}, \tau)$. We say that a temporal graph $\boldsymbol{G}$ is *connected* if its underlying graph $G_\downarrow$ is connected. Let $s, z \in V$. Throughout the whole paper we assume that the temporal input graph $\boldsymbol{G}$ is connected and that there is no time-edge between s and z. Furthermore, in accordance with Wu et al. [16] we assume that the time-edge set $\boldsymbol{E}$ is ordered by ascending labels.[1] The *concatenation* of two temporal graphs $\boldsymbol{G}_1 = (V, \boldsymbol{E}_1, \tau_1)$, $\boldsymbol{G}_2 = (V, \boldsymbol{E}_2, \tau_2)$ is denoted by $\boldsymbol{G}_1 \circ \boldsymbol{G}_2 = (V, \boldsymbol{E}_1 \cup \{(e, t + \tau_1) \mid (e, t) \in \boldsymbol{E}_2\}, \tau_1 + \tau_2)$. Furthermore, we define that $\boldsymbol{G}_1^1 = \boldsymbol{G}_1$ and $\boldsymbol{G}_1^x = \boldsymbol{G}_1^{x-1} \circ \boldsymbol{G}_1$, for all integers $x \geq 2$.

Lemma 1 ($\star$). *Given a temporal graph $\boldsymbol{G} = (V, \boldsymbol{E}, \tau)$ and two distinct vertices s and z, a temporal (s, z)-path can be computed in $\mathcal{O}(|\boldsymbol{E}|)$ time.*

Parameterized Complexity. A *parameterized problem* is in XP if there is an algorithm that solves each instance (I, r) in $|I|^{f(r)}$ time, as well as it is *fixed-parameter tractable* (in FPT) if there is an algorithm that solves each instance (I, r) in $f(r) \cdot |I|^{\mathcal{O}(1)}$ time, where f is a computable function depending only on the *parameter* r [6]. There is a hierarchy of hardness classes for parameterized problems, of which the most important one is W[1]. If a parameterized problem is W[1]-hard, then it is (presumably) not in FPT.

3   Structural Restrictions

Two approaches to define temporal graph classes are (i) restricting each layer to be contained in a specific graph class or (ii) restricting the underlying graph to be contained in a specific graph class. We point out that both are independent of the linear order of the layer and hence appear to not fully capture the temporal characteristics of a given temporal graph. Indeed, our results support this fact as we obtain intractability for many restricted graph classes.

---

[1] If this is not the case, then $\boldsymbol{E}$ can be sorted by ascending labels with bucketsort or mergesort in $\mathcal{O}(\min\{\tau, |\boldsymbol{E}| \log |\boldsymbol{E}|\})$ time.

3.1   Layer-Wise Restrictions

Restricting the layers to fall into a specific graph class neither captures any temporal aspect of the temporal graph nor the full picture drawn by all layers together. In fact, we show that such restrictions are not helpful: the problem is already NP-hard when each layer consists of at most one edge.

Lemma 2 ($\star$). *There is a polynomial-time many-one reduction that maps any instance $(G = (V, E, \tau), s, t, k)$ of* TEMPORAL (s, z)-SEPARATION *to an equivalent instance $(G' = (V, E', \tau'), s, t, k)$ such that each layer in G' has at most one edge and $\tau' \leq \tau \cdot |V|^4$.*

Lemma 2 (together with known hardness results [11,17]) implies the following.

Corollary 1. TEMPORAL (s, z)-SEPARATION *is* NP-*complete and* W[1]-*hard when parameterized by the solution size k even if each layer has at most one edge.*

Now we consider a scenario where temporal graphs have a certain geometric interpretation. For example in data sets where vertices are individuals and edges model physical proximity (see e.g. [2]), it is a reasonable assumption that the individual layers are disc intersection graphs (assuming the individuals only move in the plane). We focus on the one-dimensional case, where we get (unit) interval graphs, and investigate this restriction as a starting point for further research. We show in the following that if each layer of a given temporal graph G is restricted to be a unit interval graph and there is an ordering on the vertices that matches the relative positions of the intervals in all layers, then we can solve TEMPORAL (s, z)-SEPARATION on G in polynomial time. We first give a formal definition of the restriction.

In the following we introduce temporal interval graphs. We call a temporal graph $G = (V, E, \tau)$ a *temporal interval graph* if every layer G_i is an interval graph. We say that a temporal graph $G = (V, E, \tau)$ is a *temporal unit interval graph* if every layer G_i is a *unit* interval graph. By Lemma 2, TEMPORAL (s, z)-SEPARATION on temporal unit interval graph is NP-hard.

We call a total ordering $<_V$ on a vertex set V *compatible* with a unit interval graph $G = (V, E)$ if there are unit intervals $[a_v, a_v + 1]$ with $a_v \in \mathbb{R}$ for all vertices $v \in V$ that induce the graph G and for all $u, v \in V$ with $u <_V v$ we have that $a_u \leq a_v$. Note that for every unit interval graph there is a total ordering on the vertices that is compatible with it.

Definition 1. *A temporal graph $G = (V, E, \tau)$ is an* order-preserving temporal unit interval graph *if G is a temporal unit interval graph and there is a total ordering $<_V$ on the vertex set V that is compatible with every layer G_i.*

One can prove a number of useful properties of order-preserving temporal unit interval graphs [9], which lead to the following result.

Theorem 1 ($\star$). TEMPORAL (s, z)-SEPARATION *on order-preserving temporal unit interval graphs with given ordering $<_V$ is solvable in polynomial time.*

3.2   Underlying-Wise Restrictions

We next study temporal graphs where the underlying graph is contained in some graph class. A graph is *complete-but-one* if all but one possible edges are present.

Lemma 3 ($\star$). *There is a polynomial-time many-one reduction that maps any instance $(G = (V, E, \tau), s, t, k)$ of* TEMPORAL (s, z)-SEPARATION *to an equivalent instance $(G' = (V, E', \tau'), s, t, k)$ such that $E(G_\downarrow(G')) = \binom{V}{2} \setminus \{s, t\}$.*

Lemma 3 implies that TEMPORAL (s, z)-SEPARATION remains NP-hard on all temporal graphs where the underlying graph falls into a graph class containing all complete-but-one graphs, for instance the classes of unit interval or threshold graphs. We refer to Fig. 1 in Sect. 1 for an overview.

Note that complete-but-one graphs are no line graphs, as each complete-but-one graph (with at least five vertices) contains a $K_5 - e$ as induced subgraph, which is forbidden in line graphs [3, Graph G_3]. Hence, we next study TEMPORAL (s, z)-SEPARATION on temporal graphs where the underlying graph is a line graph.

Lemma 4 ($\star$). TEMPORAL (s, z)-SEPARATION *on temporal graphs where the underlying graph is a line graph is* NP-*complete.*

Classification Through Parameterization. An alternative way to classify an instance of a graph-theoretic problem is through its (graph) parameters. We study TEMPORAL (s, z)-SEPARATION employing several problem specific parameterizations. Any upper bound on the maximum length of a temporal (s, z)-path leads to a straightforward search-tree algorithm.

Lemma 5 ($\star$). TEMPORAL (s, z)-SEPARATION *is solvable in $\mathcal{O}(\ell^k \cdot |E|)$ time, where k is the solution size and ℓ is the maximum length of a temporal (s, z)-path.*

From Lemma 5 we can derive that TEMPORAL (s, z)-SEPARATION is linear-time solvable on temporal graph classes where the underlying graph has a constant vertex cover number[2].

Corollary 2 ($\star$). TEMPORAL (s, z)-SEPARATION *can be solved in $\mathcal{O}((2\,\mathrm{vc})^{\mathrm{vc}} \cdot |E|)$ time, where* vc *is the vertex cover number of the underlying graph.*

Another graph parameter which upper-bounds the maximum length of an (s, z)-path in the underlying graph is the well-studied *tree-depth*[3] of the underlying graph.

Corollary 3 ($\star$). TEMPORAL (s, z)-SEPARATION *can be solved in $\mathcal{O}(2^{\mathrm{td}(G_\downarrow)\cdot k} \cdot |E|)$ time, where k is the solution size and $\mathrm{td}(G_\downarrow)$ is the tree-depth of the underlying graph.*

---

[2] The vertex cover number of a graph is the smallest number of vertices such that each edges has at least one of these vertices as an endpoint.

[3] We refer to the long version [9] for details.

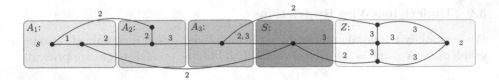

Fig. 2. The idea for the dynamic program from Theorem 2 for a temporal graph G. Vertices in S form the temporal (s, z)-separator, vertices in Z are not reachable from s in $G - S$, and vertices in A_t are not reachable from s in $G - S$ before time t.

One of the tools from the repertoire for designing fixed-parameter algorithms for (static) graph problems are tree decompositions [6]. A tree decomposition is a mapping of a graph into a related tree-like structure. For many graph problems this tree-like structure can be used to formulate a bottom-up dynamic program that starts at the leaves and ends at the root of the tree decomposition [6]. Indeed, if we parameterize by $\mathrm{tw}_\downarrow(G)$, where $\mathrm{tw}_\downarrow(G)$ is defined as the treewidth of the underlying graph $G_\downarrow(G)$, then we obtain an XP-algorithm by dynamic programming.

Theorem 2 ($\star$). TEMPORAL (s, z)-SEPARATION *is solvable in time* $\mathcal{O}((\tau + 2)^{\mathrm{tw}_\downarrow(G)+2} \cdot \mathrm{tw}_\downarrow(G) \cdot |V| \cdot |E|)$, *if a nice tree-decomposition of the underlying graph with treewidth* $\mathrm{tw}_\downarrow(G)$ *is given, and where* τ *is the maximum time label.*

Note that a tree-decomposition of width $\mathcal{O}(\mathrm{tw}(G))$ can be computed in $2^{\mathcal{O}(\mathrm{tw}(G))} \cdot n$ time [4] and can be turned into a nice tree-decomposition in polynomial-time [6, Lemma 7.4], where G is a graph with n vertices. The dynamic program is based on the fact that for each vertex $v \in V$ in a temporal graph $G = (V, E, \tau)$ there is a point of time $t \in \{1, \dots, \tau\}$ such that v cannot be reached from $s \in V$ before time t. In particular, we guess a partition $V = A_1 \uplus \dots \uplus A_\tau \uplus S \uplus Z$ such that S is the temporal (s, z)-separator and in $G - S$, no vertex contained in Z is reachable from s and no vertex $v \in A_t$ can be reached from s before time step t, where $t \in \{1, \dots, \tau\}$. See Fig. 2 for an illustrative example.

It remains open whether TEMPORAL (s, z)-SEPARATION is fixed-parameter tractable when parameterized by $\mathrm{tw}_\downarrow$ or by $k + \mathrm{tw}_\downarrow$.

4   Temporal Restrictions

In Sect. 3 we considered restrictions on the layers and the underlying graph. Observe that these restrictions do not cover the temporal aspects of a temporal graph, that is, any reordering of the layers yields a different temporal graph having the same restrictions. In this section, we study temporal graph classes whose definitions depend on the ordering of the layers. Herein, we study *monotone*, *periodic*, *consecutively connected*, and *steady* temporal graphs.

Note that *monotone*, *periodic*, and *consecutively connected* temporal graphs are quite restricted temporal graph classes [5]. Unfortunately, even on these temporal graph classes, except for trivial cases, we encounter computational

Table 1. Let τ denote the maximum time label and r the number of periods in G.

| | TEMPORAL (s, z)-SEPARATION | |
	Polynomial-time	NP-hard
p-monotone	single-peaked	$p \geq 2$
p-periodic	$p = 1$, or $r \geq n$	$p \geq 2$
T-interval connected	-	$T \geq 1$
λ-steady	$\lambda = 0$ or $(\lambda, \tau$ constant)	$\lambda \geq 1$

hardness by straightforward arguments. We refer to Table 1 for an overview on our results.

Monotone Temporal Graphs. Intuitively, a temporal graph is monotone if it can be decomposed into time-intervals on which the layers are consecutively subgraphs or supergraphs.

Definition 2. *A temporal graph $G = (V, E, \tau)$ is p-monotone if $p \in \mathbb{N}$ is the smallest number such that there are $1 = i_1 < i_2 < \ldots < i_{p+1} = \tau$ such that for all $\ell \in \{1, \ldots, p\}$ holds $E_j \subseteq E_{j+1}$ or $E_j \supseteq E_{j+1}$ for all $i_\ell \leq j < i_{\ell+1}$.*

Khodaverdian et al. [12] call a temporal graph monotone if whenever an edge is contained in a layer, this edge is contained in all succeeding layers. Their motivation is activation of proteins, or more general, temporal graphs that model activation by connected components. Casteigts et al. [5, Class 6] call this property of temporal graphs while additionally requiring the underlying graph to be connected as "recurrence of edges". Since we only consider temporal graphs with connected underlying graphs, both definitions form special cases of our 1-monotone temporal graphs where each layer is a subgraph of its successor.

A *peak* in a p-monotone temporal graph is an index $i_\ell \in \{i_1, i_2, \ldots, i_{p+1}\}$ such that there exists a $i_{\ell-1} \leq j < i_\ell$ with $E_j \subset E_{j+1}$ or a $i_\ell \leq j < i_{\ell+1}$ with $E_j \supset E_{j+1}$. As a convention, 1-monotone temporal graphs are *single-peaked*, that is, they have only one peak. Indeed, observe that for TEMPORAL (s, z)-SEPARATION only the peaks matter. Hence, we obtain the following reduction.

Observation 1. *Given an instance $\mathcal{I} = (G = (V, E, \tau), s, t, k)$ with G being p-monotone with ℓ peaks, we can compute in polynomial time an instance $\mathcal{I}' = (G' = (V, E', \tau'), s, t, k)$ such that $\mathcal{I}$ is equivalent to $\mathcal{I}'$ and $\tau' \leq \ell$.*

Observation 1 at hand, the following is straightforward:

Observation 2. TEMPORAL (s, z)-SEPARATION *is solvable in polynomial time on single-peaked temporal graphs.*

Surprisingly, the situation changes when the temporal graph is already 2-monotone but not single-peaked. We can make every temporal graph τ-monotone by simply adding edge-free layers between any two consecutive layers; formally:

Observation 3. *There is a polynomial-time many-one reduction that maps any instance* $(G = (V, E, \tau), s, t, k)$ *of* TEMPORAL (s, z)-SEPARATION *to an equivalent instance* $(G' = (V, E', 2\tau - 1), s, t, k)$ *such that for all* $i \in \{1, \ldots, \tau\}$ *it holds that* $E_{2i-1}(G') = E_i(G)$ *and for all* $i \in \{1, \ldots, \tau - 1\}$ *it holds that* $E_{2i}(G') = \emptyset$.

As TEMPORAL (s, z)-SEPARATION is already NP-complete for $\tau = 2$ [17], we get the following.

Observation 4. *For all* $p \geq 2$, TEMPORAL (s, z)-SEPARATION *on* p-*monotone temporal graphs is* NP-*complete.*

Periodic Temporal Graphs. In several real-world scenarios one observes periodicity; Indeed, whenever one observes mobile entities with periodic movements [5], as satellites or (scheduled) public transport, over longer time periods, periodic patterns appear. Such models motivate the following class of temporal graphs.

Definition 3. *A temporal graph* $G = (V, E, \tau)$ *is* p-*periodic if* $p \in \mathbb{N}$ *is the smallest number such that* $G = G'^{r}$ *where* $G' = (V, E', p)$ *and* r *denotes the number of periods.*

Different notions of periodic temporal graphs exist in the literature. Flocchini et al. [8] consider periodic temporal graphs obtained from "carriers", that is, a set of strict temporal paths define a network. Liu and Wu [14] consider delay tolerant networks where nodes have some cyclic movement pattern and get connected when they are in reach: if the time steps are large enough, periodicity is observed. In both cases, the smallest common multiple of the time spans of the entities define the length of a period. Casteigts et al. [5, Class 8] define periodic temporal graphs by periodicity of edges, that is, for all edges e, time steps t, and $c \in \mathbb{N}$, edge e is present at time step t if and only if e is present at time step $t + c \cdot p$, where p is the periodicity. They require the underlying graph to be connected, but they do not require minimality on the periodicity.

It is known that TEMPORAL (s, z)-SEPARATION is NP-complete on 2-periodic temporal graphs [17]. Contrarily, on 1-periodic temporal graphs, TEMPORAL (s, z)-SEPARATION collapses to (s, z)-SEPARATION in the underlying graph. Surprisingly, if the number of periods is large enough, then the problem becomes polynomial-time solvable.

Let P be an (s, z)-path of length ℓ in the underlying graph $G_\downarrow$ of the temporal graph $G = (V, E, \tau)$. A time-edge sequence $P' = ((e_1, t_1), \ldots, (e_\ell, t_\ell)) \in E^\ell$ is a *realization* of P $(P' \simeq P)$ if $(e_1, \ldots, e_\ell)$ is P. The *distance to temporality* of P in G is $\min_{P' \simeq P} |f_{P'}| - 1$, where $|f_{P'}|$ is the number of monotonically increasing intervals of the function $f_{P'} : \{1, \ldots, \ell\} \rightarrow \{1, \ldots, \tau\}, f_{P'}(x) = t_x$ and t_x is the label of the x-th time-edge of P'. Furthermore, the distance to temporality from s to z in G is the maximum distance to temporality over all (s, z)-paths in $G_\downarrow$.

Lemma 6 $(\star)$. *Let* $G = G'^{r}$ *be a* p-*periodic temporal graph such that the number of periods* r *is at least the distance to temporality from* s *to* z *in* G'. *Then* TEMPORAL (s, z)-SEPARATION *is solvable in polynomial time.*

Observe that in the reduction of Zschoche et al. [17] for maximum time label $\tau = 2$, the distance to temporality from s to z is two. Thus, TEMPORAL (s, z)-SEPARATION is NP-hard even if the input temporal graph $G = G'^r$ is p-periodic and the number r of periods is the distance to temporality from s to z minus one. However, the distance to temporality is clearly upper-bounded by the number of vertices. Hence, we obtain the following.

Corollary 4. *Let $G = (V, E, \tau)$ be a p-periodic temporal graph. If the number of periods $r \geq |V|$, then* TEMPORAL (s, z)-SEPARATION *is solvable in polynomial time.*

Interval-Connected Temporal Graphs. Kuhn et al. [13, Definition 2.1] introduced the following class of temporal graphs.

Definition 4. *A temporal graph $G = (V, E, \tau)$ is T-interval connected for $T \in \mathbb{N}$ if for every $t \in \{1, \ldots, \tau - T + 1\}$ the static graph $G = (V, \bigcap_{i=t}^{t+T-1} E_i(G))$ is connected.*

Kuhn et al. [13] studied T-interval connected temporal graphs in the context of counting and token dissemination in distributed systems. Note that temporal graphs where each layer is connected are 1-interval connected temporal graphs, but are not necessarily T-interval connected for some $T \geq 2$. On the contrary, for every T-interval connected temporal graph it holds true that each layer is connected.

Observation 5 ($\star$). *There is a polynomial-time many-one reduction that maps any instance $(G = (V, E, \tau), s, t, k)$ of* TEMPORAL (s, z)-SEPARATION *to an equivalent instance $(G' = (V', E', \tau), s, t, k + 1)$ such that G' is T-interval connected for every $T \geq 1$.*

Steady Temporal Graphs. When monitoring a network over time with high resolution, we expect evolutionary instead of revolutionary changes in one time step. For instance, observing any contact network per second, we do not expect many contacts to appear in the same second. More generally, in several real-world scenarios we do not expect big changes from one time step to the other. This assumption motivates the following class of temporal graphs.

Definition 5. *A temporal graph $G = (V, E, \tau)$ is λ-steady if $\lambda \in \mathbb{N} \cup \{0\}$ is the smallest number such that for each point in time $t \in \{1, \ldots, \tau - 1\}$ the size of the symmetric difference of two consecutive edge sets $|E_t \triangle E_{t+1}|$ is at most λ.*

To the best of our knowledge, this class has not yet been considered in the literature.

One can expect that hardness results for temporal graphs translate to steady temporal graphs, even if $\lambda \leq 1$.

Observation 6 ($\star$). *There is a polynomial time many-one reduction that maps any instance $(G = (V, E, \tau), s, t, k)$ of* TEMPORAL (s, z)-SEPARATION *to an equivalent instance $(G' = (V', E', \tau'), s, t, k)$ such that G' is 1-steady.*

The reduction of Observation 6 increases the maximum time label by a factor depending on the input size. Indeed, from previous results [17] it follows that TEMPORAL (s, z)-SEPARATION on λ-steady temporal graphs is fixed-parameter tractable when parameterized by τ.

Corollary 5 ($\star$). TEMPORAL (s, z)-SEPARATION *on λ-steady temporal graphs is fixed-parameter tractable when parameterized by the maximum time label τ.*

5   Conclusion

We studied TEMPORAL (s, z)-SEPARATION on different temporal graph classes—with structural and temporal restrictions on temporal graph models. We proved TEMPORAL (s, z)-SEPARATION to remain NP-complete on the majority of the considered classes of restricted temporal graphs. Polynomial-time solvability is achieved for temporal graphs where the underlying graph has bounded treewidth, on single-peaked temporal graphs, temporal graphs with many periods, and order-preserving temporal unit interval graphs.

Our results call into question to which extent currently in the literature considered notions of temporal graph classes address the features of temporal graphs and hence impose useful restrictions on temporal graphs. For instance, the introduced class of order-preserving temporal unit interval graphs is more restrictive than just requiring the layers to fall into a specific graphs class; however, also this notion does not capture temporal aspects. Exploring further, more sophisticated structural restrictions of temporal graphs, whose definitions may rely on global properties and on temporal aspects, is of particular interest when asking for computationally tractable cases of TEMPORAL (s, z)-SEPARATION.

A specific direction for future work would be to use the derived polynomial-time algorithms as a basis for distance-to-triviality parameterizations [10]. For instance, for a temporal unit interval graph one may introduce a parameter κ that upper-bounds how much the vertex orderings of two consecutive layers of a temporal unit interval graph differ. More specifically, given a temporal unit interval graph $G = (V, E, \tau)$, we define κ as the smallest integer such that there are vertex orderings $<_V^1, \ldots, <_V^\tau$ such that $<_V^t$ is compatible with layer G_t for all $t \in \{1, \ldots, \tau\}$, and the orderings of any two consecutive layers have *Kendall tau* distance[4] at most κ, that is, for all $t \in \{1, \ldots, \tau - 1\}$ we have that $K(<_V^t, <_V^{t+1}) \leq \kappa$. Clearly for order-preserving temporal unit interval graphs we have that $\kappa = 0$ and it is easy to observe (with the help of Lemma 2) that we get NP-hardness for $\kappa = 1$. We conjecture that we can achieve fixed-parameter tractability for the parameter combination (κ, τ) for TEMPORAL (s, z)-SEPARATION on temporal unit interval graphs.

---

[4] The Kendall tau distance is a metric that counts the number of inversions between two total orderings; it is also known as "bubble sort distance".

References

1. Axiotis, K., Fotakis, D.: On the size and the approximability of minimum temporally connected subgraphs. In: Proceedings of 43rd ICALP, vol. 55, pp. 149:1–149:14. Dagstuhl Publishing (2016)
2. Barrat, A., Fournet, J.: Contact patterns among high school students. PLoS ONE **9**(9), e107878 (2014)
3. Beineke, L.W.: Characterizations of derived graphs. J. Comb. Theor. **9**(2), 129–135 (1970)
4. Bodlaender, H.L., Drange, P.G., Dregi, M.S., Fomin, F.V., Lokshtanov, D., Pilipczuk, M.: A $c^k n$ 5-approximation algorithm for treewidth. SIAM J. Comput. **45**(2), 317–378 (2016)
5. Casteigts, A., Flocchini, P., Quattrociocchi, W., Santoro, N.: Time-varying graphs and dynamic networks. Int. J. Parallel Emergent Distrib. Syst. **27**(5), 387–408 (2012)
6. Cygan, M., et al.: Parameterized Algorithms. Springer, Cham (2015). https://doi.org/10.1007/978-3-319-21275-3
7. Erlebach, T., Hoffmann, M., Kammer, F.: On temporal graph exploration. In: Halldórsson, M.M., Iwama, K., Kobayashi, N., Speckmann, B. (eds.) ICALP 2015. LNCS, vol. 9134, pp. 444–455. Springer, Heidelberg (2015). https://doi.org/10.1007/978-3-662-47672-7_36
8. Flocchini, P., Mans, B., Santoro, N.: On the exploration of time-varying networks. Theor. Comput. Sci. **469**, 53–68 (2013)
9. Fluschnik, T., Molter, H., Niedermeier, R., Zschoche, P.: Temporal graph classes: a view through temporal separators. CoRR, abs/1803.00882 (2018). http://arxiv.org/abs/1803.00882. Long version of this paper
10. Guo, J., Hüffner, F., Niedermeier, R.: A structural view on parameterizing problems: distance from triviality. In: Downey, R., Fellows, M., Dehne, F. (eds.) IWPEC 2004. LNCS, vol. 3162, pp. 162–173. Springer, Heidelberg (2004). https://doi.org/10.1007/978-3-540-28639-4_15
11. Kempe, D., Kleinberg, J., Kumar, A.: Connectivity and inference problems for temporal networks. J. Comput. Syst. Sci. **64**(4), 820–842 (2002)
12. Khodaverdian, A., Weitz, B., Wu, J., Yosef, N.: Steiner network problems on temporal graphs. CoRR, abs/1609.04918v2 (2016)
13. Kuhn, F., Lynch, N., Oshman, R.: Distributed computation in dynamic networks. In: Proceedings of 42nd STOC, pp. 513–522. ACM (2010)
14. Liu, C., Wu, J.: Scalable routing in cyclic mobile networks. IEEE Trans. Parallel Distrib. Syst. **20**(9), 1325–1338 (2009)
15. Michail, O., Spirakis, P.G.: Traveling salesman problems in temporal graphs. Theor. Comput. Sci. **634**, 1–23 (2016)
16. Wu, H., Cheng, J., Ke, Y., Huang, S., Huang, Y., Wu, H.: Efficient algorithms for temporal path computation. IEEE Trans. Knowl. Data. Eng. **28**(11), 2927–2942 (2016)
17. Zschoche, P., Fluschnik, T., Molter, H., Niedermeier, R.: The complexity of finding small separators in temporal graphs. In: Proceedings of the 43rd MFCS, LIPIcs. Schloss Dagstuhl-Leibniz Center for Informatics (2018, to appear). https://arxiv.org/abs/1711.00963

Covering a Graph with Nontrivial Vertex-Disjoint Paths: Existence and Optimization

Renzo Gómez[✉][iD] and Yoshiko Wakabayashi[iD]

Instituto de Matemática e Estatística, Universidade de São Paulo,
São Paulo, Brazil
{rgomez,yw}@ime.usp.br

Abstract. Let G be a connected graph and $\mathcal{P}$ be a set of pairwise vertex-disjoint paths in G. We say that $\mathcal{P}$ is a *path cover* if every vertex of G belongs to a path in $\mathcal{P}$. In the *minimum path cover problem*, one wishes to find a path cover of minimum cardinality. In this problem, known to be NP-hard, the set $\mathcal{P}$ may contain trivial (single-vertex) paths. We study the problem of finding a path cover composed only of nontrivial paths. First, we show that the corresponding existence problem can be reduced to a matching problem on a bipartite graph via the Edmonds-Gallai Decomposition. This reduction gives, in polynomial time, a certificate for both the YES-answer and the NO-answer. When trivial paths are forbidden, for the feasible instances, one may consider either minimizing or maximizing the number of paths in the path cover. We show that the maximization problem has a close relation with the maximum matchings of a graph, and can be solved in polynomial time. For the minimization problem on feasible instances, we show that its computational complexity is equivalent to the minimum path cover problem. We also show a linear-time algorithm on (edge-weighted) trees.

1 Introduction

All graphs considered here are simple and undirected. The *length* of a path in a graph is its number of edges. If a path has length k, we say that it is a k-*path*; and when its length is zero, we say that it is *trivial*. Here, a *path cover* of a graph G means a set of pairwise vertex-disjoint paths that collectively spans $V(G)$.

The MINIMUM PATH COVER (MINPC) problem asks for a path cover of minimum cardinality. Clearly, MINPC is NP-hard on the classes of graphs for which the Hamiltonian path problem is NP-complete. This is known to hold for cubic planar 3-connected graphs [5], circle graphs, split graphs, chordal bipartite graphs [12], etc. Polynomial-time algorithms have been designed for MINPC on several classes of perfect graphs, such as interval graphs [2], cocomparability

Research supported by CNPq (Proc. 456792/2014-7, 306464/2016-0), FAPESP (Proc. 2015/11937-9), CAPES (235671298-48), MaCLinC Proj. NUMEC/USP, Brazil.

© Springer Nature Switzerland AG 2018
A. Brandstädt et al. (Eds.): WG 2018, LNCS 11159, pp. 228–238, 2018.
https://doi.org/10.1007/978-3-030-00256-5_19

graphs [3], trees [4], etc. No approximation algorithm has been designed for this problem. Some applications of MINPC include establishing ring protocols in a network, code optimization and mapping parallel processes to parallel architectures [11].

We study the problem of finding a path cover without trivial paths. First, we consider the existence problem, and then the corresponding optimization problems: the MINIMUM NONTRIVIAL PATH COVER (MINNTPC) and the MAXIMUM NONTRIVIAL PATH COVER (MAXNTPC), both for the cardinality version.

In Sect. 2 we show that the existence of a nontrivial path cover in a graph is closely related to the structure of its maximum matchings, which is in turn related to the MAXNTPC. We show a new characterization of graphs that have a nontrivial path cover, which allows us to obtain an algorithm that solves the existence problem in an interesting way: it returns in polynomial time either (a) a YES-answer which is an optimal solution to the MAXNTPC or (b) a NO-answer together with a certificate. In the case of MINNTPC, we show a polynomial-time algorithm that transforms a minimum path cover into one without trivial paths, if it exists, and with the same cardinality (and therefore optimal). This result shows that the complexity of both problems, MINPC and MINNTPC, is in some sense equivalent. We also show a linear-time algorithm for MINNTPC on trees, which can be extended to the edge-weighted version of the problem. Owing to space limitation, most of the proofs are sketched and in some cases they are omitted.

2   Forbidding Trivial Paths

This section is devoted to nontrivial path covers of a graph: both the existence and the optimization problem. To shorten notation, we write pc (resp. $ntpc$) to refer to a *path cover* (resp. *nontrivial path cover*) of a graph.

While every graph admits a pc, not every graph admits an ntpc. It is immediate that some trees do not admit an ntpc. One may naturally ask whether minimum degree 2 would suffice for a graph to admit an ntpc. This is not the case, even when we ask for a higher constant minimum degree.

The ntpc existence problem is, in fact, a special case of classic and intensively studied problems in graph theory. Given a graph G with integer functions g and f defined on its vertices, a (g, f)-*factor* of G is a spanning subgraph H of G such that each vertex x in H has degree at least $g(x)$ and at most $f(x)$. The special case in which g and f are constants, say a and b, is referred to an $[a, b]$-*factor*. Thus, the existence of an ntpc in a graph is equivalent to the existence of a $[1, 2]$-factor. Kano and Saito [7] proved that, for $r \geq 1$, if G is a graph with degree at least r and at most $r + s$, where $s \in \{1, \ldots, r\}$, then G contains a $[1, 2]$-factor. This means that all regular graphs contain $[1, 2]$-factors. These authors also show classes of complete bipartite graphs that do not contain $[1, 2]$-factors ($K_{r,r+s}$, with $s > r \geq 1$).

There is a large number of structural results on the existence of (g, f)-factors in graphs. In 1952, Tutte [13] characterized graphs that have an (f, f)-factor. Lovász [9] generalized this result characterizing graphs that have a (g, f)-factor.

Later, some results on the algorithmic aspect of this problem were obtained. Anstee [1] showed a polynomial-time algorithm that finds a (g, f)-factor, if it exists, or a negative certificate. Heinrich et al. [6] showed an algorithm that finds a (g, f)-factor when $g(x) \leq 1$ and $g(x) < f(x)$ for every vertex x in G, whose running time is better than Anstee's algorithm. Thus, the ntpc existence problem has been shown to be solvable in polynomial time.

In what follows, we adopt another approach that focuses on the close relationship between the existence of an ntpc in a graph and the structure of its maximum matchings, and derive results also on MAXNTPC. Clearly, if a graph has a perfect matching, it has an ntpc consisting solely of 1-paths. And, if a graph has an ntpc but does not have a perfect matching, we need k-paths with $k \geq 2$ to cover it. In fact, we do not need $k > 2$, as such k-paths can be broken into paths of length 1 or 2. This observation indicates that we may focus only on the problem of deciding whether a graph G admits an ntpc consisting only of 1-paths and 2-paths.

We denote by $\mathscr{P}_{1,2}(G)$, or simply $\mathscr{P}_{1,2}$, the class of these special types of ntpc in G, and by $\mathscr{P}_{1,2}^1$ the subclass of $\mathscr{P}_{1,2}$ consisting of path covers with the largest possible number of 1-paths. The following result shows how a pc in $\mathscr{P}_{1,2}^1$ and a maximum matching in G are related.

Proposition 1. *If a graph G admits an ntpc, then the cardinality of any path cover in $\mathscr{P}_{1,2}^1$ coincides with the cardinality of a maximum matching in G.*

Proof (sketch). Let $\mathcal{P}$ be a pc in $\mathscr{P}_{1,2}^1$, and denote by $\mathcal{P}_2$ the set of 2-paths in $\mathcal{P}$. Let M be a matching obtained by choosing one edge from every path in $\mathcal{P}$. We claim that M is a maximum matching in G. We may assume that $|\mathcal{P}_2| \geq 2$, otherwise the result is immediate. Suppose, by contradiction, that M is not a maximum matching, and let P be a shortest M-augmenting path. Observe that the endvertices of P are endvertices of 2-paths, say Q and R, in $\mathcal{P}_2$. From the minimality of P, it can be shown that Q and R are the only paths in $\mathcal{P}_2$ that intersect P. Let $E(P)$ (resp. $E(\mathcal{P})$) be the set of edges in P (resp. $\mathcal{P}$). Observe that $E(\mathcal{P}') \bigtriangleup E(P)$ induce a path cover of G by 1-paths and 2-paths with more 1-paths than $\mathcal{P}$, a contradiction. Therefore, M is a maximum matching in G. $\square$

Proposition 1 tells us that if a graph G admits an ntpc, then G has a maximum matching that can be extended (by adding some edges) to a path cover in $\mathscr{P}_{1,2}^1$. However, it is not true that every maximum matching has this property. In Fig. 1, we show an example of such matching (given by the wavy edges).

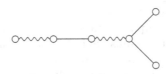

Fig. 1. A maximum matching that cannot be extended to a path cover in $\mathscr{P}_{1,2}^1$.

The idea is then to investigate whether, using information on every maximum matching of a graph, we may go further. It is well known that the Edmonds-Gallai Decomposition [10] of a graph provides the information we are seeking for. Moreover, it can be obtained in polynomial time. It defines a partition of the vertex set of a graph G into the following sets, from which we know the structure of all maximum matchings of G.

- $D(G)$, the set of vertices in G which are not covered by at least one maximum matching,
- $A(G)$, the set of vertices in $V(G) \setminus D(G)$ adjacent to at least one vertex in $D(G)$, and
- $C(G) = V(G) \setminus (A(G) \cup D(G))$.

A *near-perfect* matching of a graph G is one covering all but exactly one vertex of G. We say that a graph G is *hypomatchable* if $G - v$ has a perfect matching for every vertex v in G. Now, we state the Edmonds-Gallai Structure Theorem [10].

Theorem 1 (Edmonds, Gallai, 1965). *Let G be a graph and $A(G)$, $C(G)$ and $D(G)$ as defined above. Then*

(a) *the components of the subgraph induced by $D(G)$ are hypomatchable,*
(b) *the subgraph induced by $C(G)$ has a perfect matching,*
(c) *every maximum matching of G contains a near-perfect matching of each component of $D(G)$, a perfect matching of each component of $C(G)$ and matches all vertices of $A(G)$ with vertices in different components of $D(G)$.*

Next, from a pc $\mathcal{P}$ in $\mathscr{P}^1_{1,2}$, we define the following partition of $V(G)$.

$$L(\mathcal{P}) := \{v \in V(G) : v \text{ is an endvertex of a 2-path in } \mathcal{P}\},$$
$$R(\mathcal{P}) := \{v \in V(G) : v \text{ is an internal vertex of a 2-path in } \mathcal{P}\},$$
$$S(\mathcal{P}) := V(G) \setminus (L(\mathcal{P}) \cup R(\mathcal{P})).$$

Now, regarding this partition, we show the following result, where the sets $A(G)$, $C(G)$ and $D(G)$ are those defined above.

Proposition 2. *Let G be a graph that admits an ntpc, and let $\mathcal{P} \in \mathscr{P}^1_{1,2}(G)$. Then,*

(a) $L(\mathcal{P}) \subseteq D(G)$,
(b) *if $u \in S(\mathcal{P}) \cap A(G)$, then, the neighbor of u in $\mathcal{P}$ belongs to $D(G)$,*
(c) *let $u, v \in R(\mathcal{P}) \cup S(\mathcal{P})$ (possibly, $u = v$). Let N be the set of neighbors of u and v in $\mathcal{P}$. If $u, v \in A(G)$, then, each vertex in N belongs to a different component in $D(G)$.*

Proof (sketch). By Proposition 1, if we choose one edge from every path of $\mathcal{P}$, we obtain a maximum matching of G. Since any of the two edges of a 2-path in $\mathcal{P}$ may belong to this matching, (a) follows. By Theorem 1, in every maximum matching, the vertices in $A(G)$ are matched to vertices in different components of $D(G)$. This fact implies (b). To prove (c) we use the fact that every maximum matching in G contains a near-perfect matching of the components in $D(G)$, and analyse three cases concerning the membership of u and v in $R(\mathcal{P}) \cup S(\mathcal{P})$. □

Consider a pc in $\mathscr{P}_{1,2}^1$. Observe that Propositions 1 and 2 imply that the vertices in $C(G)$ are covered by 1-paths obtained from a perfect matching in $C(G)$. Moreover, Proposition 2 tells us how the vertices in $A(G) \cup D(G)$ are covered by a pc in $\mathscr{P}_{1,2}^1$. Observe that to cover a nontrivial component K in $D(G)$, we can take a near-perfect matching of K and add one of the edges in K incident to the uncovered vertex to obtain a pc by 1-paths and 2-paths. Since the nontrivial components of $D(G)$ are hypomatchable, to find a pc in $\mathscr{P}_{1,2}^1$, it suffices to focus on how to cover the trivial components in $D(G)$.

On the other hand, note that $\mathscr{P}_{1,2}^1$ consists of path covers of G with the maximum number of paths. Thus, MaxNtPC on G reduces to the problem of finding a pc in $\mathscr{P}_{1,2}^1$. Now, we show that the latter can be reduced to a maximum matching problem in a bipartite graph that is obtained using the Edmonds-Gallai Decomposition of G.

Let $T \subseteq D(G)$ be the set of vertices corresponding to the trivial components in $D(G)$, and let N be a set of vertices representing the nontrivial components in $D(G)$. Moreover, let A' and N' be copies of $A(G)$ and N, respectively. We define a bipartite graph $H = (U \cup W, F)$, in the following way.

$$U = A(G) \cup A' \cup N',$$
$$W = T \cup N.$$

For every component K in $D(G)$, let us denote by w_K the vertex representing this component in W. Now, we describe the set of edges F. Let u be a vertex in $A(G) \cup A'$ and w_K be a vertex in W. The edge $uw_K \in F$, if and only if, the vertex represented by u in G is adjacent to a vertex in K. Also, there is an edge linking every vertex in N to its copy in N'. Using this construction, we obtain the following result.

Theorem 2. MaxNtPC *can be solved in polynomial time.*

The classic results characterizing graphs containing (g, f)-factors, when specialized to $[1, 2]$-factors, give the following result (see Las Vergnas [8]).

Theorem 3 (Lovász 1970). *A graph G has a $[1, 2]$-factor, if and only if, for every $S \subseteq V(G)$, we have that $i(G - S) \leq 2|S|$, where $i(G - S)$ is the number of isolated vertices in $G - S$.*

Most of the characterization results, except for Anstee [1], were not concerned with an efficient way to find a NO-certificate (a set S that does not satisfy the condition stated in Theorem 3). Interestingly, our approach of searching for a pc in $\mathscr{P}_{1,2}^1$ gives an efficient way to find a $[1, 2]$-factor (and therefore an ntpc), when it exists, or to find a NO-certificate. The next theorem tells how this can be achieved.

Theorem 4. *Let G be a graph, $D(G)$ be the set given by the Edmonds-Gallai Decomposition of G, and $T \subseteq D(G)$ be the set of vertices corresponding to the trivial hypomatchable components in $D(G)$. Then the following holds:*

(i) G has a $[1,2]$-factor if and only if $|X| \le 2|N(X)|$, for every $X \subseteq T$.
(ii) If G does not have a $[1,2]$-factor, and X is a set that violates the condition stated in (i), then $S = N(X)$ is a set that violates the condition stated in Theorem 3. Moreover, S can be found in polynomial time.

In what follows, we show that the computational complexity of MINPC and of MINNTPC are closely related. More precisely, we show a polynomial-time algorithm that transforms a minimum pc of a graph into a minimum ntpc, if it exists, or exhibits a set of vertices that violates the condition given in Theorem 3.

If $\mathcal{P}$ denotes a pc of a graph $G = (V, E)$, then $\mathcal{P}_k$ denotes the set of k-paths of $\mathcal{P}$ and V_k denotes the set of vertices covered by the paths in $\mathcal{P}_k$. Furthermore, let $B \subseteq V_2$ be the set of the interior vertices of the paths in $\mathcal{P}_2$.

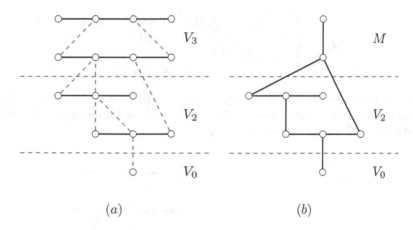

Fig. 2. (a) a graph G and a path cover $\mathcal{P}$; (b) the graph G^*.

Given a graph G and a minimum pc $\mathcal{P}$ of G, let G^* be the graph obtained from G and $\mathcal{P}$ in the following way. First, we contract every k-path in $\mathcal{P}_k$, $k \ge 3$, into a single vertex. Let M be the set of vertices obtained this way. Then, we remove every edge with one end in B and the other in $V_1 \cup B \cup M$. In Fig. 2, we show an example of a graph G^* obtained from a graph G and a pc $\mathcal{P}$ (represented by one trivial path and the paths consisting of solid edges). The following result is the core of our algorithm.

Lemma 1. *Let G be a graph and let $\mathcal{P}$ be a minimum pc of G. Consider the graph G^* as defined above. If there is a path in G^* from V_0 to M, then there is a pc $\mathcal{Q}$ such that $|\mathcal{Q}| = |\mathcal{P}|$, and $\mathcal{Q}$ has fewer trivial paths than $\mathcal{P}$; otherwise, G does not have an ntpc.*

Proof (sketch). First, suppose that there is a path in G^* between V_0 and M. Let P be a shortest path with endvertices $u \in V_0$ and $v \in M$. Consider that

$$P := \langle r_1 = u, r_2, \dots, r_s = v \rangle.$$

It can be shown that P has the following structure

(a) $s = 2k$, for some $k \geq 1$.
(b) $r_{2i} \in B$ and $r_{2i+1} \in (V_2 \setminus B)$, for $i = 1, 2 \ldots, k - 1$.

In what follows, given paths S and T such that $|V(S) \cap V(T)| = 1$ and its common vertex is an endvertex of both S and T, we denote by $S \cdot T$ the path resulting from the concatenation of S and T.

Let $P' \in \mathcal{P}$ be the path represented by the vertex $v \in M$. Let $w \in V(P')$ be a vertex adjacent to r_{s-1} in G. Consider that $P' = P'_1 \cdot P'_2$ where w is an endvertex of both P'_1 and P'_2, and $|P'_1| \leq |P'_2|$. Moreover, let $P_i \in \mathcal{P}_2$ be the path containing r_{2i} and r_{2i+1} such that $P_i := \langle s_i, r_{2i}, r_{2i+1} \rangle$, for $i = 1, 2, \ldots, k - 1$. Now, consider the paths

$$Q_i = \begin{cases} \langle r_1, r_2, s_1 \rangle, & \text{if } i = 1 \text{ and } s > 2, \\ \langle r_{2i-1}, r_{2i}, s_i \rangle, & \text{if } 1 < i < k, \\ \langle r_{s-1}, w \rangle \cdot P'_1, & \text{if } i = k, \\ P'_2, & \text{if } i = k + 1. \end{cases}$$

By replacing the paths $\langle r_1 \rangle, P_1, P_2, \ldots, P_{k-1}$ and P' with $Q_1, Q_2, \ldots, Q_{k+1}$, we obtain the desired path cover $\mathcal{Q}$. In Fig. 3 we show an example considering the path cover $\mathcal{P}$ and G^* shown in Fig. 2.

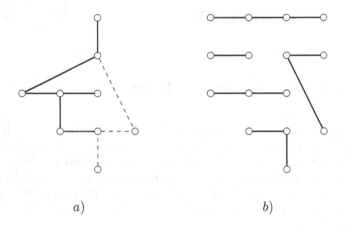

$a)$ $b)$

Fig. 3. (a) a path P and (b) the resulting path cover $\mathcal{Q}$.

Now, suppose there is no path from V_0 to M in G^*. Let u be a vertex in V_0. First, let B_u be the set of vertices in B which are reachable from u in G^*. It can be shown that, by the minimality of $\mathcal{P}$, the set $S = B_u$ violates the condition stated in Theorem 3. $\square$

By the proof of Lemma 1, given a minimum pc $\mathcal{P}$ of G, if we find a path in G^* from V_0 to M, we can obtain a minimum pc of G with one fewer trivial path than $\mathcal{P}$. Observe that, given $\mathcal{P}$, we can obtain G^* in polynomial time. Therefore, by repeating this process $|V_0|$ times, we obtain in polynomial time either an ntpc of G or a set that violates the condition given in Theorem 3.

Let $\mu(G)$ (resp. $\mu_{nt}(G)$) be the cardinality of a minimum pc (resp. ntpc) of G. By Lemma 1, the following result holds.

Corollary 1. *Let G be a graph that admits an ntpc. Then, $\mu(G) = \mu_{nt}(G)$.*

Since every minimum ntpc is also a minimum pc, and we can obtain in polynomial time a minimum ntpc given a minimum pc, we have the following result.

Corollary 2. *Let $\mathcal{G}$ be the class of graphs that admit an ntpc. Then* MINNTPC *and* MINPC *on the class $\mathcal{G}$ have the same computational complexity.*

As we mentioned, MINPC on trees can be solved in linear time. In what follows, we show a linear-time algorithm for MINNTPC, called Algorithm 1. It is for the cardinality version, but once it is understood, one can easily extend it to the edge-weighted version of the problem (in which one considers weights on the edges of T, and seeks to minimize or maximize the sum of the weights of the edges in an ntpc of T).

Let T be a tree and $\mathcal{P}$ be a minimum ntpc of T. Consider that T' is the arborescence obtained when we root T at a vertex r in T. Let T'_u be the subtree of T' rooted at u. Note that, when we consider the subgraph of T' spanned by the edges of $\mathcal{P}$, the vertex u has degree $0, 1$ or 2 in T'_u. Let $f(u, d)$ be the cardinality of a minimum ntpc in T_u where the vertex u has degree d. Since $r \in V(T)$ can be any vertex, we choose r to be a leaf of T and, therefore $f(r, 1)$ would be the cardinality of an ntpc of T.

Algorithm 1 computes the values $f(u, d)$ in post-order using a DFS traversal of T. In order to obtain the cardinality of a minimum ntpc of T, we call DFS(r, nil) where r is a leaf of T. We note that T has an ntpc if $f(r, 1) \neq +\infty$. Our algorithm can be easily modified to obtain the edges in a minimum ntpc.

Theorem 5. *Algorithm 1 correctly computes the values $f(v, d)$ for every vertex $v \in V(T)$ and $d \in \{0, 1, 2\}$.*

Proof. The proof is by induction on $n := |V(T'_v)|$, where v is any vertex in T. If $n = 1$, then v is a leaf of the tree. In this case, the algorithm correctly computes $f(v, 0) = 1$ and $f(v, 1) = f(v, 2) = +\infty$.

Now, suppose that $n \geq 2$. Let $u_1, u_2, \ldots, u_k$ be the neighbors of v in T'_v. First, if $d > k$, then there is no solution to $f(v, d)$. Therefore, we set $f(v, d) = +\infty$ in line 17, and this value is not changed afterwards. So, suppose that $d \leq k$. Since we are restricting the path covers to those in which v has degree d in T'_v, we have to choose d edges incident to v and u_i, $1 \leq i \leq k$, to belong to the pc. Observe that for the vertices u_i such that vu_i belongs to the pc, the degree in its corresponding subtree T'_{u_i} can be zero or one. In case this edge does not

---

Algorithm 1. $DFS(v, parent)$

1: $deg \leftarrow 0$
2: $m_1 \leftarrow +\infty$
3: $m_2 \leftarrow +\infty$
4: $sum \leftarrow 0$
5: **for** $u \in N(v)$:
6:      **if** $u \neq parent$:
7:          $DFS(u, v)$
8:          $Y \leftarrow \min(f(u, 1), f(u, 2))$
9:          $X \leftarrow \min(f(u, 1), f(u, 0))$
10:          $sum \leftarrow sum + Y$
11:          $deg \leftarrow deg + 1$
12:          **if** $m_1 > Y - X$:
13:             $m_1 \leftarrow Y - X$
14:          **else if** $m_2 > Y - X$:
15:             $m_2 \leftarrow Y - X$
16: **for** $d = 0$ **to** 2 :
17:      $f(v, d) \leftarrow +\infty$
18: **if** $deg \geq 0$:
19:      $f(v, 0) \leftarrow sum + 1$
20: **if** $deg \geq 1$:
21:      $f(v, 1) \leftarrow sum + m_1$
22: **if** $deg \geq 2$:
23:      $f(v, 2) \leftarrow sum + m_1 + m_2 - 1$

---

belong to the cover, its degree in T'_{u_i} must be one or two. For $i = 1, 2, \ldots, k$, let $X_{u_i} := \min\{f(u_i, 1), f(u_i, 0)\}$ and let $Y_{u_i} := \min\{f(u_i, 1), f(u_i, 2)\}$. Now, we will show how to express $f(u, d)$ in terms of X_{u_i} and Y_{u_i}. In what follows, we show this for $d = 2$.

Let u_a and u_b be neighbors of v in a minimum ntpc where v has degree two in T'_v. By the previous arguments, we have that

$$f(v, 2) = \sum_{\substack{i=1 \\ i \neq a, b}}^{k} Y_{u_i} + X_{u_a} + X_{u_b} - 1.$$

Adding and subtracting Y_{u_a} and Y_{u_b}, we get the expression

$$f(v, 2) = \sum_{i=1}^{k} Y_{u_i} + (X_{u_a} - Y_{u_a}) + (X_{u_b} - Y_{u_b}) - 1.$$

Since $\sum_{i=1}^{k} Y_{u_i}$ is a constant, to compute $f(u, 2)$ we need to find two neighbors of u that minimize $X_{u_i} - Y_{u_i}$. Observe that, at the end of the loop at line 5, the variable sum is equal to $\sum_{i=1}^{k} Y_{u_i}$, and m_1 and m_2 hold the desired minima.

By the induction hypothesis, Algorithm 1 correctly computes $f(u_i, d)$. Therefore, it correctly computes the value $f(v, 2)$. The cases in which $d = 0$ or $d = 1$ can be shown using analogous arguments.   □

Since we process each vertex v of the tree just once, and we iterate through its neighbors to compute the values of $f(u, d)$, the complexity of Algorithm 1 is $\mathcal{O}(n)$, where n is the number of vertices of the tree. Thus, we obtain the following result.

Corollary 3. MINNTPC *on trees can be solved in linear time.*

3 Concluding Remarks

As far as we know, the problems MINNTPC and MAXNTPC have not been treated in the literature. To deal with these optimization problems, we considered first the corresponding existence problem, which turns out to be a special case of the well-studied (g, f)-factor problem. Our result showing the close relation between the existence problem and the maximum matchings of the graph contributes with a characterization that gives a polynomial-time algorithm that either finds an ntpc in a graph or finds a NO-certificate. The proof of this characterization (Theorem 4) can also be seen as an alternative proof of Theorem 3. Regarding MINNTPC, we have shown that if we consider the class of graphs that admit an ntpc, then MINPC and MINNTPC have the same computational complexity. This result also shows an interesting fact about the cardinalities of these path covers. Finally, we also showed a linear-time algorithm for MINNTPC on trees.

A related optimization problem is the MINIMUM WEIGHT NONTRIVIAL PATH COVER (MINWNTPC) problem: given an edge-weighted graph, find a nontrivial path cover of minimum total weight. It should be noted that, for unit weights this problem is not equivalent to MINNTPC. In fact, we have proved that MIN-WNTPC can be solved in polynomial time. This result, as well as some integer programming formulations we have proposed for MINPC and MINNTPC are part of an ongoing research. The computational results are preliminary, but seem very promising. The design of approximation algorithms for MINPC and MINNTPC is a further interesting topic, for which finding good lower bounds is a challenging problem.

References

1. Anstee, R.: An algorithmic proof of Tutte's f-factor theorem. J. Algorithms **6**(1), 112–131 (1985)
2. Arikati, S.R., Pandu Rangan, C.: Linear algorithm for optimal path cover problem on interval graphs. Inform. Process. Lett. **35**(3), 149–153 (1990)
3. Corneil, D.G., Dalton, B., Habib, M.: LDFS-based certifying algorithm for the minimum path cover problem on cocomparability graphs. SIAM J. Comput. **42**(3), 792–807 (2013)

4. Franzblau, D.S., Raychaudhuri, A.: Optimal Hamiltonian completions and path covers for trees, and a reduction to maximum flow. ANZIAM J. **44**(2), 193–204 (2002)
5. Garey, M., Johnson, D., Tarjan, R.: The planar Hamiltonian circuit problem is NP-complete. SIAM J. Comput. **5**(4), 704–714 (1976)
6. Heinrich, K., Hell, P., Kirkpatrick, D., Liu, G.: A simple existence criterion for $(g < f)$-factors. Discrete Math. **85**(3), 313–317 (1990)
7. Kano, M., Saito, A.: $[a, b]$-factors of graphs. Discrete Math. **47**(1), 113–116 (1983)
8. Las Vergnas, M.: An extension of Tutte's 1-factor theorem. Discrete Math. **23**(3), 241–255 (1978)
9. Lovász, L.: Subgraphs with prescribed valencies. J. Comb. Theory **8**, 391–416 (1970)
10. Lovász, L., Plummer, M.: Matching Theory. North-Holland Mathematics Studies, vol. 121. North-Holland, Amsterdam (1986)
11. Moran, S., Wolfstahl, Y.: Optimal covering of cacti by vertex-disjoint paths. Theoret. Comput. Sci. **84**(2), 179–197 (1991). Algorithms Automat. Complexity Games
12. Müller, H.: Hamiltonian circuits in chordal bipartite graphs. Discrete Math. **156**(1–3), 291–298 (1996)
13. Tutte, W.: The factors of graphs. Can. J. Math. **4**, 314–328 (1952)

On the Relation of Strong Triadic Closure and Cluster Deletion

Niels Grüttemeier$^{(\boxtimes)}$ and Christian Komusiewicz

Fachbereich für Mathematik und Informatik, Philipps-Universität Marburg,
Marburg, Germany
{niegru,komusiewicz}@informatik.uni-marburg.de

Abstract. We study the parameterized and classical complexity of two
related problems on undirected graphs $G = (V, E)$. In STRONG TRIADIC
CLOSURE we aim to label the edges in E as strong and weak such
that at most k edges are weak and G contains no induced P_3 with two
strong edges. In CLUSTER DELETION we aim to destroy all induced P_3s
by a minimum number of edge deletions. We first show that STRONG
TRIADIC CLOSURE admits a $4k$-vertex kernel. Then, we study parameter-
ization by $\ell := |E| - k$ and show that both problems are fixed-parameter
tractable and unlikely to admit a polynomial kernel with respect to ℓ.
Finally, we give a dichotomy of the classical complexity of both problems
on H-free graphs for all H of order four.

1 Introduction

We study two related graph problems arising in social network analysis and
data clustering. Assume we are given a social network where vertices represent
agents and edges represent interactions between these agents, and want to predict
which of the interactions are important. In online social networks for example,
one could aim to distinguish between close friends and spurious relationships.
Sintos and Tsaparas [19] proposed to use the notion of triadic closures for this
problem. Informally, they assume that if one agent has strong relations to two
other agents, then these two should have at least a weak relation. The aim is
then to label a maximum number of edges of the social network as strong while
fulfilling this requirement. This may be defined as follows.

Definition 1. *A labeling* $L = (S_L, W_L)$ *of an undirected graph* $G = (V, E)$ *is
a partition of the edge set* E. *The edges in* S_L *are called* strong *and the edges
in* W_L *are called* weak. *A labeling* $L = (S_L, W_L)$ *is an* STC-labeling *if there
exists no pair of strong edges* $\{u, v\} \in S_L$ *and* $\{v, w\} \in S_L$ *such that* $\{u, w\} \notin E$.

For any weak (strong) edge $\{u, v\}$ we will refer to u as a weak (strong) neighbor
of v. The computational problem described informally above is now the following.

© Springer Nature Switzerland AG 2018
A. Brandstädt et al. (Eds.): WG 2018, LNCS 11159, pp. 239–251, 2018.
https://doi.org/10.1007/978-3-030-00256-5_20

STRONG TRIADIC CLOSURE (STC)
Input: An undirected graph $G = (V, E)$ and an integer $k \in \mathbb{N}$.
Question: Is there an STC-labeling L with $|W_L| \leq k$?

We call an STC-labeling $L = (S_L, W_L)$ *optimal* for a graph G, if the number $|W_L|$ of weak edges is minimal. The STC-labeling property can also be stated in terms of induced subgraphs: for every induced P_3, the path on three vertices, of G at most one edge is labeled strong. Therefore, as observed previously [13], STC is closely related to the problem of destroying induced P_3s by edge deletions. Since the graphs without an induced P_3 are exactly the disjoint union of cliques, this problem is usually formulated as follows.

CLUSTER DELETION (CD)
Input: An undirected graph $G = (V, E)$ and an integer $k \in \mathbb{N}$.
Question: Can we transform G into a *cluster graph*, that is, a disjoint union of cliques by at most k edge deletions?

More precisely, any set D of at most k edge deletions that transform G into a cluster graph, directly implies an STC-labeling $(E \setminus D, D)$ with at most k weak edges. There are, however, graphs G where the minimum number of weak edges in an STC-labeling is strictly smaller than the number of edge deletions that are needed in order to transform G into a cluster graph [13]. Due to the close relation between the two problems there are graph classes where any minimum-cardinality solution for CLUSTER DELETION directly implies an optimal STC-labeling [13].

In this work, we study the parameterized complexity of STC and CLUSTER DELETION and the classical computational complexity of both problems in graph classes that can be described by one forbidden induced subgraph of order four.

Known Results. STC is NP-hard [19], even when restricted to graphs with maximum degree four [13], or to split graphs [14]. In contrast, STC is solvable in polynomial time when the input graph is bipartite [19], subcubic [13], a proper interval graph [14], or a cograph, that is, a graph with no induced P_4 [13]. STC can be solved in $\mathcal{O}(1.28^k + nm)$ time [19] by solving VERTEX COVER in the so-called Gallai graph of the input graph. CLUSTER DELETION is NP-hard [18], even when restricted to graphs with maximum degree four [12], and solvable in polynomial-time on cographs [7] and in time $O(1.42^k + m)$ on general graphs [1].

Our Results. We provide the first polynomial kernel for STC parameterized by k. More precisely, we show that in $\mathcal{O}(n^3)$ time we can reduce an arbitrary instance of STC to an equivalent instance with at most $4k$ vertices.

Second, we initiate the study of the parameterized complexity of STC and CLUSTER DELETION with respect to the parameter $\ell := |E| - k$. Hence, in STC we are searching for an STC-labeling with at least ℓ strong edges and in CLUSTER DELETION we are searching for a cluster graph that is a subgraph of G and that has at least ℓ edges; we call these edges the *cluster edges* of this subgraph. While we present fixed-parameter algorithms for both problems and the parameter ℓ, we

Table 1. The parameterized complexity of STC and Cluster Deletion for parameters k and $\ell := |E| - k$.

Parameter	STC	Cluster Deletion
k	$\mathcal{O}(1.28^k + nm)$-time algo [19]	$\mathcal{O}(1.42^k + m)$-time algo [1]
	$4k$-vertex kernel (Theorem 1)	$4k$-vertex kernel [9]
ℓ	$\ell^{\mathcal{O}(\ell)} \cdot n$-time algo (Theorem 5)	$\mathcal{O}(9^\ell \cdot \ell n)$-time algo (Theorem 4)
	No poly kernel (Theorem 3)	No poly kernel (Corollary 1)

also show that, somewhat surprisingly, both problems do not admit a polynomial kernel with respect to ℓ, unless NP $\subseteq$ coNP/poly. Our result is obtained by polynomial parameter transformations from CLIQUE parameterized by the size of a vertex cover of the input graph to MULTICOLORED CLIQUE parameterized by the sum of the sizes of all except one color class to STC and CLUSTER DELETION parameterized by ℓ. The MULTICOLORED CLIQUE variant may be of independent interest as a suitable base problem for polynomial parameter transformations. Table 1 gives an overview of the parameterized complexity.

Independent from our work, Heggernes et al. [8] showed that STC parameterized by ℓ has no polynomial kernel unless NP $\subseteq$ coNP/poly, even when the input graph is a split graph. Moreover, they discuss the STRONG F-CLOSURE problem, which is a generalization of STC. For an arbitrary graph F, the STRONG F-CLOSURE problem asks for a labeling $L = (S_L, W_L)$ of a graph G such that there is no induced subgraph F in G that consists only of strong edges under L and where the number of strong edges is maximum under this property.

Finally, we extend the line of research studying the complexity of CLUSTER DELETION [7] and STC [13] on H-free graphs where H is a graph of order four. We present a complexity dichotomy between polynomial-time solvable and NP-hard cases for all possibilities for H. Moreover, we show for all such graphs H whether STC and CLUSTER DELETION *correspond* on H-free graphs, that is, whether every STC-labeling with at most k edges implies a CLUSTER DELETION solution with at most k edge deletions. These results are shown in Table 2.

Table 2. Complexity Dichotomy and correspondence of STC and Cluster Deletion on H-free graphs.

	STC	CD	Correspondent
$H \in \{K_4, 4K_1, C_4, 2K_2, \text{claw}, \text{co-claw}, \text{co-diamond}, \text{co-paw}\}$	NP-h	NP-h	NO
$H = \text{diamond}$	NP-h	NP-h	YES
$H \in \{\text{paw}, P_4\}$	P	P	YES

Preliminaries. We consider undirected simple graphs $G = (V, E)$ where $n := |V|$ and $m := |E|$. For any vertex $v \in V$ the open neighborhood of v is denoted by $N_G(v)$, the closed neighborhood is denoted by $N_G[v]$. The set of vertices in G

which have a distance of exactly 2 to v is denoted by $N_G^2(v)$. For any two vertex sets $V_1, V_2 \subseteq V$, we let $E_G(V_1, V_2) := \{\{v_1, v_2\} \in E \mid v_1 \in V_1, v_2 \in V_2\}$ denote the set of edges between V_1 and V_2. For any set $V' \subseteq V$ of vertices, we let $E_G(V') := E_G(V', V')$ be the set of edges between the vertices of V'. We may omit the subscript G if the graph is clear from the context.

For any $V' \subseteq V$, $G[V'] := (V', E(V'))$ denotes the *subgraph induced by* V'. A *clique* in a graph G is a set $K \subseteq V$ of vertices, such that $G[K]$ is complete. A *cut* $C = (V_1, V_2)$ is a partition of the vertex set into two parts. The *cut-set* $E_C := E_G(V_1, V_2)$ is the set of edges between V_1 and V_2. A *matching* $M \subseteq E$ is a set of pairwise disjoint edges. A matching is *maximal* if adding any edge to M does not give a matching. A matching in a graph G is *maximum* if G has no larger matching. A graph G is *H-free* if it does not contain an induced subgraph that is isomorphic to the graph H. For the definitions of single small graphs such as the $2K_2$, refer to [4] or http://graphclasses.org. For the relevant notions of parameterized complexity refer to the standard monographs [5,6]. A reduction rule is called *safe* if it produces an equivalent instance.

Due to lack of space, several proofs are deferred to a full version of this article.

2   On Problem Kernelizations

We now discuss problem kernelizations for STC parameterized by k and ℓ. First, we give a $4k$-vertex kernel and an $\mathcal{O}(\ell \cdot 2^\ell)$-size kernel. Then, we show that there is no polynomial problem kernel for ℓ unless $NP \subseteq coNP/poly$. An important concept for our kernelizations are *weak cuts* which are defined as follows.

Definition 2. *Let $G = (V, E)$ be a graph and $L = (S_L, W_L)$ an STC-Labeling for G. A weak cut for G under L is a cut C such that E_C is contained in W_L.*

Proposition 1. *Let $G = (V, E)$ be a graph and $L = (S_L, W_L)$ an STC-Labeling for G. If there is a weak cut C with cut-set E_C, then there is an STC-Labeling $L' = (S_{L'}, W_{L'})$ for $G' = (V, E \setminus E_C)$ such that $|S_{L'}| = |S_L|$.*

A $4k-vertex kernel for STC$. We now show that STC parameterized by k admits a kernel with at most $4k$ vertices. In the kernelization, we will make use of the concepts of *critical cliques* and *critical clique graphs* as introduced in [16]. These concepts were also used for a kernelization for CLUSTER EDITING [9].

Definition 3. *A critical clique of a graph G is a clique K where the vertices of K all have the same neighbors in $V \setminus K$, and K is maximal under this property.*

Definition 4. *Given a graph $G = (V, E)$, let $\mathcal{K}$ be the collection of its critical cliques. The critical clique graph $\mathcal{C}$ of G is the graph $(\mathcal{K}, E_{\mathcal{C}})$ with $\{K_i, K_j\} \in E_{\mathcal{C}} \Leftrightarrow \forall u \in K_i, v \in K_j : \{u, v\} \in E$.*

For a critical clique K we let $\mathcal{N}(K) := \bigcup_{K' \in N_C(K)} K'$ denote the union of its neighbor cliques in the critical clique graph and $\mathcal{N}^2(K) := \bigcup_{K' \in N_C^2(K)} K'$ denote the union of the critical cliques at distance exactly two from K. The critical clique graph can be constructed in $\mathcal{O}(n + m)$ time [11]. Note that the edges within a critical clique K are not part of any P_3. It is known that these kind of edges are labeled as *strong* in every optimal solution for STC [19].

In the following, we will distinguish between two types of critical cliques. We say that K has *type 1*, if $\mathcal{N}(K)$ does not form a clique in G, and that K has *type 2*, if $\mathcal{N}(K)$ forms a clique in G. We will see that the number of critical cliques of type 1 is bounded for every yes-instance of STC. The main step of the kernelization is to delete large critical cliques of type 2. Before we give the concrete rule we provide two useful properties of critical cliques of type 2.

Proposition 2. *If K_1 and K_2 are critical cliques of type 2, then $\{K_1, K_2\} \notin E_C$.*

Proposition 3. *Let K be a critical clique of type 2, $v \in \mathcal{N}(K)$ and $L = (S_L, W_L)$ an STC-labeling for G with a minimal number of weak edges. Then $E(\{v\}, K) \subseteq S_L$ or $E(\{v\}, K) \subseteq W_L$.*

Now, we may formulate the reduction rule.

Rule 1. *If G has a critical clique K of type 2 with $|K| > |E_G(\mathcal{N}(K), \mathcal{N}^2(K))|$, then remove K and $\mathcal{N}(K)$ from G and decrease k by $|E_G(\mathcal{N}(K), \mathcal{N}^2(K))|$.*

Proposition 4. *Rule 1 is safe and can be carried out in $\mathcal{O}(n^3)$ time.*

Proof. Let K be a critical clique in G with $|K| > |E_G(\mathcal{N}(K), \mathcal{N}^2(K))|$ and let G' be the reduced graph after deleting K and $\mathcal{N}(K)$ from G. We show that there is an STC-labeling $L = (S_L, W_L)$ for G with $|W_L| \leq k$ if and only if there is an STC-labeling $L' = (S_{L'}, W_{L'})$ for G' with $|W_{L'}| \leq k - |E_G(\mathcal{N}(K), \mathcal{N}^2(K))|$.

First, let $L' = (S_{L'}, W_{L'})$ be an STC-labeling, such that $|W_{L'}| \leq k - |E_G(\mathcal{N}(K), \mathcal{N}^2(K))|$ for G'. We define a labeling $L = (S_L, W_L)$ with $|W_L| \leq k$ for G by setting

$$S_L := S_{L'} \cup E_G(K) \cup E_G(K, \mathcal{N}(K)) \text{ and } W_L := W_{L'} \cup E_G(\mathcal{N}(K), \mathcal{N}^2(K)).$$

It remains to show that L is an STC-labeling. Since the edges in $S_{L'}$ do not have a common endpoint with the edges in $E_G(K) \cup E_G(K, \mathcal{N}(K))$ it suffices to show that there is no induced P_3 containing two edges $e_1, e_2 \in S_{L'}$ or $e_1, e_2 \in E_G(K) \cup E_G(K, \mathcal{N}(K))$. If $e_1, e_2 \in S_{L'}$, the edges do not form a strong P_3 since L' is an STC-labeling. Let $e_1, e_2 \in E_G(K) \cup E_G(K, \mathcal{N}(K))$. Since K is a critical clique of type 2, $K \cup \mathcal{N}(K)$ is a clique. It follows, that e_1 and e_2 are edges between vertices of a clique, so they do not form a P_3. Since there is no strong P_3 under L, it follows that L is an STC-labeling with $|W_L| \leq k$.

Conversely, let $L = (S_L, W_L)$ be an STC-labeling for G with a minimal number of weak edges. We prove the safeness by using Proposition 1. To this end, we show that $C = (K \cup \mathcal{N}(K), V \setminus (K \cup \mathcal{N}(K)))$ is a weak cut under L.

Assume there is a vertex $v \in \mathcal{N}(K)$ that has a strong neighbor $w \in \mathcal{N}^2(K)$. Then, for each $u \in K$, the edge $\{u, v\}$ is weak under L. Otherwise $\{u, v\}$ and $\{v, w\}$ would form a strong P_3, which contradicts the fact that L is an STC-labeling. Then, we have exactly $|K|$ weak edges in $E_G(\{v\}, K)$ and at most $|E_G(\mathcal{N}(K), \mathcal{N}^2(K))|$ strong edges in $E_G(\{v\}, \mathcal{N}^2(K))$. We define a new labeling $L^+ = (S_{L^+}, W_{L^+})$ by

$$S_{L^+} := S_L \cup E_G(\{v\}, K) \setminus E_G(\{v\}, \mathcal{N}^2(K)),$$
$$W_{L^+} := W_L \cup E_G(\{v\}, \mathcal{N}^2(K)) \setminus E_G(\{v\}, K).$$

From $|V(K)| > |E_G(\mathcal{N}(K), \mathcal{N}^2(K))|$ we get that $|W_{L^+}| < |W_L|$. It remains to show that L^+ is an STC-labeling, which contradicts the fact that L is an STC-labeling with a minimal number of weak edges.

Since we add edges $\{u, v\}$ with $u \in K$ to S_{L^+} we need to show that no such edge is part of a strong P_3 under L^+. Let $(\{u, v\}, e)$ with $u \in K$ be a tuple of edges that share exactly one endpoint. Consider the case $e = \{v, w\}$ with $w \in \mathcal{N}^2(K)$. It follows that $e \in W_L^+$ by the construction of L^+. So $\{u, v\}$ and e do not form a strong P_3 under L^+. Otherwise, $w \in K \cup \mathcal{N}(K)$. Then $\{u, v\}$ and e do not form an induced P_3 since $K \cup \mathcal{N}(K)$ is a clique by the definition of type-2 critical cliques.

Since there is no strong P_3 under L^+ it follows, that L^+ is an STC-labeling. In combination with the fact that $|W_{L^+}| < |W_L|$, we conclude that L^+ is an STC-labeling for G with fewer weak edges than L which contradicts the fact that L is an STC-labeling with a minimal number of weak edges. This proves the claim that $C = (K \cup \mathcal{N}(K), V \setminus (K \cup \mathcal{N}(K)))$ is a weak cut under L. By using Proposition 1, we conclude that there exists an STC-labeling $L' = (S_{L'}, W_{L'})$ with $|W_{L'}| \leq k - |E_G(\mathcal{N}(K), \mathcal{N}^2(K))|$ in G', proving the safeness of Rule 1.

The running time of Rule 1 is $\mathcal{O}(n^3)$. The critical clique graph C can be computed in $\mathcal{O}(n + m)$ time [11]. For each critical clique K the sizes of $\mathcal{N}(K)$ and $\mathcal{N}^2(K)$ can be computed in $\mathcal{O}(n^2)$ time [9]. Since every application of Rule 1 removes some vertices from G, it can be applied at most n times.   □

Theorem 1. *Rule 1 can be applied exhaustively in $\mathcal{O}(n^3)$ time and yields a problem kernel for STC with at most $4k$ vertices.*

An $\mathcal{O}(\ell \cdot 2^\ell)$ kernel for STC. We now show that STC parameterized by $\ell := |E| - k$ admits a kernel of size $\mathcal{O}(\ell \cdot 2^\ell)$. Let $G = (V, E)$ be a graph and let $M \subseteq E$ be a maximum matching in G. We partition the vertices of G into

- $M_V := \{v \in V \mid v$ is an endpoint of some $e \in M\}$,
- $I_2 := \{v \in V \mid \exists\{u, w\} \in M : u$ and w are both neighbors of $v\}$,
- $I_1 := V \setminus (I_2 \cup M_V)$.

Note that since M is maximal, $I_1 \cup I_2$ is an independent set. We will see that the number of vertices in I_2 is upper-bounded by ℓ in every STC instance. The main step of the kernelization is to delete superfluous vertices form I_1.

We will say that two vertices $v_1, v_2 \in I_1$ are *members of the same family* F, if $N(v_1) = N(v_2)$. Given a family F, we will refer to the neighborhood of the vertices in F as $N(F) := N(v)$ for some $v \in F$.

Rule 2. For every family F of vertices in I_1: If $|F| > |N(F)|$ then delete $|F| - |N(F)|$ of the vertices in F and decrease k by $(|F| - |N(F)|) \cdot |N(F)|$.

Note that Rule 2 decreases the value of k by the exact amount of edges deleted. Hence, the value of the parameter ℓ does not change.

Proposition 5. *Rule 2 is safe.*

Theorem 2. STC *admits a problem kernel of size* $\mathcal{O}(\ell \cdot 2^\ell)$ *that can be computed in time* $\mathcal{O}(\sqrt{n}m)$.

A kernel lower bound for the parameter ℓ. Above, we gave an exponential-size problem kernelization for STC parameterized by the number of strong edges ℓ. Now we will prove that STC does not admit a polynomial kernel for the parameter ℓ unless $\text{NP} \subseteq \text{coNP/poly}$ by reducing from CLIQUE.

CLIQUE
Input: $G = (V, E)$, $t \in \mathbb{N}$
Question: Is there a clique on t vertices in G?

CLIQUE parameterized by the size s of a vertex cover does not admit a polynomial kernel unless $\text{NP} \subseteq \text{coNP/poly}$ [2]. Our proof gives a polynomial parameter transformation [3] from CLIQUE parameterized by s to STC parameterized by ℓ in two steps. The first step is a reduction to the following problem.

RESTRICTED-MULTICOLORED-CLIQUE
Input: A properly t-colored graph $G = (V, E)$ with color classes $C_1, \ldots, C_t \subseteq V$ such that $|C_1| = |C_2| = \ldots = |C_{t-1}|$.
Question: Is there a clique containing one vertex from each color in G?

Proposition 6. RESTRICTED-MULTICOLORED-CLIQUE *parameterized by* $|C_1 \cup \ldots \cup C_{t-1}|$ *does not admit a polynomial kernel unless* $\text{NP} \subseteq \text{coNP/poly}$.

The next step to prove the kernel lower bound is to give a polynomial parameter transformation from RESTRICTED-MULTICOLORED-CLIQUE to STC.

Theorem 3. STC *parameterized by the number of strong edges* ℓ *does not admit a polynomial kernel unless* $\text{NP} \subseteq \text{coNP/poly}$.

Proof. We give a polynomial parameter transformation from RESTRICTED-MULTICOLORED-CLIQUE to STC. Let $G = (V, E)$ be a properly t-colored graph with color classes $C_1 = \{v_{1,1}, v_{2,1}, \ldots, v_{z,1}\}$, $C_2 = \{v_{1,2}, v_{2,2}, \ldots, v_{z,2}\}, \ldots, C_{t-1} = \{v_{1,t-1}, v_{2,t-1}, \ldots, v_{z,t-1}\}$, each of size z, and C_t. We now describe how to construct an STC-instance $(G' = (V', E'), k)$ from G such that there is an STC-labeling $L = (S_L, W_L)$ with $|W_L| \leq k$ for G' if and only if G has a multicolored clique.

For each of the first $(t-1)$ classes C_r, $r = 1, \ldots, t-1$, we define a family $\mathcal{K}_r$ of $z-1$ vertex sets $K_{1,r}, K_{2,r}, \ldots, K_{z-1,r}$, each of size t, and we add edges such that each $K \in \mathcal{K}_r$ becomes a clique. For every fixed $i = 1, \ldots, z-1$ we also add edges $\{u, v\}$ from all $u \in K_{i,r}$ to all $v \in C_r$.

Setting $k := |E| - \left(\binom{t}{2} + (t-1)(z-1)\binom{t+1}{2}\right)$ gives us $\ell = \binom{t}{2} + (t-1) \cdot (z-1)\binom{t+1}{2}$. Obviously, ℓ is polynomially bounded in $|C_1 \cup \ldots \cup C_{t-1}| = (t-1) \cdot z$. This completes the construction. The correctness proof is deferred. $\square$

The proof of Theorem 3 also implies that CLUSTER DELETION has no kernel with respect to the parameter $\ell := |E| - k$: The strong edges in the STC-labeling obtained in the forward direction of the proof form a disjoint union of cliques and the converse direction follows from the fact that a cluster subgraph with at least ℓ cluster edges implies an STC-labeling with at least ℓ strong edges which then implies that the MULTICOLORED CLIQUE instance is a yes-instance.

Corollary 1. CLUSTER DELETION *parameterized by the number of cluster edges* $\ell := |E| - k$ *does not admit a polynomial kernel unless* $NP \subseteq coNP/poly$.

3   Fixed-Parameter Algorithms for the Parameterization by the Number of Strong or Cluster Edges

For CLUSTER DELETION, we obtain a fixed-parameter algorithm by a simple dynamic programming algorithm.

Theorem 4. CLUSTER DELETION *can be solved in* $\mathcal{O}(9^\ell \cdot \ell n)$ *time.*

For STC, we combine a branching on the graph that is induced by a maximal matching with a dynamic programming over the vertex sets of this graph.

Theorem 5. STC *can be solved in* $\ell^{\mathcal{O}(\ell)} \cdot n$ *time.*

Proof. The initial step of the algorithm is to compute a maximal matching M in G. If $|M| \geq \ell$, then answer yes. Otherwise, the endpoints of M are a vertex cover of size less than 2ℓ since M is maximal. Let C denote this vertex cover and let $I := V \setminus C$ denote the independent set consisting of the vertices that are not an endpoint of M. The algorithm now has two further main steps. First, try all STC-labelings of $G[C]$ with at most ℓ strong edges. If there is one STC-labeling with ℓ strong edges, then answer yes. Otherwise, compute for each STC-labeling of $G[C]$ with fewer than ℓ edges, whether it can be extended to an STC-labeling of G with ℓ strong edges by labeling sufficiently many edges of $E(C, I)$ as strong.

Observe that $G[C]$ has $\ell^{\mathcal{O}(\ell)}$ STC-labelings with at most ℓ strong edges and that they can be enumerated in $\ell^{\mathcal{O}(\ell)}$ time: The set of edges in $G[C]$ has size less than $\binom{2\ell}{2} = \mathcal{O}(\ell^2)$ and we enumerate all subsets of size at most ℓ of this set. Now consider one such set E_C. In $\mathcal{O}(\ell^2)$ time, we can check whether $(E_C, E(C) \setminus E_C)$ is a valid STC-labeling. If this is not the case, then discard the current set. Otherwise, compute whether this labeling can be extended into a labeling of G with at least ℓ strong edges by using dynamic programming over subsets of C.

Assume in the following that $I := \{1, \ldots, n - |C|\}$. The dynamic programming table T has entries of the type $T[i, C']$ for all $i \in \{1, \ldots, n - |C|\}$ and all $C' \subseteq C$. Each entry stores the maximum number of strong edges in an STC-labeling of $G[C \cup \{1, \ldots, i\}]$ in which the strong edges of $E(C)$ are exactly those of E_C and in which the strong neighbors of the vertices in $\{1, \ldots, i\}$ are exactly from C'. Observe that the set of strong neighbors $N_S(i)$ of each vertex i has to fulfill three properties:

- $N_S(i)$ is a clique.
- No vertex of $N_S(i)$ has a strong neighbor in $C \setminus N(i)$.
- No vertex of $N_S(i)$ has a strong neighbor in $I \setminus \{i\}$.

We call a set that fulfills the first two properties *valid* for i. We ensure the third property by the recurrence in the dynamic programming.

After filling this table completely, we have a yes-instance if $T[n - |C|, C] \geq \ell$ and a no-instance otherwise. The entries are computed for increasing values of i and subsets C' of increasing size. The basic entry is $T[0, \emptyset]$ which is set to $|E_C|$. The recurrence to compute an entry for $i \geq 1$ is

$$T[i, C'] = \max_{C'' \subseteq C' : C'' \text{is valid for} \{i\}} T[i - 1, C' \setminus C''] + |C''|.$$

The correctness follows from the observation that we consider all valid sets for strong neighbors and that in the optimal solution for $G[i - 1, C' \setminus C'']$ no vertex from $\{1, \ldots, i - 1\}$ has strong neighbors in C''.

The running time of the algorithm can be seen as follows. A maximal matching can be computed greedily in linear time. If the matching has size less than ℓ, we fill the dynamic programming table as defined above. The number of partial labelings E_C is $\ell^{\mathcal{O}(\ell)}$. For each of them, in $\mathcal{O}(2^{2\ell} \cdot \ell n)$ time, we can compute for each i the subsets of C which are valid for i. The number of terms that are subsequently evaluated in the recurrences is $3^{|C|}$ as each term corresponds to one partition of C into $C \setminus C'$, $C' \setminus C''$, and C''. For each term, one needs to evaluate the equation in $\mathcal{O}(1)$ time. Hence, the overall time needed to fill T for one partial labeling E_C is $\mathcal{O}(3^{2\ell} \cdot n) = \mathcal{O}(9^\ell \cdot n)$; the overall running time follows. $\square$

4   STC and Cluster Deletion on H-free Graphs

Recall that every solution for CLUSTER DELETION provides an STC-labeling $L = (S_L, W_L)$ by defining S_L as the set of edges inside the cliques in the resulting graph. We call such L a *cluster labeling*. However, this solution is not necessarily an optimal one [14].

In this section we discuss the complexity and the solution structure if the input for STC and CLUSTER DELETION is limited to H-free graphs, that is, graphs that do not have an induced subgraph H. We give a dichotomy for all classes of H-free graphs, where H is a graph on four vertices, that is $H \in \{K_4, 4K_1, C_4, 2K_2, \text{diamond}, \text{co-diamond}, \text{claw}, \text{co-claw}, \text{paw}, \text{co-paw}, P_4\}$.

a)   b)   c)

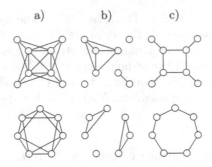

Fig. 1. Two graphs where no cluster labeling is an optimal STC-labeling. Column (a) shows the input graph, column (b) shows an optimal cluster labeling, and column (c) shows the strong edges in an optimal STC-labeling.

The correspondence between STC and Cluster Deletion on H-free graphs. We say that the two problems *correspond* on a graph class Π if for every graph in Π we can find a cluster labeling that is an optimal STC-labeling.

Figure 1 shows two examples, where a cluster labeling is not an optimal solution for STC. The upper example, taken from [14], is C_4-, $2K_2$-, co-paw-, and co-diamond-free; an optimal STC-labeling has eight strong edges, while the best cluster labeling has only seven cluster edges. The second example is the complement of a C_7. It is K_4-, $4K_1$-, claw- and co-claw free; the optimal STC-labeling has seven strong edges, while the best cluster labeling has six cluster edges. The examples give the cases where STC and CLUSTER DELETION do not correspond.

Theorem 6. *The problems* CLUSTER DELETION *and* STC

- *do not correspond on the class of H-free graphs, for $H \in \{C_4, 2K_2,$ co-paw, co-diamond, $K_4, 4K_1,$ claw, co-claw$\}$, and*
- *correspond on the class of H-free graphs, for $H \in \{P_4,$ diamond, paw$\}$.*

The complexity of STC and Cluster Deletion on H-free graphs. We first identify the cases where both problems are solvable in polynomial time.

Proposition 7. *If $H \in \{P_4,$ paw$\}$, STC and* CLUSTER DELETION *are solvable in polynomial time on H-free graphs.*

In all other possible cases for H, both problems remain NP-hard on H-free graphs. To show this, we first use the following construction:

Definition 5. *Let $G = (V, E)$ be a graph. The* expanded graph $\tilde{G}$ *of G is the graph obtained by adding a clique $\tilde{K} = \{v_1, \ldots, v_{|V|^3}\}$ and all edges such that every $v \in V$ is adjacent to all vertices in $\tilde{K}$.*

Obviously, we can construct $\tilde{G}$ from G in polynomial time. We use this construction to give a reduction from CLIQUE to STC and CLUSTER DELETION. The construction also transfers certain H-freeness properties from G to $\tilde{G}$.

Proposition 8. *Let* $(G = (V, E), t)$ *be a* CLIQUE *instance.*

(a) There is a clique of size at least t in G if and only if there is an STC-labeling $L = (S_L, W_L)$ for $\tilde{G}$ such that $|S_L| \geq \binom{n^3}{2} + t \cdot n^3$.

(b) There is a clique of size at least t in G if and only if $\tilde{G}$ has a solution for CLUSTER DELETION *with at least $\binom{n^3}{2} + t \cdot n^3$ cluster edges.*

Proposition 9. *Let* $H \in \{2K_2, \text{co-diamond}, \text{co-paw}, 4K_1\}$. *If a graph G is H-free, then the expanded graph $\tilde{G}$ is H-free as well.*

As mentioned above, the Propositions 8 and 9 deliver the following.

Proposition 10. STC *and* CLUSTER DELETION *remain NP-hard on H-free graphs if $H \in \{2K_2, \text{co-diamond}, \text{co-paw}, 4K_1, \text{claw}, C_4, \text{diamond}, K_4\}$.*

Proof. **Case 1:** $H \in \{2K_2, \text{co-diamond}, \text{co-paw}, 4K_1\}$. CLIQUE remains NP-hard on $2K_2$-, co-diamond-, co-paw- and $4K_1$-free graphs since INDEPENDENT SET is NP-hard on the complement graphs: C_4-, diamond-, paw- and K_4-free graphs [15]. By Proposition 8, $(G, k) \mapsto (\tilde{G}, m - (\binom{n^3}{2} + k \cdot n^3))$ is a polynomial time reduction from CLIQUE to STC and CLUSTER DELETION. From Proposition 9 we know, that if G is $2K_2$-, co-diamond-, co-paw- or $4K_1$-free, so is $\tilde{G}$. Thus, STC and CLUSTER DELETION remain NP-hard on H-free graphs.

Case 2: $H = \text{claw}$. Analogously to Case 1, we use the reduction $(G, k) \mapsto (\tilde{G}, m - (\binom{n^3}{2} + k \cdot n^3))$ to show that STC and CLUSTER DELETION remain NP-hard on claw-free graphs. It is known that CLIQUE remains NP-hard on $3K_1$-free graphs, since INDEPENDENT SET is NP-hard on triangle-free graphs [15]. If G is $3K_1$-free, then $\tilde{G}$ is claw-free: Assuming $\tilde{G}$ has a claw as induced subgraph, there are three vertices in $\tilde{G}$, which are pairwise non-adjacent. Since G is $3K_1$-free, one of those vertices must lie in $\tilde{K}$. This contradicts the fact, that every vertex in $\tilde{K}$ is adjacent to every other vertex in $\tilde{G}$ by construction.

Case 3: $H \in \{C_4, \text{diamond}, K_4\}$. There is a reduction from 3SAT to CLUSTER DELETION producing a C_4-, K_4-, and diamond-free CLUSTER DELETION instance [12]. By Theorem 6, there is an optimal cluster labeling for STC on diamond-free graphs, so the reduction works also for STC. Thus, STC and CLUSTER DELETION remain NP-hard on C_4-, K_4-, and diamond-free graphs. □

It remains to show NP-hardness on co-claw-free graphs. Since CLIQUE can be solved in polynomial time on co-claw-free graphs [17] we cannot reduce from CLIQUE in this case. Instead, we reduce from the following problem.

3-CLIQUE COVER
Input: A graph $G = (V, E)$
Question: Can V be partitioned into three cliques K_1, K_2, and K_3?

3-CLIQUE COVER is NP-hard on co-claw-free graphs [10].

Proposition 11. STC *and* CLUSTER DELETION *remain NP-hard on co-claw-free graphs.*

Proof. We give a reduction from 3-CLIQUE COVER on co-claw-free graphs to STC and CLUSTER EDITING on co-claw free graphs. Let $G = (V, E)$ be a co-claw-free instance for 3-CLIQUE COVER. We describe how to construct a co-claw-free STC-/CLUSTER DELETION- instance $(G' = (V', E'), k)$. We define three vertex sets K_1, K_2, and K_3. Every K_i consists of exactly n^3 vertices $v_{1,i}, \ldots, v_{n^3,i}$. We set $V' := V \cup K_1 \cup K_2 \cup K_3$. Moreover, we define edges from every vertex in $K_1 \cup K_2 \cup K_3$ to every vertex in V and edges of the form $\{v_{c,i}, v_{d,j}\}$, where $c \neq d$. The set E' is the union of those edges and E. Note that this makes each K_i a clique of size n^3. We set $k := |E'| - (3 \cdot \binom{n^3}{2} + n^4)$. This completes the construction; the correctness proof is deferred. □

Theorem 7. *The problems* CLUSTER DELETION *and* STC *are*

- *solvable in polynomial time on H-free graphs, if $H \in \{P_4, paw\}$, and*
- *NP-hard on H-free graphs, if $H \in \{K_4, 4K_1, C_4, 2K_2, diamond, co\text{-}diamond, claw, co\text{-}claw, co\text{-}paw\}$.*

References

1. Böcker, S., Damaschke, P.: Even faster parameterized cluster deletion and cluster editing. Inf. Process. Lett. **111**(14), 717–721 (2011)
2. Bodlaender, H.L., Jansen, B.M.P., Kratsch, S.: Kernelization lower bounds by cross-composition. SIAM J. Discret. Math. **28**(1), 277–305 (2014)
3. Bodlaender, H.L., Thomassé, S., Yeo, A.: Kernel bounds for disjoint cycles and disjoint paths. Theor. Comput. Sci. **412**(35), 4570–4578 (2011)
4. Brandstädt, A., Le, V.B., Spinrad, J.P.: Graph Classes: A Survey, vol. 3. SIAM, Philadelphia (1999)
5. Cygan, M., et al.: Parameterized Algorithms. Springer, Cham (2015). https://doi.org/10.1007/978-3-319-21275-3
6. Downey, R.G., Fellows, M.R.: Fundamentals of Parameterized Complexity. Texts in Computer Science. Springer, London (2013). https://doi.org/10.1007/978-1-4471-5559-1
7. Gao, Y., Hare, D.R., Nastos, J.: The cluster deletion problem for cographs. Discret. Math. **313**(23), 2763–2771 (2013)
8. Golovach, P.A., Heggernes, P., Konstantinidis, A.L., Lima, P.T., Papadopoulos, C.: Parameterized Aspects of Strong Subgraph Closure. ArXiv e-prints, abs/1802.10386, February 2018
9. Guo, J.: A more effective linear kernelization for cluster editing. Theor. Comput. Sci. **410**(8–10), 718–726 (2009)
10. Holyer, I.: The NP-completeness of edge-coloring. SIAM J. Comput. **10**(4), 718–720 (1981)
11. Hsu, W.-L., Ma, T.-H.: Substitution decomposition on chordal graphs and applications. In: Hsu, W.-L., Lee, R.C.T. (eds.) ISA 1991. LNCS, vol. 557, pp. 52–60. Springer, Heidelberg (1991). https://doi.org/10.1007/3-540-54945-5_49
12. Komusiewicz, C., Uhlmann, J.: Cluster editing with locally bounded modifications. Discret. Appl. Math. **160**(15), 2259–2270 (2012)
13. Konstantinidis, A.L., Nikolopoulos, S.D., Papadopoulos, C.: Strong triadic closure in cographs and graphs of low maximum degree. In: Cao, Y., Chen, J. (eds.) COCOON 2017. LNCS, vol. 10392, pp. 346–358. Springer, Cham (2017). https://doi.org/10.1007/978-3-319-62389-4_29

14. Konstantinidis, A.L., Papadopoulos, C.: Maximizing the strong triadic closure in split graphs and proper interval graphs. In: Proceedings of the 28th ISAAC. LIPIcs, Dagstuhl, Germany, vol. 92, pp. 53:1–53:12. Schloss Dagstuhl-Leibniz-Zentrum fuer Informatik (2017)

15. Poljak, S.: A note on stable sets and colorings of graphs. Commentationes Mathematicae Universitatis Carolinae **15**(2), 307–309 (1974)

16. Protti, F., da Silva, M.D., Szwarcfiter, J.L.: Applying modular decomposition to parameterized cluster editing problems. Theory Comput. Syst. **44**(1), 91–104 (2009)

17. Sbihi, N.: Algorithme de recherche d'un stable de cardinalite maximum dans un graphe sans etoile. Discret. Math. **29**(1), 53–76 (1980)

18. Shamir, R., Sharan, R., Tsur, D.: Cluster graph modification problems. Discret. Appl. Math. **144**(1–2), 173–182 (2004)

19. Sintos, S., Tsaparas, P.: Using strong triadic closure to characterize ties in social networks. In: Proceedigns of the 20th KDD, pp. 1466–1475. ACM, New York (2014)

On Perfect Linegraph Squares

Meike Hatzel[(✉)] and Sebastian Wiederrecht

Institut für Softwaretechnik und Theoretische Informatik, TU Berlin, Berlin,
Germany
{meike.hatzel,sebastian.wiederrecht}@tu-berlin.de

Abstract. A strong edge colouring is a proper colouring of the edges of
a graph such that no two edges that are incident with a common edge
receive the same colour. The square of a graph G is obtained from G by
adding edges between vertices of distance exactly 2. Therefore the strong
edge colouring problem can be transformed to the problem of finding
a proper vertex colouring of the squared linegraph. In this paper we
characterise families of graphs whose squared linegraphs exclude induced
paths of a fixed length. As an example, we give a characterisation of
graphs with P_4-free linegraph squares by a finite family of forbidden
induced subgraphs. Our main result is a characterisation of graphs with
perfect linegraph squares by providing forbidden induced subgraphs. In
addition we are able to observe that all of these classes are χ-bounded.

Keywords: Perfect graphs · Graph powers · χ-bounded

1 Introduction

The colouring of graphs is a well known and highly active area of research that
has many applications. A *proper colouring* (VCol) of a graph is an assignment of
colours to its vertices such that no two adjacent vertices receive the same colour.
Similarly a *proper edge colouring* (ECol) is an assignment of colours to the edges
of a graph, such that two edges with a common vertex receive different colours.

The concept of colourings for both, vertices and edges, can be generalised
by adding a distance constraint. A *strong vertex colouring* (SVCol) is a proper
colouring such that vertices within distance 2 of each other also receive different
colours and a *strong edge colouring* (SECol) is a proper edge colouring where
every two edges having an adjacent edge in common are also coloured differently.
The colour classes of SECol form an *induced matching*.

M. Hatzel and S. Wiederrecht—Both authors are supported by the European
Research Council (ERC) under the European Union's Horizon 2020 research and
innovation programme (ERC Consolidator Grant DISTRUCT, grant agreement No
648527).

© Springer Nature Switzerland AG 2018
A. Brandstädt et al. (Eds.): WG 2018, LNCS 11159, pp. 252–265, 2018.
https://doi.org/10.1007/978-3-030-00256-5_21

The SECol-problem and the related *maximum induced matching*(MIM)-problem have received a lot of attention from the network community as this problem appears in the context of interference-free channel assignments [1,11,15–17].

The SVCol-problem (see [13]), the SECol-problem (see [14]) and the MIM-problem [2,21] are all NP-complete. This is especially surprising for the MIM-problem, as maximum matching is long known to be in P. Moreover, MIM stays NP-hard even on bipartite graphs [2] and on planar graphs of maximum degree 4 [12].

While many problems are solvable with more reasonable running time if parametrised, SVCol is still $W[1]$-hard, even when parametrised by treewidth [8].

By adding edges between vertices of distance at most l, we obtain the k-*th distance power* G^k of G. So the problem of finding a SVCol of G can be transformed into finding a VCol of G^2. If we were able to ensure G^2 to be in a class of graphs where the VCol-problem is known to be in P, we could use this for the construction of new algorithms for the SVCol-problem.

A classical example of graphs on which many NP-complete problems, such as the VCol-problem, become easy to solve are chordal graphs. While squares of chordal graphs are not again chordal in general, Duchet proved the following result:

Theorem 1 (Duchet [7]). *If G^k is chordal, so is G^{k+2}.*

A ECol corresponds to a VCol of the vertices of the linegraph. In the same fashion a SECol corresponds to a VCol of the squared linegraph, so in order to find graphs on which the SECol-problem becomes accessible we are interested in $L(G)^2$ to belong to a class of graphs that we can colour in polynomial time. Again chordal graphs provide a nice example.

Theorem 2 (Cameron [2]). *If G is a chordal graph, then $L(G)^2$ is chordal.*

In this paper we investigate the structure of graphs whose squared linegraphs have nice properties with respect to the VCol-problem. Namely we find families A and B of forbidden induced subgraphs in order to make sure that for any graph G excluding the graphs of A, $L(G)^2$ excludes the graphs of B as induced subgraphs. This leads in particular to a characterisation of graphs with perfect linegraph squares and therefore yields a class of graphs on which the strong edge colouring problem can be solved in polynomial time. We extend the results from [19] by using similar techniques which are also used in [3].

2 Preliminaries

All graphs considered in this paper are finite, simple and undirected. If $G = (V, E)$ is a graph and $H = (V', E')$ such that $V' \subseteq V$ and $E' = \{e \in E \mid e \subseteq V'\}$, we call H an *induced subgraph* of G and H is said to be *contained* in G. We denote the subgraph induced by the vertex set V' by $G[V']$. If H is not contained in

G, we call G *H-free*, respectively if $H_1, \ldots, H_t$ are graphs not contained in G we say that G is $(H_1, \ldots, H_t)$-*free*.

Let x, y be two vertices of a graph G, the *distance* $\text{dist}_G (x, y)$ between x and y is the length of a shortest path in G with endpoints x and y. The neighbourhood $N_G(v)$ of a vertex v in G is defined as the set $\{y \in V(G) \mid vy \in E(G)\}$. The set $N_G[v] := N_G(v) \cup \{v\}$ is the closed neighbourhood of v in G. If G is understood from the context, we write $\text{dist}(x, y)$. The *linegraph* of $G = (V, E)$ is the graph $L(G) := (E, \{e_1 e_2 \mid e_1 \cap e_2 \neq \emptyset,\ e_1, e_2 \in E\})$. The *k-th distance power*, or simply *k-th power* of G, for $k \geq 1$, is the graph $G^k := (V, \{xy \mid \text{dist}(x, y) \leq k\})$. For $t \geq 1$, $k \geq 3$ we denote the path on t vertices by P_t and the cycle on k vertices by C_k. The *complement* of G is defined as $\overline{G} := \left(V, \binom{V}{2} \setminus E\right)$, where $\binom{V}{2}$ denotes the set of all 2-element subsets of V. For $k \geq 5$ the complement of a cycle C_k, namely $\overline{C_k}$, is called an *antihole*. If we are given a set S of vertices and a vertex $x \in S$, we sometimes refer to x as a *S-vertex*, the same occurs for edges. As reference for other definitions including those of paths and cycles the reader is referred to the book by Diestel [6].

3   Induced Graphs in $L(G)^2$

The Strong Perfect Graph Theorem by Chudnovsky et al. [4] allows us to describe *perfect graphs* in terms of forbidden induced subgraphs.

A graph G is *perfect* if $\omega(H) = \chi(H)$ holds for all induced subgraphs $H \subseteq G$. Here ω denotes the size of a maximum clique and χ the minimum number of colours required for a VCol of a given graph.

Theorem 3 (Strong Perfect Graph Theorem, Chudnovsky et al. [4]). *A graph is perfect if and only if it contains neither C_k nor $\overline{C}_k$ for odd $k \geq 5$.*

We are aiming to forbid induced cycles and antiholes on an odd number of vertices ≥ 5 in the squared linegraph. A possible generalisation of perfect graphs is the concept of χ-boundedness. A class of graphs $\mathcal{C}$ is χ-bounded if there is a function $f : \mathbb{N} \to \mathbb{N}$ such all $G \in \mathcal{C}$ satisfy $\chi(G) \leq f(\omega(G))$. A famous result by Gyárfás [10] states that the class of P_t-free graphs is χ-bounded for every $t \geq 1$.

Inspired by this, we start our discussion on induced graphs in G^2 by investigating graphs with P_t-free squares.

3.1   Induced Paths in G^2

The goal of this subsection is to provide some insight in the technical descriptions of the structures, i.e. subgraphs, that are responsible for the existence of induced paths in the square of a graph G. Our main technique is quite technical and uses a lot of case distinctions. The proof of Lemma 5, which characterises the structures that give rise to induced paths in G^2, is presented as an example.

Definition 4 (Spire). *A spire of order n is a graph $\mathsf{S}_n = (U \cup W, E)$ with $U = \{u_1, \ldots, u_n\}$ and $W = \{w_1, \ldots, w_q\}$ such that the following conditions hold:*

(S1) S_n contains a path P with endpoints u_1 and u_n such that all u_i and w_j are ordered by their appearance along P.

(S2) No three U-vertices form a subpath of length 2 on P.

(S3) No two W-vertices are consecutive on P.

(S4) The graph $G[U]$ only contains edges that lie on P, additionally: $W \subseteq V(P)$ and $U \subseteq V(P)$.

(S5) Either $u_i u_{i+1} \in E(S_n)$, or $u_i w_j u_{i+1}$ is an induced subpath of P for some j.

(S6) Each $w_j \in W$ is adjacent to exactly two U-vertices and those are consecutive on P.

If S_n is contained in some graph G and there exists an additional vertex v with $vu_i, vu_j \in E(G)$ for some $j \neq i \pm 1$, then S_n is called a withered spire or just withered.

We refer to U as the base of S_n and if $w_j u_i w_{j+1} \subseteq P$ and $w_j w_{j+1} \in E(S_n)$, u_i is called pending.

Lemma 5. Let G be a graph and $t \geq 1$. There is a set of vertices $U \subseteq V(G)$ such that $G^2[U] \cong P_t$ if and only if G contains an unwithered spire of order t with base U.

Proof. Let $U = \{u_1, \ldots, u_t\} \subseteq V(G)$ such that $G^2[U] \cong P_t$. Assume that the vertices of U are ordered by their appearance along P_t. If $\mathrm{dist}_G(u_i, u_{i+1}) = 2$ for some $i \in \{1, \ldots, t-1\}$, there is a vertex $w \in V(G)$ with $u_i, u_{i+1} \in N(w)$. Suppose there is some $j \in \{1, \ldots, t\} \setminus \{i, i+1\}$ with $u_j \in N(w)$. Then $\mathrm{dist}_G(u_i, u_j) \leq 2$ and $\mathrm{dist}_G(u_{i+1}, u_j) \leq 2$ which contradicts P_t being an induced path. Hence we can collect a set W of such vertices by choosing exactly one such w for every pair of vertices u_i, u_{i+1} of distance 2 in G. This set W immediately satisfies Property (S6). In addition, we now have a path P on the vertices of U and W in which two consecutive vertices u_i and u_{i+1} are not adjacent if and only if there is some $w \in W$ adjacent to both of them. Thus, Properties (S1) to (S3) hold as well.

As we have seen, all W-vertices connect U-vertices of distance exactly 2 in G, so if $e \in E(G[U]) \setminus E(P)$, then e joins two U-vertices that do not have a common neighbour in W and are not consecutive on P. This contradicts P_n being induced, so such edges do not exists and Property (S4) holds. Property (S5) holds by construction and thus we are done.

For the reverse direction of the proof let $S_t = (U \cup W, E)$ be an unwithered spire in G and P its corresponding path. By Property (S5), the $u_i u_{i+1} \in E(G^2)$ for all $1 \leq i \leq t-1$, so $G^2[U]$ contains a path of length $t-1$ on the vertices of U. Assume towards a contradiction that this path is not induced. Hence, there are two U-vertices u_i and u_j with $j \neq i \pm 1$ and $\mathrm{dist}_G(u_i, u_j) \leq 2$. By Properties (S2) and (S4), u_i and u_j cannot be adjacent and since S_t is not withered, there is no $v \in V(G) \setminus V(S_t)$ adjacent to both. So there is some $x \in V(S_t)$ with $u_i, u_j \in N(x)$. By Properties (S2) and (S4), $x \notin U$ and thus $x \in W$. But by Property (S6), u_i and u_j now have to be consecutive, a contradiction. $\square$

Consider the class $\mathcal{S}_t$ of graphs G such that G^2 excludes a fixed P_t. There has to be an additional vertex in the graph, and not within the spire, in order for the spire to be withered. So our result does not allow us to conclude the χ-boundedness of $\mathcal{S}_t$. However, by Gyárfás' Theorem, for every $t \geq 1$ the class $\mathcal{S}_t^2$ of graphs G for which some graph H exists with $H^2 = G$ and all S_n in H are withered is χ-bounded.

3.2   Linegraphs

The problem of spires and other structures responsible for the existence of pre-scribed induced subgraphs in G^2 is that an additional vertex in the graph G is required to render them withered.

Flotow (see [9]) gives the following construction to show that there is no finite family of forbidden induced subgraphs describing a class $\mathcal{C}'$ such that $G^2 \in \mathcal{C}$ if and only if $G \in \mathcal{C}'$ if we put no bound on the clique number. Suppose there was a finite family $\mathcal{F}$ characterising $\mathcal{C}$ in this way. We construct the graph G by taking a copy of every graph in $\mathcal{F}$ together with one additional vertex v adjacent to everything else. Thus G^2 is complete, so $G^2 \in \mathcal{C}$, contradicting the $\mathcal{F}$-free graphs to be exactly those with squares in $\mathcal{C}$.

This construction exploits the problem of additional vertices responsible for the existence of paths of length 2. However, if we consider linegraphs this problem does not occur. In order for a vertex to be responsible for a structure being withered in $L(G)$, an edge in G has to contain endpoints of two other edges that are not supposed to be connected in $L(G)^2$. Hence, the edge is part of the subgraph induced by the vertex set of the edges producing the structure in $L(G)$ (Fig. 1).

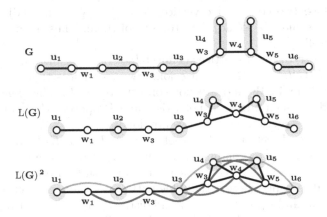

Fig. 1. A graph G containing an unwithered spire in its linegraph together with the induced P_6 in $L(G)^2$. (Color figure online)

Definition 6 (Sprout [19]). *A* sprout *of order n is defined as a graph* $\mathsf{ST}_n = (V, U \cup W \cup E)$ *with* $|U| = n$ *and* $|W| = q$, U, W *and* E *having a pairwise empty intersection and* $\lceil \frac{n}{2} \rceil \le q \le n$ *satisfying the following conditions:*

(ST1) *There is a cycle C with $E(C) \supseteq W$ containing the edges of W in the order $w_1, \ldots, w_q$.*

(ST2) *The elements of $U = \{u_1, \ldots, u_n\}$ are sorted such that they appear along C in order with $u_1 \cap w_q \neq \emptyset$ and $u_2 \cap w_1 \neq \emptyset$. In addition $u_i \cap u_j = \emptyset$ for all $j \neq i \pm 1 \,(\mathrm{mod}\ n)$.*

(ST3) *If $w_i \cap w_{i+1} \neq \emptyset$, then there is exactly one $u \in U$ with $(w_i \cap w_{i+1}) \cap u \neq \emptyset$. These edges are called* pending.

(ST4) *If $w_i \cap w_{i+1} = \emptyset$, then there either is one $u \in U$ connecting w_i and w_{i+1} in C, or there are exactly two edges $t, t' \in U$, such that the graph induced by t and t' is a path starting on w_i, ending on w_{i+1} and being part of C.*

(ST5) *The pending U-edges are pairwise non-adjacent and every U-edge that is not pending is an edge in $E(C)$.*

If a sprout $\mathsf{ST}_n = (V, U \cup W \cup E)$ contains an edge $e \in E$ connecting two non-consecutive u-edges, we say ST_n is infertile, *otherwise ST_n is called* fertile.

Theorem 7 (Scheidweiler and Wiederrecht [19]). *Let G be a graph and $n \in \mathbb{N}$. Then $\mathrm{L}(G)^2$ contains a cycle of length n if and only if G contains a fertile sprout of order n.*

Induced Paths in Linegraph Squares

In order to forbid P_t in $\mathrm{L}(G)^2$ we translate spires into the world of linegraphs. Similar to sprouts, the first definition is very technical but for $t = 4$ we will be able to make use of it in order find a small family of graphs to forbid in G.

Definition 8 (Plantlet). *A* plantlet *of order n* $\mathsf{PL}_n = (V, U \cup W \cup R)$ *is a graph with $U = \{u_1, \ldots, u_n\}$, $W = \{w_1, \ldots, w_q\}$ and U, W and R pairwise disjoint such that the following conditions hold:*

(P1) $V = \bigcup_{u \in U} u$.

(P2) *There is a path P with $E(P) \subseteq U \cup W$ and $W \subseteq E(P)$ such that for every $u_i \in U \setminus E(P)$ there is a unique vertex $\{v\} = u_i \cap V(P)$. U-edges with a vertex that is not on P are called* pending. *Every pending U-edge is adjacent to two W-edges. Both, U- and W-edges, are ordered by their appearance along P.*

(P3) *If $u \in U$ is pending, then none of its endpoints belongs to another U-edge.*

(P4) *No U-edge contains vertices of two other U-edges.*

(P5) *Either $u_i \cap u_{i+1} \neq \emptyset$, or there exists a unique $w_j \in W$ such that $u_i \cap w_j \neq \emptyset$ and $u_{i+1} \cap w_j \neq \emptyset$.*

If there is an edge $e \in R$ and $i, j \in \{1, \ldots, n\}$ with $|i - j| \ge 2$ such that $e \cap u_i \neq \emptyset$ and $e \cap u_j \neq \emptyset$, PL_n is called infertile, *otherwise it is called* fertile.

Lemma 9. *A graph G contains a fertile $\mathsf{PL}_n = (V, U \cup W \cup R)$ if and only if $\mathrm{L}(G)$ contains an unwithered spire $\mathsf{S}_n = (U \cup W, E)$ for some $E \subseteq E(\mathrm{L}(G))$.*

To reach a characterisation of a graph class excluding an induced path of fixed length in $L(G)$ in terms of a succinct list of forbidden subgraphs, one usually needs a huge case distinction. Excluding the P_4 results in a class of perfect graphs (see [20]). Furthermore, excluding any single induced subgraph not contained in the P_4 results in a class of graphs for which no linear χ-bounding function can exist (see [18]). So, considering the class of graphs whose squared linegraphs exclude P_4 seems natural in the context of investigating generally perfect linegraph squares. We also obtain the following general observations on graph classes whose linegraph squares exclude induced paths of a certain length.

Lemma 10. *Let $n \geq 2$ and G be a graph such that $L(G)^2$ is P_n-free. Then G is $P_{\lceil \frac{3}{2}n \rceil}$-free.*

There are two main characteristics by which we can distinguish different plantlets of the same order: The length of the path P of Property (P2), and the number and position of pending edges along P. By Lemma 10, we have a lower bound of $\lceil \frac{3}{2}n \rceil - 1$ on the length of P in a plantlet of order n that does not have any pending U-edges. Next, we show that $2n - 1$ is an upper bound. As long as there are no pending U-edges these bounds are strict. By allowing pending edges we obtain a lower bound of $n + 2$. This is summarised in the following corollary.

Corollary 11. *Let $n \in \mathbb{N}$, then there is no plantlet of order n such that its Property (P2)-path P is isomorphic to P_i for some $i \in \mathbb{N} \setminus \{n + 2, \ldots, 2n\}$.*

Let us return to the case where $n = 4$. In this case the Property (P2)-paths of a plantlet has between 6 and 8 vertices. We categorise these plantlets into seven different types as depicted in Fig. 2.

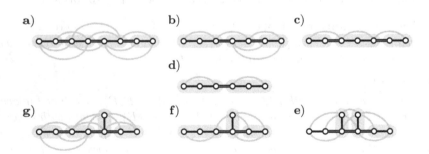

Fig. 2. The seven types of plantlets of order 4. The yellow marked edges are the U-edges, the thicker violet edges belong to W and the grey edges are those that may exist in R, but are not necessary for the graph to be a fertile plantlet. (Color figure online)

We claim that, up to isomorphism, these are all possible plantlets of order 4. To prove this, we partition them into three families based on the length of their Property (P2)-path. By Corollary 11, there are no other plantlets of order 4, so this case distinction suffices.

Still, it is clear that some of these fertile plantlets of order 4 are subgraphs of plantlets of the other types, so this family is not minimal with respect to excluding all fertile plantlets of order 4. The next step is to reduce the number of forbidden subgraphs. Figure 3 depicts a smaller family of forbidden subgraphs for excluding plantlets of order 4.

Type A plantlets correspond exactly to the type d plantlets of Fig. 2. The type B plantlets are a specified version of type b plantlets where the v_3v_5-edge has to exist in order to distinguish them from plantlets of type A. Type C is obtained from type a by requiring the additional edge v_3v_6 and allowing all possible combinations of all other allowed edges except for the existence of v_3v_5 and v_4v_6 at the same time, while type D is obtained by requiring v_3v_5 and v_4v_6 to exist at the same time, distinguishing them from plantlets of type B.

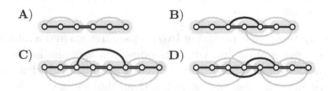

Fig. 3. The four minimal types of plantlets of order 4. Unmarked black edges must necessarily exist, all other colours/patterns are chosen as in Fig. 2. (Color figure online)

To summarise, we obtain the following theorem for graphs without an induced P_4 in their squared linegraph.

Theorem 12. *Let G be a graph. Then $\mathrm{L}(G)^2$ is P_4-free if and only if G does not contain a graph of type A, B, C, or D.*

Induced Odd Antiholes and Perfect Linegraph Squares

In order to describe the class of graphs with perfect linegraph squares, we need to find a structure similar to sprouts. The edges that will become the vertices of an induced antihole in $\mathrm{L}(G)^2$ are ordered in a cyclic fashion and we use a cycle to represent this ordering. When deleting a vertex from this cycle we obtain a path, hence a bipartite graph. Whenever we talk about the colour classes of a path obtained this way, we refer to the two classes of the unique 2-colouring of this path.

Definition 13 (Meristem). *A meristem of order $n = 2k+1$, $k \geq 2$, is a graph $\mathsf{M}_n = (V, U \cup W \cup R)$ with $U = \{u_1, \ldots, u_n\}$ such that the following conditions hold:*

(M1) $V = \bigcup_{u \in U} u$.
(M2) *There is a cyclic ordering on the edges of U such that $u_{n+1} = u_1$. We represent this ordering by a cycle C_U whose vertices correspond to the edges in U.*

(M3) There exists a family $\{U_1, \ldots, U_n\} \subseteq \binom{U}{k}$ such that the edges in U_i are identified with the vertices of $C_U - u_i$ from the same colour class as u_{i+1}.

(M4) For all $u \in U$ and $v, v' \in U$ with $v \cap u \neq \emptyset$ and $v' \cap u \neq \emptyset$ there is some $j \in \{1, \ldots, n\}$ with $v, v' \in U_j$.

(M5) There are sets $W_1, \ldots, W_n$, possibly empty and not necessarily disjoint, with $\bigcup_{i=1}^n W_i = W$ and for all $w = xy \in W_i$ there are $u, v \in U_i$ such that $x \in u$, $y \in v$, $u \cap v = \emptyset$ and $e \not\subseteq u \cup v$ for all e in $(U \cup W) \setminus \{u, v, w\}$.

(M6) For all $w \in W$ and $u, v \in U$ with $u \cap w \neq \emptyset$ and $v \cap w \neq \emptyset$ there is some j such that $u, v \in U_j$ and $w \in W_j$.

(M7) For all $u, u' \in U_i$ with $u \cap u' = \emptyset$ there is a $w \in W_i$ with $u \cap w \neq \emptyset$ and $u' \cap w \neq \emptyset$.

We call U the base of M_n. If $u_i \cap u = \emptyset$ for all $u \in U \setminus \{u_i\}$, u_i is called pending. If there is an edge $e \in R$ with $e \cap u_i \neq \emptyset$ and $e \cap u_{i+1} \neq \emptyset$ for some $i \in \{1, \ldots, n\}$, M_n is called infertile.

The following theorem is proved in two steps. First we find a family of graphs whose squares contain antiholes, which is done in a way similar to spires, complete with a notion of being withered. Such a graph is called a *thornbush* of order n, where n is the size of the induced antihole it generates when being squared. Then, as the second step, we show that a fertile meristem of order n in G corresponds to an unwithered thornbush in $\mathrm{L}(G)$.

Theorem 14. *A graph G contains a fertile meristem of order n if and only if $\mathrm{L}(G)^2$ contains an antihole of size n.*

Theorem 14 allows us to state a first straight forward characterisation of graphs with perfect linegraph squares by combining it with Theorem 7.

Corollary 15. *Let G be a graph, then $\mathrm{L}(G)^2$ is perfect if and only if G does not contain a fertile sprout or fertile meristem of order $n = 2k + 1$ for any $k \geq 2$.*

We now further refine this result by taking a closer look at the structure of sprouts and meristems. We need the following lemmata.

Lemma 16 (Scheidweiler and Wiederrecht [19]). *Let G be a graph and $k \geq 2$ an integer. If C is an induced cycle in G^k, then G^r cannot contain two consecutive edges of C for all $r \leq \lfloor \frac{k}{2} \rfloor$.*

Lemma 17 (Scheidweiler and Wiederrecht [19]). *A graph G contains a cycle of length at least four as a, not necessarily induced, subgraph if and only if $\mathrm{L}(G)$ contains an induced cycle of the same length.*

Lemma 18 (Scheidweiler and Wiederrecht [19]). *Let $k \geq 4$ be an integer. If a graph G does not contain induced cycles of length $\ell \geq k$, then $\mathrm{L}(G)^2$ contains no induced cycles of length $\ell \geq k$.*

Lemma 19. *For all $n \geq 4$ the cycle C_j with $n + \lceil \frac{n}{2} \rceil \leq j \leq 2n$ is a fertile sprout of order n.*

Proof. By Lemma 16 there are at most $\left\lfloor \frac{n}{2} \right\rfloor$ pairs of adjacent u-edges in a fertile sprout of size n.

First, consider $j = n + \left\lceil \frac{n}{2} \right\rceil$. If there are exactly $\left\lfloor \frac{n}{2} \right\rfloor$ pairs of adjacent u-edges, then the remaining $\left\lceil \frac{n}{2} \right\rceil - \left\lfloor \frac{n}{2} \right\rfloor \in \{0, 1\}$ U-edges cannot be adjacent to any other U-edge. Hence, we need exactly $\left\lceil \frac{n}{2} \right\rceil$ W-edges to complete the sprout. By alternating between U-edge pairs, W-edges and possibly one single U-edge we obtain a cycle of length $n + \left\lceil \frac{n}{2} \right\rceil$. The sprout definition allows some additional chords, but no such edge is necessary, so the C_j is a fertile sprout of size n.

Second, we note that each of those $\left\lfloor \frac{n}{2} \right\rfloor$ pairs of U-edges may be split by an additional W-edge, hence with $k \leq \left\lfloor \frac{n}{2} \right\rfloor$ such splits we can produce a cycle of length $n + \left\lceil \frac{n}{2} \right\rceil + k \leq 2n$ which again is a fertile sprout of size n. $\qquad\square$

Corollary 20. *A graph G with an induced cycle of length $\ell \geq 8$ contains a fertile sprout of odd order.*

Proof. By Lemma 19 cycles of length $8, 9$ or 10 contain a fertile sprout of order 5. So consider $\ell \geq 11$. We observe that for $n \geq 7$ we have $n + \left\lceil \frac{n}{2} \right\rceil \leq 2(n-2)+1$ and $7 + \left\lceil \frac{7}{2} \right\rceil = 11$. So by this observation and Lemma 19, if ℓ is odd, then the cycle is a fertile sprout of order $\frac{\ell-1}{2} + 2$. If ℓ is even, then the cycle is a fertile sprout of order $\frac{\ell}{2}$. $\qquad\square$

As the squared linegraph of C_7 yields an antihole of size 7, the following lemma reduces the length of allowed induced cycles even further.

Lemma 21. *If a graph G contains a C_7, $\mathrm{L}(G)^2$ contains an antihole of the same size.*

Proof. If G contains a C_7 Lemma 17 yields the existence of an induced cycle C of the same length in $\mathrm{L}(G)$. For each pair of non-adjacent vertices $u, w \in V(\mathrm{L}(G))$ of C that do not have a common neighbour on C, $\mathrm{dist}_{\mathrm{L}(G)}(u, w) \geq 3$ holds, since a path of length 2 between two such vertices would correspond to a chord in the C_7, which does not exist. In $\mathrm{L}(G)^2$ each vertex of C is adjacent to its four nearest vertices on C and not adjacent to the two opposite vertices of C. By reordering the vertices C^2 is an antihole. $\qquad\square$

Lemma 18 implies that a graph G without induced cycles of length $\ell \geq 7$ does not contain a fertile sprout of order $\ell \geq 7$, hence $\mathrm{L}(G)^2$ only contains holes of size at most 5. In order to forbid the holes of size 5, we exclude all fertile sprouts of order 5 in G. Since we can exclude the existence of induced cycles of length $\ell \geq 7$, it suffices to consider sprouts of order 5 with a longest induced cycle, or base cycle, of length 5 and 6. Figure 4 depicts the three possible types of sprouts of order 5 with a base cycle of length 5. We proceed by discussing the case of a base cycle of length 6.

Lemma 22. *Let* $\mathsf{ST}_5 = (V, U \cup W \cup R)$ *be a fertile sprout of order 5 with a base cycle C of length 6 and $E(C) \cap R = \emptyset$. Then* ST_5 *has either three or four pending edges, which are incident with consecutive vertices of C.*

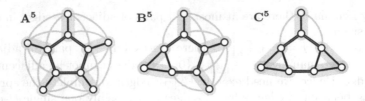

Fig. 4. The sprouts of order 5 with a base cycle of length 5. Colours are chosen as in Figs. 2 and 3. (Color figure online)

Proof. Because ST_5 is a fertile sprout of order 5, the number of pending edges is at least 0 and at most 5. Assume that ST_5 has 5 pending edges. Then the cycle consists of exactly one U-edge and five W-edges. Since pending U-edges can only be incident to vertices on C that are not contained in any other U-edge, the number of pending edges in ST_5 is at most four contradicting the assumption.

Suppose ST_5 has no pending edge, then C contains five U-edges. But, by Lemma 16 there are at least $\lceil \frac{5}{2} \rceil = 3$ W-edges, which must be contained in C as well since C is the base cycle of ST_5. This contradicts the fact that $|C| = 6$. For the same reason ST_5 does not have exactly one pending edge. There remain 4 U-edges and at least 3 W-edges on C, which contradicts $|C| = 6$ again.

Next, suppose there are exactly two pending edges. There are at least three W-edges necessary for the pending edges and since three U-edges may not form a path on C we need at least one additional W-edge, thus C must now consist of 3 U-edges and at least 4 W-edges, again exceeding the length of C.

So there only are two possibilities: Either ST_5 has three or four pending U-edges. These are incident to consecutive vertices of C. Suppose they were not, then the base cycle must contain at least 6 W-edges and an additional U-edge and if there are just three pending edges not being adjacent to consecutive vertices of C, 5 W-edges and 2 additional U-edges are required.   □

Notice that every fertile sprout of type B^6 contains a type A^6-sprout. Hence the fertile ST_5 not containing a type A^6-sprout but still having a base cycle of length 6 are the types C^6. D^6 and E^6 from Fig. 5.

Corollary 23. *Let $\mathsf{ST}_5 = (V, U \cup W \cup R)$ be a fertile sprout of order 5 with a longest induced cycle C of length 6, then ST_5 contains a fertile sprout ST_5' of type A^6, C^6, D^6, or E^6 (see Fig. 5).*

As a last observation on sprouts one can see that the sprouts of type C^6 are also of type B^6, and the sprouts of type D^6 or E^6 certainly contain type A^6 sprouts. The complement of a C_5 is again a C_5, hence the family of meristems of order 5 is exactly the family of sprouts of order 5. With these last observations we can further reduce the number of obstructions to perfect linegraph squares to the statement of our main result.

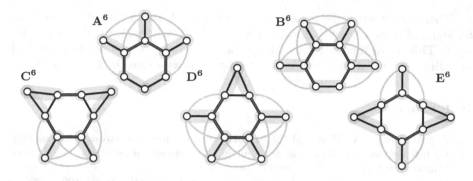

Fig. 5. The sprouts of order 5 with a base cycle of length 6. Colours are chosen as in Figs. 2 and 3. (Color figure online)

Theorem 24. *Let G be a graph. Then $L(G)^2$ is perfect if and only if G does not contain a cycle of length $\ell \geq 7$, a fertile sprout of type A^5, B^5, or C^5 (see Fig. 4), a fertile sprout of type A^6 (see Fig. 5), or a fertile meristem of order $n = 2k + 1$ with $k \geq 3$.*

4   Conclusion (χ-Boundedness)

Theorem 24 states that the class of graphs with perfect linegraph squares excludes induced cycles of length $\ell \geq 7$. Similarly, if we exclude induced cycles of certain length in the squared linegraph of a graph G, by Lemma 19 G itself also excludes cycles of some length. Hence, if $\mathcal{C}_n$ is the class of graphs G such that $L(G)^2$ does not contain an induced cycle of length $\ell \geq n$, the graphs in $\mathcal{C}$ also exclude cycles of length $\ell \geq n + \lceil \frac{n}{2} \rceil$.

Something similar holds for graph classes excluding induced paths of a certain length in their squared linegraphs, see Lemma 10. With Gyárfás' Theorem on classes excluding induced paths and the following theorem by Chudnovsky et al. we are able to deduce the χ-boundedness of graph classes with certain excluded induced subgraphs in their linegraph squares.

Theorem 25. (Chudnovsky et al. [5]). *If $\mathcal{C}_n$ is a class excluding induced cycles of length $\ell \geq n$, $\mathcal{C}_n$ is χ-bounded.*

Theorem 26. *The following classes are χ-bounded: 1. the class $\mathcal{C}_{chordal}$ of graphs with $L(G)^2$ chordal, 2. the class $\mathcal{C}_{perfect}$ of graphs with $L(G)^2$ perfect, 3. the class $\mathcal{C}_{P_t}$ of graphs with $L(G)^2$ P_t-free, $t \geq 1$, and 4. the class $\mathcal{C}_n$ of graphs with $L(G)^2$ excluding induced cycles of length $\ell \geq n$.*

The families of forbidden induced subgraphs for all four of these classes contain many more graphs than just cycles and paths of a certain length. In fact, we forbid plantlets, sprouts or meristems in all of them, which have a far more complicated structure.

Therefore, it might be possible to derive much better χ-bounding functions for some of those classes than those provided by Theorem 25 and Gyárfás' Theorem. This seems especially possible for classes like $\mathcal{C}_{P_4}$, which can be described by a finite family of forbidden induced subgraphs as we saw in Theorem 12.

References

1. Arputhamary, I.A., Mercy, M.H.: An analytical discourse on strong edge coloring for interference-free channel assignment in interconnection networks. Wirel. Pers. Commun. **94**(4), 2081–2094 (2017)
2. Cameron, K.: Induced matchings. Discret. Appl. Math. **24**(1–3), 97–102 (1989)
3. Cameron, K., Sritharan, R., Tang, Y.: Finding a maximum induced matching in weakly chordal graphs. Discret. Math. **266**(1–3), 133–142 (2003)
4. Chudnovsky, M., Robertson, N., Seymour, P., Thomas, R.: The strong perfect graph theorem. Ann. Math. **164**, 51–229 (2006)
5. Chudnovsky, M., Scott, A., Seymour, P.: Induced subgraphs of graphs with large chromatic number. iii. long holes. Combinatorica **121**, 1–16 (2016)
6. Diestel, R.: Graph Theory, vol. 5. Springer, Heildelberg (2010)
7. Duchet, P.: Classical perfect graphs: an introduction with emphasis on triangulated and interval graphs. North-Holland Math. Stud. **88**, 67–96 (1984)
8. Fiala, J., Golovach, P.A., Kratochvíl, J.: Parameterized complexity of coloring problems: treewidth versus vertex cover. Theor. Comput. Sci. **412**(23), 2513–2523 (2011)
9. Flotow, C.: Graphs whose powers are chordal and graphs whose powers are interval graphs. J. Graph Theory **24**(4), 323–330 (1997)
10. Gyárfás, A.: Problems from the world surrounding perfect graphs. Applicationes Mathematicae **19**(3–4), 413–441 (1987)
11. Janssen, J., Narayanan, L.: Approximation algorithms for channel assignment with constraints. Theor. Comput. Sci. **262**(1), 649–667 (2001)
12. Ko, C., Shepherd, F.: Adding an identity to a totally unimodular matrix (1994)
13. Lloyd, E.L., Ramanathan, S.: On the complexity of distance-2 coloring. In: Proceedings of the Fourth International Conference on Computing and Information, ICCI 1992, pp. 71–74. IEEE (1992)
14. Mahdian, M.: On the computational complexity of strong edge coloring. Discret. Appl. Math. **118**(3), 239–248 (2002)
15. Nandagopal, T., Kim, T.E., Gao, X., Bharghavan, V.: Achieving MAC layer fairness in wireless packet networks. In: Proceedings of the 6th Annual International Conference on Mobile Computing and Networking, pp. 87–98. ACM (2000)
16. Ramanathan, S.: A unified framework and algorithm for (T/F/C) DMA channel assignment in wireless networks. In: INFOCOM 1997, Sixteenth Annual Joint Conference of the IEEE Computer and Communications Societies. Driving the Information Revolution, Proceedings IEEE, vol. 2, pp. 900–907. IEEE (1997)
17. Ramanathan, S., Lloyd, E.L.: Scheduling algorithms for multihop radio networks. IEEE/ACM Trans. Netw. (TON) **1**(2), 166–177 (1993)
18. Randerath, B., Schiermeyer, I.: Vertex colouring and forbidden subgraphs-a survey. Graphs Combin. **20**(1), 1–40 (2004)

19. Scheidweiler, R., Wiederrecht, S.: On chordal graph and line graph squares. Discret. Appl. Math. **243**, 239–247 (2018)
20. Seinsche, D.: On a property of the class of n-colorable graphs. J. Comb. Theory, Ser. B **16**(2), 191–193 (1974)
21. Stockmeyer, L.J., Vazirani, V.V.: NP-completeness of some generalizations of the maximum matching problem. Inf. Process. Lett. **15**(1), 14 19 (1982)

On Weak Isomorphism of Rooted
Vertex-Colored Graphs

Lars Jaffke$^{(\boxtimes)}$ and Mateus de Oliveira Oliveira

University of Bergen, Bergen, Norway
{lars.jaffke,mateus.oliveira}@uib.no

Abstract. In this work we consider a notion of isomorphism of rooted vertex-colored graphs which allows not only vertices, but also colors to be permuted. Here, a prospective color permutation must be chosen from a group specified at the input. We call this notion *weak isomorphism*. It turns out that already for severely restricted classes of graphs, the corresponding WEAK GRAPH ISOMORPHISM problem is as hard as the well studied STRING ISOMORPHISM problem. Our main result states that weak isomorphism can be solved in FPT time when simultaneously parameterized by three graph invariants: maximum degree, BFS color number, and BFS width. Intuitively, the second parameter quantifies the number of colors that cross a level of a breadth first search (BFS) tree of the corresponding graph. The third parameter is a width measure based on a BFS-based decomposition introduced independently by Yamazaki et al. [Algorithmica '99] and by Chepoi and Dragan [Eur. J. Comb. '00]. We show that the resulting parameterized problem has close relations to the notion of (strong) isomorphism of bounded color class hypergraphs. Our algorithm can be used to solve the latter problem in FPT time. Another consequence is that isomorphism of hypergraphs implicitly represented by ordered decision diagrams (ODD's) can be solved in FPT time if the width of the involved ODD's is an additional parameter.

Keywords: Weak Graph Isomorphism
Implicit hypergraph representation · Fixed-parameter tractability

1 Introduction

In the GRAPH ISOMORPHISM problem (GI), we are given vertex-colored graphs G and G', each with n vertices, and the goal is to determine if there is a bijection between the vertex sets of G and G' that preserves both adjacencies and vertex-colors. It has been known for a long time that GI can be decided in time $2^{O(\sqrt{n \log n})}$ [4]. In a recent work, Babai proposed an algorithm that improves this upper bound to $2^{O(\log^k n)}$ for some constant k [2].

From the perspective of parameterized algorithms [10], graph isomorphism is solvable in time $f(k) \cdot n^{O(1)}$ (that is, FPT time) whenever the parameter k

This work was supported by the Bergen Research Foundation.

© Springer Nature Switzerland AG 2018
A. Brandstädt et al. (Eds.): WG 2018, LNCS 11159, pp. 266–278, 2018.
https://doi.org/10.1007/978-3-030-00256-5_22

stands for eigenvalue multiplicity [3], treewidth [16], feedback vertex-set number [15], or size of the largest color class [11] of the involved graphs. On the other hand, GI can be solved in time $f_1(k) \cdot n^{f_2(k)}$ (that is, in XP time), whenever the parameter k stands for genus [18], rankwidth [14], maximum degree [17], size of an excluded topological subgraph [13] or size of an excluded minor [12].

We note that Babai's recent breakthrough [2] does not immediately imply an improvement on the running time of any of the parameterized algorithms mentioned above. On the one hand, some of these parameterized algorithms, such as those in [12–14,18], are obtained using techniques which are very distinct from Luks' group-theoretic framework, which is crucially employed in [2]. On the other hand, the reduction to Johnson schemes, which is the central technique introduced by Babai in [2] is only guaranteed to work when applied to groups of order at least $n^{1+\log n}$, while the size of the primitive groups that arise as the bottleneck for the running time of algorithms for bounded degree graph isomorphism [17] and bounded color-class graph isomorphism [11] is bounded by $n^{f(k)}$ where k is the parameter.

In this work we consider a relaxation of isomorphism of vertex-colored graphs in which not only vertices, but also colors can be permuted. Here, a prospective permutation of the set of colors must be chosen from a permutation group specified at the input. We call this notion *weak isomorphism*[1] and the corresponding decision problem WEAK GRAPH ISOMORPHISM (WGI). To avoid confusion, we may call *strong isomorphism* the usual notion of isomorphism for colored graphs (and hypergraphs), in which colors must be preserved.

It turns out that WGI for severely restricted classes of graphs, such as stars[2] is as hard as the well studied STRING ISOMORPHISM (SI) problem, and therefore as hard as the GROUP GRAPH ISOMORPHISM problem (GGI)[3]. Note that stars are graphs of genus 0, treewidth 1, which have feedback vertex-set number 0 and that exclude triangles as minors and topological minors. Therefore, the existence of XP algorithms for WGI when parameterized by the magnitude of these invariants would imply that SI can be solved in polynomial time. Contrast this with the fact that GI has been shown to be in FPT or in XP for each of these parameters.

Main Result: Our main result (Theorem 3) is an algorithm that solves WGI on rooted graphs in FPT time when the problem is simultaneously parameterized by three invariants: maximum degree (d), BFS color number (α) and BFS width (β). In other words, we show that this problem can be solved in time $f(d, \alpha, \beta) \cdot n^{O(1)}$ for some function f. While maximum degree does not require introduction, the

---

[1] If one is allowed to permute the colors arbitrarily, according to the symmetric group acting on the set of colors, then the corresponding isomorphism problem is sometimes termed *isomorphism of labeled graphs* [8].

[2] A *star* is a graph with a root r and n other vertices which are connected to r and only to r.

[3] SI was shown to be polynomially many-one equivalent to GGI in [4]. Nevertheless it is a longstanding open problem to determine whether any of these problems can be reduced to GI even via polynomial-time Turing reductions.

last two parameters are based on a breadth first search (BFS) of the graph starting in its root. Intuitively, the BFS color number of a rooted graph quantifies the number of colors that cross a level of such a BFS. The third parameter is a width measure based a notion of BFS-decomposition introduced in two distinct contexts by Yamazaki et al. [24] and by Chepoi and Dragan [9]. These parameters will be properly defined in Sect. 2.

Hardness Results: It is worth noting that proving the existence of an FPT algorithm for WGI in the case that either BFS color (α) number or BFS width (β) is unbounded would solve major open problems in algorithmics. First, we note that trees (and in particular stars) have BFS width 1. Therefore, by the discussion above, WGI for graphs of BFS width 1 is as hard as SI. On the other hand, WGI on graphs of BFS color number 1 is as hard as GI. Additionally, WGI for bounded degree trees and for bounded degree graphs of BFS color number 1 is as hard as bounded degree graph isomorphism. This shows that developing algorithms running in time $f(d, \alpha) \cdot n^{g(\alpha)}$ or running in time $f(d, \beta) \cdot n^{g(\beta)}$ would imply that bounded degree graph isomorphism is in FPT, solving a major open problem in the field of parameterized algorithmics [10, p. 680]. We refer to the full version for proofs of the statements discussed in this paragraph.

Bounded Color Class Hypergraph Isomorphism: Weak isomorphism for rooted graphs where the degree, BFS color number and BFS width are bounded is intimately connected with strong isomorphism of bounded color class hypergraphs. In particular, we show that bounded color-class hypergraph isomorphism is reducible in FPT time to weak isomorphism of rooted colored trees (BFS width 1), bounded degree and bounded BFS color number. In this sense, our main result (Theorem 3) may be regarded as a generalization of an FPT algorithm due to Arvind, Das, Köbler and Toda for bounded color class hypergraph isomorphism parameterized by maximum color-class size [1].

Implicitly Represented Hypergraph Isomorphism: Considering weak isomorphism of rooted graphs of higher BFS width, we are able to address the strong isomorphism problem for *implicitly represented* bounded color class hypergraphs containing an exponential number of hyperedges. A popular way of implicitly representing graphs and hypergraphs of exponential size is via the notion of ordered decision diagrams, or equivalently, via the notion of leveled finite automata (LFA's) [6,19]. Within this formalism, vertices and edges are encoded as strings of symbols. Since LFA's with poly(n) states may accept exponentially many strings, such LFA's can be used to represent graphs with exp(n) edges. The goal then is to devise algorithms for computational problems on graphs that operate directly on these implicit representations without ever having to construct the underlying graph or hypergraph explicitly. Problems that have been addressed from this symbolic perspective include ALL-PAIRS SHORTEST PATHS [20], MINIMUM SPANNING TREE [5], MAXIMUM MATCHING [7] and TOPOLOGICAL SORTING [23]. In this work we use LFA's to provide implicit representations of bounded color class hypergraphs. Our main result in this respect states that the problem of deciding strong isomorphism between bounded color class hypergraphs H and H' represented by LFA's $\mathcal{A}$ and $\mathcal{A}'$ respectively can be solved in time $f(w, b) \cdot |m|^{O(1)}$ where w is the width of the input automata, m is the number of color classes, and b is the size of the largest color class in the graphs represented by $\mathcal{A}$ and $\mathcal{A}'$ respectively.

Throughout the text, proofs of statements marked with '★' are deferred to the full version.

2 Problem Definition and Statement of the Main Result

Let Γ be a finite set of colors. A *rooted Γ-colored graph* is a tuple $G = (V, E, \gamma, r)$ where V is a set of vertices, $E \subseteq \binom{V}{2}$ is a set of undirected edges, $\gamma : V \to \Gamma$ is a function that colors vertices in V with elements from Γ, and r is a special vertex which is called the *root* of G. (For a formal explanation of the following group-theoretic notions, see the beginning of Sect. 3.) Let $\mathfrak{R}$ be a permutation group on Γ. An $\mathfrak{R}$-weak isomorphism between rooted Γ-colored graphs $G = (V, E, \gamma, r)$ and $G' = (V', E', \gamma', r')$ is a pair (φ, ψ) where $\varphi : V \to V'$ is a bijection between V and V', and ψ is a permutation in $\mathfrak{R}$ satisfying the following conditions.

(i) $\varphi(r) = r'$.
(ii) For each $u, v \in V$, $\{u, v\} \in E$ if and only if $\{\varphi(u), \varphi(v)\} \in E'$.
(iii) For each $v \in V$, $\psi(\gamma(v)) = \gamma'(\varphi(v))$.

We refer to the computational problem of deciding whether two Γ-colored rooted graphs are $\mathfrak{R}$-weakly isomorphic as WEAK ISOMORPHISM.

We observe that by setting ψ in the definition above to be the identity on Γ we recover the traditional notion of isomorphism for vertex-colored graphs. In this case, we say that the pair (φ, ψ) is a *strong isomorphism*.

We parameterize WEAK ISOMORPHISM for vertex-colored graphs with respect to three parameters. The first is simply the maximum degree of the involved graphs. The other two, the *BFS color number* and the *BFS-width* which we define next, are based on a breadth first search (BFS) starting at the root vertex.

Let $G = (V, E, \gamma, r)$ be a rooted Γ-colored graph. For each $i \in \{0, \ldots, |V|\}$, we let $B_G(i)$ be the set of vertices of G at distance at most i from r, $C_G(i)$ be the set of vertices of G at distance at least i from r, and $S_G(i)$ be the set of vertices of G at distance precisely i from r.

BFS Color Number. The *BFS color number* of G, $\alpha(G)$, is defined as follows.

$$\alpha(G) = \max_{i \geq 0} |\gamma(B_G(i)) \cap \gamma(C_G(i))|.$$

In other words, $\alpha(G)$ is the maximum among all $i \in \{0, 1, \ldots, |V|\}$ of colors that occur, at the same time, in some vertex at distance at most i from v and in some vertex at distance at least i from v. For a given i, we call the set of such colors the *colors crossing level i*.

BFS-Width. We now define the *BFS-decomposition* of a rooted graph, introduced by Yamazaki et al. [24] in the study of graph isomorphism, and by Chepoi and Dragan in the context of distance-approximating trees [9].

Definition 1 (BFS-decomposition). *Let G be rooted graph with root r, and let m be the maximum distance of any vertex in $V(G)$ to r. For $i \in [m]$, we say that a set $X \subseteq S_G(i)$ is i-normal if for each pair $u, u' \in X$, there is a path from u to u' that is contained in $C_G(i)$.*

The BFS-decomposition of G is a pair (T, X) where T is a rooted tree and $X = \{X_u\}_{u \in V(T)}$ is a family of subsets of vertices of G, called the bags, satisfying the following properties.

(i) *For each $S \subseteq V(G)$, $S = X_u$ for some $u \in V(T)$ if and only if S is i-normal for some $i \in [m]$.*

(ii) *There is an edge $(u, u') \in E(T)$ if and only if there are vertices $v \in X_u$ and $v' \in X_{u'}$ such that $\{v, v'\} \in E(G)$.*

The width *of (T, X) is defined as $\max_{t \in V(T)} |X_t|$.*

We note that unlike other notions of graph decompositions, each rooted graph G has a unique BFS-decomposition (T, X). Hence, we refer to the width of the BFS-decomposition as the *BFS-width of G*. Note that the BFS-width of G is simply the largest size of an i-normal set, over all $i \in [m]$. Chepoi and Dragan observed that the BFS-decomposition of a rooted graph can be constructed in time linear in the size of the input graph.

Proposition 2 ([9]). *Let G be a rooted graph on n vertices and m edges. There is an algorithm that computes the BFS-decomposition of G in time $\mathcal{O}(n + m)$.*

Intuitively, the BFS-width of a rooted graph measures how close the oriented graph induced by the ordering obtained from a BFS started in the root is to being an outbranching tree. It is not difficult to see that a rooted graph has BFS-width 1 if and only if it is a rooted tree and that the clique on n vertices has BFS-width $n - 1$. However, taking edge subgraphs can increase the BFS-width.

The notion of BFS width of a rooted graph G described above coincides with the notion of *rooted tree distance width* defined in [24] in the special case in which a single root is fixed[4]. A comparison between *rooted tree distance width* and some well known width measures for graphs can be found in [24]. We compare BFS width with some well known width measures in Fig. 1 and we refer to the full version for the details. It is known [16] that strong isomorphism parameterized by treewidth, and hence by BFS-width, can be solved in FPT time. However, an fpt-algorithm for *weak* isomorphism parameterized by BFS-width would give a polynomial-time algorithm for STRING ISOMORPHISM. Hence we turn to a more fine-grained parameterization.

Theorem 3 (Main Theorem). *Let Γ be a finite set of colors and G and G' be rooted Γ-colored graphs with n vertices, maximum degree d, BFS color number α and BFS-width β. Let $\mathfrak{R}$ be a subgroup of $\mathrm{Sym}(\Gamma)$ specified by a generating set of*

---

[4] The work [24] also considers the setting in which several roots are fixed. They showed that graph isomorphism is solvable in XP time in the number of fixed roots.

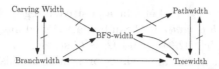

Fig. 1. Relationship between BFS-width and some well-known graph width measures. An arrow $A \rightarrow B$ translates to 'if A is bounded, then B is bounded' and an arrow $A \nrightarrow B$ means that there are graphs where A is bounded and B is unbounded.

$size^5 |\Gamma|^2$. Then, for some function f, one can determine in time $f(d, \alpha, \beta) \cdot n^{O(1)}$ whether G and G' are $\mathfrak{R}$-weakly isomorphic.

3 FPT-Algorithm for Weak Isomorphism

This section is devoted to the proof of Theorem 3. Before we proceed, we review some key standard group-theoretic concepts (see e.g. [21]), and set some notation concerning rooted trees.

Group-Theoretic Background. Let Ω be a finite set. A *permutation* $\pi \colon \Omega \rightarrow \Omega$ is a bijection from Ω to itself. We denote by $\mathrm{Sym}(\Omega)$ the *symmetric group* on Ω, i.e. the group of all permutations of Ω with function composition. We use the shorthand S_n for $\mathrm{Sym}([n])$. For two groups G and H, we denote by $H \sqsubseteq G$ that H is a subgroup of G. For each subgroup H of G and each element $r \in G$, we say that $Hr = \{hr \mid h \in H\}$ is a (right) coset of H in G. We say that r is a coset representative of the set Hr. A group $G \sqsubseteq \mathrm{Sym}(\Omega)$ is called a *permutation group* on Ω. For a set $\Delta \subseteq \Omega$, we denote by $\mathrm{Stab}(G, \Delta)$ the *pointwise stabilizer subgroup* of Δ in G, i.e. the subgroup of G that fixes each element in Δ. For an element $\omega \in \Omega$, we use the shorthand $\mathrm{Stab}(G, \omega)$ for $\mathrm{Stab}(G, \{\omega\})$. We furthermore denote by $\mathrm{SetStab}(G, \Delta)$, the *setwise stabilizer* of Δ in G.

Let G be a group and Ω a finite set. An *action* of G on Ω is a homomorphism $\varphi \colon G \rightarrow \mathrm{Sym}(\Omega)$ and we say that G *acts on* Ω via φ. Suppose G acts on Ω via φ and let $\omega \in \Omega$ and $g \in G$. For notational convenience, we denote by $\omega^g \in \Omega$ the element $\varphi(g)(\omega)$, i.e. the element of Ω to which ω is mapped by the permutation $\varphi(g)$. For an element $\omega \in \Omega$, we call the set $\omega^G := \{\omega^g \mid g \in G\}$ the *orbit* of ω. We remark that the set of orbits forms a partition of Ω.

A set of permutations $S \subseteq \mathrm{Sym}(\Omega)$ is a *generating set* for a permutation group G, if G is the smallest subgroup of $\mathrm{Sym}(\Omega)$ containing S and we write $G = \langle S \rangle$ to denote that G is generated by S.

Theorem 4 ([22]). *Let $G \sqsubseteq S_n$ given by a set of t generators. There is an algorithm that computes a generating set for G of size at most n^2 in time $\mathcal{O}(t \cdot n^2)$.*

For a permutation $\pi \in \mathrm{Sym}(\Omega)$ and a set $\Delta \subseteq \Omega$, we denote by Δ^π the set $\{\omega^\pi \mid \omega \in \Delta\}$. For a set of permutations $P \subseteq \mathrm{Sym}(\Omega)$, we let $P_\Delta = \{\pi \in P \mid \Delta^\pi = \Delta\}$.

---

[5] By Theorem 4 one may assume that the input generating set has at most $|\Gamma|^2$ elements.

Δ is called P-*stable* if $P_\Delta = P$. Let b be a positive integer. A permutation group $G \sqsubseteq \mathrm{Sym}(\Omega)$ is called b-*bounded* if Ω can be partitioned into $\Omega = \Omega_1 \cup \cdots \cup \Omega_r$ such that for each $i \in [r]$, $|\Omega_i| \le b$ and Ω_i is G-stable.

An important tool in the algorithms presented in this section is an FPT-algorithm due to Arvind et al. [1] for the following problem: Given two groups $G, H \sqsubseteq \mathrm{Sym}(\Omega)$ such that $\langle G \cup H \rangle$ is b-bounded, and two permutations $x, y \in \mathrm{Sym}(\Omega)$, compute a set of generators and a representative for the coset $Gx \cap Hy$ (if nonempty). This problem is called COLORED COSET INTERSECTION.

Theorem 5 (Theorem 4 in [1]). *There is an algorithm that solves* COLORED COSET INTERSECTION *in time* $\mathcal{O}(17^b n^9 + sn^5)$ *where* $n = |\Omega|$ *and* s *is the size of the generating sets for* G *and* H.

We will also need the following lemma which is well known in the computational group theory literature. A proof of this lemma can be found in [1].

Lemma 6. *Let* $H_i = \langle S_i \rangle$, $1 \le i \le t$, *be subgroups of* $\mathrm{Sym}(\Omega)$ *given by generating sets* S_i, *and* y_i, $1 \le i \le t$, *be permutations in* $\mathrm{Sym}(\Omega)$ *such that* $\bigcup_{i=1}^t H_i y_i$ *is a coset* Hy *of some subgroup* H *of* $\mathrm{Sym}(\Omega)$. *Then* $Hy = \langle S \rangle y_1$, *where*

$$S = \cup_{i=1}^t S_i \cup \{y_i y_1^{-1} \mid 1 \le i \le t\}.$$

Rooted Trees. Let T be a rooted tree. The *depth* of a vertex $v \in V(T)$ is defined as the distance between v and the root of T. The *height* of v is defined as the minimum distance between v and some leaf of T plus one. In particular, each leaf of T has height one. If e is an edge in $E(T)$, then we let $\mathfrak{t}(e)$ be the vertex in e which is closest from the root of T, and $\mathfrak{b}(e)$ be the vertex in e that is closest to a leaf of T. The depth of e is defined as the depth of $\mathfrak{t}(e)$, while the height of e is defined as the height of $\mathfrak{b}(e)$. If u and u' are vertices in $V(T)$, then we say that u is a child of u' if there exists an edge $e \in E(T)$ such that $u = \mathfrak{b}(e)$ and $u' = \mathfrak{t}(e)$. For each vertex $u \in V(T)$ we let T^u denote the subtree of T rooted at u. Analogously, for each edge $e \in E(T)$ we let T^e denote the subtree of T rooted at e. More precisely, T^e is the minimal subtree of T that contains the subtree $T^{\mathfrak{b}(e)}$, the vertex $\mathfrak{t}(e)$, and the edge e.

Proof of Theorem 3. Let G be a rooted Γ-colored graph of BFS color number α and BFS width β, and let (T, X) be the unique BFS decomposition of G. For each vertex $u \in V(T)$ we let $G(u) := G[X_u]$ be the Γ-colored subgraph of G induced by X_u, and $\widehat{G}(u) := G[\bigcup_{u' \in V(T^u)} X_{u'}]$ be the Γ-colored subgraph of G induced by the vertices occurring in bags of the tree T^u. Analogously, for each edge $e \in V(T)$ we let $G(e) := G[X_{\mathfrak{t}(e)} \cup X_{\mathfrak{b}(e)}]$ and $\widehat{G}(e) := G[\bigcup_{u' \in V(T^e)} X_{u'}]$.

Let m be the depth of the tree T. For each $i \in \{0, \dots, m\}$, let Γ_i be the set of colors crossing level i, $L_i = \mathrm{SetStab}(\mathrm{Sym}(\Gamma), \Gamma_i)$ be the set of permutations in $\mathrm{Sym}(\Gamma)$ that map colors in Γ_i to colors in Γ_i, and let $\mathfrak{R}_i = \mathfrak{R} \cap L_i$. Finally, we let $\widehat{\mathfrak{R}} = \mathfrak{R} \cap \bigcap_{j=0}^m L_j$. We note that for any root preserving weak isomorphism (φ, ψ) from T_1 to T_2, ψ must necessarily belong to $\mathfrak{R}$. Furthremore, since for each $i \in [m]$, $|\Gamma_i| \le \alpha$, $\widehat{\mathfrak{R}}$ is α-bounded.

We say that two vertices u and u' in $V(T)$ are *matchable* if they have the same number of children, the same depth and the same height. We say that two edges e and e' in $E(T)$ are matchable if the vertices $\mathfrak{b}(e)$ and $\mathfrak{b}(e')$ are matchable. If u is matchable to u' then we let $Iso(u, u')$ be the set of weak isomorphisms from $G(u)$ to $G(u')$, and $\widehat{Iso}(u, u')$ the set of all level-preserving[6] weak isomorphisms from $\widehat{G}(u)$ to $\widehat{G}(u')$. Analogously, if e and e' are matchable edges in T, then we let $Iso(e, e')$ be the set of weak isomorphisms from $G(e)$ to $G(e')$, and $\widehat{Iso}(e, e')$ the set of weak isomorphisms from $\widehat{G}(e)$ to $\widehat{G}(e')$. Note that for each $u \in V(T)$ and $e \in E(T)$, $\widehat{Iso}(u, u)$ and $\widehat{Iso}(e, e)$ are the sets of weak automorphisms of $\widehat{G}(u)$ and $\widehat{G}(e)$ respectively. In particular, $\widehat{G}(r)$ is the group of root preserving weak automorphisms of G.

For each pair u and u' of matchable vertices of T, and for each weak isomorphism (φ, ψ) of $G(u)$ to $G(u')$, we define the following sets.

$$\Pi(u, u', \varphi, \psi) := \left\{ \Psi \in \hat{\mathfrak{R}} \mid \psi = \Psi|_{V(G(u))} \right\}$$

$$\widehat{\Pi}(u, u', \varphi, \psi) := \left\{ \Psi \in \hat{\mathfrak{R}} \mid \exists (\widehat{\varphi}, \widehat{\psi}) \in \widehat{Iso}(u, u'), \varphi = \widehat{\varphi}|_{V(G(u))}, \right.$$
$$\left. \psi = \Psi|_{V(G(u))}, \widehat{\psi} = \Psi|_{V(\widehat{G}(u))} \right\}$$

Intuitively, the set $\Pi(u, u', \varphi, \psi)$ is the set of all color permutations in $\hat{\mathfrak{R}}$ that are compatible with the coloring ψ and $\widehat{\Pi}(u, u', \varphi, \psi)$ is the set of all colorings in $\hat{\mathfrak{R}}$ that are compatible with the coloring $\widehat{\psi}$ of some weak isomorphism $(\widehat{\varphi}, \widehat{\psi})$ of $\widehat{Iso}(u, u')$ that extends (φ, ψ). Going further, we define the following sets.

$$\Pi(u, u') := \bigcup_{(\varphi, \psi) \in Iso(u, u')} \Pi(u, u', \varphi, \psi)$$

$$\widehat{\Pi}(u, u') := \bigcup_{(\varphi, \psi) \in Iso(u, u')} \widehat{\Pi}(u, u', \varphi, \psi)$$

Intuitively, $\Pi(u, u')$ (resp. $\widehat{\Pi}(u, u')$) is the set of colorings in $\hat{\mathfrak{R}}$ that extend the coloring of some weak isomorphism from $G(u)$ to $G(u')$ (resp. from $\widehat{G}(u)$ to $\widehat{G}(u')$). If e, e' are matchable edges of T and $(\varphi, \psi) \in Iso(e, e')$ then we define the sets $\Pi(e, e', \varphi, \psi)$, $\widehat{\Pi}(e, e', \varphi, \psi)$, $\Pi(e, e')$, $\widehat{\Pi}(e, e')$ analogously.

Observation 7. *Let u and u' be a pair of matchable vertices of T. Then there is a weak isomorphism from $\widehat{G}(u)$ to $\widehat{G}(u')$ if and only if $\widehat{\Pi}(u, u')$ is non-empty.*

The following lemma, which is the main technical result of this section, provides us with an efficient way of deciding whether $\widehat{\Pi}(u, u') \neq \emptyset$ for any given pair of matchable vertices u and u'.

Lemma 8. *Let x and x' be either a pair of matchable vertices or a pair of matchable edges of T. Let (φ, ψ) be a weak isomorphism in $Iso(x, x')$. Then the set $\widehat{\Pi}(x, x', \varphi, \psi)$ is either empty or a coset of some subgroup $\widehat{\Pi}(x, \psi)$ of*

---

[6] Level preserving means that vertices at level i are mapped to vertices at level i.

$\mathfrak{R}$. *Furthermore in the second case, a generator set for the group* $\widehat{\Pi}(x, \psi)$ *and a coset representative* $\Psi(x, x', \varphi, \psi)$ *for* $\widehat{\Pi}(x, x', \varphi, \psi)$ *can be computed in time* $f(d, \alpha, \beta) \cdot n^{\mathcal{O}(1)}$ *for some computable function* f.

Proof. We prove the lemma by induction on the height of the pair x, x'.

Base Case: In the base case, both u and u' are leaves of T. For each $(\varphi, \psi) \in Iso(u, u')$, let X denote the domain of ψ and $\Psi(u, u', \varphi, \psi) \in \widehat{\mathfrak{R}}$ be any permutation such that $\Psi(u, u', \varphi, \psi)|_X = \psi$. We then have that

$$\widehat{\Pi}(u, u', \varphi, \psi) = \Pi(u, u', \varphi, \psi) = \text{Stab}(\widehat{\mathfrak{R}}, X)\Psi(u, u', \varphi, \psi).$$

It follows that we can use $\text{Stab}(\widehat{\mathfrak{R}}, X)$ as $\widehat{\Pi}(x, \psi)$ for which we can compute a generating set of size at most $|\Gamma|^2$ in polynomial time using standard methods.

Inductive Step for Edges: Now let $e \in E(T)$ and $e' \in E(T)$ be matchable edges of height h and let (φ, ψ) be a weak isomorphism in $Iso(e, e')$. Assume that generators for $\widehat{\Pi}(u, \cdot)$ have been computed for every vertex u of height at most h and that $\Psi(u, u', \varphi', \psi')$ has been computed for every pair of matchable vertices u, u' of height at most h, and every $(\varphi', \psi') \in Iso(u, u')$.

Let $u := \mathfrak{b}(e)$ and $u' := \mathfrak{b}(e')$. Furthermore, let X denote the domain of ψ, then we have that $\Pi(e, e', \varphi, \psi) = \text{Stab}(\widehat{\mathfrak{R}}, X)\Psi$ where $\Psi \in \widehat{\mathfrak{R}}$ is such that $\Psi|_X = \psi$. Let $\varphi_u := \varphi|_{V(G(u))}$ and $\psi_u := \psi|_{V(G(u))}$ be the restrictions of φ and ψ, respectively, to $V(G(u))$. Then,

$$\widehat{\Pi}(e, e', \varphi, \psi) = \widehat{\Pi}(u, u', \varphi_u, \psi_u) \cap \Pi(e, e', \varphi, \psi).$$

Since $\widehat{\Pi}(u, u', \varphi_u, \psi_u)$ and $\Pi(e, e', \varphi, \psi)$ are cosets of subgroups of $\widehat{\mathfrak{R}}$ (which is α-bounded), we can apply the algorithm of Theorem 5 to compute $\widehat{\Pi}(e, e', \varphi, \psi)$ in time $17^{\alpha} \cdot |\Gamma|^{\mathcal{O}(1)}$. The output of the algorithm of Theorem 5 (if nonempty) is a coset, given by a set of $|\Gamma|^2$ generators for the corresponding subgroup of $\widehat{\mathfrak{R}}$ and a coset representative which we use as $\widehat{\Pi}(e, \psi)$ and $\Psi(e, e', \varphi, \psi)$, respectively.

Inductive Step for Vertices: Now suppose that u and u' are matchable non-leaf vertices of height $h + 1$ and let (φ, ψ) be a weak isomorphism in $Iso(u, u')$. Suppose the node u has degree d^*, and note that $d^* \leq d \cdot \beta$. (The max. degree of the BFS-decomposition is at most the max. degree of the underlying graph times the width.) Since u and u' are matchable, we know that u' has degree d^* as well. Let $u_1, \ldots, u_{d^*}$ and $u'_1, \ldots, u'_{d^*}$ be the children of u and u' respectively, and $e_1 = \{u, u_1\}, \ldots, e_{d^*} = \{u, u_{d^*}\}$ and $e'_1 = \{u', u'_1\}, \ldots, e'_{d^*} = \{u', u'_{d^*}\}$ be the corresponding edges. Assume that for every pair of matchable edges e, e' of height at most h and every weak isomorphism $(\varphi', \psi') \in Iso(e, e')$, the coset $\widehat{\Pi}(e, e', \varphi', \psi')$ has been computed. We show how to compute $\widehat{\Pi}(u, u', \varphi, \psi)$. (Keeping in mind that $(\varphi, \psi) \in Iso(u, u')$) for a pair of edges e_i, e'_j as above, we define

$$\widehat{\Pi}_{(\varphi, \psi)}(e_i, e'_j) := \left\{ \Psi \in \widehat{\Pi}(e_i, e'_j, \varphi', \psi') \;\middle|\; (\varphi', \psi')|_{V(G(u))} = (\varphi, \psi) \right\}.$$

Hence, $\widehat{\Pi}_{(\varphi,\psi)}(e_i, e'_j)$ is a subset of $\widehat{\Pi}(e_i, e'_j)$ where we restrict the indexing weak isomorphisms in $Iso(e_i, e'_j)$ to act on $V(G(u))$ in the same way as (φ, ψ). Suppose $\widehat{\psi} \in \widehat{\Pi}(u, u', \varphi, \psi)$ and let $\widehat{\varphi} \colon V(\widehat{G}(u)) \to V(\widehat{G}(u'))$ be a bijection such that

(i) $\widehat{\varphi}|_{V(G(u))} = \varphi$ (while we already have that $\widehat{\psi}|_{V(G(u))} = \psi$),

(ii) $(\widehat{\varphi}, \widehat{\psi})$ is a weak isomorphism from $\widehat{G}(u)$ to $\widehat{G}(u')$.

Then $\widehat{\varphi}$ induces a permutation $\tau \in S_{d^*}$ from the children of u to the children of u' (depending on where the corresponding vertices are mapped), and we have that $\widehat{\psi}$ belongs to the intersection $\bigcap_{j \in [d^*]} \widehat{\Pi}_{(\varphi,\psi)}(e_j, e'_{\tau(j)})$, implying that

$$\widehat{\Pi}(u, u', \varphi, \psi) \subseteq \bigcup_{\tau \in S_{d^*}} \bigcap_{j \in [d^*]} \widehat{\Pi}_{(\varphi,\psi)}(e_j, e'_{\tau(j)}).$$

For the converse direction, suppose that $\widehat{\psi} \in \bigcup_{\tau \in S_{d^*}} \bigcap_{j \in [d^*]} \widehat{\Pi}_{(\varphi,\psi)}(e_j, e'_{\tau(j)})$. This means that there is a permutation $\tau \in S_{d^*}$ from the children of u to the children of u' such that for each $j \in [d^*]$, $\widehat{\psi} \in \widehat{\Pi}_{(\varphi,\psi)}(e_j, e'_{\tau(j)})$.

We can conclude that there is a bijection $\widehat{\varphi} \colon V(\widehat{G}(u)) \to V(\widehat{G}(u'))$ such that $(\widehat{\varphi}, \widehat{\psi})$ satisfies (i) and (iii) above. In particular the definitions guarantee the existence of weak isomorphisms $(\widehat{\varphi}_j, \widehat{\psi}_j)$ from $\widehat{G}(e_j)$ to $\widehat{G}(e_{\tau(j)})$ such that $(\widehat{\varphi}_j, \widehat{\psi}_j)$ acts in the same way as (φ, ψ) on $V(G(u))$ for all $j \in [d^*]$. Hence,

$$\widehat{\Pi}(u, u', \varphi, \psi) = \bigcup_{\tau \in S_{d^*}} \bigcap_{j \in [d^*]} \widehat{\Pi}_{(\varphi,\psi)}(e_j, e'_{\tau(j)}).$$

For each pair of edges e_i, e'_j as above, we can compute the coset $\widehat{\Pi}_{(\varphi,\psi)}(e_i, e'_j)$ using Lemma 6 a number of times that only depends on α and β. Again since all $\widehat{\Pi}_{(\varphi,\psi)}(e_i, e'_j)$ are cosets of subgroups of $\mathfrak{R}$, $\widehat{\Pi}(u, u', \varphi, \psi)$ can be computed using $2^{\mathcal{O}(d^* \log d^*)}$ calls to the algorithm of Theorem 5 taking in total FPT in $d + \alpha + \beta$ time. We use Lemma 6 at most $d^*!$ times to compute the union over all $\tau \in S_{d^*}$. So, the algorithm takes time $f(d, \alpha, \beta) \cdot n^{\mathcal{O}(1)}$ for some computable f. $\square$

Isomorphism Construction: Now, to finalize the proof of Theorem 3 we need to show how Lemma 8 can be used to compute a weak isomorphism between rooted Γ-colored graphs H and H'. Let r be the root of H and r' the root of H'. Let G be the graph obtained by taking the disjoint union of H with H', creating a new root r'', and adding new edges $\{r'', r\}$ and $\{r'', r'\}$. Then in the tree T of the BFS-decomposition (T, X) of G, the root vertex v has two children u and u' with $X_u = \{r\}$ and $X_{u'} = \{r'\}$. Furthermore, $\widehat{G}(u) = H$ and $\widehat{G}(u') = H'$. Therefore, we have a weak isomorphism from H to H' if and only if $\widehat{\Pi}(u, u', \varphi, \psi)$ is non-empty for some weak isomorphism $(\varphi, \psi) \in Iso(u, u')$ which can be decided in time $f(d, \alpha, \beta) \cdot n^{\mathcal{O}(1)}$ by Lemma 8.

Now suppose that $\widehat{\Pi}(u, u', \varphi, \psi)$ is not empty, and let $\Psi \in \widehat{\Pi}(u, u', \varphi, \psi)$. Then there exists a mapping $\Phi \colon V(H) \to V(H')$ such that (Φ, Ψ) is a weak isomorphism from H to H'. We explain how to determine such a mapping Φ. Note that it is sufficient to determine Φ for each bag in T^u. For the vertices in

X_u, we can use $\Phi|_{X_u} = \varphi$. Let $u_1, \ldots u_{d^*}$ be the children of u and $u'_1, \ldots, u'_{d^*}$ the children of u'. Then (as used in our algorithm) there is a permutation $\tau \in S_{d^*}$ such that for each $i \in [d^*]$, $\Psi \in \widehat{\Pi}(u_i, u'_{\tau(i)}, \varphi_i, \psi_i)$ for some weak isomorphism (φ_i, ψ_i) from $G(u_i)$ to $G(u'_{\tau(i)})$ (with $\Psi|_{X_{u_i}} = \psi_i$). To determine τ and all φ_i's, we look up a number of table entries that only depends on d, α and β. For each $i \in [d^*]$, we let $\Phi|_{X_{u_i}} = \varphi_i$. We continue this process inductively until we reached the leaves of T and hence have determined Φ on all vertices of H. Clearly, this reconstruction process also takes time $f(d, \alpha, \beta) \cdot n^{O(1)}$.

4   Bounded Color Class Hypergraph Isomorphism

In this section we show that weak isomorphism for rooted graphs of simultaneously bounded degree, BFS number and BFS width is intimately connected with the notion of strong isomorphism for bounded color class hypergraphs.

We now show that the algorithm of Theorem 3 can be used to solve strong isomorphism of colored hypergraphs. A colored hypergraph is a hypergraph with a partition of its vertex set the blocks of which we call the *color classes*. In the corresponding BOUNDED COLOR CLASS HYPERGRAPH ISOMORPHISM problem we are given two colored hypergraphs and the question is whether there is a color-class preserving isomorphism from one to the other. The parameter in this problem is the maximum size of any color class in the input hypergraphs.

Proposition 9 (★). BOUNDED COLOR CLASS HYPERGRAPH ISOMORPHISM *can be reduced in time $m^{O(1)}$ to* WEAK ISOMORPHISM *of colored trees of bounded degree and bounded BFS color number, where m denotes the size of the input hypergraphs.*

We now address the question of determining whether two hypergraphs *implicitly* represented by leveled finite automata (LFA, formally defined below) are isomorphic. We note that a set of strings can be represented by an LFA if and only if it can be represented by an ordered decision diagram of similar size.

For each set X we let $\mathcal{P}(X)$ be the set of all subsets of X. Let H be a hypergraph with color classes $C_1, \ldots, C_n$, each of size at most b. A hyperedge in H may be represented as a sequence $h = X_1 X_2 \ldots X_n$ where for each $i \in [n]$, $X_i \subseteq C_i$. Fixing a numbering on the vertices in each C_i, we can then represent a hyperedge as a string of length n over the alphabet $\Sigma_b := \mathcal{P}([b])$. This way, a hypergraph with n color classes and color class size b can be represented by an LFA $\mathcal{A}$ over Σ_b, such that the set of hyperedges of H corresponds to the language $\mathcal{L}(\mathcal{A})$.

Let Σ be a finite set of symbols. A *leveled finite automaton* (LFA) over Σ is a tuple $\mathcal{A} = (Q, \Sigma, \mathcal{R}, I, F)$ where Q is a set of states with partition $Q_0, Q_1, \ldots, Q_n$, $I \subseteq Q$ is a set of initial states, $F \subseteq Q$ is a set of final states, and $\mathcal{R} \subseteq \bigcup_{i \in [n]} Q_{i-1} \times \Sigma \times Q_i$ is a transition relation. We say that $\mathcal{A}$ accepts a string $w_1 w_2 \ldots w_n \in \Sigma^*$ if there exists a sequence $(q_0, w_1, q_1)(q_1, w_2, q_2) \ldots (q_{n-1}, w_n, q_n)$ of transitions in $\mathcal{R}$ such that $q_0 \in I$ and

$q_n \in F$. We denote by $\mathcal{L}(\mathcal{A})$ the set of all strings accepted by $\mathcal{A}$. We say that n is the length of $\mathcal{A}$. The width of $\mathcal{A}$ is defined as $w(\mathcal{A}) = \max_{n \leq i \leq n} |Q_i|$. LFA's can only represent finite languages, but the number of strings in these languages may be exponential in n.

Theorem 10 (★). *Let H and H' be hypergraphs with n color classes and color class size b. Let $\mathcal{A}$ and $\mathcal{A}'$ be LFA's of width w and length n over Σ_b representing H and H' respectively. Then one can determine in time $f(w,b) \cdot n^{O(1)}$ whether H is isomorphic to H'.*

Acknowledgements. We would like to thank Daniel Lokshtanov for helpful discussions and Laszlo Babai for clarifying many aspects of his algorithm during a workshop on Symmetry in Finite and Infinite Structures.

References

1. Arvind, V., Das, B., Köbler, J., Toda, S.: Colored hypergraph isomorphism is fixed parameter tractable. Algorithmica **71**, 120–138 (2015)
2. Babai, L.: Graph isomorphism in quasipolynomial time [extended abstract]. In: STOC, pp. 684–697. ACM (2016)
3. Babai, L., Grigoryev, D.Y., Mount, D.M.: Isomorphism of graphs with bounded eigenvalue multiplicity. In: STOC, pp. 310–324. ACM (1982)
4. Babai, L., Kantor, W.M., Luks, E.M.: Computational complexity and the classification of finite simple groups. In: FOCS, pp. 162–171. IEEE (1983)
5. Bollig, B.: On symbolic obdd-based algorithms for the minimum spanning tree problem. Theor. Comput. Sci. **447**, 2–12 (2012)
6. Bollig, B., Bury, M.: On the OBDD representation of some graph classes. Discret. Appl. Math. **214**, 34–53 (2016)
7. Bollig, B., Pröger, T.: On efficient implicit obdd-based algorithms for maximal matchings. Inf. Comput. **239**, 29–43 (2014)
8. Booth, K.S., Colbourn, C.J.: Problems polynomially equivalent to graph isomorphism. Computer Science Department, Univ. Waterloo (1979)
9. Chepoi, V., Dragan, F.: A note on distance approximating trees in graphs. Eur. J. Combin. **21**(6), 761–766 (2000)
10. Downey, R.G., Fellows, M.R.: Fundamentals of Paramterized Complexity. Springer, London (2013). https://doi.org/10.1007/978-1-4471-5559-1
11. Furst, M., Hopcroft, J., Luks, E.: Polynomial-time algorithms for permutation groups. In: FOCS, pp. 36–41. IEEE (1980)
12. Grohe, M.: Fixed-point definability and polynomial time on graphs with excluded minors. J. ACM **59**(5), 27 (2012)
13. Grohe, M., Marx, D.: Structure theorem and isomorphism test for graphs with excluded topological subgraphs. SIAM J. Comp. **44**(1), 114–159 (2015)
14. Grohe, M., Schweitzer, P.: Isomorphism testing for graphs of bounded rank width. In: FOCS, pp. 1010–1029. IEEE (2015)
15. Kratsch, S., Schweitzer, P.: Isomorphism for graphs of bounded feedback vertex set number. In: Kaplan, H. (ed.) SWAT 2010. LNCS, vol. 6139, pp. 81–92. Springer, Heidelberg (2010). https://doi.org/10.1007/978-3-642-13731-0_9

16. Lokshtanov, D., Pilipczuk, M., Pilipczuk, M., Saurabh, S.: Fixed-parameter tractable canonization and isomorphism test for graphs of bounded treewidth. In: FOCS, pp. 186–195. IEEE (2014)
17. Luks, E.M.: Isomorphism of graphs of bounded valence can be tested in polynomial time. J. Comput. Syst. Sci. **25**(1), 42–65 (1982)
18. Miller, G.: Isomorphism testing for graphs of bounded genus. In: STOC, pp. 225–235. ACM (1980)
19. Nunkesser, R., Woelfel, P.: Representation of graphs by OBDDs. In: Deng, X., Du, D.-Z. (eds.) ISAAC 2005. LNCS, vol. 3827, pp. 1132–1142. Springer, Heidelberg (2005). https://doi.org/10.1007/11602613_112
20. Sawitzki, D.: A symbolic approach to the all-pairs shortest-paths problem. In: Hromkovič, J., Nagl, M., Westfechtel, B. (eds.) WG 2004. LNCS, vol. 3353, pp. 154–167. Springer, Heidelberg (2004). https://doi.org/10.1007/978-3-540-30559-0_13
21. Seress, A.: Permutation Group Algorithms. Cambridge University Press, Cambridge (2003)
22. Sims, C.C.: Computational methods in the study of permutation groups. In: Computational Problems in Abstract Algebra, pp. 169–183 (1970)
23. Woelfel, P.: Symbolic topological sorting with OBDDs. J. Discret. Algorithms **4**(1), 51–71 (2006)
24. Yamazaki, K., Bodlaender, H.L., de Fluiter, B., Thilikos, D.M.: Isomorphism for graphs of bounded distance width. Algorithmica **24**(2), 105–127 (1999)

Connected Vertex Cover
for $(sP_1 + P_5)$-Free Graphs

Matthew Johnson, Giacomo Paesani$^{(\boxtimes)}$, and Daniël Paulusma

Department of Computer Science, Durham University, Durham, UK
{matthew.johnson2,giacomo.paesani,daniel.paulusma}@durham.ac.uk

Abstract. The CONNECTED VERTEX COVER problem is to decide if a graph G has a vertex cover of size at most k that induces a connected subgraph of G. This is a well-studied problem, known to be NP-complete for restricted graph classes, and, in particular, for H-free graphs if H is not a linear forest. On the other hand, the problem is known to be polynomial-time solvable for sP_2-free graphs for any integer $s \geq 1$. We prove that it is also polynomial-time solvable for $(sP_1 + P_5)$-free graphs for every integer $s \geq 0$.

1 Introduction

A set S of vertices in a graph G forms a *vertex cover* of G if every edge of G is incident with a vertex of S. The set S is an *independent set* if no two vertices in S are adjacent. These definitions lead to two classical graph problems, which are both NP-complete: the VERTEX COVER problem is to decide if a given graph G has a vertex cover of size at most k for a given integer k; the INDEPENDENT SET problem is to decide if a given graph G has an independent set of size at least ℓ for a given integer ℓ. A set S of at least k vertices of a graph G on n vertices is a vertex cover if and only if $V_G \setminus S$ is an independent set (of size at most $n - k$). Hence VERTEX COVER and INDEPENDENT SET are polynomially equivalent. A vertex cover of a graph G is connected if it induces a connected subgraph of G. In our paper, we focus on the corresponding decision problem.

CONNECTED VERTEX COVER
Instance: a graph G and an integer k.
Question: does G have a connected vertex cover S with $|S| \leq k$?

In 1977, Garey and Johnson [9] proved that CONNECTED VERTEX COVER is NP-complete for planar graphs of maximum degree 4. More recently, Priyadarsini and Hemalatha [18] and Fernau and Manlove [8] strengthened this result to 2-connected planar graphs of maximum degree 4 and planar bipartite graphs of maximum degree 4, respectively. Wanatabe et al. [22] proved that CONNECTED VERTEX COVER is NP-complete even for 3-connected graphs. Very recently,

This work was supported by The Leverhulme Trust (Grant RPG-2016-258).

A. Brandstädt et al. (Eds.): WG 2018, LNCS 11159, pp. 279–291, 2018.
https://doi.org/10.1007/978-3-030-00256-5_23

Munaro [16] proved the same for line graphs of planar cubic bipartite graphs and for planar bipartite graphs of arbitrarily large girth, and Li et al. [13] showed NP-completeness for 4-regular graphs.

We now turn to tractable cases. Ueno et al. [21] proved that CONNECTED VERTEX COVER is polynomial-time solvable for graphs of maximum degree at most 3. Escoffier et al. [7] proved the same result for chordal graphs. As VERTEX COVER is also polynomial-time solvable for chordal graphs [10], the authors of [7] proposed a general study on the complexity of CONNECTED VERTEX COVER on graph classes for which VERTEX COVER is polynomial-time solvable. This leads us to the research question of our paper:

For which classes of graphs do the complexities of VERTEX COVER *and* CONNECTED VERTEX COVER *coincide?*

This question was addressed by Chiarelli et al. [6] who considered classes of graphs characterized by a single forbidden induced subgraph H. Such graphs are called H-free. They observed that the results of Munaro [16] imply that CONNECTED VERTEX COVER is NP-complete for H-free graphs if H contains a cycle or a claw. Using Poljak's construction [17], VERTEX COVER is readily seen to be NP-complete for graphs of arbitrarily large girth and thus for H-free graphs whenever H contains a cycle. When H is the claw, VERTEX COVER becomes polynomial-time solvable for H-free graphs [15,20]. Hence, there exist graphs H such that CONNECTED VERTEX COVER and VERTEX COVER have different complexities when restricted to H-free graphs (assuming P $\neq$ NP).

So the complexity of CONNECTED VERTEX COVER is known for H-free graphs unless H is a linear forest (the disjoint union of one or more paths). Even the case where H is a single path on r vertices (denoted P_r) is settled neither for VERTEX COVER nor for CONNECTED VERTEX COVER; it is not known if there exists an integer r such that VERTEX COVER or CONNECTED VERTEX COVER is NP-complete for P_r-free graphs. Lokshtanov et al. [14] proved that INDEPENDENT SET, and thus VERTEX COVER, is polynomial-time solvable for P_5-free graphs. Recently, Grzesik et al. [11] extended this to P_6-free graphs. We also note that if VERTEX COVER is polynomial-time solvable on H-free graphs for some graph H, then it is polynomial-time solvable on $(P_1 + H)$-free graphs. This follows from the folklore observation that to solve the complementary problem of INDEPENDENT SET on a $(P_1 + H)$-free graph one solves the problem on each H-free graph obtained by removing a vertex and all its neighbours.

Theorem 1 ([11]). *For every $s \geq 0$,* VERTEX COVER *can be solved in polynomial time for $(sP_1 + P_6)$-free graphs.*

By using the concept of the price of connectivity [3,5,12], Chiarelli et al. [6] proved that CONNECTED VERTEX COVER is polynomial-time solvable for sP_2-free graphs for any integer $s \geq 1$. For VERTEX COVER this follows by combining two classical results [2,19] (as is well-known). No other complexity results are known for CONNECTED VERTEX COVER for H-free graphs if H is a linear forest.

Our Contribution. We continue the study of [6,7] and prove the following result, which includes polynomial-time solvability for P_5-free graphs.

Theorem 2. *For every $s \geq 0$,* CONNECTED VERTEX COVER *can be solved in polynomial time for $(sP_1 + P_5)$-free graphs.*

Our Method. It is easy to construct graphs with a minimum connected vertex cover that do not contain a minimum vertex cover; see the graph G_1 in Fig. 1. We also note that the difference between a minimum vertex cover and a minimum connected vertex cover in an $(sP_1 + P_5)$- free graph is at most 3 if $s = 0$ and at most $3s + 10$ if $s \geq 1$ [12]. We cannot exploit this property directly as that would require an algorithm to enumerate all minimum vertex covers in polynomial time. Moreover, the graph G_2 in Fig. 1 shows that even if this were possible, it is not immediately obvious how to proceed; one cannot necessarily hope to find a minimum connected vertex cover by extending a minimum vertex cover. As an extra complication, for CONNECTED VERTEX COVER one cannot extend results on H-free graphs to results on $(sP_1 + H)$-free graphs in a straightforward way (certainly one cannot use the technique for VERTEX COVER described before Theorem 1).

Our method is based on an analysis of the structure of dominating sets in $(sP_1 + P_5)$-free graphs using a characterization of P_5-free graphs due to Bacsó and Tuza [1]. We translate the problem into a problem in which we try to extend a partial vertex cover into a full connected vertex cover. We solve this extension variant of CONNECTED VERTEX COVER by using Theorem 1 (applied to the smaller class of $(sP_1 + P_5)$-free graphs). We show how to do this in Sect. 3 and then show how to use this result to prove Theorem 2 in Sect. 4. An important ingredient of our proof is to reduce the size of the input graph by contracting an edge between two vertices u and v whenever we detect that u and v will belong to the connected vertex cover. This idea stems from the observation that a connected graph G on n vertices has a connected vertex cover of size k if and only if G contains the star $K_{1,n-k}$ on $n - k + 1$ vertices as a contraction.

2 Preliminaries

Let $G = (V, E)$ be a graph. For a set $S \subseteq V$, the graph $G[S]$ denotes the subgraph of G induced by S, and we say that S is *connected* if $G[S]$ is connected. We write $G - S = G[V \setminus S]$, and if $S = \{u\}$ we may simply write $G - u$. For a vertex $u \in V$, we write $N_G(u) = \{v \mid uv \in E\}$ to denote the neighbourhood of u. For a set $S \subseteq V$, we write $N_G(S) = (\bigcup_{u \in S} N_G(u)) \setminus S$. A subset $D \subseteq V$ is a *dominating* set of G if every vertex of $V \setminus D$ is adjacent to at least one vertex of D. An edge uv of a graph $G = (V, E)$ is *dominating* if $\{u, v\}$ is dominating. The *contraction* of an edge $uv \in E$ is the operation that replaces u and v by a new vertex adjacent to precisely those vertices of $V \setminus \{u, v\}$ adjacent to u or v in G. Recall that for a graph H, we say that another graph G is *H-free* if it does not contain an induced subgraph isomorphic to H. The *disjoint union* $G + H$ of two

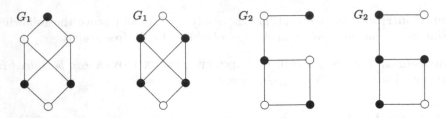

Fig. 1. An example of a P_5-free graph G_1 with a minimum connected vertex cover (coloured black in the right-hand drawing) that contains no minimum vertex cover (there are exactly two, indicated by the sets of black and white vertices in the left-hand drawing). The graph G_2 is an example of a $(P_1 + P_5)$-free graph with a minimum vertex cover (coloured black in the left hand drawing) that is not contained in any minimum connected vertex cover; clearly any connected vertex cover that contains it has at least five vertices and an example of a minimum connected vertex cover on four vertices is indicated by the vertices coloured black in the right-hand drawing.

vertex-disjoint graphs G and H is the graph $(V_G \cup V_H, E_G \cup E_H)$. The disjoint union of r copies of a graph G is denoted by rG. A *linear forest* is the disjoint union of one or more paths. The following, straightforward lemma holds for any linear forest.

Lemma 1. *Let G be a connected $(sP_1 + P_5)$-free graph for some $s \geq 0$. The graph obtained from G after contracting an edge is also connected and $(sP_1 + P_5)$-free.*

We will use the following result of Bacsó and Tuza [1] as a lemma.

Lemma 2. ([1]). *Every connected P_5-free graph G has a dominating set D, computable in $O(n^3)$ time, that induces either a P_3 or a complete graph.*

Note that it is not difficult to compute the set D in polynomial time; this also follows from a more general result of Camby and Schaudt [4] for P_r-free graphs $(r \geq 1)$.

Proofs of some lemmas are omitted due to space restrictions.

3   An Auxiliary Problem

In this section we prove that a variant of CONNECTED VERTEX COVER can be solved in polynomial time for $(sP_1 + P_5)$-free graphs for every integer $s \geq 0$.

To prove Theorem 2 we will solve a polynomial number of instances of this variant, which we show can be solved in polynomial time for $(sP_1 + P_5)$-free graphs for every $s \geq 0$. We introduce the variant by first describing its input. Let G be a connected graph, let $J \subseteq V_G$ be a subset of the vertex set of G and let y be a vertex of J. We call the triple (G, J, y) *cover-complete* if it has the following properties (see also Fig. 2):

(A) J is an independent set;
(B) y is adjacent to every vertex of $G - J$;
(C) the neighbours of each vertex in $J \setminus \{y\}$ form an independent set in $G - J$.

We now describe the problem.

CONNECTED VERTEX COVER COMPLETION
   Instance: a cover-complete triple (G, J, y).
      Goal: find a smallest connected vertex cover S of G such that $J \subseteq S$.

We will show how to solve this problem in polynomial time for $(sP_1 + P_5)$-free graphs for any $s \geq 0$.

Let (G, J, y) be a cover-complete triple, where G is a connected $(sP_1 + P_5)$-free graph. For a vertex $w \in N_G(J \setminus \{y\})$, we write $J_w = N_G(w) \cap J$. Note that, by (B), $y \in J_w$. Let G' be the graph obtained from G by contracting every edge of $G[J_w \cup \{w\}]$. As $G[J_w \cup \{w\}]$ is connected, contracting its edges reduces it to a single vertex which we denote y_w. We say that we have *set-contracted* G into G' via w and that we *contracted* $J_w \cup \{w\}$ into y_w.

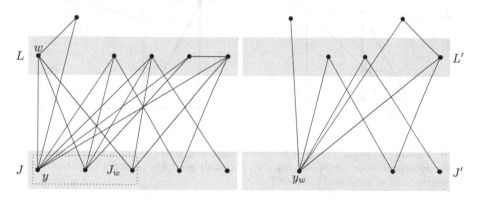

Fig. 2. An example of a cover-complete triple (G, J, y) and the cover-complete triple (G', J', y_w) obtained from set-contracting G via vertex w. The sets $J' = (J \setminus J_w) \cup \{y_w\}$, $L = N_G(J \setminus \{y\})$ and $L' = N_{G'}(J' \setminus \{y_w\})$ are also displayed (the latter two sets will be formally introduced later).

Lemma 3. *Let (G, J, y) be a cover-complete triple, where G is a connected $(sP_1 + P_5)$-free graph for some $s \geq 0$. Let $w \in N_G(J \setminus \{y\})$, and let G' be the graph obtained from G after set-contracting via w. Let $J' = (J \setminus J_w)) \cup \{y_w\}$ and $y' = y_w$. Then the following hold:*

1. *G' is a connected $(sP_1 + P_5)$-free graph;*
2. *(G', J', y') is a cover-complete triple;*
3. *A set $S \subseteq V_G$ is a (smallest) connected vertex cover of G that contains $J \cup \{w\}$ if and only if $(S \setminus (J \cup \{w\}) \cup J'$ is a (smallest) connected vertex cover of G' that contains J'.*

Let (G, J, y) be a cover-complete triple. We define $L_J = N_G(J \setminus \{y\})$. If there is no ambiguity, we will just write $L = L_J$. Note that, by (C), L is the union of a number of independent sets, but L itself might not be independent. However we can deduce the following lemma, which follows immediately from property (C).

Lemma 4. *Let (G, J, y) be a cover-complete triple. If w_1 and w_2 are two adjacent vertices in L, then no vertex of $J \setminus \{y\}$ is adjacent to both w_1 and w_2.*

We introduce two key definitions. Two vertices $w_1, w_2 \in L$ form a *pseudo-dominating pair* if w_1 and w_2 are non-adjacent; w_1 has a neighbour $x_1 \in J$ not adjacent to w_2; and w_2 has a neighbour $x_2 \in J$ not adjacent to w_1. Three vertices $w_1, w_2, w_3 \in L$ form a *pseudo-dominating triple* if w_1 is adjacent to neither w_2 nor w_3; w_2 and w_3 are adjacent; J contains two distinct vertices x_1 and x_2 such that $x_1 \in N_G(w_1) \setminus N_G(\{w_2, w_3\})$ and $x_2 \in (N_G(w_1) \cap N_G(w_2)) \setminus N_G(w_3)$. See the illustrations in Fig. 3, from which we also observe that no pseudo-dominating pair or pseudo-dominating triple can be found in a P_5-free graph.

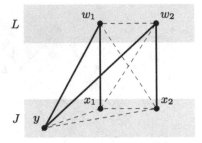 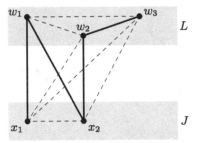

Fig. 3. Examples, on the left, of a pseudo-dominating pair (w_1, w_2), and, on the right, of a pseudo-dominating triple (w_1, w_2, w_3). As easily seen, the presence of either implies the existence of at least one induced P_5.

Let S be a connected vertex cover of G that contains J. Recall that J is an independent set. A subset $L^* \subseteq L \cap S$ is a *connector* of S if $J \cup L^*$ is connected.

Lemma 5. *Let (G, J, y) be a cover-complete triple, where G is an $(sP_1 + P_5)$-free graph for some $s \geq 0$. Let S be a connected vertex cover of G that contains J. If S contains both vertices of a pseudo-dominating pair w_1, w_2, then S has a connector of size at most $s + 1$ that contains both w_1 and w_2.*

Lemma 6. *Let (G, J, y) be a cover-complete triple, where G is an $(sP_1 + P_5)$-free graph for some $s \geq 0$. Let S be a connected vertex cover of G that contains J. If S contains all three vertices of a pseudo-dominating triple w_1, w_2, w_3, then S has a connector of size at most $s + 2$ that contains $\{w_1, w_2, w_3\}$.*

Let (G, J, y) be a cover-complete triple. Let S be a connected vertex cover of G that contains J. If S contains both vertices of some pseudo-dominating pair of

G or all three vertices of some pseudo-dominating triple of G, then S is of *type 1*. Otherwise S must contain at most one vertex of any pseudo-dominating pair and at most two vertices of any pseudo-dominating triple of G. In that case we say that S is of *type 2*. We observe that G might have connected vertex covers of only one type.

We will now see, in Lemma 8, how to find a smallest type 1 connected vertex cover of a graph G of a cover-complete triple (G, J, y) in polynomial time (if it exists). After that we shall prove how to find a smallest type 2 connected vertex cover of G in polynomial time (if it exists). To compute these sets we need the following lemma, which uses Theorem 1 in its proof.

Lemma 7. *Let $(G, \{y\}, y)$ be a cover-complete triple, where G is an $(sP_1 + P_5)$-free graph for some $s \geq 0$. Then it is possible to compute a smallest connected vertex cover of G that contains y in polynomial time.*

Using Lemmas 5–7, we can now prove the following lemma.

Lemma 8. *Let (G, J, y) be a cover-complete triple. Then it is possible to find in polynomial time a smallest type 1 connected vertex cover of G.*

Let (G, J, y) be a cover-complete triple. Using Lemma 8 we can find a smallest type 1 connected vertex cover of G. However, it might be possible that G has a smaller connected vertex cover of type 2. To investigate this, we introduce two reduction rules that will transform a cover-complete triple (G, J, y) into a triple (G', J', y') with $|J'| < |J|$. We say that such a rule is *safe* if the following holds:

1. If G is $(sP_1 + P_5)$-free and connected, then G' is $(sP_1 + P_5)$-free and connected.
2. (G', J', y') is cover-complete.
3. Given a smallest connected vertex cover S' of G' that contains J', it is possible, in polynomial time, to find a smallest connected vertex cover S of G that contains J.

Rule 1. Set-contract via x whenever x is a vertex in $L \cap N_G(w_1) \cap N_G(w_2)$ for some pseudo-dominating pair (w_1, w_2).

Rule 2. For any vertex $w_5 \in L$ that is not adjacent to any vertex of a clique of four vertices w_1, w_2, w_3, w_4 in L, delete w_5 and set-contract via u for every $u \in L \cap N_G(w_5)$.

Lemma 9. *Rules 1 and 2 are safe.*

We call a cover-complete triple (G, J, y) *free* if G has no pseudo-dominating pair with a common neighbour in L, and moreover, $G[L]$ is $(P_1 + K_4)$-free. By exhaustively applying Rules 1 and 2 in arbitrary order, which we may safely do due to Lemma 9, we have the following lemma.

Lemma 10. *A cover-complete triple (G, J, y) can be modified, in polynomial time, into a free cover-complete triple (G', J', y) with the following properties:*

1. *If G is (sP_1+P_5)-free and connected, then G' is (sP_1+P_5)-free and connected.*
2. *Given a smallest connected vertex cover S' of G' that contains J', it is possible to find in polynomial time a smallest connected vertex cover S of G that contains J.*

Let (G, J, y) be a free cover-complete triple. A connector of a connected vertex cover S of G is *minimal* if it does not properly contain a smaller connector of S.

Lemma 11. *Let (G, J, y) be a free cover-complete triple that has a pseudo-dominating pair (w_1, w_2). Then every minimal connector L^* of every type 2 connected vertex cover S of G has size at most 5.*

Lemma 12. *Let (G, J, y) be a free cover-complete triple that has no pseudo-dominating pair. It is possible to find in polynomial time a clique $K \subseteq L$ with $N_G(K) \cap J = J$.*

We are now ready to prove the following theorem.

Theorem 3. *For every $s \geq 0$, CONNECTED VERTEX COVER COMPLETION can be solved in polynomial time for $(sP_1 + P_5)$-free graphs.*

Proof. Let $s \geq 0$ and let (G, J, y) be a cover-complete triple, where G is an $(sP_1 + P_5)$-free graph. We first apply Lemma 10 to obtain a free cover-complete triple (G', J', y') in polynomial time. By the same lemma, G' is $(sP_1 + P_5)$-free. Our aim is to find a smallest connected vertex cover of G' that contains J' in polynomial time, so that we can apply statement 2 of Lemma 10. We first compute in polynomial time a smallest type 1 connected vertex cover S^* of G' using Lemma 8. We now need to compute a smallest type 2 connected vertex cover S' of G' and compare $|S'|$ with $|S^*|$.

First suppose that G' contains a pseudo-dominating pair. We guess a minimal connector of size at most 5 and apply Lemma 3 on its vertices. (By guess, we mean choose a set of up to 5 vertices and test to see if they form a minimal connector. We eventually look at all such sets.) If we obtain an instance of the form $(G'', \{y''\}, y'')$, then we apply Lemma 7. Then we uncontract all contracted edges to get a connected vertex cover of G' of type 2. By Lemma 11, doing this for every guessed minimal connector of size at most 5 gives us a smallest type 2 connected vertex cover S' of G'. As we process each guess in polynomial time and there are at most $O(n^5)$ guesses, we find S' in polynomial time. We compare S' and S^* and choose the smaller of the two.

Now suppose that G' has no pseudo-dominating pair. Let $L' = N_{G'}(J' \backslash \{y'\})$. By Lemma 12, we can obtain in polynomial time a clique $K \subseteq L'$ with $N_{G'}(K) \cap J' = J'$. Let $K = \{w_1, \ldots, w_r\}$ for some $r \geq 1$. As K is a clique, every vertex cover contains at least $r - 1$ vertices of K. We will do as follows: first we will find in polynomial time a smallest connected vertex cover of G' that contains $J' \cup K$, and then we will find in polynomial time, for $i = 1, \ldots, r$, a smallest connected vertex cover of G' that contains $J' \cup (K \setminus \{w_i\})$ and that does not contain w_i. As there are $O(n)$ cases, the total time is polynomial.

We start by computing a smallest connected vertex cover of G' that contains $J' \cup K$ by set-contracting via each vertex of K. By Lemma 3, this yields a cover-complete triple $(G'', \{y''\}, y'')$ to which we apply Lemma 7. Then we uncontract all contracted edges in polynomial time. By Lemma 3, this yields a smallest connected vertex cover S_K of G' that contains $J' \cup K$.

We now show how to compute, in polynomial time, a smallest connected vertex cover of G' that contains $J' \cup (K \setminus \{w_1\})$ and that does not contain w_1. The case $i \geq 2$ is done in the same way.

Let $A = L' \setminus N_{G'}(w_1)$ consist of all non-neighbours of w_1 in L'. As $G'[L']$ is $(K_4 + P_1)$-free by definition, we find that $G'[A]$ is K_4-free. As w_1 is not in the connected vertex cover we are looking for we remove w_1, and we set-contract via each neighbour of w_1 in L. By Lemma 3, we may now consider the resulting cover-complete triple (G'', J'', y'') where G'' is connected and $(sP_1 + P_5)$-free. As G' had no pseudo-dominating pairs, we have that G'' has no pseudo-dominating pairs. We write $L'' = N_{G''}(J'' \setminus \{y''\})$. As $L'' \subseteq A$, we find that $G''[L'']$ is K_4-free.

Claim. Every minimal connector L^* of every connected vertex cover of G'' that contains J'' has size at most 3.

We prove the claim by showing that L^* is a clique, which implies that L^* has size at most 3, as $G''[L'']$ is K_4-free. Suppose instead that L^* is not a clique. Then L^* contains two non-adjacent vertices w_1 and w_2. As L^* is a minimal connector, w_1 has a neighbour in J'' not adjacent to w_2, and vice versa. But then (w_1, w_2) is a pseudo-dominating pair of G'': this is not possible, as G'' has no pseudo-dominating pairs. This contradiction proves the claim.

We now guess a minimal connector by considering all subsets in L'' that have size at most 3. For each guess we apply Lemma 3 on its vertices. If we obtain an instance $(G''', \{y'''\}, y''')$, then we apply Lemma 7. Then we uncontract all contracted edges to obtain in polynomial time a connected vertex cover of G'' that contains J''. We take the smallest one of these connected vertex covers of G''. For this connected vertex cover of G'', we uncontract all contracted edges again to obtain in polynomial time a smallest connected vertex cover S_{w_1} of G' that contains $J' \cup (K \setminus \{w_1\})$ and that does not contain w_1.

As mentioned, we pick the smallest one out of the connected vertex covers S_K and S_{w_i}, $1 \leq i \leq r$, to obtain a smallest type 2 connected vertex cover of G', the size of which we compare with the size of S^*. We pick the smallest one.

Thus we obtain in polynomial time a smallest connected vertex cover of G' that contains J' (both in the case where G' has a pseudo-dominating pair and in the case where G' has no pseudo-dominating pair). As stated, it remains to apply statement 2 of Lemma 10 to find in polynomial time a smallest connected vertex cover of G that contains J. The correctness of our algorithm follows immediately from the above case analysis and the description of the cases.   □

4 Our Main Result

In this section we prove Theorem 2. We need two more lemmas (we use Lemma 2 to prove the first one).

Lemma 13. *Let $s \geq 0$ and let G be a connected $(sP_1 + P_5)$-free graph. Then G has a connected dominating set D that is either a clique or has size at most $2s^2 + s + 3$. Moreover, D can be found in $O(n^{2s^2+s+3})$ time.*

Lemma 14. *Let J be an independent set in a connected graph G such that J has a vertex y that is adjacent to every vertex of $G - J$. Let J' consist of those vertices of $J \setminus \{y\}$ that have two adjacent neighbours in $G - J$ (or equivalently, in G). Then a subset S is a connected vertex cover of G that contains J if and only if $S \setminus J'$ is a connected vertex cover of $G - J'$ that contains $J \setminus J'$.*

We are now ready to prove our main result.

Theorem 2 (Restated). *For every $s \geq 0$, CONNECTED VERTEX COVER can be solved in polynomial time for $(sP_1 + P_5)$-free graphs.*

Proof. Let G be an (sP_1+P_5)-free graph for some $s \geq 0$. We may assume without loss of generality that G is connected. By Lemma 13 we can first compute in $O(n^{2s^2+s+3})$ time a connected dominating set D that either has size at most $2s^2 + s + 3$ or is a clique. We note that, if D is a clique, any vertex cover of G contains all but at most one vertex of D. This leads to a case analysis where we guess the subset $D^* \subseteq D$ of vertices not in a minimum connected vertex cover of G. Because $|D^*| \leq 2s^2 + s + 3$, the number of guesses is polynomial. For each guess of D^*, we compute a smallest connected vertex cover S_{D^*} that contains all vertices of $D \setminus D^*$ and no vertex of D^*. Then, in the end, we return one that has minimum size overall.

Let D^* be a guess. We first show the following claim (proof omitted).

Claim 1. We may assume without loss of generality that $D \setminus D^*$ is connected.

Case 1. $D^* = \emptyset$.
We compute a minimum vertex cover S' of $G - D$ in polynomial time by Theorem 1. Clearly $S' \cup D$ is a vertex cover of G. As D is a connected dominating set, $S' \cup D$ is a connected vertex cover of G. Let $S_\emptyset = S' \cup D$. As S' is a minimum vertex cover of $G - D$, $S_\emptyset$ is a smallest connected vertex cover of G that contains all vertices of D. We remember $S_\emptyset$, which we found in polynomial time.

Case 2. $1 \leq |D^*| \leq |D|$ (recall that $|D| \leq 2s^2 + s + 3$).
Recall that we are looking for a smallest connected vertex cover of G that contains every vertex of $D \setminus D^*$ but does not contain any vertex of D^*. Hence D^* must be an independent set and $G - D^*$ must be connected (if one of these conditions is false, then we stop considering the guess D^*). Moreover, a vertex cover that contains no vertex of D^* must contain all vertices of $N_G(D^*)$. Hence we can safely contract not only any edge between two vertices of $D \setminus D^*$, but also any edge between two vertices in $N_G(D^*)$ or between a vertex of $D \setminus D^*$ and a vertex in $N_G(D^*)$. We perform edge contractions recursively and as long as possible while remembering all the edges that we contract. Let G^* be the resulting graph.

Note that the set D^* still exists in G^*, as we did not contract any edges with an endpoint in D^*. By Claim 1, the set $D \setminus D^*$ in G corresponds to exactly one vertex of G^*. We denote this vertex by y. We observe the following equivalence.

Claim 2. Every smallest connected vertex cover of G^* that contains y and that does not contain any vertex of D^* corresponds to a smallest connected vertex cover of G that contains $D \setminus D^*$ and that does not contain any vertex of D^*, and vice versa.

As we obtained G^* in polynomial time, and we can uncontract all contracted edges in polynomial time as well, Claim 2 tells us that we may consider G^* instead of G. As G is connected and $(sP_1 + P_5)$-free, G^* is connected and $(sP_1 + P_5)$-free as well by Lemma 1.

We write $J^* = N_{G^*}(D^*)$ and note that y belongs to J^* as D is connected in G. We now consider the graph $G^* - D^*$. As $G - D^*$ is connected, $G^* - D^*$ is connected. By Claim 2, our new goal is to find a smallest connected vertex cover of $G^* - D^*$ that contains J^*. By our procedure, J^* is an independent set of $G^* - D^*$. As D dominates G, we find that $D \setminus D^*$ dominates every vertex of $G - D^*$ that is not adjacent to a vertex of D^*. Hence the vertex y, which corresponds to the set $D \setminus D^*$, is adjacent to every vertex of $(G^* - D^*) - J^*$ in the graph $G^* - D^*$.

Let $J \subseteq J^*$ consist of y and those vertices in J^* whose neighbourhood in $G^* - D^*$ is an independent set. As y is adjacent to every vertex of $(G^* - D^*) - J^*$ in $G^* - D^*$, and we can remember the set $J^* \setminus J$, we can apply Lemma 14 and remove $J^* \setminus J$. That is, it suffices to find a smallest connected vertex cover of the graph $G' = (G^* - D^*) - (J^* \setminus J)$ that contains J.

As J^* is an independent set of $G^* - D^*$, we find that J is an independent set of G'. By definition, $y \in J$. As y is adjacent to every vertex of $(G^* - D^*) - J^*$ in $G^* - D^*$, we find that y is adjacent to every vertex in $G' - J$. By definition, the neighbours of each vertex in $J \setminus \{y\}$ form an independent set in $G' - J$. Hence the triple (G', J, y) is cover-complete. This means that we can apply Theorem 3 to find in polynomial time a smallest connected vertex cover S' of G' that contains J.

We translate S' in polynomial time into a smallest connected vertex cover S^* of $G^* - D^*$ that contains J^* by adding $J^* \setminus J$ to S'. We translate S^* in polynomial time into a smallest connected vertex cover S_{D^*} of G that contains no vertex of D^* by uncontracting any contracted edges.

As mentioned, in the end we pick, in polynomial time, a smallest set of the sets S_{D^*}. This set is then a minimum connected vertex cover of G, which is obtained in polynomial time. We have not sought to optimize the running time of the algorithm so do not provide a detailed analysis, but observe that, for sufficiently large s, it is $n^{O(s^3)}$. The running time is dominated by obtaining a connected $D \setminus D^*$ (in Claim 1). As $D \setminus D^*$ has $O(n^{2s^2+s+3})$ components and the paths required to join them each have $O(s)$ vertices, the time required to find them is $n^{O(s^3)}$. The correctness of our algorithm follows immediately from the above case analysis and the description of the cases.   □

5   Future Work

We pose two open problems. First, determine the complexity of CONNECTED VERTEX COVER for P_6-free graphs. Second, is there an integer r such that CONNECTED VERTEX COVER is NP-complete for P_r-free graphs?

References

1. Bacsó, G., Tuza, Zs.: Dominating cliques in P_5-free graphs, Periodica Mathematica Hungarica **21**, 303–308 (1990)
2. Balas, E., Yu, C.S.: On graphs with polynomially solvable maximum-weight clique problem. Networks **19**, 247–253 (1989)
3. Camby, E., Cardinal, J., Fiorini, S., Schaudt, O.: The price of connectivity for vertex cover. Discret. Math. Theor. Comput. Sci. **16**, 207–224 (2014)
4. Camby, E., Schaudt, O.: A new characterization of P_k-free graphs. Algorithmica **75**, 205–217 (2016)
5. Cardinal, J., Levy, E.: Connected vertex covers in dense graphs. Theor. Comput. Sci. **411**, 2581–2590 (2010)
6. Chiarelli, N., Hartinger, T.R., Johnson, M., Milanic, M., Paulusma, D.: Minimum connected transversals in graphs: new hardness results and tractable cases using the price of connectivity. Theor. Comput. Sci. **705**, 75–83 (2018)
7. Escoffier, B., Gourvès, L., Monnot, J.: Complexity and approximation results for the connected vertex cover problem in graphs and hypergraphs. Theor. Comput. Sci. **8**, 36–49 (2010)
8. Fernau, H., Manlove, D.: Vertex and edge covers with clustering properties: complexity and algorithms. J. Discret. Algorithms **7**, 149–167 (2009)
9. Garey, M.R., Johnson, D.S.: The rectilinear Steiner tree problem is NP-complete. SIAM J. Appl. Math. **32**, 826–834 (1977)
10. Gavril, F.: The intersection graphs of subtrees in trees are exactly the chordal graphs. J. Comb. Theory, Ser. B **16**, 47–56 (1974)
11. Grzesik, A., Klimošová, T., Pilipczuk, M., Pilipczuk, M.: Polynomial-time algorithm for maximum weight independent set on P_6-free graphs. Manuscript (2017)
12. Hartinger, T.R., Johnson, M., Milanic, M., Paulusma, D.: The price of connectivity for transversals. Eur. J. Comb. **58**, 203–224 (2016)
13. Li, Y., Yang, Z., Wang, W.: Complexity and algorithms for the connected vertex cover problem in 4-regular graphs. Appl. Math. Comput. **301**, 107–114 (2017)
14. Lokshtanov, D., Vatshelle, M., Villanger, Y.: Independent set in P_5-free graphs in polynomial time. In: Proceedings of SODA 2014, pp. 570–581 (2014)
15. Minty, G.J.: On maximal independent sets of vertices in claw-free graphs. J. Comb. Theory, Ser. B **28**, 284–304 (1980)
16. Munaro, A.: Boundary classes for graph problems involving non-local properties. Theor. Comput. Sci. **692**, 46–71 (2017)
17. Poljak, S.: A note on stable sets and colorings of graphs. Commentationes Mathematicae Universitatis Carolinae **15**, 307–309 (1974)
18. Priyadarsini, P.K., Hemalatha, T.: Connected vertex cover in 2-connected planar graph with maximum degree 4 is NP-complete. Int. J. Math. Phys. Eng. Sci. **2**, 51–54 (2008)
19. Tsukiyama, S., Ide, M., Ariyoshi, H., Shirakawa, I.: A new algorithm for generating all the maximal independent sets. SIAM J. Comput. **6**, 505–517 (1977)

20. Sbihi, N.: Algorithme de recherche d'un stable de cardinalité maximum dans un graphe sans étoile. Discret. Math. **29**, 53–76 (1980)
21. Ueno, S., Kajitani, Y., Gotoh, S.: On the nonseparating independent set problem and feedback set problem for graphs with no vertex degree exceeding three. Discret. Math. **72**, 355–360 (1988)
22. Wanatabe, T., Kajita, S., Onaga, K.: Vertex covers and connected vertex covers in 3-connected graphs. In: Proceedings of IEEE ISCAS 1991, pp. 1017–1020 (1991)

Structurally Parameterized
d-Scattered Set

Ioannis Katsikarelis[✉], Michael Lampis, and Vangelis Th. Paschos

Université Paris-Dauphine, PSL Research University, CNRS, UMR 7243 LAMSADE,
75016 Paris, France
{ioannis.katsikarelis,michail.lampis,paschos}@lamsade.dauphine.fr

Abstract. In d-SCATTERED SET we are given an (edge-weighted) graph and are asked to select at least k vertices, so that the distance between any pair is at least d, thus generalizing INDEPENDENT SET. We provide upper and lower bounds on the complexity of this problem with respect to various standard graph parameters. In particular, we show the following:

- For any $d \geq 2$, an $O^*(d^{tw})$-time algorithm, where tw is the treewidth of the input graph and a tight SETH-based lower bound matching this algorithm's performance. These generalize known results for INDEPENDENT SET.
- d-SCATTERED SET is W[1]-hard parameterized by vertex cover (for edge-weighted graphs), or feedback vertex set (for unweighted graphs), even if k is an additional parameter.
- A single-exponential algorithm parameterized by vertex cover for unweighted graphs, complementing the above-mentioned hardness.
- A $2^{O(td^2)}$-time algorithm parameterized by tree-depth (td), as well as a matching ETH-based lower bound, both for unweighted graphs.

We complement these mostly negative results by providing an *FPT approximation scheme* parameterized by treewidth. In particular, we give an algorithm which, for any error parameter $\epsilon > 0$, runs in time $O^*((tw/\epsilon)^{O(tw)})$ and returns a $d/(1 + \epsilon)$-scattered set of size k, if a d-scattered set of the same size exists.

1 Introduction

In this paper we study the d-SCATTERED SET problem: given graph $G = (V, E)$ and a metric weight function $w : E \mapsto \mathbb{N}^+$ that gives the length of each edge, we are asked if there exists a set K of at least k *selections* from V, such that the distance between any pair $v, u \in K$ is at least $d(v, u) \geq d$, where $d(v, u)$ denotes the shortest-path distance from v to u under weight function w. If w assigns weight 1 to all edges, the variant is called *unweighted*.

The problem can already be seen to be hard, as it generalizes INDEPENDENT SET (for $d = 2$), even to approximate (under standard complexity assumptions), i.e. the optimal k cannot be approximated to $n^{1-\epsilon}$ in polynomial time [18], while an alternative name is DISTANCE-d INDEPENDENT SET [12,13,28]. This hardness prompts the analysis of the problem when the input graph is of restricted

© Springer Nature Switzerland AG 2018
A. Brandstädt et al. (Eds.): WG 2018, LNCS 11159, pp. 292–305, 2018.
https://doi.org/10.1007/978-3-030-00256-5_24

structure, our aim being to provide a comprehensive account of the complexity of d-SCATTERED SET through various upper and lower bound results. Our viewpoint is parameterized: we consider the well-known structural parameters treewidth **tw**, tree-depth **td**, vertex cover number **vc** and feedback vertex set number **fvs**, that comprehensively express the intended restrictions on the input graph's structure (as they range in size and applicability), while we examine both the edge-weighted and unweighted variants of the problem.

Our Contribution: First, in Sect. 3 we present a lower bound of $(d - \epsilon)^{\mathrm{tw}} \cdot n^{O(1)}$ on the complexity of any algorithm solving d-SCATTERED SET parameterized by tw, based on the Strong Exponential Time Hypothesis (SETH [19,20]). This result can be seen as a non-trivial extension of the bound of $(2 - \epsilon)^{\mathrm{tw}} \cdot n^{O(1)}$ for INDEPENDENT SET [25] for larger values of d, for which the construction is required to be much more compact in terms of encoded information per unit of treewidth.

In Sect. 4 we provide a dynamic programming algorithm of running time $O^*(d^{\mathrm{tw}})$, matching this lower bound, over a given tree decomposition of width tw. The algorithm actually solves the counting version of d-SCATTERED SET, making use of standard techniques (dynamic programming on tree decompositions), with an application of the fast subset convolution technique of [2] (or *state changes* [7,30]) to bring the running time down to match the size of the dynamic programming tables.

Having thus identified the complexity of the problem with respect to tw, we next focus on the more restrictive parameters vc and fvs and we show in Sect. 5 that the edge-weighted d-SCATTERED SET problem parameterized by $vc + k$ is W[1]-hard. If, on the other hand, all edge-weights are set to 1, then d-SCATTERED SET (the unweighted variant) parameterized by $fvs + k$ is W[1]-hard. Our reductions also imply lower bounds based on the Exponential Time Hypothesis (ETH [19,20]), yet we do not believe these to be tight, due to the quadratic increase in parameter size (as the construction's focus lies on the edges). One observation we can make is that there are few cases where we can expect to obtain an FPT algorithm without bounding the value of d.

We complement these results with a single-exponential algorithm for the unweighted variant, of running time $O^*(3^{\mathrm{vc}})$ for the case of even d, while for odd d the running time is $O^*(4^{\mathrm{vc}})$. The algorithm is based on defining a sub-problem based on a variant of SET PACKING that we solve via dynamic programming. The difference in running times, depending on the parity of d, is due to the number of possible situations for a vertex with respect to potential candidates for selection.

Further, for the unweighted variant we also show in Sect. 7 the existence of an algorithm parameterized by td of running time $O^*(2^{O(\mathrm{td}^2)})$, as well as a matching ETH-based lower bound. The upper bound follows from known connections between the tree-depth of a graph and its diameter, while the lower bound comes from a reduction from 3-SAT.

Finally, we turn again to tw in Sect. 8 and we present a fixed-parameter-tractable approximation scheme (FPT-AS) on d of running time

$O^*((\text{tw}/\epsilon)^{O(\text{tw})})$, that finds a $d/(1+\epsilon)$-scattered set of size k, if a d-scattered set of the same size exists. The algorithm is based on a rounding technique introduced in [24] and can be much faster than any exact algorithm for the problem (for large d, i.e. $d \geq O(\log n)$), even for the unweighted case and more restricted parameters. Figure 1 illustrates the relationships between considered parameters and summarizes our results, while we refer the reader to the full version [22] for all omitted definitions, constructions and proofs.

Related Work: Our work can be considered as a continuation of the investigations in [21], where the (k,r)-CENTER problem is similarly studied with respect to several well-known structural parameters and a number of fine-grained upper/lower bounds is presented, while some of the techniques employed for our SETH lower bound are also present in [8].

The SETH-based lower bound of $(2-\epsilon)^{\text{tw}} \cdot n^{O(1)}$ on the running time of any algorithm for INDEPENDENT SET parameterized by tw comes from [25]. For d-SCATTERED SET, Halldórsson et al. [17] showed a tight inapproximability ratio of $n^{1-\epsilon}$ for even d and $n^{1/2-\epsilon}$ for odd d, while Eto et al. [13] showed that on r-regular graphs the problem is APX-hard for $r, d \geq 3$, while also providing polynomial-time $O(r^{d-1})$-approximations and a polynomial-time approximation scheme (PTAS) for planar graphs. For a class of graphs with at most a polynomial (in n) number of minimal separators, d-SCATTERED SET can be solved in polynomial time for even d, while it remains NP-hard on chordal graphs (contained in the class) and any odd $d \geq 3$ [28]. It remains NP-hard even for planar bipartite graphs of maximum degree 3, while a 1.875-approximation is available on cubic graphs [14]. Several hardness results for planar and chordal (bipartite) graphs can be found in [12], while [16] shows the problem admits an EPTAS on (apex)-minor-free graphs, based on the theory of bidimensionality. Finally, on a related result, Marx and Pilipczuk recently offered an $n^{O(\sqrt{k})}$-time algorithm for planar graphs, making use of Voronoi diagrams and based on ideas previously used to obtain geometric QPTASs [27].

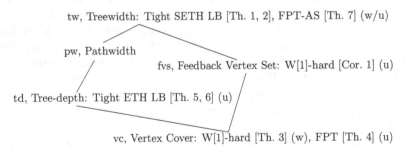

tw, Treewidth: Tight SETH LB [Th. 1, 2], FPT-AS [Th. 7] (w/u)

pw, Pathwidth

fvs, Feedback Vertex Set: W[1]-hard [Cor. 1] (u)

td, Tree-depth: Tight ETH LB [Th. 5, 6] (u)

vc, Vertex Cover: W[1]-hard [Th. 3] (w), FPT [Th. 4] (u)

Fig. 1. Relationships between parameters and an overview of our results (with theorem numbers, for weighted/unweighted variants). In the downwards direction (from tw to vc) parameter size increases and algorithmic results are inherited, while hardness results are inherited in the upwards direction.

2   Definitions and Preliminaries

We use standard graph-theoretic notation. For a graph $G = (V, E)$, $n = |V|$ denotes the number of vertices, $m = |E|$ the number of edges, an edge $e \in E$ between $u, v \in V$ is denoted by (u, v), and for a subset $X \subseteq V$, $G[X]$ denotes the graph induced by X. The functions $\lfloor x \rfloor$ and $\lceil x \rceil$, for $x \in \mathbb{R}$, denote the maximum integer that is not larger and the minimum integer that is not smaller than x, respectively. Further, we assume the reader has some familiarity with standard definitions from parameterized complexity theory (see [10,11,15]).

For a parameterized problem with parameter k, an FPT-AS is an algorithm which for any $\epsilon > 0$ runs in time $O^*(f(k, \frac{1}{\epsilon}))$ (i.e. FPT time when parameterized by $k + \frac{1}{\epsilon}$) and produces a $(1 + \epsilon)$-approximation (see [26]). We use $O^*(\cdot)$ to imply omission of factors polynomial in n. In this paper we present approximation schemes with running times of the form $(\log n / \epsilon)^{O(k)}$. These can be seen to imply an FPT running time by a well-known win-win argument: *If a parameterized problem with parameter k admits, for some $\epsilon > 0$, an algorithm running in time $O^*((\log n / \epsilon)^{O(k)})$, then it also admits an algorithm running in time $O^*((k/\epsilon)^{O(k)})$.*

Treewidth and *pathwidth* are standard notions in parameterized complexity that measure how close a graph is to being a tree or path [3,4,23]. We will also use the parameters *vertex cover number* and *feedback vertex set number* of a graph G, which are the sizes of the minimum vertex set whose deletion leaves the graph edgeless, or acyclic, respectively. Finally, we will consider the related notion of *tree-depth* [29], which is defined as the minimum height of a rooted forest whose completion (the graph obtained by connecting each node to all its ancestors) contains the input graph as a subgraph. We will denote these parameters for a graph G as $\operatorname{tw}(G), \operatorname{pw}(G), \operatorname{vc}(G), \operatorname{fvs}(G)$, and $\operatorname{td}(G)$, and will omit G if it is clear from the context. We recall the following well-known relations [5,9] between these parameters which justify the hierarchy given in Fig. 1: *For any graph G we have* $\operatorname{tw}(G) \leq \operatorname{pw}(G) \leq \operatorname{td}(G) \leq \operatorname{vc}(G)$, $\operatorname{tw}(G) \leq \operatorname{fvs}(G) \leq \operatorname{vc}(G)$.

We also recall here the two main complexity assumptions used in this paper [19,20]. The Exponential Time Hypothesis (ETH) states that 3-SAT cannot be solved in time $2^{o(n+m)}$ on instances with n variables and m clauses. The Strong Exponential Time Hypothesis (SETH) states that for all $\epsilon > 0$, there exists an integer q such that q-SAT (where q is the maximum size of any clause) cannot be solved in time $O((2 - \epsilon)^n)$.

3   Treewidth: SETH Lower Bound

In this section we show that for any fixed $d > 2$, the existence of any algorithm for the d-SCATTERED SET problem of running time $O^*((d - \epsilon)^{\operatorname{tw}})$, for some $\epsilon > 0$, would imply the existence of some algorithm for q-SAT on instances with n variables, of running time $O^*((2 - \delta)^n)$, for some $\delta > 0$ and any $q \geq 3$. First, let us briefly discuss the reduction for the SETH lower bound of $(2 - \epsilon)^{\operatorname{tw}}$ for INDEPENDENT SET from [25]. The reduction is based on the construction of

n paths (one for each variable) on $2m$ vertices each, conceptually divided into m pairs of vertices (one for each clause), with each vertex signifying assignment of value 0 or 1 to the corresponding variable. A gadget is introduced for each clause, connected to the vertex of some path that signifies the assignment to the corresponding variable that would satisfy the clause. The pathwidth of the graph (and thus also its treewidth) is (roughly) equal to the number of paths and so a correspondence between a satisfying assignment and an independent set can be established, meaning an $O^*((2-\epsilon)^{\mathrm{tw}})$-time algorithm for INDEPENDENT SET would imply an $O^*((2-\epsilon)^n)$-time algorithm for SAT, for any $\epsilon > 0$.

Intuitively, the reduction for INDEPENDENT SET needs to "embed" the 2^n possible variable assignments into the 2^{tw} states of some optimal dynamic program for the problem, while in our lower bound construction for d-SCATTERED SET we need to be able to encode these 2^n assignments by d^{tw} states and thus there can be no one-to-one correspondence between a variable and only one vertex in some bag of the tree decomposition (that the optimal dynamic program might assign states to); instead, every vertex included in some bag must carry information about the assignment for a *group* of variables. Furthermore, as now $d > 2$, in order to make the converse direction of our reduction to work, we need to make our paths sufficiently long to ensure that any solution will eventually settle into a pattern that encodes a consistent assignment, as the optimal d-scattered set may "cheat" by not selecting the same vertex from each part of some long path (periodically), a situation that would imply a different assignment for the appearances of the same variable for two different clauses (see also [8] and the SETH-based lower bound for DOMINATING SET from [25]).

Clause Gadget $\hat{C}$: We first describe the construction of our clause gadget $\hat{C}$: this gadget has N *input* vertices and its purpose is to only allow for selection of one of these in any d-scattered set, along with another, standard selection. Given vertices $v_1, \ldots, v_N$, we first make N paths $A_i = (a_i^1, \ldots, a_i^{\lfloor d/2 \rfloor - 1}), \forall i \in [1, N]$ on $\lfloor d/2 \rfloor - 1$ vertices. We connect vertices a_i^1 to inputs v_i, while only for even d, we also make all vertices $a_i^{\lfloor d/2 \rfloor - 1}$ into a clique (all other endpoints of each path). We then make a path $B = (b^1, \ldots, b^{\lceil d/2 \rceil + 1})$ and we connect its endpoint $b^{\lceil d/2 \rceil + 1}$ to all $a_i^{\lfloor d/2 \rfloor - 1}$. Observe that any d-scattered set can only include one of the input vertices (as the distance between them is $d - 1$) and the vertex b^1, being the only option at distance d from all inputs.

Construction: We will describe the construction of a graph G, given some $\epsilon < 1, q \geq 3, d > 2$ and an instance ϕ of q-SAT with n variables, m clauses and at most q variables per clause. We first choose an integer $p = \lceil \frac{1}{(1-\lambda)\log_2(d)} \rceil$, for $\lambda = \log_d(d-\epsilon) < 1$ (i.e. p depends only on d and ϵ) and then group the variables of ϕ into $t = \lceil \frac{n}{\gamma} \rceil$ groups $F_1, \ldots, F_t$, for $\gamma = \lfloor \log_2(d)^p \rfloor$, being also the maximum size of any such group.

For each group F_τ of variables of ϕ, with $\tau \in [1, t]$, we make a simple gadget $\hat{G}_\tau^1$ that consists of p paths $P_\tau^l = (p_1^l, \ldots, p_d^l)$ on d vertices each, for $l \in [1, p]$. We then make $m(tp(d-1)+1)$ copies of this "column" of t gadgets $\hat{G}_1^1, \ldots, \hat{G}_t^1$ (i.e. t vertically arranged gadgets), that we connect horizontally (so that we have tp

"long paths"): we connect each last vertex p_d^l from a gadget $\hat{G}_\tau^j$ to vertex p_1^l from the following gadget $\hat{G}_\tau^{j+1}$, for all $l \in [1,p]$, $\tau \in [1,t]$ and $j \in [1, m(tp(d-1))]$ (see Fig. 2(b) for an example).

Next, for every clause C_μ, with $\mu \in [1,m]$, we make $tp(d-1)+1$ copies of the clause gadget $\hat{C}^j$, for $j \in [1, m(tp(d-1)+1)]$, where for each $\mu \in [1,m]$, the number of inputs in the $tp(d-1)+1$ copies is $N = q_\mu d^p/2$, where q_μ is the number of literals in clause C_μ. One clause is assigned to each column of gadgets, so that the first m columns correspond to one clause each, with $tp(d-1)+1$ repetitions of this pattern giving the complete association. Then, for every $\tau \in [1,t]$ we associate a set $S_\tau \subset \bigcup_{l \in [1,p]} P_\tau^l$, that contains exactly one vertex from each of the p paths in $\hat{G}_\tau^j$, with an assignment to the variables in group F_τ. As there are at most $2^\gamma = 2^{\lfloor \log_2(d)^p \rfloor}$ assignments to the variables in F_τ and $d^p \geq 2^\gamma$ such sets S_τ, the association can be unique for each τ (i.e. for each *row* of gadgets). Now, for every literal appearing in clause C_μ, exactly half of the partial assignments to the group F_τ in which the literal's variable appears will satisfy it and thus, each of the $q_\mu d^p/2$ input vertices of the clause gadget will correspond to one literal and one assignment to the variables of the group that satisfy it.

Let v be an input vertex of a clause gadget $\hat{C}^j$, corresponding to a literal of clause C_μ that is satisfied by a partial assignment to the variables of group F_τ that is associated with set $S_\tau \subset \bigcup_{l \in [1,p]} P_\tau^l$, containing exactly one vertex from each path $P_\tau^l, l \in [1,p]$, from gadget $\hat{G}_\tau^j$. For even d, we then make a path $w_1, \ldots, w_{d/2-1}$ on $d/2-1$ vertices, connecting vertex w_1 to v and for each vertex $p_i^l \notin S$ of each path $P_\tau^l \in \hat{G}_\tau^j$ we also make a path $y_1, \ldots, y_{d/2-1}$ on $d/2-1$ vertices, attaching endpoint y_1 to its corresponding path vertex p_i^l, while the other endpoints $y_{d/2-1}$ are all attached to vertex $w_{d/2-1}$ and to each other (into a clique). For odd d, we make a similar construction for each such v, only the number of vertices in constructed paths is now $\lfloor d/2 \rfloor$ instead of $d/2-1$ and vertices $y_{\lfloor d/2 \rfloor}$ are not made into a clique. Thus every input vertex v of some clause gadget is at distance exactly $d-1$ from every path vertex that does not belong to the set associated with its corresponding partial assignment (and thus exactly d from the only vertex per path that is), while the distances between any pair of other (i.e. intermediate) vertices via these paths are $\leq d-1$. This concludes our construction, while Fig. 2 provides illustrations of the above.

In this way, a satisfying assignment for ϕ would correspond to a d-scattered set that selects the vertices in each gadget $\hat{G}_\tau$ that match the partial assignment S_τ for that group's variables F_τ in all $m(tp(d-1)+1)$ columns, along with the corresponding input vertex from each clause gadget (implying the existence of a satisfied literal within the clause). On the other hand, for any d-scattered set of size $(tp+2)m(tp(d-1)+1)$ in G, the maximum number of times it can "cheat" by not periodically selecting the "same" vertices in each column is $tp(d-1)$. The number of columns being $m(tp(d-1)+1)$, by the pigeonhole principle, there will always exist m consecutive columns for which the selection pattern does not change, from which a consistent assignment for all clauses can be extracted.

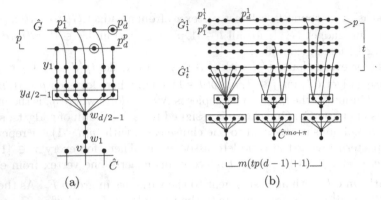

(a)   (b)

Fig. 2. (a): The connection of an input vertex v of a clause gadget $\hat{C}$ to its corresponding path vertices in some $\hat{G}$, where vertices of set S_τ are circled and boxed vertices form a clique (for even d). (b): A simplified picture of the global construction, with some exemplative connecting paths between clause gadgets and path vertices shown as edges.

Theorem 1. *For any fixed $d > 2$, if d-SCATTERED SET can be solved in $O^*((d-\epsilon)^{tw(G)})$ time for some $\epsilon > 0$, then there exists some $\delta > 0$, such that q-SAT can be solved in $O^*((2-\delta)^n)$ time, for any $q \geq 3$.*

4   Treewidth: Dynamic Programming Algorithm

In this section we present an $O^*(d^{tw})$-time dynamic programming algorithm for the counting version of the d-SCATTERED SET problem. The input is a graph $G = (V, E)$, a nice tree decomposition $(\mathcal{X}, T)$ for G, where $T = (I, F)$ is a tree and $\mathcal{X} = \{X_i | i \in I\}$ is the set of bags, while $\max_{i \in I} |X_i| - 1 = \text{tw}$, along with two numbers $k \in \mathbb{N}^+, d \geq 2$, while the output is the number of d-scattered sets of size k in G.

There is a table D_i associated with every node i of the tree decomposition with $X_i = \{v_0, \ldots, v_t\}$, while each table entry $D_i[\kappa, s_0, \ldots, s_t]$ contains the number of (distinct) d-scattered sets K of size $|K| = \kappa$ (its *partial solution*) and is indexed by a number $\kappa \in [1, k]$ and a $t + 1$-sized tuple $(s_0, \ldots, s_t)$ of *state-configurations*, assigning a state $s_j \in [0, d - 1]$ to each vertex v_j in the bag. There are d possible states for each vertex, designating its distance to the closest selection for the d-scattered set at the "current" stage of the algorithm (i.e. within the graph defined by each node of the tree decomposition), with vertices of *zero* state $s_j = 0$ being included in K, vertices of *low* state $s_j \in [1, \lfloor d/2 \rfloor]$ being at distance at least s_j from their closest selection and $d - s_j$ from the second closest, while vertices of *high* state $s_j \in [\lfloor d/2 \rfloor + 1, d - 1]$ are at distance at least s_j from K. That is, each partial solution is described by a given budget for selections (up to k) and the minimum distances of all vertices in the bag to an already selected vertex.

Based on this scheme, the inductive computations for each type of node of the nice tree decomposition are straightforward to obtain, yet a direct implementation would not lead to an algorithm of running time that matches the size of the constructed tables (d^{tw}): using the above state-representation, the computations at a join node would require an additional 2^{tw} factor, as this is the number of possible combinations of previously computed partial solutions (from the tables of its children) that could be combined to give a partial solution for the new node. This can be avoided by an application of the *state changing* technique (or fast subset convolution, see [2, 7, 30] and Chap. 11 from [10]), for which it is also more convenient to *count* the number of solutions of each size $\kappa \in [1, k]$, instead of computing the maximum k for which a solution satisfying the given state-configuration exists.

Theorem 2. *Given graph G, along with $d \in \mathbb{N}^+$ and nice tree decomposition $(\mathcal{X}, T)$ of width tw for G, there exists an algorithm to solve the counting version of the d-SCATTERED SET problem in $O^*(d^{tw})$ time.*

5   Vertex Cover, Feedback Vertex Set: W[1]-Hardness

In this section we show that the edge-weighted variant of the d-SCATTERED SET problem parameterized by vc $+ k$ is W[1]-hard via a reduction from k-MULTICOLORED INDEPENDENT SET, a well-known W[1]-complete problem (see [10]: given a graph $G = (V, E)$, with V partitioned into k independent sets $V = V_1 \uplus \cdots \uplus V_k$, $|V_i| = n, \forall i \in [1, k]$, where E only contains edges between vertices of sets V_i, V_j with $i \neq j$, we are asked to find a subset $S \subseteq V$, such that $G[S]$ forms an independent set and $|S \cap V_i| = 1, \forall i \in [1, k]$.

Construction: Given an instance $[G = (V, E), k]$ of k-MULTICOLORED INDEPENDENT SET, we construct an instance $[G' = (V', E'), k']$ of edge-weighted d-SCATTERED SET where $d = 6n$. First, for every color class $V_i \subseteq V$ we create a set $P_i \subseteq V'$ of n vertices $p_l^i, \forall l \in [1, n], \forall i \in [1, k]$ (that directly correspond to the vertices of V_i). Next, for each $i \in [1, k]$ we make a pair of vertices a_i, b_i, connecting a_i to each vertex p_l^i by an edge of weight $n + l$, while b_i is connected to each vertex p_l^i by an edge of weight $2n - l$. Next, for every non-edge $e \in \bar{E}$ (i.e. $\bar{E}$ contains all pairs of vertices from V that are not connected by an edge from E) between two vertices from different V_{i_1}, V_{i_2} (with $i_1 \neq i_2$), we make a vertex u_e that we connect to vertices a_{i_1}, b_{i_1} and a_{i_2}, b_{i_2}. We set the weights of these edges as follows: suppose that e is a non-edge between the j_1-th vertex of V_{i_1} and the j_2-th vertex of V_{i_2}. We then set $w(u_e, a_{i_1}) = 5n - j_1$, $w(u_e, b_{i_1}) = 4n + j_1$ and $w(u_e, a_{i_2}) = 5n - j_2$, $w(u_e, b_{i_2}) = 4n + j_2$. Next, for every pair of i_1, i_2 we make two vertices $g_{i_1,i_2}, g'_{i_1,i_2}$. We connect g_{i_1,i_2} to all vertices u_e that correspond to non-edges e between vertices of the same pair V_{i_1}, V_{i_2} by edges of weight $(6n - 1)/2$ and also g_{i_1,i_2} to g'_{i_1,i_2} by an edge of weight $(6n + 1)/2$. In this way, a k-multicolored independent set in G corresponds to a $6n$-scattered set in G' of size k^2. This concludes the construction of G', with Fig. 3 providing an illustration.

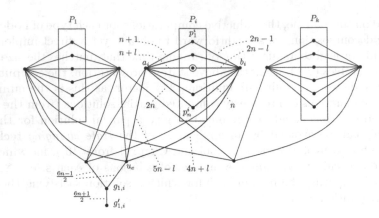

Fig. 3. A general picture of graph G', where the circled vertex is p_l^i and dotted lines match weights to edges.

Theorem 3. *The edge-weighted d-SCATTERED SET problem is $W[1]$-hard parameterized by $vc + k$. Furthermore, if there is an algorithm for edge-weighted d-SCATTERED SET running in time $n^{o(\sqrt{vc}+\sqrt{k})}$ then the ETH is false.*

Using essentially the same reduction (with minor modifications) we also obtain similar hardness results for unweighted d-SCATTERED SET parameterized by fvs:

Corollary 1. *The unweighted d-SCATTERED SET problem is $W[1]$-hard parameterized by $fvs + k$. Furthermore, if there is an algorithm for unweighted d-SCATTERED SET running in time $n^{o(fvs+\sqrt{k})}$ then the ETH is false.*

6 Vertex Cover: FPT Algorithm

We next show that unweighted d-SCATTERED SET admits an FPT algorithm parameterized by vc, in contrast to its weighted version (Theorem 3). Given graph G along with a vertex cover C of G and $d \geq 3$, our algorithm first defines an instance of PARTIAL SET PACKING, where elements may be *partially* included in some sets and then solves the problem by dynamic programming. In this variant, any element has a *coefficient* of inclusion in each set and a collection of sets is a solution if there is no pair of sets for which the sum of any element's coefficients is >1.

We make a set for each vertex and an element for each vertex of C. Our aim is to identify two vertices (sets) as incompatible selections if there is some third "middle" vertex from C (elements), whose sum of distances to the other two is $<d$, based on the observation that for any vertex not belonging to the d-scattered set, only one selection can be at distance $<d/2$, yet any number of selections can be at distance $\geq d/2$ (consider a star as an example).

These coefficients of inclusion are then used to assign vertices of C to their closest possible selections, with complete inclusion (i.e. coefficient equal to 1) implying the distance is $<d/2$ and no inclusion (equal to 0) that it is $>d/2$. For the middle vertices, depending on the parity of d (and causing the difference in running times), we require either one (i.e. $1/2$) or two ($1/3$ and $2/3$) extra coefficients to be able to determine the exact position of a possible middle vertex from C (element) on the path between two potential selections (sets). If the sum of coefficients is ≤ 1, the vertex from C is either a middle vertex on the path between the two selections or at distance $<d/2$ from only one of them. On the other hand, if the sum of coefficients is >1, then the sum of distances from the vertex to the two selections is $<d$ and the incompatibility of the sets implies the corresponding vertices cannot both belong in the d-scattered set.

Theorem 4. *Given graph G, along with $d > 2$ and a vertex cover of size vc of G, there exists an algorithm solving the unweighted d-SCATTERED SET problem in $O^*(3^{vc})$ time for even d and $O^*(4^{vc})$ time for odd d.*

7 Tree-Depth: Tight ETH Lower Bound

In this section we consider the unweighted version of the d-SCATTERED SET problem parameterized by td. We first show the existence of an FPT algorithm of running time $O^*(2^{O(\text{td}^2)})$ and then a tight ETH-based lower bound. We begin with a simple upper bound argument, making use of the following fact on tree-depth, while the algorithm then follows from the dynamic programming procedure of Theorem 2 and the relationship between d, td and tw:

Lemma 1. *For any graph $G = (V, E)$ we have $D(G) \leq 2^{td+1} - 2$, where $D(G)$ denotes the graph's diameter.*

Theorem 5. *Unweighted d-SCATTERED SET can be solved in time $O^*(2^{O(\text{td}^2)})$.*

Next we show a lower bound matching Theorem 5, based on the ETH, using a reduction from 3-SAT and a construction similar to the one used in Sect. 5.

Theorem 6. *If unweighted d-SCATTERED SET can be solved in $2^{o(\text{td}^2)} \cdot n^{O(1)}$ time, then 3-SAT can be solved in $2^{o(n)}$ time.*

8 Treewidth Revisited: FPT-AS

In this section we present an FPT approximation *scheme* (FPT-AS) for d-SCATTERED SET parameterized by tw. Given as input an edge-weighted graph $G = (V, E)$, $k \in \mathbb{N}^+$, $d \geq 2$ and an arbitrarily small error parameter $\epsilon > 0$, our algorithm is able to return a set K, such that any $v, u \in K$ are at distance $d(v, u) \geq \frac{d}{1+\epsilon}$, in time $O^*((\text{tw}/\epsilon)^{O(\text{tw})})$, if G has a d-scattered set of size $|K|$.

Our algorithm makes use of a technique introduced in [24] (see also [1,21]) for approximating problems that are W-hard by treewidth. If the hardness of the problem arises from the need of the dynamic programming table to store tw large numbers (in our case, the distances of the vertices in the bag from the closest selection), we can significantly speed up the algorithm by replacing all values by the closest integer power of $(1 + \delta)$, for some appropriately chosen δ, thus reducing the table size from d^{tw} to $(\log_{(1+\delta)} d)^{\text{tw}}$. Of course, the calculations may result in values that are not integer powers of $(1 + \delta)$ that will thus have to be "rounded" to maintain the table size. This might introduce the accumulation of rounding errors, yet we are able to show that the error on any rounded value can be bounded by a function of the height of its corresponding bag and then make use of a theorem from [6] stating that any tree decomposition can be balanced so that its width remains almost unchanged and its total height becomes $O(\log n)$.

The rounding technique as applied in [24] employs randomization and an extensive analysis to procure the bounds on the propagation of error, while we only require a deterministic adaptation of the rounding process without making use of the advanced machinery there introduced, as for our particular case, the bound on the rounding error can be straightforwardly obtained. The main tool we require is the following definition of an addition-rounding operation, denoted by $\oplus$: for two non-negative numbers x_1, x_2, we define $x_1 \oplus x_2 := 0$, if $x_1 = x_2 = 0$. Otherwise, we set $x_1 \oplus x_2 := (1 + \delta)^{\lfloor \log_{(1+\delta)}(x_1+x_2) \rfloor}$.

The integers we would like to approximately store are the states $s_j \in [1, d-1]$, representing the distance of a vertex v_j in bag X_i of the tree decomposition to the closest selection in the d-scattered set K, during computation of the dynamic programming algorithm. Let $\Sigma_\delta := \{0\} \cup \{(1 + \delta)^l | l \in \mathbb{N}\}$. Intuitively, Σ_δ is the set of rounded states that our modified algorithm may use. Of course, Σ_δ as defined is infinite, but we will only consider the set of values that are at most d, denoted by Σ_δ^d. In this way, the size of Σ_δ^d is reduced to $\log_{(1+\delta)}(d)$, that for $\delta = \frac{\epsilon}{O(\log n)}$, gives $|\Sigma_\delta^d| = O(\log(d) \log(n)/\epsilon)$ and we then rely on the well-known win-win parameterized argument given in Sect. 2 to get a running time of $O^*((\text{tw}/\epsilon)^{O(\text{tw})})$.

Modifications: First, we make use of an adaptation of the algorithm of Theorem 2 that works for the maximization version of the problem (albeit not optimally). We next explain the necessary modifications to the exact algorithm for use of the rounded states $\sigma \in \Sigma_\delta^d$. Consider a node i introducing vertex v_{t+1}: for a new entry to describe a proper extension to some previously computed partial solution, if the new vertex is of state $s_{t+1} \in [1, d-1]$ in the new entry, then there must be some vertex $v_j \in X_i$, such that $s_{t+1} \leq d(v_{t+1}, v_j) + s_j$ (the one for which this sum is minimized), i.e. we require that the new state of the introduced vertex matches its distance to some other vertex in the bag plus the state of that vertex (being the one "responsible" for connecting v_{t+1} to the partial solution). The rounded state σ_{t+1} for v_{t+1} must now satisfy: $\sigma_{t+1} \leq d(v_{t+1}, v_j) \oplus \sigma_j$, where $\oplus$ is the operator defined above. Further, we define the symmetrical (around $d/2$) state $\bar{\sigma}$ for a given low state σ as the minimum state σ' for which $\sigma + \sigma' \geq \frac{d}{(1+\epsilon)}$ and we arbitrarily choose the computed states for the table of one of the children nodes to represent the new entries and $\bar{\sigma}$ to identify the symmetrical of each low state (from the other node's table).

Moreover, we require that the tree decompositions on which our algorithm is to be applied are rooted and of maximum depth $O(\log n)$. In [6] (Lemma 1), it is shown that any tree decomposition of width tw can be converted to a rooted and binary tree decomposition of depth $O(\log n)$ and width at most $3\text{tw} + 2$ in $O(\log n)$ time and $O(n)$ space. The following lemma employs the transformation to bound the error of any value calculated in this way, based on an appropriate choice of δ and therefore set Σ_δ^d of available values, by relating the rounded states σ computed at any node to the states s that the exact algorithm would use at the same node instead.

Lemma 2. *Given ϵ and a tree decomposition $(\mathcal{X}, T)$ with $T = (I, F), \mathcal{X} = \{X_i | i \in I\}$, where T is rooted, binary and of depth $O(\log n)$, there exists a constant C, such that for all rounded states $\sigma_j \in \Sigma_\delta^d$ it is $\sigma_j \geq \frac{s_j}{(1+\epsilon)}, \forall v_j \in X_i, \forall i \in I$, where $\delta = \frac{\epsilon}{C \log n}$.*

Theorem 7. *There is an algorithm which, given an edge-weighted instance of d-*SCATTERED SET $[G, k, d]$, *a tree decomposition of G of width tw and a parameter $\epsilon > 0$, runs in time $O^*((\text{tw}/\epsilon)^{O(\text{tw})})$ and finds a $d/(1 + \epsilon)$-scattered set of size k, if a d-scattered set of the same size exists in G.*

References

1. Angel, E., Bampis, E., Escoffier, B., Lampis, M.: Parameterized power vertex cover. In: Heggernes, P. (ed.) WG 2016. LNCS, vol. 9941, pp. 97–108. Springer, Heidelberg (2016). https://doi.org/10.1007/978-3-662-53536-3_9
2. Björklund, A., Husfeldt, T., Kaski, P., Koivisto, M.: Fourier meets möbius: fast subset convolution. In: STOC, pp. 67–74 (2007)
3. Bodlaender, H.L.: The algorithmic theory of treewidth. Electron. Notes Discret. Math. **5**, 27–30 (2000)
4. Bodlaender, H.L.: Treewidth: characterizations, applications, and computations. In: Fomin, F.V. (ed.) WG 2006. LNCS, vol. 4271, pp. 1–14. Springer, Heidelberg (2006). https://doi.org/10.1007/11917496_1
5. Bodlaender, H.L., Gilbert, J.R., Hafsteinsson, H., Kloks, T.: Approximating treewidth, pathwidth, frontsize, and shortest elimination tree. J. Algorithms **18**(2), 238–255 (1995)
6. Bodlaender, H.L., Hagerup, T.: Parallel algorithms with optimal speedup for bounded treewidth. SIAM J. Comput. **27**(6), 1725–1746 (1998)
7. Bodlaender, H.L., van Leeuwen, E.J., van Rooij, J.M.M., Vatshelle, M.: Faster algorithms on branch and clique decompositions. In: Hliněný, P., Kučera, A. (eds.) MFCS 2010. LNCS, vol. 6281, pp. 174–185. Springer, Heidelberg (2010). https://doi.org/10.1007/978-3-642-15155-2_17
8. Borradaile, G., Le, H.: Optimal dynamic program for r-domination problems over tree decompositions. In: IPEC, vol. 63, pp. 8:1–8:23 (2016)
9. Courcelle, B., Olariu, S.: Upper bounds to the clique width of graphs. Discret. Appl. Math. **101**(1–3), 77–114 (2000)

10. Cygan, M., et al.: Parameterized Algorithms. Springer, Cham (2015). https://doi. org/10.1007/978-3-319-21275-3

11. Downey, R.G., Fellows, M.R.: Fundamentals of Parameterized Complexity. Texts in Computer Science. Springer, London (2013). https://doi.org/10.1007/978-1-4471-5559-1

12. Eto, H., Guo, F., Miyano, E.: Distance- d independent set problems for bipartite and chordal graphs. J. Comb. Optim. **27**(1), 88–99 (2014)

13. Eto, H., Ito, T., Liu, Z., Miyano, E.: Approximability of the distance independent set problem on regular graphs and planar graphs. In: Chan, T.-H.H., Li, M., Wang, L. (eds.) COCOA 2016. LNCS, vol. 10043, pp. 270–284. Springer, Cham (2016). https://doi.org/10.1007/978-3-319-48749-6_20

14. Eto, H., Ito, T., Liu, Z., Miyano, E.: Approximation algorithm for the distance-3 independent set problem on cubic graphs. In: Poon, S.-H., Rahman, M.S., Yen, H.-C. (eds.) WALCOM 2017. LNCS, vol. 10167, pp. 228–240. Springer, Cham (2017). https://doi.org/10.1007/978-3-319-53925-6_18

15. Flum, J., Grohe, M.: Parameterized Complexity Theory. Texts in Theoretical Computer Science. An EATCS Series. Springer, Heidelberg (2006). https://doi.org/10. 1007/3-540-29953-X

16. Fomin, F.V., Lokshtanov, D., Raman, V., Saurabh, S.: Bidimensionality and EPTAS. In: SODA, pp. 748–759. SIAM (2011)

17. Halldórsson, M.M., Kratochvíl, J., Telle, J.A.: Independent sets with domination constraints. Discret. Appl. Math. **99**(1–3), 39–54 (2000)

18. Håstad, J.: Clique is hard to approximate within $n^{1-\epsilon}$. Acta Mathematica **182**, 105–142 (1999)

19. Impagliazzo, R., Paturi, R.: On the complexity of k-sat. J. Comput. Syst. Sci. **62**(2), 367–375 (2001)

20. Impagliazzo, R., Paturi, R., Zane, F.: Which problems have strongly exponential complexity? J. Comput. Syst. Sci. **63**(4), 512–530 (2001)

21. Katsikarelis, I., Lampis, M., Paschos, V.Th.: Structural parameters, tight bounds, and approximation for (k, r)-center. CoRR, abs/1704.08868 (2017)

22. Katsikarelis, I., Lampis, M., Paschos, V.Th.: Structurally parameterized d-scattered set. CoRR, abs/1709.02180 (2017)

23. Kloks, T. (ed.): Treewidth, Computations and Approximations. LNCS, vol. 842. Springer, Heidelberg (1994). https://doi.org/10.1007/BFb0045375

24. Lampis, M.: Parameterized approximation schemes using graph widths. In: Esparza, J., Fraigniaud, P., Husfeldt, T., Koutsoupias, E. (eds.) ICALP 2014. LNCS, vol. 8572, pp. 775–786. Springer, Heidelberg (2014). https://doi.org/10. 1007/978-3-662-43948-7_64

25. Lokshtanov, D., Marx, D., Saurabh, S.: Known algorithms on graphs on bounded treewidth are probably optimal. In: SODA, pp. 777–789 (2011)

26. Marx, D.: Parameterized complexity and approximation algorithms. Comput. J. **51**(1), 60–78 (2008)

27. Marx, D., Pilipczuk, M.: Optimal parameterized algorithms for planar facility location problems using voronoi diagrams. In: Bansal, N., Finocchi, I. (eds.) ESA 2015. LNCS, vol. 9294, pp. 865–877. Springer, Heidelberg (2015). https://doi.org/10. 1007/978-3-662-48350-3_72

28. Montealegre, P., Todinca, I.: On distance-d independent set and other problems in graphs with "few" minimal separators. In: Heggernes, P. (ed.) WG 2016. LNCS, vol. 9941, pp. 183–194. Springer, Heidelberg (2016). https://doi.org/10.1007/978-3-662-53536-3_16

29. Nesetril, J., Ossona de Mendez, P.: Tree-depth, subgraph coloring and homomorphism bounds. Eur. J. Comb. **27**(6), 1022–1041 (2006)

30. van Rooij, J.M.M., Bodlaender, H.L., Rossmanith, P.: Dynamic programming on tree decompositions using generalised fast subset convolution. In: Fiat, A., Sanders, P. (eds.) ESA 2009. LNCS, vol. 5757, pp. 566–577. Springer, Heidelberg (2009). https://doi.org/10.1007/978-3-642-04128-0_51

Popular Matchings of Desired Size

Telikepalli Kavitha[✉]

Tata Institute of Fundamental Research, Mumbai, India
kavitha@tcs.tifr.res.in

Abstract. Our input is an instance of the stable marriage problem with strict and possibly incomplete lists, i.e., it is a bipartite graph $G = (A \cup B, E)$ where each vertex has a strict preference list ranking its neighbors. We consider a generalization of stable matchings called *popular matchings*: a matching M in G is popular if there is no matching M' such that the vertices that prefer M' to M outnumber those that prefer M to M'.

There are linear time algorithms to compute a *min-size* popular matching and a *max-size* popular matching in G. The following question is a natural variant of the min-size and max-size popular matching problems:

– given a parameter k, is there a popular matching of size k in G?

Here min $< k <$ max, where min and max are the sizes of a min-size and a max-size popular matching in G. We show the above problem is NP-hard. For any min $< k <$ max, we also show a linear time algorithm to construct a matching of size k whose *unpopularity factor* is at most 2.

1 Introduction

Let $G = (A \cup B, E)$ be a bipartite graph where every vertex has a strict preference list ranking its neighbors. A matching M is *stable* if there is no pair (a, b) such that a and b prefer each other to their respective assignments in M. The classical result of Gale and Shapley [6] shows that stable matchings always exist in G and such a matching can be computed in linear time.

Popularity is a notion of *global stability* that was proposed by Gärdenfors [8] in 1975. We say a vertex u prefers matching M to matching M' if (i) u is either matched in M and unmatched in M' or (ii) u is matched in both M, M' and u prefers its partner in M to its partner in M'. Let $\phi(M, M')$ be the number of vertices that prefer M to M'.

Definition 1. *A matching M is popular if $\phi(M, M') \geq \phi(M', M)$ for all matchings M' in G.*

Thus a matching M is popular if M never loses an election against any matching M' in G: here each vertex casts a vote for the matching in $\{M, M'\}$ that it prefers. Popular matchings always exist in $G = (A \cup B, E)$ since every stable matching is popular [8]. In fact, all stable matchings have the same size [7] and every stable matching is a min-size popular matching [10].

© Springer Nature Switzerland AG 2018
A. Brandstädt et al. (Eds.): WG 2018, LNCS 11159, pp. 306–317, 2018.
https://doi.org/10.1007/978-3-030-00256-5_25

The main incentive to relax stability to popularity is to obtain larger matchings. Max-size popular matchings are useful in applications such as matching students to projects or trainees to posts, where one wants a globally stable matching with large size. A max-size popular matching in $G = (A \cup B, E)$ can be computed in linear time [11].

A natural variant of the max-size popular matching problem is one where we seek a popular matching of size k, for a given parameter k. So the input is an instance $G = (A \cup B, E)$ along with an integer k, where min $< k <$ max: here min (similarly, max) is the size of a stable (resp., max-size popular) matching in G. This problem arises in applications such as assigning internships to visiting students where our resources are constrained. Hence rather than a max-size popular matching that may be too large to suit our constraints, we seek a popular matching of size k, for an appropriate parameter k.

Interestingly, G need not admit a popular matching of size k for every k sandwiched between min and max. Consider the following instance on 8 vertices a_i, b_i, c_i, d_i for $i = 0, 1$. For $i = 0, 1$:

- the preference list of a_i is $b_i \succ b_{1-i} \succ c_i$
- the preference list of b_i is $a_{1-i} \succ a_i \succ d_i$.

That is, the vertex a_i's top choice is b_i, second choice is b_{1-i}, and third choice is c_i while the vertex b_i's top choice is a_{1-i}, second choice is a_i, and third choice is d_i. Each of the vertices c_0, c_1, d_0, d_1 has only 1 neighbor. That is, for $i = 0, 1$, c_i's only neighbor is a_i and d_i's is b_i.

This instance has 2 stable matchings: these are $S = \{(a_0, b_0), (a_1, b_1)\}$ and $S' = \{(a_0, b_1), (a_1, b_0)\}$. The matching $M = \{(a_i, c_i), (b_i, d_i) : i = 0, 1\}$ is a (max-size) popular matching of size 4. It can be checked that this instance has *no* popular matching of size 3. We show the following result here.

Theorem 1. *The problem of deciding whether* $G = (A \cup B, E)$ *admits a popular matching of size* k, *for a given parameter* k, *is* NP-*hard*.

While min-size and max-size popular matchings in G can be computed in linear time, surprisingly, computing one of size k, for a given k, is NP-hard. We show that a general graph H admits a vertex cover of size at most t if and only if our bipartite instance G has a popular matching of size $k \in \{s+1, \ldots, s+t\}$, where $s = 3mn + 2m + n$ (m and n are the number of edges and vertices in H).

The *unpopularity factor* $u(M)$ of a matching M in G is defined below. Let

$$u(M) = \max_{N \neq M} \frac{\phi(N, M)}{\phi(M, N)}. \tag{1}$$

If M is popular, then $\phi(N, M) \leq \phi(M, N)$ for all matchings N in G. Thus $u(M) \leq 1$ for a popular matching M. A matching M with $u(M) = O(1)$ can be regarded as a *near-popular* matching since in an election between M and any matching N, the ratio of the number of votes for N and the number of votes for M is $O(1)$. We show the following result here.

Theorem 2. *For any k, where* min $< k <$ max, *a matching M of size k in* $G = (A \cup B, E)$ *such that $u(M) \leq 2$ can be computed in linear time.*

Background. Algorithms for computing popular matchings were first studied in instances $G = (A \cup B, E)$ where vertices in A have preferences over their neighbors while the vertices in B have no preferences. Popular matchings need not always exist in such an instance and there is a polynomial time algorithm [1] to determine if such an instance admits a popular matching or not.

The notion of unpopularity factor was introduced in [15] and it was shown that computing a least unpopularity factor matching in the above model is NP-hard. Popular *fractional* matchings always exist here and such a fractional matching can be computed in polynomial time by linear programming [14].

When every vertex in $G = (A \cup B, E)$ has a strict preference list, popular matchings always exist in G since stable matchings are popular [8]. When ties are allowed in preference lists, the problem of deciding whether G admits a popular matching or not is NP-hard [2, 3].

The first polynomial time algorithm to compute a max-size popular matching in $G = (A \cup B, E)$ with strict preference lists was given in [10]. It was observed in [10] that G need not admit popular matchings of all sizes between min and max. The hardness result that we show here has been improved very recently in [5] to show that deciding if G admits a popular matching of *any* intermediate size between min and max (rather than a particular size k) is NP-hard.

Techniques. Our NP-hardness proof uses the LP framework for popular matchings from [14]. Every popular matching M is a max-weight $(A \cup B)$-complete matching in a bipartite graph $\tilde{G}$ with an edge weight function wt_M. An optimal solution to the dual LP will be called a *witness* to M's popularity.

Witnesses will play an important role in our hardness proof. Witnesses were also recently used in [13] to show that computing a max-utility popular matching in $G = (A \cup B, E)$ with edge utilities is NP-hard. Our algorithm to construct a size k matching with unpopularity factor at most 2 is based on the linear time algorithm from [11] to construct a max-size popular matching in $G = (A \cup B, E)$.

Organization of the paper. Section 2 describes the linear programming framework for popular matchings. Section 3 proves the NP-hardness of the size k popular matching problem and Sect. 4 has our algorithm to construct a *near-popular* matching of size k.

2   Preliminaries

Let $\tilde{G}$ be the graph G augmented with *last resort vertices*, that is, add the set $\{\ell(u) : u \in A \cup B\}$ to the vertex set of G. The vertex $\ell(u)$ will be at the bottom of u's preference list. The edge set $\tilde{E}$ of $\tilde{G}$ is $E \cup \{(u, \ell(u)) : u \in A \cup B\}$. Corresponding to any matching M in G, define the matching $\tilde{M}$ in $\tilde{G}$ as follows:

$$\tilde{M} = M \cup \{(u, \ell(u)) : u \text{ is unmatched in } M\}.$$

Thus $\tilde{M}$ is the $(A \cup B)$-complete matching in $\tilde{G}$ that corresponds to M in G.

To define the edge weight function wt_M in $\tilde{G}$, we will first define the function $\mathsf{vote}_u(v, v')$ for any vertex u and neighbors v, v' of u. This is as follows:

$$\mathsf{vote}_u(v, v') = \begin{cases} 1 & \text{if } u \text{ prefers } v \text{ to } v' \\ 1 & \text{if } u \text{ prefers } v' \text{ to } v \\ 0 & \text{otherwise, i.e., } v = v'. \end{cases}$$

For any edge (a, b) in G, let $\mathsf{wt}_M(a, b)$ be $\mathsf{vote}_a(b, \tilde{M}(a)) + \mathsf{vote}_b(a, \tilde{M}(b))$. Observe that if $(a, b) \in M$ then $\mathsf{wt}_M(a, b) = 0$.

For any $u \in A \cup B$, if $\tilde{M}(u) = \ell(u)$ then let $\mathsf{wt}_M(u, \ell(u))$ be 0; else let $\mathsf{wt}_M(u, \ell(u))$ be -1. Thus $\mathsf{wt}_M(u, \ell(u))$ is u's vote for $\ell(u)$ versus its partner in $\tilde{M}$.

Let N be any matching in G. $\tilde{N}$ is the $(A \cup B)$-complete matching in $\tilde{G}$ corresponding to N. We have:

$$\mathsf{wt}_M(\tilde{N}) = \sum_{(a,b)\in N} \mathsf{wt}_M(a, b) + \sum_{(u,\ell(u))\in \tilde{N}} \mathsf{wt}_M(u, \ell(u))$$

$$= \sum_{u\in A\cup B} \mathsf{vote}_u(\tilde{N}(u), \tilde{M}(u)) = \phi(N, M) - \phi(M, N).$$

Since $\phi(N, M) - \phi(M, N) = \mathsf{wt}_M(\tilde{N})$, the matching M is popular if and only if $\mathsf{wt}_M(\tilde{N}) \leq 0$ for all matchings N in G. Observe that $\mathsf{wt}_M(\tilde{M}) = 0$. So M is popular if and only if the optimal value of the max-weight $(A \cup B)$-complete matching in $\tilde{G}$ with edge weight function wt_M is 0.

The linear program LP1 given below is the dual of the max-weight $(A \cup B)$-complete matching LP in $\tilde{G}$. It follows from LP-duality that M is popular if and only if the optimal value of LP1 is 0.

$$\text{minimize} \sum_{u\in A\cup B} \alpha_u \qquad \text{(LP1)}$$

subject to

$$\alpha_u + \alpha_v \geq \mathsf{wt}_M(u, v) \quad \forall (u, v) \in E$$
$$\alpha_u \geq \mathsf{wt}_M(u, \ell(u)) \quad \forall u \in A \cup B.$$

Definition 2. *For any popular matching M, an optimal solution α to LP1 above is called a* witness *of M.*

The vector $\mathbf{0}$ is a witness of popularity of any stable matching. Note that a popular matching may have several witnesses.

Lemma 1 ([12]). *Every popular matching M in $G = (A \cup B, E)$ has a witness in $\{0, \pm 1\}^n$, where $n = |A| + |B|$.*

Let M be a popular matching and let α be a witness to M's popularity. Let N be another popular matching in G. Then $\phi(M, N) = \phi(N, M)$. So $\mathsf{wt}_M(\tilde{N}) = 0$, thus $\tilde{N}$ is an optimal solution to the max-weight $(A \cup B)$-complete matching linear program in $\tilde{G}$ with edge weight function wt_M. Recall that α is an optimal solution to the dual LP, i.e., LP1.

Call $(a, b) \in E$ a *popular edge* if there is a popular matching in G that contains (a, b), i.e., $(a, b) \in N$ for some popular matching N. Lemma 2 follows from complementary slackness. A vertex u is *unstable* if u is left unmatched in any stable matching. The second part of Lemma 2 follows from the observation that $(u, \ell(u)) \in \tilde{S}$ for any unstable vertex u and stable matching S.

Lemma 2 ([13]). *If (a, b) is a popular edge then $\alpha_a + \alpha_b = \mathsf{wt}_M(a, b)$. Also, $\alpha_u = \mathsf{wt}_M(u, \ell(u))$ for any unstable vertex u.*

3   The NP-Hardness proof

Let $H = (V_H, E_H)$ be an instance of the vertex cover problem on n vertices and m edges. Let $V_H = [n]$. We will now build a bipartite instance $G = (A \cup B, E)$.

Let $e = (i, j) \in E_H$, where $i < j$. Corresponding to e, we will have a gadget C_e on $6n + 4$ vertices in G. The gadget C_e will consist of a 4-cycle on vertices x_e, y_e, x'_e, y'_e along with n gadgets D_e^t for $1 \leq t \leq n$. Each gadget D_e^t consists of 6 vertices: p_e^t, r_e^t, u_e^t in A and q_e^t, s_e^t, w_e^t in B.

For any $e \in E_H$ and $1 \leq t \leq n$, the preference lists of the 6 vertices in the gadget D_e^t are as follows. These preference lists are inspired by the gadget used in [13] to show the NP-hardness of the max-utility popular matching problem in bipartite instances.

- The preference list of p_e^t is $q_e^t \succ w_e^t$ and that of q_e^t is $p_e^t \succ u_e^t$.
- The preference list of r_e^t is $s_e^t \succ w_e^t$ and that of s_e^t is $r_e^t \succ u_e^t$.

Thus the vertices p_e^t and q_e^t are each other's top choice neighbors. Similarly, the vertices r_e^t and s_e^t are each other's top choice neighbors. The last choice of p_e^t (also, r_e^t) is w_e^t and the last choice of q_e^t (also, s_e^t) is u_e^t.

- The preference list of u_e^t is $q_e^t \succ y_e \succ s_e^t \succ y'_e$.
- The preference list of w_e^t is $r_e^t \succ x_e \succ p_e^t \succ x'_e$.

Thus there are edges between u_e^t and y_e, y'_e, and similarly, between w_e^t and x_e, x'_e, for every t. The preference lists of the vertices x_e, y_e, x'_e, y'_e are as follows.

- The preference lists of x_e and x'_e are very similar: while x_e's list consists of $y_e \succ y'_e \succ w_e^1 \succ \cdots \succ w_e^n$, x'_e's list is exactly the same as x_e's list with b'_i at the bottom of its list.
- Similarly, the preference lists of y_e and y'_e are very similar: while y_e's list consists of $x_e \succ x'_e \succ u_e^1 \succ \cdots \succ u_e^n$, y'_e's list is exactly the same as y_e's list with a'_j at the bottom of its list.

Thus the last choices of x'_e and y'_e are b'_i and a'_j, respectively. Recall that $e = (i, j)$, the vertices b'_i and a'_j belong to the gadgets corresponding to vertices i and j, respectively. We describe these vertex gadgets now.

Corresponding to each $i \in V_H$, there will be 4 vertices a_i, b_i, a'_i, b'_i in G. Their preference lists are:

- a_i's preference list is $b_i \succ b'_i$ while b_i's is $a_i \succ a'_i$,
- a'_i's preference list is $b_i \succ y'_{e_1} \succ \cdots \succ y'_{e_h}$, and
- b'_i's preference list is $a_i \succ x'_{e'_1} \succ \cdots \succ x'_{e'_\ell}$.

Here $e_1, \ldots, e_h$ are edges incident to i in H with i as their higher valued endpoint and $e'_1, \ldots, e'_\ell$ are edges incident to i in H with i as their lower valued endpoint. The order among $y'_{e_1}, \ldots, y'_{e_h}$ (similarly, $x'_{e'_1}, \ldots, x'_{e'_\ell}$) in the preference list of a'_i (resp., b'_i) does not matter.

It is known [10] that every popular matching matches all *stable* vertices (those matched in a stable matching). For each $e \in E_H$ and $t \in [n]$, the vertices u^t_e and w^t_e are *unstable* since any stable matching in G includes the edges (p^t_e, q^t_e), (r^t_e, s^t_e) for all t, and also (x_e, y_e), (x'_e, y'_e). Similarly, for any $i \in V_H$, the vertices a'_i and b'_i are unstable.

Vertex preferences have been set such that the following lemma holds for any popular matching in G.

Lemma 3. *Let M be any popular matching in G. Then for every $e \in E_H$, either $\{(x_e, y_e), (x'_e, y'_e)\} \subseteq M$ or $\{(x_e, y'_e), (x'_e, y_e)\} \subseteq M$.*

Since x_e, y_e, x'_e, y'_e have to be matched in any popular matching in G, Claims 1 and 2 given below immediately imply Lemma 3.

Claim 1. *For any $e = (i, j) \in E_H$, where $i < j$, neither (x'_e, b'_i) nor (a'_j, y'_e) is a popular edge.*

Proof. In order to show that neither (x'_e, b'_i) nor (a'_j, y'_e) is a popular edge, consider the following matching M:

$$M = \cup_e \{(x_e, y_e), (x'_e, y'_e)\} \cup_{e,t} \{(p^t_e, q^t_e), (r^t_e, s^t_e)\} \cup_i \{(a_i, b'_i), (a'_i, b_i)\}.$$

The matching M is popular as witnessed by the following vector α:

- for all $1 \leq i \leq n$: set $\alpha_{a_i} = \alpha_{b_i} = 1$ and $\alpha_{a'_i} = \alpha_{b'_i} = -1$.
- for all $e \in E_H$: set $\alpha_v = 0$ for all vertices v in the gadget C_e, i.e., for $v \in \{x_e, y_e, x'_e, y'_e\} \cup \{p^t_e, q^t_e, r^t_e, s^t_e, u^t_e, w^t_e : 1 \leq t \leq n\}$.

It is easy to check that the above vector α satisfies all the constraints of LP1, i.e., for each edge (v, v'), we have $\alpha_v + \alpha_{v'} \geq \mathrm{wt}_M(v, v')$. Also $\alpha_v \geq \mathrm{wt}_M(v, \ell(v))$ for all v. Moreover, $\sum_{v \in V_G} \alpha_v = 0$. Thus M is a popular matching.

For each $e = (i, j) \in E_H$, $\mathrm{wt}_M(x'_e, b'_i) = \mathrm{wt}_M(a'_j, y'_e) = -2$ while $\alpha_{x'_e} = \alpha_{y'_e} = 0$ and $\alpha_{a'_j} = \alpha_{b'_i} = -1$. Thus $\alpha_{x'_e} + \alpha_{b'_i} > \mathrm{wt}_M(x'_e, b'_i)$ and $\alpha_{a'_j} + \alpha_{y'_e} > \mathrm{wt}_M(a'_j, y'_e)$. It follows from Lemma 2 that neither (x'_e, b'_i) nor (a'_j, y'_e) is a popular edge. $\square$

Claim 2. *For all $e \in E_H, t \in [n]$, neither (x_e, w_e^t) nor (x'_e, w_e^t) is a popular edge; similarly, neither (u_e^t, y_e) nor (u_e^t, y'_e) is a popular edge.*

The proof of Claim 2 is similar to the proof of Claim 1. Let M be any popular matching in G. Since all stable vertices have to be matched in every popular matching in G, all the vertices $p_e^t, q_e^t, r_e^t, s_e^t$ for all $t \in [n]$ in any edge gadget C_e have to be matched in M.

Lemma 5 shows that for any $e \in E_H$, a popular matching in G either matches *no* unstable vertex in C_e or it matches *all* the $2n$ unstable vertices u_e^t, w_e^t for $1 \leq t \leq n$ in C_e. Before we show Lemma 5, we will show the following useful lemma. Let M be a popular matching in G and let $\boldsymbol{\alpha} \in \{0, \pm 1\}^{|V_G|}$ be a witness of M.

Lemma 4. *For any $e \in E_H$, if M matches at least one unstable vertex in C_e, then at least one of $\alpha_{x'_e}, \alpha_{y'_e}$ is -1.*

Proof. Let u_e^t be an unstable vertex in C_e that is matched in M for some $t \in \{1, \ldots, n\}$. So by Lemma 3, either $(u_e^t, q_e^t), (p_e^t, w_e^t)$ are in M or $(u_e^t, s_e^t), (r_e^t, w_e^t)$ are in M. So one of the following cases holds:

- $(p_e^t, w_e^t) \in M$ in which case $\mathsf{wt}_M(x_e, w_e^t) = 0$ (since w_e^t prefers x_e to p_e^t)
- $(u_e^t, s_e^t) \in M$ in which case $\mathsf{wt}_M(u_e^t, y_e) = 0$ (since u_e^t prefers y_e to s_e^t).

So either $\mathsf{wt}_M(x_e, w_e^t) = 0$ or $\mathsf{wt}_M(u_e^t, y_e) = 0$. We know that $\alpha_{x_e} + \alpha_{w_e^t} \geq \mathsf{wt}_M(x_e, w_e^t)$ and $\alpha_{u_e^t} + \alpha_{y_e} \geq \mathsf{wt}_M(u_e^t, y_e)$.

We know from Lemma 2 that $\alpha_{u_e^t} = \mathsf{wt}_M(u_e^t, \ell(u_e^t))$. Since u_e^t is matched in M, we have $\mathsf{wt}_M(u_e^t, \ell(u_e^t)) = -1$. Similarly, $\alpha_{w_e^t} = -1$. Since $\alpha_{u_e^t} = \alpha_{w_e^t} = -1$, in order to maintain the above constraints for edges (x_e, w_e^t) and (u_e^t, y_e), at least one of $\alpha_{x_e}, \alpha_{y_e}$ has to be 1.

We will now show that $\alpha_{x'_e} = -\alpha_{y_e}$ and $\alpha_{y'_e} = -\alpha_{x_e}$. We know that either (i) $(x_e, y_e), (x'_e, y'_e)$ are in M or (ii) $(x_e, y'_e), (x'_e, y_e)$ are in M (by Lemma 3). In both cases, note that $\mathsf{wt}_M(x_e, y'_e) = \mathsf{wt}_M(x'_e, y_e) = 0$. In case (ii), we have $\alpha_{x_e} + \alpha_{y'_e} = 0$ and $\alpha_{x'_e} + \alpha_{y_e} = 0$ (by Lemma 2).

In case (i), $\alpha_{x_e} + \alpha_{y'_e} \geq 0$ and $\alpha_{x'_e} + \alpha_{y_e} \geq 0$; however $\alpha_{x_e} + \alpha_{y_e} = \alpha_{x'_e} + \alpha_{y'_e} = 0$. This means $\alpha_{x_e} + \alpha_{y_e} + \alpha_{x'_e} + \alpha_{y'_e} = 0$ and so it has to be the case that $\alpha_{x_e} + \alpha_{y'_e} = \alpha_{x'_e} + \alpha_{y_e} = 0$.

So in both cases we have $\alpha_{x'_e} = -\alpha_{y_e}$ and $\alpha_{y'_e} = -\alpha_{x_e}$. Since at least one of $\alpha_{x_e}, \alpha_{y_e}$ has to be 1, it follows that at least one of $\alpha_{x'_e}, \alpha_{y'_e}$ is -1. $\square$

Lemma 5. *For any edge e, if some unstable vertex in C_e is matched in a popular matching M then all unstable vertices in C_e are matched in M.*

Proof. Let u_e^t be an unstable vertex in C_e that is matched in M for some $t \in \{1, \ldots, n\}$. We know from Lemma 4 that at least one of $\alpha_{x'_e}, \alpha_{y'_e}$ is -1, where $\boldsymbol{\alpha} \in \{0, \pm 1\}^{|V_G|}$ is a witness of M.

Suppose some unstable vertices u_e^h, w_e^h are *unmatched* in M for some $h \in \{1, \ldots, n\}$. Then $\mathsf{wt}_M(u_e^h, y_e') = 0$, since u_e^h prefers to be matched to y_e' than be unmatched while y_e' prefers its partner in M (either x_e or x_e') to u_e^h. Similarly, $\mathsf{wt}_M(x_e', w_e^h) = 0$.

Since $\mathsf{wt}_M(u_e^h, \ell(u_e^h)) = \mathsf{wt}_M(w_e^h, \ell(w_e^h)) = 0$, it follows from Lemma 2 that $\alpha_{u_e^h} = \alpha_{w_e^h} = 0$. Also $\alpha_{x_e'} = -1$ or $\alpha_{y_e'} = -1$. Thus either $\alpha_{u_e^h} + \alpha_{y_e'} < 0$ or $\alpha_{x_e'} + \alpha_{w_e^h} < 0$.

Hence $\boldsymbol{\alpha}$ violates some constraint of LP1, a contradiction to $\boldsymbol{\alpha}$ being a feasible solution to LP1. So every unstable vertex in C_e has to be matched in M. $\qquad \square$

Lemma 6. *Let M be a popular matching in G of size more than $m \cdot (3n+2) + n$, where $|E_H| = m$. The set $U = \{i \in [n] : (a_i, b_i') \in M\}$ is a vertex cover of H.*

Proof. The matching M has to match at least one unstable vertex in every C_e. Otherwise C_e would contribute only $2n + 2$ edges to M and this would make $|M| \le (2n+2) + (3n+2) \cdot (m-1) + 2n = m \cdot (3n+2) + n$.

Since $|M| > m \cdot (3n+2) + n$, M has to match at least one unstable vertex in every C_e. It now follows from Lemma 4 that for every $e \in E_H$, at least one of $\alpha_{x_e'}, \alpha_{y_e'}$ is -1, where $\boldsymbol{\alpha} \in \{0, \pm 1\}^{|V_G|}$ is a witness to M's popularity.

Let $(i, j) \in E_H$, where $i < j$. We need to show that either i or j is in U. Suppose not. Then both (a_i, b_i) and (a_j, b_j) are in M. This means $\alpha_{b_i'} = \alpha_{a_j'} = 0$ (by Lemma 2). Also $\mathsf{wt}_M(x_e', b_i') = 0$ since b_i' prefers x_e' to being unmatched, while x_e' prefers both y_e and y_e' (its possible partners in M) to b_i'. Similarly, $\mathsf{wt}_M(a_j', y_e') = 0$.

At least one of $\alpha_{x_e'}, \alpha_{y_e'}$ is -1 (by Lemma 4), so either $\alpha_{x_e'} + \alpha_{b_i'} < \mathsf{wt}_M(x_e', b_i')$ or $\alpha_{a_j'} + \alpha_{y_e'} < \mathsf{wt}_M(a_j', y_e')$. This is a contradiction to $\boldsymbol{\alpha}$ being a witness to M's popularity. Thus either i or j is in U for every edge $(i, j) \in E_H$. $\qquad \square$

Theorem 3. *For any integer $c \ge 1$, the graph H has a vertex cover of size c if and only if G has a popular matching of size $m \cdot (3n+2) + n + c$.*

Proof. Suppose G has a popular matching M of size $m \cdot (3n+2) + n + c$, where $c \ge 1$. We know from Lemma 6 that the set $U = \{i \in [n] : (a_i, b_i') \in M\}$ is a vertex cover of H. We will now show that $|U| = c$.

Since $|M| > m \cdot (3n+2) + n$, it follows from the proof of Lemma 6 that M has to match at least one unstable vertex in C_e, for every $e \in E_H$. This means that M has to match all vertices in C_e for all $e \in E_H$ (by Lemma 5). Edges within C_e for all $e \in E_H$ account for $m \cdot (3n+2)$ many edges in M.

Corresponding to each $i \in U$, there are 2 edges (a_i, b_i') and (a_i', b_i) in M and corresponding to each $i \notin U$, there is 1 edge (a_i, b_i) in M. Hence $|M| = m \cdot (3n+2) + 2|U| + n - |U| = m \cdot (3n+2) + n + |U|$. Thus $|U| = c$.

We now show the converse. Let $U \subseteq [n]$ be a vertex cover of size c in H. We will use U to build a popular matching M of size $m \cdot (3n+2) + n + c$ as follows:

- for every $i \in U$ do: include edges (a_i, b_i') and (a_i', b_i) in M.
- for every $i \notin U$ do: include the edge (a_i, b_i) in M.
- for every $e = (i, j)$, where $i < j$, in E_H do: include edges (x_e, y_e) and (x_e', y_e') in M;

- if $i \in U$ then add $(p_e^t, q_e^t), (r_e^t, w_e^t), (u_e^t, s_e^t)$ to M for all $1 \leq t \leq n$
- if $i \notin U$ then add $(p_e^t, w_e^t), (r_e^t, s_e^t), (u_e^t, q_e^t)$ to M for all $1 \leq t \leq n$

The size of M is $m \cdot (3n+2) + 2|U| + n - |U| = m \cdot (3n+2) + n + c$. In order to prove the popularity of M, consider the vector $\boldsymbol{\alpha}$ defined as follows:

- for every $i \in U$ do: set $\alpha_{a_i} = \alpha_{b_i} = 1$ and set $\alpha_{a_i'} = \alpha_{b_i'} = -1$
- for every $i \notin U$ do: set $\alpha_{a_i} = \alpha_{b_i} = \alpha_{a_i'} = \alpha_{b_i'} = 0$
- for every $e = (i, j)$, where $i < j$, in E_H do:
  - if $i \in U$ then set $\alpha_{x_e} = \alpha_{x_e'} = -1$ and $\alpha_{y_e} = \alpha_{y_e'} = 1$;
    set $\alpha_{q_e^t} = \alpha_{r_e^t} = \alpha_{s_e^t} = 1$ and $\alpha_{p_e^t} = \alpha_{u_e^t} = \alpha_{w_e^t} = -1$ for all $t \in [n]$.
  - if $i \notin U$ then set $\alpha_{x_e} = \alpha_{x_e'} = 1$ and $\alpha_{y_e} = \alpha_{y_e'} = -1$;
    set $\alpha_{p_e^t} = \alpha_{q_e^t} = \alpha_{r_e^t} = 1$ and $\alpha_{s_e^t} = \alpha_{u_e^t} = \alpha_{w_e^t} = -1$ for all $t \in [n]$.

It can be checked that $\boldsymbol{\alpha}$ is a witness to M's popularity. Thus M is a popular matching in G of size $m \cdot (3n+2) + n + c$. □

This finishes the NP-hardness proof of the size k popular matching problem in $G = (A \cup B, E)$. Thus we have proved Theorem 1 stated in Sect. 1.

Note that it is easy to check if a given matching in $G = (A \cup B, E)$ is popular or not [2,10]. Hence the problem of deciding whether G admits a size k popular matching is NP-complete.

4   A Near-Popular Matching of Size k

Let $G = (A \cup B, E)$ be our input instance with strict preference lists. Given a parameter k sandwiched between min and max, we consider the problem of computing a matching M of size k in G with $u(M) \leq 2$ (see (1) in Sect. 1 for the definition of $u(M)$). Recall that min and max are the sizes of a stable matching and a max-size popular matching in G.

Our algorithm is an adaptation of the linear time algorithm called the *2-level Gale-Shapley algorithm* from [11] to compute a max-size popular matching in G. Running the 2-level Gale-Shapley algorithm in G is equivalent to running the Gale-Shapley algorithm in a related graph G', as shown in [4].

It is known that all max-size popular matchings match the same subset of vertices [9], let Pop denote this subset of $A \cup B$. Let Stab be the set of vertices matched in a stable matching of G. It is known [7] that all stable matchings in G match the same subset Stab of vertices. Note that $|\text{Stab} \cap A| = \text{min}$ and $|\text{Pop} \cap A| = \text{max}$.

Our algorithm computes the sets Stab and Pop using the linear time algorithms in [6,11], respectively. It then picks any subset X of A such that $\text{Stab} \cap A \subseteq X \subseteq \text{Pop} \cap A$ and $|X| = k$.

A new graph G_X on vertex set $A_X \cup B'$ is constructed. Define A_X and B' as follows: $A_X = \{a_0 : a \in A\} \cup \{a_1 : a \in X\}$ and $B' = B \cup \{d(a) : a \in A\}$. Thus for each $a \in A$, a new vertex $d(a)$ is introduced in the set B'.

The vertices in $\{a_0 : a \in A\}$ will be called *level 0* vertices and those in $\{a_1 : a \in X\}$ will be called *level 1* vertices.

For every $a \in A$, the edge set E_X of G_X consists of the edges $(a_0, d(a))$ and (a_0, b) for every neighbor b of a in G. For $a \in X$, the edges $(a_1, d(a))$ and (a_1, b) for every neighbor b also belong to the edge set of G_X. Thus $E_X = \{(a_i, b) : (a, b) \in E \text{ and } a_i \in A_X\} \cup \{(a_i, d(a)) : a_i \in A_X\}$.

The preference lists of all vertices in G_X are as follows:

- For any $a \in A$, the preference list of a_0 is the same as a's preference list in G with the vertex $d(a)$ as its *least* preferred neighbor.
- For any $a \in X$, the preference list of a_1 is the same as a's preference list in G with the vertex $d(a)$ as its *most* preferred neighbor.
- For $a \in X$, the preference list of $d(a)$ is $a_0 \succ a_1$, i.e., $d(a)$'s top choice is a_0 and second choice is a_1.
- For each $b \in B$, the preference list of b in G_X is all its *level 1* neighbors as per their order of preference in G followed by all its *level 0* neighbors in their original order of preference.

For instance, let b's preference list in G be $a \succ a' \succ a''$ and let a, a'' belong to X and $a' \notin X$. Then b's preference list in G'_X is $a_1 \succ a_1'' \succ a_0 \succ a_0' \succ a_0''$.

The main step of our algorithm is to run Gale-Shapley algorithm [6] in the instance G_X: so vertices in A_X propose and those in B' dispose. Let M_X be the stable matching in G_X computed by this algorithm.

Let M be the matching obtained by deleting all $(a_i, d(a))$ edges from M_X (for $i = 0, 1$) and projecting any (a_j, b) edge in M_X, where $j \in \{0, 1\}$ and $b \in B$, to (a, b). Lemmas 7 and 8 show that M is the matching we seek.

Lemma 7. *The size of M is k.*

Proof. We will show that every vertex in X is matched in M and no vertex in $A \setminus X$ is matched in M. Thus $|M| = |X| = k$.

Since a_0 is $d(a)$'s most preferred neighbor, every vertex in $\{a_0 : a \in A\}$ has to be matched in any stable matching in G_X. Let $a \in A \setminus X$. We claim that a_0 is matched to $d(a)$ in the matching M_X. In the Gale-Shapley algorithm in G_X, the vertex a_0 would be rejected by all its neighbors in B. This is because every vertex in $A \setminus X$ is *unstable* in G (since $X \supseteq \text{Stab} \cap A$). Thus $d(a)$ would be the only neighbor to accept a_0's proposal. So $(a_0, d(a)) \in M_X$ for every $a \in A \setminus X$. Corresponding to any $a \in A \setminus X$, there is no level 1 vertex a_1; thus the vertex a is unmatched in M.

We will now show that every vertex in X is matched in M. Suppose $a \in X$ is not matched in M, i.e., a_1 is unmatched in M_X. Then a_1 would also be unmatched in the matching M^* obtained by running Gale-Shapley algorithm in the graph G', which is the graph G_X when $X = A$. The graph G' is obtained by adding the vertices a_1 for $a \in A \setminus X$ along with appropriate edges (a_1, b) to G_X.

It was shown in [4] that running the Gale-Shapley algorithm in G' computes a *max-size* popular matching in G. Since every vertex in X has to be matched in a max-size popular matching (recall that $X \subseteq \text{Pop}$), the vertex a_1 is matched in M^*. This is a contradiction to a being unmatched in M. Hence every $a \in X$ is matched in M. $\qquad\square$

We will partition the vertex set of G as follows: we already have $A = X \cup (A \setminus X)$ and now we will further partition X into X_0 and X_1. Let $X_0 = \{a : (a_0, b) \in M_X$ for some b in $B\}$ and $X_1 = \{a : (a_1, b) \in M_X$ for some b in $B\}$.

The set B will be partitioned into $B_0 \cup B_1$ as follows: let $B_1 = \{b \in B : (a_1, b) \in M_X$ for some a in $A\}$ and let $B_0 = B \setminus B_1$.

Observe that $M \subseteq (X_0 \times B_0) \cup (X_1 \times B_1)$. The following properties were proved for $X = A$ in [11] and it can be shown that they hold for any $X \subseteq A$:

(∗) For any $(a, b) \in X_1 \times B_0$, a and b prefer their partners in M to each other.

(∗∗) For any $(a, b) \in (X_0 \times B_0) \cup (X_1 \times B_1)$, either a prefers $M(a)$ to b or b prefers $M(b)$ to a. For any $(a, b) \in (A \setminus X) \times B_0$, b prefers $M(b)$ to a.

It follows from (∗) and (∗∗) that any *blocking edge* (a, b) to M, i.e., where both a and b prefer each other to their assignments in M, has to satisfy $b \in B_1$ and $a \in A \setminus X_1$, i.e., $a \in X_0$ or $a \in A \setminus X$.

We are ready to prove the following lemma. The proof of Lemma 8 is similar to the proof of correctness of the max-size popular matching algorithm in [11].

Lemma 8. *The unpopularity factor of M is at most 2.*

Proof. We will show the following:

(i) for any alternating path ρ with respect to M, $\phi(M \oplus \rho, M) \leq 2\phi(M, M \oplus \rho)$.

(ii) for any alternating cycle ρ with respect to M, $\phi(M \oplus \rho, M) \leq \phi(M, M \oplus \rho)$.

Let ρ be an alternating path with respect to M. Consider the edges in $\rho \setminus M$: suppose $\rho \setminus M$ has x edges from $((A \setminus X_1) \times B_0) \cup (X_1 \times B_1)$, y edges from $(A \setminus X_1) \times B_1$, and z edges from $X_1 \times B_0$.

We know from (∗) that for each edge $(a, b) \in X_1 \times B_0$, both a and b prefer M to $M \oplus \rho$. We know from (∗∗) that for every edge $(a, b) \in ((A \setminus X_1) \times B_0) \cup (X_1 \times B_1)$, at least one of a, b prefers M to $M \oplus \rho$. An edge in $(A \setminus X_1) \times B_1$ may be a *blocking edge* to M. Hence among the vertices of ρ, the number of votes for $M \oplus \rho$ versus M is at most $x + 2y$.

Suppose exactly one endpoint of ρ is unmatched in M. Then among the vertices of ρ, the number of votes for M versus $M \oplus \rho$ is at least $x + 2z + 1$: here the "1" counts the matched endpoint v of ρ. This vertex v prefers M to $M \oplus \rho$ since it is matched in M but unmatched in $M \oplus \rho$.

The crucial point here is that since $M \subseteq (X_0 \times B_0) \cup (X_1 \times B_1)$, we have $z \geq y - 1$. This is because between 2 occurrences of edges from $(A \setminus X_1) \times B_1$ in $\rho \setminus M$, there needs to be an edge from $X_1 \times B_0$.

Thus $\phi(M, M \oplus \rho) \geq x + 2z + 1 \geq x + 2y - 1$. So we have:

$$\frac{\phi(M \oplus \rho, M)}{\phi(M, M \oplus \rho)} \leq \frac{x + 2y}{x + 2y - 1} \leq 2.$$

Suppose both endpoints of ρ are unmatched in M. Thus one endpoint of ρ is in $A \setminus X$ and the other endpoint is in B_0. This implies $z = y$. Hence the number of votes for M versus $M \oplus \rho$ is at least $x + 2y$. Thus we have $\phi(M \oplus \rho, M) \leq \phi(M, M \oplus \rho)$ here.

The proof when both endpoints of ρ are matched in M is similar to the first case: here $z \geq y - 1$, however *both* the endpoints of ρ prefer M to $M \oplus \rho$. Thus $\phi(M, M \oplus \rho) \geq x + 2(y-1) + 2 = x + 2y$. The proof in the case of alternating cycle is the same as in the second case above: here $z = y$ and so $\phi(M, M \oplus \rho) \geq x + 2y$. Thus in both these cases as well, we have $\phi(M \oplus \rho, M) \leq \phi(M, M \oplus \rho)$.

Let $N \neq M$ be any matching in $G - (A \cup B, E)$. So $N \oplus M$ is a collection of alternating paths and cycles with respect to M. For every $\rho \in N \oplus M$, we know from (i) and (ii) above that $\phi(M \oplus \rho, M) \leq 2\phi(M, M \oplus \rho)$. This immediately implies that $\phi(N, M)/\phi(M, N) \leq 2$. Thus $u(M) \leq 2$.   □

Since our algorithm runs in linear time, this proves Theorem 2 from Sect. 1.

References

1. Abraham, D.J., Irving, R.W., Kavitha, T., Mehlhorn, K.: Popular matchings. SIAM J. Comput. **37**(4), 1030–1045 (2007)
2. Biró, P., Irving, R.W., Manlove, D.F.: Popular matchings in the marriage and roommates problems. In: Proceedings of 7th International Conference on Algorithms and Complexity (CIAC), pp. 97–108 (2010)
3. Cseh, Á., Huang, C.-C., Kavitha, T.: Popular matchings with two-sided preferences and one-sided ties. SIAM J. Discret. Math. **31**(4), 2348–2377 (2017)
4. Cseh, Á., Kavitha, T.: Popular edges and dominant matchings. In: Louveaux, Q., Skutella, M. (eds.) IPCO 2016. LNCS, vol. 9682, pp. 138–151. Springer, Cham (2016). https://doi.org/10.1007/978-3-319-33461-5_12
5. Faenza, Y., Kavitha, T., Powers, V., Zhang, X.: Popular matchings and limits to tractability. http://arxiv.org/abs/1805.11473 (2018)
6. Gale, D., Shapley, L.S.: College admissions and the stability of marriage. Am. Math. Mon. **69**, 9–15 (1962)
7. Gale, D., Sotomayor, M.: Some remarks on the stable matching problem. Discret. Appl. Math. **11**, 223–232 (1985)
8. Gärdenfors, P.: Match making: assignments based on bilateral preferences. Behav. Sci. **20**, 166–173 (1975)
9. Hirakawa, M., Yamauchi, Y., Kijima, S., Yamashita, M.: On the structure of popular matchings in the stable marriage problem - who can join a popular matching? In: The 3rd International Workshop on Matching Under Preferences (2015)
10. Huang, C.-C., Kavitha, T.: Popular matchings in the stable marriage problem. Inf. Comput. **222**, 180–194 (2013)
11. Kavitha, T.: A size-popularity tradeoff in the stable marriage problem. SIAM J. Comput. **43**(1), 52–71 (2014)
12. Kavitha, T.: Popular half-integral matchings. In: Proceedings of 43rd International Colloquium on Automata, Languages, and Programming (ICALP), pp. 22.1–22.13 (2016)
13. Kavitha, T.: Max-size popular matchings and extensions (2018). http://arxiv.org/abs/1802.07440
14. Kavitha, T., Mestre, J., Nasre, M.: Popular mixed matchings. Theor. Comput. Sci. **412**, 2679–2690 (2011)
15. McCutchen, R.M.: The least-unpopularity-factor and least-unpopularity-margin criteria for matching problems with one-sided preferences. In: Laber, E.S., Bornstein, C., Nogueira, L.T., Faria, L. (eds.) LATIN 2008. LNCS, vol. 4957, pp. 593–604. Springer, Heidelberg (2008). https://doi.org/10.1007/978-3-540-78773-0_51

Convexity-Increasing Morphs of Planar Graphs

Linda Kleist[1], Boris Klemz[2(✉)], Anna Lubiw[3], Lena Schlipf[4], Frank Staals[5], and Darren Strash[6]

[1] Technische Universität Berlin, Berlin, Germany
[2] Freie Universität Berlin, Berlin, Germany
`klemz@inf.fu-berlin.de`
[3] University of Waterloo, Waterloo, Canada
[4] FernUniversität in Hagen,
Hagen, Germany
[5] Utrecht University, Utrecht, The Netherlands
[6] Hamilton College, Clinton, USA

Abstract. We study the problem of *convexifying* drawings of planar graphs. Given any planar straight-line drawing of a 3-connected graph G, we show how to morph the drawing to one with convex faces while maintaining planarity at all times. Furthermore, the morph is *convexity increasing*, meaning that angles of inner faces never change from convex to reflex. We give a polynomial time algorithm that constructs such a morph as a composition of a linear number of steps where each step either moves vertices along horizontal lines or moves vertices along vertical lines.

1 Introduction

A *morph* between two planar straight-line drawings Γ_0 and Γ_1 of a graph G is a continuous movement of the vertices from one to the other, with the edges following along as straight-line segments between their endpoints. A morph is *planar* if it preserves planarity of the drawing at all times.

Motivated by applications in animation and in reconstruction of 3D shapes from 2D slices, the study of morphing has focused on finding a morph between two given planar drawings. The existence of planar morphs was established long ago [5,24], followed by algorithms that produce good visual results [11,13], and algorithms that find "piece-wise linear" morphs with a linear number of steps [2].

Our focus is somewhat different, and more aligned with graph drawing goals—our input is a planar graph drawing and our aim is to morph it to a better drawing, in particular to a convex drawing. A morph *convexifies* a given straight-line graph drawing if the end result is a *convex graph drawing*, i.e. a planar straight-line graph drawing in which every face is a convex polygon. For a survey on convex graph drawing, see [22].

Due to space constraints, some proofs in this manuscript are only sketched or omitted entirely. Full proofs of all claims can be found in the full preprint version [16].

© Springer Nature Switzerland AG 2018
A. Brandstädt et al. (Eds.): WG 2018, LNCS 11159, pp. 318–330, 2018.
https://doi.org/10.1007/978-3-030-00256-5_26

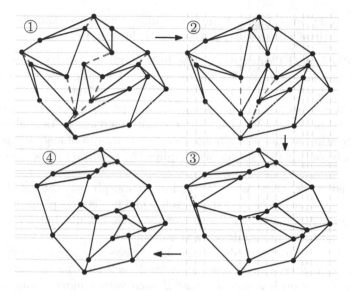

Fig. 1. A sequence of convexity-increasing morphs (horizontal, vertical, horizontal) that morph a straight-line drawing of a graph G (drawn in black) into a strictly convex drawing of G.

We first observe that it is easy, using known results, to find a planar morph that convexifies a given graph drawing—we can just create a convex drawing with the same faces (assuming such a drawing exists), and morph to that specific drawing using the known planar morphing algorithms. In this paper, we impose a stronger condition: we want to find a confexifying morph which is also *convexity-increasing*, meaning that an angle of an inner face never switches from convex to reflex. Besides the theoretical goal of studying continuous motion that is monotonic in some measure (e.g. edge lengths [15]), another motivation comes from visualization—a morph of a graph drawing should maintain the user's "mental map" [20] which means changing as little as possible, and making observable progress towards a goal. Previous morphing algorithms fail to provide convexity-increasing morphs even if the target is a convex drawing because they all start by triangulating the drawing. This means that an original convex angle may be subdivided by new triangulation edges, so there is no constraint that keeps it convex.

Our main result is an algorithm to convexify any straight-line planar drawing of a 3-connected graph via a planar convexity-increasing morph. In fact, we show a surprising stronger property—that the morph can be composed of a linear number of *horizontal* and *vertical* morphs. A *horizontal* morph moves all vertices at constant speeds along horizontal lines, and a *vertical* morph is defined similarly. See Fig. 1 for an illustration. Orthogonality is a very desirable and well-studied criterion for graph drawing [9], in part because there is evidence that the human visual cortex comprehends orthogonal lines more easily [4,21]. Similarly, it seems natural that orthogonal *motion* should be easier to comprehend, though morphing algorithms have so far not explored this criterion.

Related Work and Concepts. To the best of our knowledge there is no previous work on convexity-increasing morphs except for the case when the input graph is a path or a cycle. Connelly et al. [7] and Canterella et al. [6] gave algorithms to convexify a simple polygon while preserving edge lengths. Since their motions are "expansive", they are convexity-increasing. Aichholzer et al. [1] gave an algorithm to find a "visibility-increasing" morph of a simple polygon to a convex polygon; this condition also implies the condition of being convexity-increasing.

In related work, there is an algorithm to morph one convex drawing to another convex drawing of the same graph while preserving planarity and convexity [3]. Such morphs are convexity-increasing by default, but do not address our problem since our initial drawing is not convex.

Many previous morphing algorithms find "piece-wise linear" morphs, where the morph is composed of discrete steps and each step moves vertices along straight lines. A morph is called *linear* if each vertex moves along a straight line at constant speed; different vertices are allowed to move at different speeds, and some may remain stationary. A linear morph is completely specified by the initial and final drawings. If, in addition, all the lines along which vertices move are parallel, then the morph is called *unidirectional* [2]. Horizontal and vertical morphs are a special kind of unidirectional morphs.

Alamdari et al. [2] gave an algorithm with runtime $O(n^3)$ that takes as input two n-vertex planar straight-line drawings that are combinatorially the same, and constructs a planar morph between them that consists of a sequence of $O(n)$ unidirectional morphs.

Our Results. Our main result is the following theorem.

Theorem 1. *Let Γ be a planar straight-line drawing of a 3-connected graph G on n vertices. Then Γ can be morphed to a strictly convex drawing via a sequence of convexity-increasing planar morphs each of which is either a horizontal morph or a vertical morph. If Γ has a convex outer face then the number of morphs in the sequence is at most $r+1$, where r is the number of internal reflex angles in Γ. In general, the number of morphs in the sequence is at most $1.5n$. Furthermore, there is an $O(n^{1+\omega/2})$ time algorithm to find the sequence of morphs, where ω is the matrix multiplication exponent.*

The run time is $O(n^{2.5})$ with Gaussian elimination, improved to $O(n^{2.1865})$ using the current fastest matrix multiplication method with $\omega \approx 2.3728639$ [17]. Our model of computation is the real-RAM—we do not have a polynomial bound on the bit-complexity of the coordinates of the vertices in the sequence of drawings that specify the morph. However, previous morphing algorithms had no such bounds either.

In terms of visualization, our algorithm has an advantage over the general morphing algorithm of Alamdari et al. [2]. That algorithm "almost contracts" vertices, which destroys the user's mental model of the graph. We do not use contractions, and therefore believe our morphs to be useful for visualizations.

As with some previous morphing results [2,3] a main ingredient of our proof of Theorem 1 is a result of Hong and Nagamochi [14] that gives conditions (and an algorithm) for redrawing a planar straight-line graph drawing to have convex faces, while preserving the y-coordinates of the vertices ("level planar drawings of hierarchical-st plane -graphs," in their terminology). Angelini et al. [3] strengthened Hong and Nagamochi's result to strictly convex faces. We give a new proof of the strengthened result for 3-connected graphs using Tutte's graph drawing algorithm. This speeds up Hong and Nagamochi's algorithm from $O(n^2)$ to $O(n^{\omega/2})$. Our improvement in run-time also speeds up the run-time of the morphing algorithm of Alamdari et al. [2] from $O(n^3)$ to $O(n^{1+\omega/2})$.

Organization. The proof of Theorem 1 is in Sect. 3, and the result about redrawing with convex faces is in Sect. 4. We begin with preliminaries in Sect. 2.

2   Preliminaries

Two planar drawings of a graph G have the *same combinatorial embedding* if they have the same clockwise cyclic ordering of edges around the outer face and around each inner face.

Given a planar straight-line drawing Γ of a graph, its *angles* are formed by pairs of consecutive edges around a face, with the angle measured inside the face. An *internal angle* is an angle of an inner face. A *reflex* angle is one that exceeds π. A *convex* angle is at most π, and a *strictly convex* angle is less than π. A drawing Γ is *convex* if all its faces are drawn as convex polygons, i.e., angles of the inner faces are convex and angles of the outer face are reflex or flat (of angle 180°). The drawing is *strictly convex* if all faces are strictly convex.

A face of a planar graph drawing is *y-monotone* if the boundary of the face consists of two y-monotone chains. A chain is *y-monotone* if the y-coordinates of points along the chain are strictly increasing. These definitions apply to general planar graph drawings, not just straight-line drawings. (We note that the directed graphs that have drawings with y-monotone faces are the *st-planar* graphs, which are well-studied [8].)

A linear morph is completely specified by the initial and final drawings. We use the notation $\langle \Gamma_1, \Gamma_2 \rangle$ to denote the linear morph from drawing Γ_1 to Γ_2.

Unidirectional Morphs. Restricting to linear morphs seems like a sensible way to discretize morphs—essentially, it asks for the vertex trajectories to be piecewise linear. At first glance, the further restriction to unidirectional morphs seems arbitrary and restrictive. However, as discovered by Alamdari et al. [2], it is easier to prove the existence of unidirectional morphs. Also, unidirectional morphs have many nice properties, as we explain in this section. Suppose we perform a horizontal morph. Then every vertex must keep its y-coordinate. Alamdari et al. gave conditions on the initial and final drawing that guarantee that the horizontal morph between them is planar:

Lemma 1. *[2, Lemma 13] If Γ and Γ' are two planar straight-line drawings of the same graph such that every line parallel to the x-axis crosses the same ordered sequence of edges and vertices in both drawings, then the linear morph from Γ to Γ' is planar.*

Observe that the conditions of the lemma imply that every vertex is at the same y-coordinate in Γ and Γ' so the linear morph between them is horizontal. Also note that the lemma generalizes in the obvious way to any direction, not just the direction of the x-axis. We note several useful consequences of Lemma 1.

Lemma 2. *Let $\Gamma_1, \Gamma_2, \Gamma_3$ be three planar straight-line drawings where the linear morphs $\langle \Gamma_1, \Gamma_2 \rangle$ and $\langle \Gamma_2, \Gamma_3 \rangle$ are horizontal and planar. Then the linear morph $\langle \Gamma_1, \Gamma_3 \rangle$ is a horizontal planar morph.*

Lemma 3. *During a horizontal morph, the convexity-status of an angle changes at most once. If $\langle \Gamma_1, \Gamma_2 \rangle$ is a horizontal morph and every convex internal angle of Γ_1 is also convex in Γ_2 then the morph is convexity-increasing.*

Alamdari et al. gave a further condition that implies the hypothesis of Lemma 1, and applies even to drawings that are not straight-line planar:

Observation 1. *[2, Lemma 13] Let Γ be a planar drawing of a graph G in which all faces (including the outer face) are y-monotone and let Γ' be another planar drawing of G that has the same combinatorial embedding, the same y-coordinates of vertices, and y-monotone edges. Then every line parallel to the x-axis crosses the same ordered sequence of edges and vertices in both drawings.*

Redrawing with Convex Faces while Preserving y-Coordinates. We build upon an $O(n^2)$ time algorithm due to Hong and Nagamochi [14] that redraws a planar graph to have convex faces while preserving the y-coordinates of the vertices. Angelini et al. [3] strengthened the result to strictly convex faces by perturbing vertices to avoid angles of 180°. They did not analyze the runtime. Both [3,14] expressed their results in terms of *level planar drawings* of *hierarchical-st plane graphs*, and handled more generally the class of graphs that have (strictly) convex drawings.

We limit ourselves to 3-connected graphs and improve the running time:

Lemma 4 (based on [3,14]). *Let Γ be a planar drawing of a 3-connected graph G such that every face is y-monotone (including the outer face). Let C be a strictly convex straight-line drawing of the outer face of G such that every vertex of C has the same y-coordinate as in Γ. Then there is a straight-line strictly convex drawing Γ' of G that has C as the outer face and such that every vertex of Γ' has the same y-coordinate as in Γ. Furthermore, Γ' can be found in time $O(n^{\omega/2})$, where ω is the matrix multiplication exponent.*

We prove Lemma 4 in Sect. 4 using Tutte's graph drawing algorithm. This is quite different from the previous approaches, and gives the improved run-time. Our run-time is $O(n^{1.5})$ without fast matrix multiplication, $O(n^{1.1865})$ with.

3   Computing Convexity-Increasing Morphs

To give some intuition about the proof of Theorem 1, we first consider an easy case where the outer face C of Γ is a strictly convex polygon and all faces are y-monotone. In this case, we can immediately apply Lemma 4 with the outer face fixed to obtain a new straight-line strictly convex drawing Γ' with all vertices at the same y-coordinates. By Observation 1 every line parallel to the x-axis crosses the same ordered sequence of edges and vertices in Γ and in Γ'. Then by Lemma 1 the morph from Γ to Γ' is planar. Also it is a horizontal morph. Thus we have a morph from Γ to a strictly convex drawing Γ' by way of a single horizontal morph. Furthermore, the morph is convexity-increasing by Lemma 3.

Morphing Drawings with a Convex Outer Face. When the outer face is convex, but the other faces are not necessarily y-monotone, we will still begin with one horizontal morph. Assume Γ has no horizontal edges (we show how to ensure this later on). A face f is y-monotone if and only if it has only one *local maximum* and only one *local minimum*, where a vertex v is a *local minimum* (*local maximum*) of face f if the neighbors of v in f lie above v (below v, respectively). A *local extremum* refers to a local minimum or a local maximum. Observe that a horizontal morph preserves the local extrema and does not change their convex/reflex status. Thus the only reflex angles that can be made convex via a horizontal morph are the *h-reflex* angles, where an angle of inner face f is called *h-reflex* if it is reflex and occurs at a vertex that has one neighbor in f above and the other below—equivalently, the angle is reflex and is not a local extremum of f.

To obtain a horizontal morph we will apply Lemma 4, and to do that, we must first augment Γ to a drawing with y-monotone faces by inserting y-monotone edges (not necessarily straight-line). For an example see Fig. 2. This is a standard operation in upward planar (or "monotone") drawing [8, Lemma 4.1] [19, Lemma 3.1], but we need the stronger property that new edges are only incident to local extrema (otherwise we would relinquish control of convexity at that vertex). We will use:

Proposition 1. *Any straight-line planar drawing of a 3-connected graph can be augmented to have y-monotone inner faces by adding edges such that each edge can be drawn as a y-monotone curve joining two local extrema in some face. Furthermore, these edges can be found in time $O(n \log n)$.*

The proof idea is illustrated in Fig. 2. Proposition 1 allows us to prove the following:

Lemma 5. *Let Γ be a straight-line planar drawing of a 3-connected graph with a convex outer face and no horizontal edge. There exists a horizontal planar morph to a straight-line drawing Γ' such that Γ' has a strictly convex outer face and every internal angle that is not a local extremum is strictly convex in Γ'. Furthermore, the morph is convexity-increasing, and can be found in time $O(n^{\omega/2})$, where ω is the matrix multiplication exponent.*

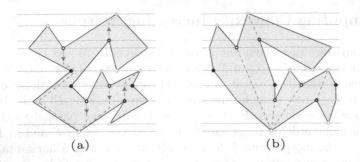

Fig. 2. (a) A face (in gray) that is not y-monotone. Gray vertices are convex; black vertices are reflex. Local minima are drawn with thin-bordered hollow vertices, and local maxima with thick-bordered hollow vertices; other vertices are drawn solid, and those that are reflex (black) will be convexified in the next morph. Red dashed edges inside the face are added by Proposition 1. (b) The face after application of Lemma 5. Angles at black solid vertices have become convex.

Proof. Use Proposition 1 to augment Γ with a set of edges A such that $\Gamma \cup A$ is a planar drawing in which all faces are y-monotone, and any edge of A goes between two local extrema in some inner face. This takes $O(n \log n)$ time. Let C be the outer face of Γ. Create a new drawing of C, call it C', that is strictly convex, but preserves the y-coordinates of vertices.

By Lemma 4 with the outer face C' we obtain (in time $O(n^{\omega/2})$) a new straight-line strictly convex drawing $\Gamma' \cup A'$ with all vertices at the same y-coordinates as in Γ. (Here A' is a set of straight-line edges corresponding to A.) By Observation 1 every line parallel to the x-axis crosses the same ordered sequence of edges and vertices in $\Gamma \cup A$ and in $\Gamma' \cup A'$. Then by Lemma 1 the morph from Γ to Γ' is a planar horizontal morph.

Any internal angle of Γ that is not a local extremum has no edge of A incident to it, and thus becomes strictly convex in Γ'. Any internal angle of Γ that is a local extremum maintains its convex/reflex status in Γ'. Thus by Lemma 3 the morph is convexity-increasing. The run-time to find the morph (i.e., to find Γ') is $O(n^{\omega/2})$. $\square$

Lemma 5 generalizes to directions d other than the horizontal direction. In order to prove Theorem 1, the plan is to apply Lemma 5 and, then, to conceptually "turn the paper" by 90° and perform a vertical morph to make any *v-reflex* angle convex, where an angle of inner face f is called a *v-reflex* angle if it is reflex and occurs at a vertex that has one neighbor in f to the left and the other to the right. Under the assumption that each such horizontal or vertical morphing step convexifies at least one angle, the proof of Theorem 1 is obtained by simply continuing to apply Lemma 5 alternately in the horizontal and vertical direction. To ensure that after each step there is at least one h-reflex or v-reflex vertex, we provide the following strengthened version of Lemma 5.

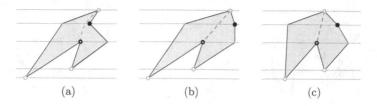

(a)   (b)   (c)

Fig. 3. (a) A face that is not y-monotone. (b) The face after application of Lemma 5. There is a vertical edge and the single reflex vertex is not v-reflex. (c) After applying a horizontal shear transformation, the reflex vertex is v-reflex and there are no vertical edges.

Lemma 6. *Let Γ be a straight-line planar drawing with a convex outer face and no horizontal edge. There exists a horizontal planar morph to a straight-line drawing Γ'' such that*

(i) the outer face of Γ'' is strictly convex,
(ii) every internal angle that is not a local extremum is convex in Γ'',
(iii) Γ'' has no vertical edge, and
(iv) if Γ'' is not convex, then it has at least one v-reflex angle.

Furthermore, the morph is convexity-increasing, and can be found in time $O(n^{\omega/2})$.

Proof sketch. We apply Lemma 5 to obtain a morph from Γ to a drawing Γ' that satisfies (i) and (ii). We apply a shearing transformation (along the x-axis) to Γ' to obtained a drawing Γ'' which satisfies all four conditions, see Fig. 3(b,c). By Observation 1 and Lemma 1 the horizontal morph $\langle \Gamma', \Gamma'' \rangle$ is planar. Thus, by Lemma 2 the horizontal morph $\langle \Gamma, \Gamma'' \rangle$ is planar. Since shearing is an affine transformation, Γ' and Γ'' have the same sets of convex/reflex angles. Hence, $\langle \Gamma, \Gamma'' \rangle$ is convexity-concreasing. □

Each application of Lemma 6 convexfies at least one reflex angle and, thus, after at most r application we obtain a strictly convex drawing. If the initial drawing contains a horizontal edge, we use one additional vertical morph to a sheared drawing, resulting in a total of $r + 1 \leq n$ morphs. This concludes the proof sketch of Theorem 1 for the case that the outer face is already realized as a convex polygon.

Morphing Drawings with a Non-convex Outer Face. To prove the general case of Theorem 1, where the outer face is not convex, we first augment the outer face of Γ with edges from its convex hull to obtain a drawing of an augmented graph with a convex outer face. We apply the algorithm for the convex case to morph to a strictly convex drawing and then remove the extra edges one-by-one. After each edge is removed we morph to a strictly convex drawing of the reduced graph using at most three horizontal or vertical morphs. We now describe these steps in more detail.

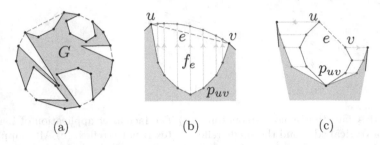

Fig. 4. (a) Schematic of the convex drawing of $G \cup A$. Graph G is depicted in gray, edges of A are dashed, and the pockets are white. (b)–(c) Cases 1 and 2 for Lemma 7, where faint gray arrows indicate explicit placements on the convex hull.

Augmenting the Outer Face. Compute the convex hull of Γ. Any segment of the convex hull that does not correspond to an edge of G becomes a new edge that we add to G. Let A denote the new edges and $G \cup A$ denote the augmented graph with straight-line planar drawing $\Gamma \cup \Gamma_A$. Each edge $e \in A$ is part of the boundary of an inner face f_e of $\Gamma \cup \Gamma_A$. We call f_e the *pocket* of e. We apply the algorithm for the convex case to obtain a strictly convex drawing of $G \cup A$, see Fig. 4(a). Note that said algorithm is guaranteed to to produce drawings without horizontal or vertical edges.

Popping a Pocket Outward. We now describe a way to remove an edge of A and "pop" out the vertices of its pocket so that they become part of the convex hull. Lemma 4 serves once again as an important subroutine. We make ample use of the fact that we may freely specify the desired subdrawing of the outer face after each application of Lemma 4, as long as we maintain either the x-coordinates or the y-coordinates of all vertices.

Lemma 7. *Let Γ be a strictly convex drawing of graph G, with an edge e on the outer face. Suppose that $G - e$ is 3-connected. Then $\Gamma - e$ can be morphed to a strictly convex drawing of $G - e$ via at most three convexity-increasing morphs, each of which is horizontal or vertical. Furthermore, the morphs can be found in time $O(n^{\omega/2})$.*

Proof. Our morph will be specified by a sequence of drawings, Γ, Γ_1, Γ_2, Γ_3, where the first and last morphs are vertical and the second morph, which we can sometimes skip, is horizontal.

Let $e = (u, v)$. We first perform a vertical morph from Γ to a drawing Γ_1 in which vertex u is top-most or bottom-most. This can be done by choosing some strictly convex drawing of the outer face in which u is extreme while maintaining the x-coordinates of all vertices, and then using one application of Lemma 4. For the remainder of the proof, assume without loss of generality that u is the top-most vertex and that v lies to the right of u in Γ_1. The other cases are symmetric. Let p_{uv} denote the path from u to v in $f_e - e$. We distinguish two cases depending on the shape of f_e in Γ_1.

Case 1: The path $p_{u,v}$ is an x-monotone chain in Γ_1, see Fig. 4(b). In this case we can skip the second step of the morph sequence. We will remove e and compute a vertical morph from $\Gamma_1 - e$ to a strictly convex drawing Γ_3 of $G - e$. This can be done by Lemma 4 as long as we can specify a strictly convex drawing of the outer face of $\Gamma_1 - e$ in which the x-coordinates match those of Γ_1. It suffices to compute a suitable new reflex chain for p_{uv}, see Fig. 4(b).

Case 2: The path p_{uv} is not an x-monotone chain. In this case we will compute a horizontal morph from Γ_1 to a strictly convex drawing Γ_2 in which u and v are the unique left-most and the unique right-most vertices. This can be done by Lemma 4 as long as we can specify a strictly convex drawing of the outer face of Γ_1 in which u and v are at the left and right and the y-coordinates match those of Γ_1. This is possible because u is top-most, see Fig. 4(c).

In the drawing Γ_2 the pocket f_e is convex with extreme points u and v so the path p_{uv} is x-monotone and, hence, by case 1, there is a vertical morph from $\Gamma_2 - e$ to a strictly convex drawing Γ_3 of $G - e$.   □

Each application of Lemma 7 increases the number of vertices of G on the convex hull. Thus, by induction on said number we obtain the proof of Theorem 1. We have used a constant number of morphs per pockets and, thus, $O(n)$ morphs in total. In [16] we improve this bound to $1.5n$.

4 Using Tutte's Algorithm to Find Convex Drawings Preserving y-Coordinates

In this section, we sketch our proof of Lemma 4. In his paper, "How to Draw a Graph," [25] Tutte showed that any 3-connected planar graph $G = (V, E)$ with a fixed convex drawing C of its outer face has a convex drawing with outer face C that can be obtained by solving a system of linear equations. Let V_I be the internal vertices of G and for $v \in V_I$ let variables (x_v, y_v) represent the coordinates of vertex v. For each vertex $v \in C$ let (x_v, y_v) be its (fixed) coordinates. Let d_v be the degree of vertex v. Consider the system of equations:

$$\forall u \in V_I \quad (x_u, y_u) = \sum_{(u,v) \in E} \frac{1}{d_u}(x_v, y_v).$$

Tutte proved that this system of equations has a solution and that the solution gives a convex drawing of G with outer face C. The drawing is strictly convex so long as C is strictly convex.

In fact, Tutte's proof carries over to more general "barycenter" weights other than $1/d_u$. The following result is proved in [10] (or see [12]). Assign a weight $w_{u,v} > 0$ to each ordered pair u, v with $(u, v) \in E$ such that for each u, $\sum_v w_{u,v} = 1$. Then the system of equations

$$\forall u \in V_I \quad (x_u, y_u) = \sum_{(u,v) \in E} w_{u,v}(x_v, y_v) \tag{1}$$

has a unique solution that gives a convex drawing of G with outer face C (and a strictly convex drawing of G if C is strictly convex).

The idea for our proof of Lemma 4 is as follows: As a preprocessing step, we compute barycenter weights $w_{u,v}$ which force the vertices to lie at the desired y-coordinates. To this end, we solve the equation system (1) restricted to the y-coordinates. Here, we consider the y's as fixed vales, while the $w_{u,v}$'s are the variables. We add further constraints that ensure that for every internal vertex u the obtained values $w_{u,v}$ sum up to 1. As a second step, we apply the generalized version of Tutte's algorithm to obtain the desired drawing using the weights obtained in the preprocessing step. The efficient runtime is obtained by using the generalized nested dissection method by Lipton et al. [18] (also see [23]).

5   Conclusions

In [16] we extend our results to *internally* 3-connected graphs. This property is known to characterize the plane graphs which admit strictly convex drawings [3,14]. We also show that our algorithm is worst-case optimal in the sense that a linear number of morphing steps may be required to convexify a given drawing. The following questions remain open: (1) Recall that during a convexity-increasing morph, the set of *internal* convex angles never decreases. We conjecture that every straight-line planar drawing of a (internally) 3-connected graph admits a convexity-increasing morph to a convex drawing such that during the morph the set of *external* reflex angles also never decreases. (2) Design piece-wise linear morphs with a polynomial bound on the bit complexity of the intermediate drawings. This would be a step towards having intermediate drawings that lie on a polynomial-sized grid, i.e. with a logarithmic number of bits for each vertex's coordinates. This is open both for our problem of morphing to a convex drawing and for the problem of morphing between two given planar straight-line drawings.

Acknowledgments. We thank André Schulz for helpful discussions on generalizations of Tutte's algorithm. This work was begun at Dagstuhl workshop 17072, "Applications of Topology to the Analysis of 1-Dimensional Objects." We thank Dagstuhl, the organizers, and the other participants for a stimulating workshop. In particular, we thank Carola Wenk and Regina Rotmann for joining some of our discussions, and Irina Kostitsyna for contributing many valuable ideas.

References

1. Aichholzer, O., Aloupis, G., Demaine, E.D., Demaine, M.L., Dujmovic, V., Hurtado, F., Lubiw, A., Rote, G., Schulz, A., Souvaine, D.L., Winslow, A.: Convexifying polygons without losing visibilities. In: Canadian Conference on Computational Geometry (CCCG) (2011)

2. Alamdari, S., Angelini, P., Barrera-Cruz, F., Chan, T.M., Da Lozzo, G., Di Battista, G., Frati, F., Haxell, P., Lubiw, A., Patrignani, M., Roselli, V., Singla, S., Wilkinson, B.T.: How to morph planar graph drawings. SIAM J. Comput. **46**(2), 29 (2017)
3. Angelini, P., Da Lozzo, G., Frati, F., Lubiw, A., Patrignani, M., Roselli, V.: Optimal morphs of convex drawings. In: Arge, L., Pach, J. (eds.) Proceedings of the 31st International Symposium on Computational Geometry (SoCG 2015), Leibniz International Proceedings in Informatics (LIPIcs), vol. 34, pp. 126–140. Dagstuhl, Wadern (2015)
4. Appelle, S.: Perception and discrimination as a function of stimulus orientation: the "oblique effect" in man and animals. Psychol. Bull. **78**(4), 266 (1972)
5. Cairns, S.: Deformations of plane rectilinear complexes. Am. Math. Monthly **51**(5), 247–252 (1944)
6. Cantarella, J.H., Demaine, E.D., Iben, H.N., O'Brien, J.F.: An energy-driven approach to linkage unfolding. In: Proceedings of the 20th Annual Symposium on Computational Geometry (SoCG), pp. 134–143. ACM (2004)
7. Connelly, R., Demaine, E.D., Rote, G.: Straightening polygonal arcs and convexifying polygonal cycles. Discret. Comput. Geom. **30**, 205–239 (2003)
8. Di Battista, G., Tamassia, R.: Algorithms for plane representations of acyclic digraphs. Theor. Comput. Sci. **61**(2–3), 175–198 (1988)
9. Eiglsperger, M., Fekete, S.P., Klau, G.W.: Orthogonal graph drawing. In: Kaufmann, M., Wagner, D. (eds.) Drawing Graphs. LNCS, pp. 121–171. Springer, Heidelberg (2001). https://doi.org/10.1007/3-540-44969-8_6
10. Floater, M.S.: Parametric Tilings and scattered data approximation. Int. J. Shape Model. **4**(03n04), 165–182 (1998)
11. Floater, M.S., Gotsman, C.: How to morph tilings injectively. J. Comput. Appl. Math. **101**(1–2), 117–129 (1999)
12. Gortler, S.J., Gotsman, C., Thurston, D.: Discrete one-forms on meshes and applications to 3D mesh parameterization. Comput. Aided Geomet. Des. **23**(2), 83–112 (2006)
13. Gotsman, C., Surazhsky, V.: Guaranteed intersection-free polygon morphing. Comput. Graph. **25**(1), 67–75 (2001)
14. Hong, S.-H., Nagamochi, H.: Convex drawings of hierarchical planar graphs and clustered planar graphs. J. Discrete Algorithms **8**(3), 282–295 (2010)
15. Iben, H.N., O'Brien, J.F., Demaine, E.D.: Refolding planar polygons. Discret. Comput. Geomet. **41**(3), 444–460 (2009)
16. Kleist, L., Klemz, B., Lubiw, A., Schlipf, L., Staals, F., Strash, D.: Convexity-increasing morphs of planar graphs. CoRR, abs/1802.06579 (2018)
17. Le Gall, F.: Powers of tensors and fast matrix multiplication. In: Proceedings of the 39th International Symposium on Symbolic and Algebraic Computation, pp. 296–303. ACM (2014)
18. Lipton, R.J., Rose, D.J., Tarjan, R.E.: Generalized nested dissection. SIAM J. Numer. Anal. **16**(2), 346–358 (1979)
19. Pach, J., Tóth, G.: Monotone drawings of planar graphs. J. Graph Theory **46**(1), 39–47 (2004)
20. Purchase, H.C., Hoggan, E., Görg, C.: How important is the "Mental map"? – An empirical investigation of a dynamic graph layout algorithm. In: Kaufmann, M., Wagner, D. (eds.) GD 2006. LNCS, vol. 4372, pp. 184–195. Springer, Heidelberg (2007). https://doi.org/10.1007/978-3-540-70904-6_19
21. Purchase, H.C., Pilcher, C., Plimmer, B.: Graph drawing aesthetics–created by users, not algorithms. IEEE Trans. Vis. Comput. Graph. **18**(1), 81–92 (2012)

22. Rahman, S.: Convex graph drawing. In: Kao, M.-Y. (ed.) Encyclopedia of Algorithms, pp. 1–7. Springer, Heidelberg (2015). https://doi.org/10.1007/978-3-642-27848-8

23. Ribó Mor, A., Rote, G., Schulz, A.: Small grid embeddings of 3-polytopes. Discret. Comput. Geomet. **45**(1), 65–87 (2011)

24. Thomassen, C.: Deformations of plane graphs. J. Combin. Theory, Ser. B **34**(3), 244–257 (1983)

25. Tutte, W.T.: How to draw a graph. Proc. Lond. Math. Soc. **3**(1), 743–767 (1963)

Treedepth Bounds in Linear Colorings

Jeremy Kun[1], Michael P. O'Brien[2(✉)], and Blair D. Sullivan[2]

[1] Google, Raleigh, USA
[2] North Carolina State University, Raleigh, USA
mpobrie3@ncsu.edu

Abstract. *Low-treedepth* colorings are an important tool for algorithms that exploit structure in classes of bounded expansion; they guarantee subgraphs that use few colors have *bounded treedepth*. These colorings have an implicit tradeoff between the total number of colors used and the treedepth bound, and prior empirical work suggests that the former dominates the run time of existing algorithms in practice. We introduce *p-linear colorings* as an alternative to the commonly used p-centered colorings. They can be efficiently computed in bounded expansion classes and use at most as many colors as p-centered colorings. Although a set of $k < p$ colors from a p-centered coloring induces a subgraph of treedepth at most k, the same number of colors from a p-linear coloring may induce subgraphs of larger treedepth. A simple analysis of this treedepth bound shows it cannot exceed 2^k, but no graph class is known to have treedepth more than $2k$. We establish polynomial upper bounds via constructive coloring algorithms in trees and intervals graphs, and conjecture that a polynomial relationship is in fact the worst case in general graphs. We also give a co-NP-completeness reduction for recognizing p-linear colorings and discuss ways to overcome this limitation in practice.

1 Introduction

Algorithms for graph classes that exhibit *bounded expansion* structure [2,9–11] offer a promising framework for efficiently solving many NP-hard problems on real-world networks. The structural restrictions of bounded expansion, which allow for pockets of localized density in globally sparse graphs, are compatible with properties of many real-world networks such as clustering and heavy-tailed degree distributions. Moreover, multiple random graph models designed to mimic these properties have been proven to asymptotically almost surely belong to classes of bounded expansion [3]. From a theoretical perspective, graphs belonging to classes of bounded expansion can be characterized by *low-treedepth colorings* of bounded size, i.e. using only a small number of colors. Roughly speaking, a low-treedepth coloring is one in which the subgraphs induced on each small set of colors have small *treedepth*, a structural property stronger than treewidth. This definition naturally implies an algorithmic pipeline [3,4,10] for classes of bounded expansion involving four stages: computing a low-treedepth coloring, using the coloring to decompose the graph into subgraphs of small treedepth,

© Springer Nature Switzerland AG 2018
A. Brandstädt et al. (Eds.): WG 2018, LNCS 11159, pp. 331–343, 2018.
https://doi.org/10.1007/978-3-030-00256-5_27

solving the problem efficiently on each such subgraph, and combining the subsolutions to construct a global solution. The complexities of algorithms using this paradigm often are of the form $O(\binom{k}{p} 2^{d \log d} \cdot n^c)$ where k is the coloring size and d is the treedepth of the subgraphs.

A recent implementation [12] and experimental evaluation [13] of this pipeline has identified that the coloring size has a much larger effect on the run time than the treedepth in practice. Although graphs in classes of bounded expansion are guaranteed to admit colorings of constant size with respect to the number of vertices, the only known polynomial-time algorithms for computing these colorings are approximations [2]. Consequently it is unclear to what extent our current coloring algorithms can be altered to reduce the coloring size. A more viable approach to improving the performance of the algorithmic pipeline without significant high-level changes would be to develop a new type of low-treedepth coloring that uses fewer colors but potentially has weaker guarantees about the treedepth of the subgraphs.

The traditional low-treedepth colorings for classes of bounded expansion are known as *p-centered colorings*. This name stems from the property that on any subgraph H, a p-centered coloring either uses at least p colors or is a *centered coloring*, which restricts the multiplicity of colors in induced subgraphs. In this paper we introduce an alternative that closely mirrors this paradigm but only extends the color multiplicity guarantees to path subgraphs. For this reason we refer to them as *p-linear colorings* and *linear colorings*. We identify that p-linear colorings share three important properties with p-centered colorings that allow them to be used in the bounded expansion algorithmic pipeline.

1. The minimum coloring size is constant in graphs of bounded expansion.
2. A coloring of bounded size can be computed in polynomial time.
3. Small sets of colors induce graphs of small treedepth.

The third of these properties is of particular interest, since understanding the tradeoffs between coloring size and treedepth in switching between p-centered and p-linear colorings fundamentally depends on bounding the maximum treedepth of a graph that admits a linear coloring with k colors. Equivalently, we frame this problem as determining the gap between the minimum number of colors needed for a linear versus a centered coloring in any given graph. A naïve analysis does not exclude the possibility that this gap is exponentially large, despite the fact that the largest known difference is a constant multiplicative factor. We conjecture that our proven constant factor lower bound is also the upper bound; as evidence, we prove that in trees and interval graphs the difference is polynomially bounded (in the coloring size) and give polynomial time algorithms (with respect to the graph size) to certify this difference. Surprisingly, we also prove that some p-linear colorings cannot be verified in polynomial time unless $P = \text{co-NP}$ and discuss the practical implications of these findings. In the interest of space, the proofs of all lemmas are omitted from the main text and can be found in [7].

2   Definitions and Background

In this section we detail the background and terminology necessary to understand p-linear colorings.

2.1   Graph Terminology

We denote the vertices and edges of a graph G as $V(G)$ and $E(G)$, respectively, and assume all graphs are simple and undirected except where specifically noted otherwise. The *open neighborhood* of a vertex v, denoted $N(v)$, is the set of vertices u such that $uv \in E(G)$, while the *closed neighborhood*, $N[v]$ is defined as $N(v) \cup \{v\}$. Vertex a is an *apex* with respect to a subgraph H if $V(H) \subseteq N(a)$.

We say P is a $v_1 v_\ell$-*path* if $V(P) = \{v_1, \ldots, v_\ell\}$ and $E(P) = \{v_i v_{i+1} : 1 \leq i \leq \ell - 1\}$; we will notate this as $P = v_1, \ldots, v_\ell$. Given disjoint paths $P = v_1, \ldots, v_\ell$ and $Q = u_1, \ldots, u_{\ell'}$, the path $P \cdot Q = v_1, \ldots, v_\ell, u_1, \ldots, u_{\ell'}$ is the *concatenation* of P and Q if v_ℓ is adjacent to u_1. A path is *Hamiltonian* with respect to subgraph H if $V(P) = V(H)$.

In a rooted tree T, we let T_v be the subtree of T rooted at v and the *leaf paths* of T_v be the set of paths from a leaf of T_v to v. Vertices u and v are *unrelated* in T if u is neither an ancestor nor a descendant of v.

A *coloring* ϕ of a graph G is a mapping of the vertices of G to *colors* $1, \ldots, k$ and has *size* $|\phi| = k$. A coloring is *proper* if no pair of adjacent vertices have the same color. For any subgraph H and color c, if there is exactly one vertex $v \in H$ such that $\phi(v) = c$ we say c *appears uniquely* in H and v is a *center* of H. A subgraph with no unique color is said to be *non-centered*.

2.2 p-Centered Colorings and Bounded Expansion

Definition 1. *A p-centered coloring ϕ of graph G is a coloring such that for every connected subgraph H, H has a center or $\phi|_H$ uses at least p colors.*

Nešetřil and Ossana de Mendez established that bounding the minimum size of a p-centered coloring is a necessary and sufficient condition for a graph class to have bounded expansion.

Proposition 1 ([9]). *A class of graphs $\mathcal{C}$ has bounded expansion iff there exists a function f such that for all $G \in \mathcal{C}$ and all $p \geq 1$, G admits a p-centered coloring with $f(p)$ colors.*

There are varying methods to compute p-centered colorings, such as transitive-fraternal augmentations [5,9] and generalized coloring numbers [17], we focus here on *distance-truncated transitive-fraternal augmentations* (DTFAs) [14], which iteratively augment the graph with additional edges to impose constraints on proper colorings. This linear time algorithm guarantees that after $(2 \log p)^p$ DTFA iterations, any proper coloring of the augmented graph is a p-centered coloring whose size is bounded in classes of bounded expansion.

2.3   Centered Colorings and Treedepth

Note that if ϕ is a p-centered coloring of G and H is a subgraph of G whose vertices use at most $p-1$ colors in ϕ, H must have a center. This relates p-centered colorings to a more restricted class of graphs defined by *centered colorings*.

Definition 2. *A* centered coloring ϕ *of graph* G *is a coloring such that every connected subgraph has a center. The minimum size of a centered coloring of* G *is denoted* $\chi_{cen}(G)$.

Note that a centered coloring is also proper, or else there would be a connected subgraph of size two with no center. Observe that if X is the set of all centers of G, then $G\backslash X$ must either be empty or disconnected. This implies that if $|G| \gg \chi_{cen}(G)$, then G breaks into many components after only a few vertex deletions. This property is captured by *treedepth decompositions*.

Definition 3. *A* treedepth decomposition $\mathcal{T}$ *of graph* G *is a rooted forest with the same vertex set as* G *such that* $uv \in E(G)$ *implies* u *is an ancestor of* v *in* $\mathcal{T}$ *or vice versa. The* depth *of* $\mathcal{T}$ *is the length of the longest path from a leaf in* $\mathcal{T}$ *to a root in its component. The* treedepth *of* G, $\mathrm{td}(G)$, *is the minimum depth of a treedepth decomposition of* G.

Given a centered coloring of size k, we can generate a treedepth decomposition of depth at most k by choosing any center v to be the root and setting the children of v to be the roots of the treedepth decompositions of the components of $G\backslash\{v\}$. Likewise, given a treedepth decomposition of depth k, we can generate a centered coloring using k colors by bijectively assigning the colors to levels of the tree and coloring vertices according to their level. We refer to the colorings and decompositions resulting from these procedures as *canonical*; together they imply that the treedepth and centered coloring numbers are equal for all graphs.

3 p-Linear and Linear Colorings

We introduce *p-linear colorings* as an alternative to p-centered colorings.

Definition 4. *A* p-linear coloring *is a coloring* ψ *of a graph* G *such that for every path[1]* P, *either* P *has a center or* $\psi|_P$ *uses at least* p *colors.*

It is proven in [14] that after performing 2^p DTFA iterations, any proper coloring of the augmented graph is a p-linear coloring. This implies that p-linear colorings indeed have constant size in bounded expansion classes and can be constructed in polynomial time (like p-centered colorings).

In the interest of maintaining consistency with prior terminology, we define *linear colorings* analogously to centered colorings.

---

[1] This includes non-induced paths.

Definition 5. *A* linear coloring *is a coloring ψ of a graph G such that every path has a center. The* linear coloring number *is the minimum number of colors needed for a linear coloring and is denoted $\chi_{lin}(G)$.*

Note that linear colorings must also be proper. A simple recursive argument shows that every path of length d requires at least $\log_2(d+1)$ colors in a linear coloring; thus a graph of linear coloring number k has no path of length 2^k. Because every depth-first search tree is a treedepth decomposition, $\mathrm{td}(G) \leq 2^{\chi_{lin}(G)}$, proving that small numbers of colors in p-linear colorings induce graphs of bounded treedepth[2].

Our study of the divergence between linear and centered coloring numbers will naturally focus on linear colorings that are not also centered colorings. We say ψ is a *non-centered linear coloring* (NCLC) of graph G if G contains a connected induced subgraph with no center. For NCLC ψ, we say a connected induced subgraph H is a *witness* to ψ if H is non-centered but every proper connected subgraph of H has a center. For the sake of completeness, we prove in Lemma 1 that many simple graph classes do not admit NCLCs.

Lemma 1. *If G is a path, star, cycle, complete graph, or complete bipartite graph, any linear coloring of G is also a centered coloring.*

4   Treedepth Lower Bounds

To understand the tradeoff between the number of colors and treedepth of small color sets when using p-linear colorings in lieu of p-centered colorings, it is important to know the maximum treedepth of a graph of fixed linear coloring number k, $t_{\max}(k)$. In Lemmas 3 and 4, we prove lower bounds on $t_{\max}(k)$ through explicit constructions of graph families. In order to show that these graphs have large treedepth, we first establish assumptions about the structure of treedepth decompositions that can be made without loss of generality.

Lemma 2. *Let G be a graph and $S \subset V(G)$ such that $G[S]$ is connected and with respect to some component $C \in G \backslash S$, every vertex in S is an apex of C. Then for any treedepth decomposition T of depth k, we can construct a treedepth decomposition T' such that:*

1. *$\mathrm{depth}(T') \leq k$*
2. *Each vertex in S is an ancestor of every vertex in C in T'*
3. *For each pair of vertices $\{u, w\} \subseteq V(C)$ or $\{u, w\} \subseteq V(G \setminus C)$, u is an ancestor of w in T' iff it is an ancestor of w in T.*

Using Lemma 2, we now show that $t_{\max}(k) \geq 2k$.

Lemma 3. *There exists an infinite sequence of graphs $R_1, R_2, \ldots$ such that*

$$\lim_{i \to \infty} \frac{\chi_{cen}(R_i)}{\chi_{lin}(R_i)} = 2.$$

---

[2] This tightens a bound in [14] from double to single exponential.

The graphs in Lemma 3 contain large cliques. We now show that this is not a necessary condition for the linear and centered coloring numbers to diverge (Fig. 1).

Lemma 4. *Let B_ℓ be the complete binary tree with ℓ levels. Then*

$$\lim_{\ell \to \infty} \frac{\chi_{cen}(B_\ell)}{\chi_{lin}(B_\ell)} \geq \frac{3}{2}$$

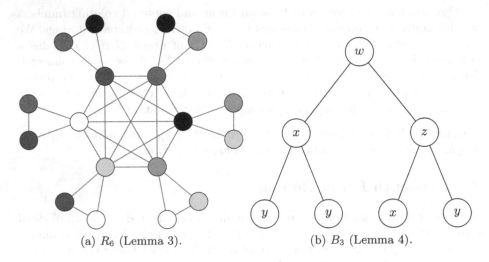

(a) R_6 (Lemma 3).   (b) B_3 (Lemma 4).

Fig. 1. Linear colorings of graphs in Lemmas 3–4. (Color figure online)

We conjecture that the construction in Lemma 3 is optimal.

Conjecture 1. For any graph G, $\chi_{cen}(G) \leq 2\chi_{lin}(G)$.

While the exclusion of a path of length 2^k indicates $t_{max}(k) \leq 2^k$, this nonetheless leaves a large gap between the upper and lower bounds on $t_{max}(k)$. To move towards a proof of Conjecture 1, we consider two restricted graph classes—namely, trees and interval graphs—in the next two sections and establish polynomial upper bounds on $t_{max}(k)$ for graphs in these classes.

5   Treedepth Upper Bounds on Trees

Schäffer proved that there is a linear time algorithm for finding a minimum-sized centered coloring of a tree T [16]. In this section we prove the following theorem by showing a correspondence between the centered coloring from Schäffer's algorithm and colors on paths in any linear coloring of T.

Theorem 1. *There exists a polynomial time algorithm that takes as input a tree T and a linear coloring ψ of T with size k and outputs a centered coloring of T whose size is at most $O(k^3)$.*

Schäffer's algorithm finds a particular centered coloring whose colors are ordered in a way that reflects their roles as centers. For this reason, the coloring is called a *vertex ranking* and the colors are referred to as *ranks*; it guarantees that in each subgraph, the vertex of maximum rank is also a center. We will use this terminology in this section to clearly distinguish between the ranks in the vertex ranking and colors in the linear coloring. Note that the canonical centered coloring of a treedepth decomposition is a vertex ranking if the colors are ranked decreasing from the root downwards, which implies that every centered coloring can be converted to a vertex ranking of the same size. Of central importance to Schäffer's algorithm are what we will refer to as *rank lists*.

Definition 6. *For a vertex ranking r of tree T, the* rank list *of T, denoted $L(T)$, can be defined recursively as $L(T) = L(T \backslash T_v) \cup \{r(v)\}$ where v is the vertex of maximum rank in T.*

Schäffer's algorithm arbitrarily roots T and builds the ranking from the leaves to the root of T, computing the rank of each vertex from the rank lists of each of its children.

Proposition 2 ([16]). *Let r be a vertex ranking of T produced by Schäffer's algorithm and let $v \in T$ be a vertex with children $u_1, \ldots, u_\ell$. If x is the largest integer appearing on rank lists of at least two children of v (or 0 if all such rank lists are pairwise disjoint) then $r(v)$ is the smallest integer satisfying $r(v) > x$ and $r(v) \notin \bigcup_{i=1}^{\ell} L(u_i)$.*

Our proof of Theorem 1 is based on tracking sets of colors of ψ on leaf paths as Schäffer's algorithm moves up the rooted tree. Define the *color vector* of a path P with respect to linear coloring ψ to be the set of colors from ψ appearing on P. Let $S(v)$ be the set of all color vectors of all leaf paths in T_v. Let $\kappa(v)$ be the maximum cardinality of any color vector in $S(v)$ and $S_\kappa(v) = \{C \in S(v) : |C| = \kappa(v)\}$. We show below that every vertex v has a corresponding vertex u that is "similar" in rank but "dissimilar" in values of κ and/or S_κ.

Lemma 5. *Let v be a vertex of rank $i > k$. There exists a vertex $u \in T_v$ such that*

- *$i - \kappa(v) - 1 < r(u) < i$ and*
- *Either $\kappa(u) < \kappa(v)$ or $|S_\kappa(u)| \leq \lfloor \frac{1}{2}|S_\kappa(v)| \rfloor$.*

Using Lemma 5 as a recursive step, we prove Theorem 1 by tracing the values of functions κ and S down towards the leaves.

Proof (Theorem 1). Let u_i be the vertex of maximum rank in T. There is a maximal sequence of vertices $u_2, \ldots, u_q$ such that u_{i+1} satisfies the properties of

Lemma 5 with respect to T_{u_i}. Note that the ranks of $u_1, \ldots, u_q$ are monotonically decreasing and $r(u_i) - r(u_{i+1}) \leq \kappa(u_i)$. Moreover, every vertex in T satisfies $1 \leq |S_\kappa(v)| \leq \binom{k}{\kappa(v)}$ and $1 \leq \kappa(v) \leq k$. Since the only vertices with $\kappa(v) = 1$ are the leaves,

$$r(u_i) \leq \sum_{i=1}^{k} i \left(\log_2 \binom{k}{i} + 1 \right) \leq O(k^3).$$

Consequently, r is a centered coloring of size at most $O(k^3)$ that can be computed in linear time.   □

6   Treedepth Upper Bounds on Interval Graphs

Because linear colorings are equivalent to centered colorings when restricted to paths, we turn our attention to the linear coloring numbers of "pathlike" graphs. We investigate a particular class of "pathlike" graphs in this section and prove a polynomial relationship between their centered and linear coloring numbers.

Definition 7. *A graph G is an* interval graph *if there is an injective mapping f from $V(G)$ to intervals on the real line such that $uv \in E(G)$ iff $f(u)$ and $f(v)$ overlap.*

We refer to the mapping f as the *interval representation* of G. Since the overlap between intervals $f(u)$ and $f(v)$ is independent of the interval representations of the other vertices, every subgraph of an interval graph is also an interval graph. The interval representation of G implies a natural "left-to-right" layout that gives it the "pathlike" qualities, which are manifested in restrictions on the length of induced cycles (*chordal*) and paths between vertex triples (*AT-free*).

Definition 8. *A graph is* chordal *if it contains no induced cycles of length ≥ 4.*

Definition 9. *Vertices u, v, w are an* asteroidal triple *(AT) if there exist uv-, vw-, and wu-paths P_{uv}, P_{vw}, and P_{wu}, respectively, such that $N[w] \cap P_{uv} = N[u] \cap P_{vw} = N[v] \cap P_{uv} = \emptyset$. A graph with no AT is called* AT-free.

Proposition 3 ([8]). *Graph G is an interval graph iff G is chordal and AT-free.*

Intuitively, Definition 9 is a set of three vertices such that every pair is connected by a path that avoids the neighbors of the third. Roughly speaking, in the context of linear colorings, Proposition 3 indicates that if w is a center of a "long" uv-path P in G, any vertex w' such that $\psi(w) = \psi(w')$ must have a neighbor on P. We devote the rest of this section proving Theorem 2.

Theorem 2. *There exists a polynomial time algorithm that takes as input an interval graph G and a linear coloring of G with size k and outputs a centered coloring of G with size at most k^2.*

Our algorithm makes extensive use of the following well-known property of maximal cliques in interval graphs.

Proposition 4 ([8]). *If G is an interval graph, its maximal cliques can be linearly ordered in polynomial time such that for each vertex v, the cliques containing v appear consecutively.*

In particular, we identify a *prevailing path* in G whose vertices "span" the maximal cliques and a *prevailing subgraph* that consists of the prevailing path as well as vertices in maximal cliques "between" consecutive vertices on the prevailing path. We will show that any linear coloring is a centered coloring when restricted to the prevailing subgraph and that after removing the prevailing subgraph, the remaining components each use fewer colors.

Let $C_1, \ldots C_m$ be an ordering of the maximal cliques of G that satisfies Proposition 4. We say vertex v is *introduced* in C_i if $v \in C_i$ but $v \notin C_{i-1}$, and denote this as $I(v) = i$. Likewise, v is *forgotten* in C_j if $v \in C_j$ but $v \notin C_{j+1}$, and denote this as $F(v) = j$. The procedure for constructing a prevailing subgraph and prevailing path is described in Algorithm 1. This algorithm selects the vertex v from the current maximal clique that is forgotten "last" and adds v to the prevailing path and $C_{F(v)}$ to the prevailing subgraph. We prove in Lemma 6 that if P, Q are a prevailing path and subgraph, the vertices in $Q \backslash P$ can be inserted between vertices of P to form a Hamiltonian path of Q.

Algorithm 1. Construction of a prevailing path and subgraph.

Input: interval graph G
Output: prevailing path P and prevailing subgraph Q
 1: $C_1, \ldots, C_m \leftarrow$ maximal cliques of G labeled in accordance with Proposition 4
 2: $P \leftarrow \emptyset$
 3: $V_Q \leftarrow \emptyset$
 4: $i \leftarrow 1$
 5: $j \leftarrow 1$
 6: **while** $i < m$ **do**
 7:     $v_j \leftarrow \arg\max_{u \in C_i} F(u)$
 8:     $P \leftarrow P \cdot \{v_j\}$
 9:     $i \leftarrow F(v)$
10:     $V_Q \leftarrow V_Q \cup V(C_i)$
11:     $j \leftarrow j + 1$
12: **end while**
13: $Q \leftarrow G[V_Q]$
14: **return** P, Q

Lemma 6. *Every prevailing subgraph has a Hamiltonian path.*

Although the fact that the prevailing subgraph Q has a Hamiltonian path implies Q has a center with respect to ψ, we must ensure that the proper subgraphs of Q also have a center. In Lemma 7, we prove $\psi|_Q$ is centered by showing every proper connected subgraph of Q also has a Hamiltonian path.

Lemma 7. *If Q is a prevailing subgraph of an interval graph G and ψ a linear coloring of G, $\psi|_Q$ is a centered coloring.*

Since any linear coloring ψ of the prevailing subgraph Q must also be a centered coloring, $\mathrm{td}(Q) \leq |\psi|$. To get a bound on the treedepth of G, we focus on the relationship between Q and $G \backslash Q$. In particular, we show that the components of $G \backslash Q$ use fewer than $|\psi|$ colors by proving that each such component has an apex in the prevailing path.

Lemma 8. *Let P, Q be a prevailing path and subgraph of an interval graph G. For each component X of $G \backslash Q$, there is a vertex $a \in P$ such that $X \subseteq N(a)$.*

We can now establish a polynomial upper bound on the treedepth of interval graphs, proving Theorem 2.

Proof (Theorem 2). Let $\mathcal{A}$ be the algorithm that constructs a treedepth decomposition $\mathcal{T}$ of G by finding a prevailing subgraph Q (Algorithm 1), using $\psi|_Q$ to create a treedepth decomposition of Q, and recursively constructing treedepth decompositions of $G \backslash Q$. If $\mathrm{depth}(\mathcal{T}) \leq k^2$ and $\mathcal{A}$ runs in polynomial time, then the canonical centered coloring of $\mathcal{T}$ is a centered coloring of G of size at most k^2. We prove $\mathcal{A}$ satisfies these requirements by induction on $k = |\psi|$. At $k = 1$, the graph consists of isolated vertices and $\mathcal{A}$ trivially constructs a treedepth decomposition of G of depth 1 in polynomial time.

Assume $\mathcal{A}$ has the desired properties for linear colorings of size at most $k-1$. Because the maximal cliques of an interval graph can be enumerated and ordered in polynomial time (Proposition 4), identifying Q via Algorithm 1 can be done in polynomial time. By Lemma 7, the canonical treedepth decomposition of Q has depth at most k. Since every component X of $G \backslash Q$ has an apex a in P (Lemma 8), we can assume a is an ancestor in $\mathcal{T}$ of each vertex in X (Lemma 2). Because ψ is proper, $\psi(a)$ does not appear in $\psi|_X$ and since induced subgraphs of interval graphs are themselves interval graphs, $\mathcal{A}$ finds a treedepth decomposition of X whose depth is at most $(k-1)^2$. Thus $\mathcal{T}$ has depth $k + (k-1)^2 \leq k^2$. The recursion only lasts $k \leq n$ steps, so $\mathcal{A}$ runs in polynomial time. $\square$

7   Hardness of Recognizing Linear Colorings

Based on the similarity in definition between linear and centered colorings, one might assume that computing them should be roughly equally difficult. Finding a centered coloring of a fixed size is NP-hard [1], but given a coloring of a graph, we can *recognize* whether it is centered in polynomial time by attempting to create the canonical treedepth decomposition; this procedure will identify a non-centered subgraph if the coloring is not centered. To the contrary, we will prove that LINEAR COLORING RECOGNITION, the problem of recognizing whether a coloring is linear, is co-NP-complete. In order to prove the hardness of LINEAR COLORING RECOGNITION, we first define a dual problem. The NON-CENTERED PATH problem takes a graph G and coloring ψ as input and decides whether G

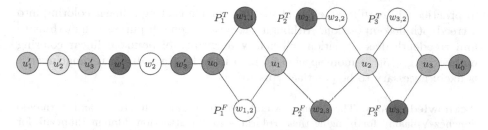

Fig. 2. The graph G and coloring ψ for $\Phi = (x_1 \lor x_2 \lor \neg x_3) \land (\neg x_1 \lor x_2 \lor x_3) \land (\neg x_2)$. (Color figure online)

has a non-centered path P. We focus on proving the hardness of NON-CENTERED PATH because a certificate to that problem is easily definable: a path where every color appears at least twice (Fig. 2).

Theorem 3. NON-CENTERED PATH *is NP-complete.*

Corollary 1. LINEAR COLORING RECOGNITION *is co-NP-complete.*

The co-NP-hardness of recognizing linear colorings is compounded by three stronger hardness implications. First, the coloring ψ given in Theorem 3 has size $m+n+1$, which means that unless the exponential time hypothesis [6] fails, there is no $2^{o(k)}$ algorithm to recognize a linear coloring of size k. Second, the graph G is outerplanar with pathwidth two, which implies that neither treewidth-style dynamic programming nor a Baker-style layering approach is likely to solve this problem efficiently. Finally, by subdividing each edge and coloring all subdivision vertices with a (single) new color, we obtain a bipartite graph with degeneracy two, proving hardness for each of those classes. Nonetheless, the fact that $\chi_{\text{cen}}(G) = O(\log m + \log n)$ while $|\psi| = m + n + 1$ leaves open the possibility that LINEAR COLORING RECOGNITION becomes easier for colorings of minimum size.

8   Conclusion

We have introduced p-linear and linear colorings as an alternative to p-centered and centered colorings for use in algorithms for classes of bounded expansion. The p-linear colorings are computable in polynomial time and require a constant number of colors in classes of bounded expansion, while inducing graphs of bounded treedepth for all small sets of colors, allowing direct substitution in existing algorithmic pipelines. A major direction for future work is to bring the upper bound on $t_{\max}(k)$ of 2^k closer to the lower bound of $2k$. In particular, it appears our current toolkit for analyzing linear colorings must be expanded in order to prove (or disprove) Conjecture 1. We also believe it is worth studying whether recognizing linear colorings can be done in polynomial time if we assume the coloring is of size $\chi_{\text{lin}}(G)$. Finally, using p-linear colorings in practice will require an efficient method for translating a linear coloring into a treedepth decomposition. Although there exist general-purpose algorithms to find treedepth decompositions efficiently in graphs of bounded linear coloring number (e.g. [15]), a more specialized algorithm that avoids "heavy machinery" is likely necessary to be practically useful.

Acknowledgments. The authors would like to thank Felix Reidl and Fernando Sánchez-Villaamil for bringing these colorings to our attention, Marcin Pilipczuk for his assistance in refining our upper and lower bounds on trees, and the anonymous reviewers for their helpful suggestion. This work was supported in part by the DARPA GRAPHS Program and the Gordon & Betty Moore Foundation's Data-Driven Discovery Initiative through Grants SPAWARN66001-14-1-4063 and GBMF4560 to Blair D. Sullivan.

References

1. Bodlaender, H.L.: Rankings of graphs. SIAM J. Discrete Math. **11**(1), 168–181 (1998)
2. Nešetřil, J., Ossona de Mendez, P.: Sparsity. Algorithms and Combinatorics, vol. 28. Springer, Heidelberg (2012). https://doi.org/10.1007/978-3-642-27875-4
3. Demaine, E.,D., et al.: Structural sparsity of complex networks: random graph models and linear algorithms. CoRR, abs/1406.2587 (2015)
4. Dvořák, Z., Král, D., Thomas, R.: Testing first-order properties for subclasses of sparse graphs. JACM **60**(5), 36:1–36:24 (2013)
5. Grohe, M., Kreutzer, S., Siebertz, S.: Deciding first-order properties of nowhere dense graphs. J. ACM (JACM) **64**(3), 17 (2017)
6. Impagliazzo, R., Paturi, R.: On the complexity of k-SAT. J. Comput. Syst. Sci. **62**(2), 367–375 (2001)
7. Kun, J., O'Brien, M.P., Sullivan, B.D.: Treedepth bounds in linear colorings. CoRR, abs/1802.09665, May 2018
8. Lekkeikerker, C., Boland, J.: Representation of a finite graph by a set of intervals on the real line. Fundam. Math. **51**(1), 45–64 (1962)
9. Nešetřil, J., de Mendez, P.O.: Grad and classes with bounded expansion I. Decompositions. Eur. J. Comb. **29**(3), 760–776 (2008)
10. Nešetřil, J., de Mendez, P.O.: Grad and classes with bounded expansion II. Algorithmic aspects. Eur. J. Comb. **29**(3), 777–791 (2008)
11. Nešetřil, J., de Mendez, P.O.: Grad and classes with bounded expansion III. Restricted graph homomorphism dualities. Eur. J. Comb. **29**(4), 1012–1024 (2008). Homomorphisms: structure and highlights
12. O'Brien, M.P., Hobbs, C.G., Jasnick, K., Reidl, F., Rodrigues, N.G., Sullivan, B.D.: CONCUSS, v2.0., June 2016. https://doi.org/10.5281/zenodo.30281
13. O'Brien, M.P., Sullivan, B.D.: An experimental evaluation of a bounded expansion algorithmic pipeline. CoRR, abs/1712.06690, December 2017
14. Reidl, F.: Structural sparseness and complex networks. Dr. Aachen, Techn. Hochsch., Aachen, 2015. Aachen, Techn. Hochsch., Dissertation (2015)
15. Reidl, F., Rossmanith, P., Villaamil, F.S., Sikdar, S.: A faster parameterized algorithm for treedepth. In: Esparza, J., Fraigniaud, P., Husfeldt, T., Koutsoupias, E. (eds.) ICALP 2014. LNCS, vol. 8572, pp. 931–942. Springer, Heidelberg (2014). https://doi.org/10.1007/978-3-662-43948-7_77

16. Schäffer, A.A.: Optimal node ranking of trees in linear time. Inf. Process. Lett. **33**(2), 91–96 (1989)
17. Zhu, X.: Colouring graphs with bounded generalized colouring number. Discrete Math. **309**(18), 5562–5568 (2009)

An Improved FPT Algorithm
for Independent Feedback Vertex Set

Shaohua Li and Marcin Pilipczuk$^{(\boxtimes)}$

Institute of Informatics, University of Warsaw, Warsaw, Poland
{Shaohua.Li,malcin}@mimuw.edu.pl

Abstract. We study the INDEPENDENT FEEDBACK VERTEX SET problem—a variant of the classic FEEDBACK VERTEX SET problem where, given a graph G and an integer k, the problem is to decide whether there exists a vertex set $S \subseteq V(G)$ such that $G \backslash S$ is a forest and S is an independent set of size at most k. We present an $\mathcal{O}^*((1+\varphi^2)^k)$-time FPT algorithm for this problem, where $\varphi < 1.619$ is the golden ratio, improving the previous fastest $\mathcal{O}^*(4.1481^k)$-time algorithm given by Agrawal et al. [1]. The exponential factor in our time complexity bound matches the fastest deterministic FPT algorithm for the classic FEEDBACK VERTEX SET problem.

On the technical side, the main novelty is a refined measure of an input instance in a branching process, that allows for a simpler and more concise description and analysis of the algorithm.

Keywords: Independent feedback vertex set · FPT algorithm

1 Introduction

Given a graph G, a feedback vertex set of G is a set of vertices $S \subseteq V(G)$ such that $G \backslash S$ is a forest. The FEEDBACK VERTEX SET problem (FVS) asks to find a feedback vertex set of the minimum size. This problem is a classic NP-hard problem which has been studied extensively in many fields of complexity and algorithms [2].

In this work, we take the point of view of *parameterized complexity*, where every instance I of a problem at hand is accompanied with a *parameter k*, intended to represent the complexity of the instance at hand. We ask for a *fixed-parameter algorithm* (*FPT algorithm* for short) that solves an instance I with parameter k in time $f(k)|I|^c$ for some computable function f and a

This research is a part of projects that have received funding from the European Research Council (ERC) under the European Union's Horizon 2020 research and innovation programme under grant agreements No. 714704 .

European Research Council
Established by the European Commission

© Springer Nature Switzerland AG 2018
A. Brandstädt et al. (Eds.): WG 2018, LNCS 11159, pp. 344–355, 2018.
https://doi.org/10.1007/978-3-030-00256-5_28

constant c. That is, the exponential blow-up in the running time bound, probably unavoidable for NP-hard problems, is confined to be a function of the parameter only. For more on parameterized complexity, we refer to a recent textbook [3].

In the context of parameterized complexity of the FEEDBACK VERTEX SET problem, there is a long line of work improving the upper bound of the FPT algorithm for the standard parameterization of the solution size [4–11] (i.e., the input consists of a graph G and a parameter k, and the goal is to find a feedback vertex set of size at most k or show that no such set exists). The fastest randomized FPT algorithm for FEEDBACK VERTEX SET, which runs in time $\mathcal{O}^*(3^k)$, is given by Cygan et al. [12].[1,2] If one asks for a deterministic FPT algorithm, the champion runs in $O^*(3.619^k)$ and is due to Kociumaka and Pilipczuk [11].

At the same time, many variants of FEEDBACK VERTEX SET received significant attention, including SUBSET FVS [13–15], GROUP FVS [14,16–18], or SIMULTANEOUS FVS [19].

In this paper, we focus on the parameterized version of the INDEPENDENT FEEDBACK VERTEX SET problem (IFVS), which is to decide if there exists a feedback vertex set S of size at most k such that no two vertices of S are adjacent in G. Misra et al. gave the first FPT algorithm running in time $\mathcal{O}(5^k n^{\mathcal{O}(1)})$ and an $\mathcal{O}(k^3)$ kernel for IFVS [20].[3] Agrawal et al. presented an improved FPT algorithm running in time $\mathcal{O}^*(4.1481^k)$ for IFVS [1]. In this paper, we propose a faster FPT algorithm.

Theorem 1. *The* INDEPENDENT FEEDBACK VERTEX SET *problem, parameterized by the solution size, can be solved in* $\mathcal{O}^*((1 + \varphi^2)^k) \leq \mathcal{O}^*(3.619^k)$ *time, where* $\varphi = \frac{1+\sqrt{5}}{2} < 1.619$ *is the golden ratio.*

We remark here that Theorem 1 is not "just another" improvement in the base of the exponential function, but in some sense "the end of the road". The exponential function of the time bound of Theorem 1 matches the one of the algorithm of Kociumaka and Pilipczuk [11] for the classic FEEDBACK VERTEX SET problem. Since FEEDBACK VERTEX SET trivially reduces to INDEPENDENT FEEDBACK VERTEX SET (subdivide each edge once), any (deterministic) improvement to the base of the exponential function of Theorem 1 would give a similar improvement for FEEDBACK VERTEX SET.

On the technical side, we follow the standard approach of iterative compression as in [1] to reduce to a "disjoint" version of the problem. Here, our approach diverges from the one of [1]. We follow a modified measure for the subsequent branching process, somewhat inspired by the work of Kociumaka and

---

[1] The $\mathcal{O}^*$-notation suppresses factors that are polynomial in the input size.

[2] Actually in the randomized FPT algorithm for FVS, the parameter is the treewidth of the graph. Since the treewidth of a yes-instance (G, k) to FVS is at most $k + 1$, the randomized algorithm for FVS runs in time $\mathcal{O}^*(3^k)$.

[3] A *kernel* of size $g(k)$ for some computable function g is a polynomial-time procedure that reduces an instance I with parameter k to an equivalent instance with size and parameter value bounded by $g(k)$.

Pilipczuk [11]. This improved measure, together with a number of new notions (generalized W-degree, potential nice vertices and tents), allow us to simplify the algorithm and analysis as compared to [1].

2   Preliminaries

The graphs in our paper are all undirected and may contain multiple edges or loops. For a graph G, we denote its vertex set by $V(G)$ and edge multiset by $E(G)$. For a vertex $v \in V(G)$, we use $N(v) = \{u \in V(G) : uv \in E(G)\}$ to denote the *neighborhood* of v; note that $N(v)$ is a set, containing a vertex u only once even in the presence of multiple edges uv. We define the *closed neighborhood* of v as $N[v] = N(v) \cup \{v\}$. For a vertex set $A \subseteq V(G)$, the neighborhood of A is $N(A) = \bigcup_{v \in A} N(v) \backslash A$. For a vertex set $X \subseteq V(G)$, we denote the *induced subgraph* of X by $G[X]$. For simplicity, we use $G\backslash X$ to denote $G[V(G)\backslash X]$. For a vertex set $X \subseteq V(G)$ and $v \in V(G)$, we define X-*degree* of v as the number of edges with one endpoint being v and the other lying in X, and we denote it by $\deg_X(v)$. Note that the X-degree counts edges with multiplicities. A *connected component* is a maximal connected subgraph. Contracting a connected subgraph H is the operation of replacing the subgraph H with a vertex v_H and every edge xy with $x \in V(H)$ and $y \in V(G)\backslash V(H)$ with an edge $v_H y$ (keeping multiplicities).

3   An Algorithm for Independent Feedback Vertex Set

Given an instance (G, k), we first invoke the $\mathcal{O}^*((1 + \varphi^2)^k)$-time FPT algorithm for the classic FEEDBACK VERTEX SET problem [11]. If the algorithm returns NO, we conclude that there is no independent feedback vertex set of size at most k since an independent feedback vertex set is also a feedback vertex set. Otherwise, the algorithm returns a feedback vertex set Z such that $|Z| \leq k$. Obviously, $F = G\backslash Z$ is a forest.

Suppose there is a solution S for the input instance (G, k). The algorithm branches into $2^{|Z|}$ directions, guessing a subset Z' of Z such that $S \cap Z = Z'$. Let $W = Z\backslash Z'$. If $G[Z']$ is not an independent set or $G[W]$ is not a forest, the algorithm rejects this guess. Hence, we can assume that $G[Z']$ is an independent set and $G[Z\backslash Z']$ is a forest. Let $R = N(Z') \cap F$. Since the solution S is an independent set and $Z' \subseteq S$, we have $R \cap S = \emptyset$. Then the algorithm tries to find an independent feedback vertex set $S' \subseteq F$ for $G\backslash Z'$ such that $S' \cap R = \emptyset$ and $|S'| \leq k - |Z'|$. Following [1], we call this subproblem DISJOINT INDEPENDENT FEEDBACK VERTEX SET (DIS-IFVS for short). We give a faster FPT algorithm for DIS-IFVS in the next section. The algorithm tries every possible $Z' \subseteq Z$ and solves the corresponding subproblem of DIS-IFVS. If the algorithm finds a YES instance of DIS-IFVS, then it returns YES for the instance (G, k) of IFVS. Otherwise, if the algorithm tries every possible $Z' \subseteq Z$ and obtains a NO answer for every corresponding instance of DIS-IFVS, it reports that (G, k) is a NO instance.

3.1   Disjoint Independent Feedback Vertex Set

We start with a formal definition of the problem.

DISJOINT INDEPENDENT FEEDBACK VERTEX SET
Input: An undirected (multi)graph G, a feedback vertex set W of G,
$R \subseteq V(G) \backslash W$, and an integer k.
Question: Is there an independent feedback vertex set $X \subseteq V(G) \backslash (W \cup R)$
for G such that $|X| \leq k$?

Let $F = V(G \backslash W)$. Obviously, $G[F]$ is a forest since W is a feedback vertex set
of G. A vertex $v \in F \backslash R$ is a *nice vertex* if $\deg_W(v) = 2$ and v has no neighbors
in F. A vertex $v \in F \backslash R$ is a *tent* if $\deg_W(v) = 3$ and v has no neighbors in F.

As mentioned earlier, we rely on a measure different from the one in [1]. The
measure μ of an instance (G, W, R, k) is defined as

$$\mu = k + \rho - (\eta + \tau).$$

Here, ρ represents the number of connected components of $G[W]$, η is the number
of nice vertices in $F \backslash R$ and τ is the number of tents in $F \backslash R$.

We remark that the distinction between sets W and R is purely for the
sake of complexity of the algorithm. The set of feasible solutions to a DISJOINT
INDEPENDENT FEEDBACK VERTEX SET instance (G, W, R, k) would be the same
if we move vertices from R to W. However, the notions of tents, nice vertices,
and the measure μ strongly depends on the distinction between the sets W and
R. The algorithm maintains this distinction to ensure the promised running time
bound.

Our main technical result is the following.

Lemma 1. *A* DISJOINT INDEPENDENT FEEDBACK VERTEX SET *instance* I
with measure μ *can be solved in time* $\mathcal{O}^*(\varphi^\mu)$, *where* $\varphi = \frac{1+\sqrt{5}}{2}$ *is the golden
ratio.*

Theorem 1 follows by standard analysis as in [1]:

Proof (Proof of Theorem 1). The algorithm for FVS of [11] runs in time $\mathcal{O}^*((1 + \varphi^2)^k)$. In a branch with a set $Z' \subseteq Z$ the routine for DIS-IFVS is passed an
instance with both $W = Z \backslash Z'$ and the parameter bounded by $k - |Z'|$, and
hence with measure bounded by $2(k - |Z'|)$. Since the algorithm for DIS-IFVS
runs in time $O^*(\varphi^\mu)$, the total running time of its applications is bounded by

$$\sum_{i=0}^{k} \binom{k}{i} \mathcal{O}^*(\varphi^{2(k-i)}) = \mathcal{O}^*((1 + \varphi^2)^k) \leq \mathcal{O}^*(3.619^k).$$

This completes the proof.   □

The remainder of this section is devoted to the proof of Lemma 1. We start
with showing that μ is nonnegative on YES instances.

Lemma 2. *Let* $I = (G, W, R, k)$ *be a* YES *instance of* DISJOINT FEEDBACK VERTEX SET. *Then* $\mu \geq 0$.

Proof. Let X be a solution to the instance I. Thus $G' = G\backslash X$ is a forest. Let $N \subseteq V(G)\backslash(W \cup R)$ be the set of nice vertices and $T \subseteq V(G)\backslash(W \cup R)$ be the set of tents. Since $X \cap W = \emptyset$, we have that $H := G[W \cup (N\backslash X) \cup (T\backslash X)]$ is a forest. Now we contract each component in $H[W]$ into a single vertex and get a forest $\tilde{H}$. Since there are at most $\rho + |N\backslash X| + |T \backslash X|$ vertices in $\tilde{H}$, there are at most $\rho + |N\backslash X| + |T \backslash X| - 1$ edges in $\tilde{H}$. According to the definition of tents and nice vertices, $(N \cup T)\backslash X$ is an independent set. Moreover, since the degree of any vertex in $N\backslash X$ and $T\backslash X$ is 2 and 3, respectively, we get the following inequality:

$$2|N\backslash X| + 3|T\backslash X| \leq |E(\tilde{H})| \leq \rho + |N\backslash X| + |T\backslash X| - 1.$$

It follows that:

$$|N\backslash X| + |T\backslash X| \leq |N\backslash X| + 2|T\backslash X| \leq \rho.$$

Hence, as $|X| \leq k$,

$$|N| + |T| \leq \rho + k.$$

As a result, $\mu = \rho + k - (\eta + \tau) \geq 0$. □

A small comment is in place. Our measure μ is different from the one of [1]: $\mu' = 2k + \rho - (\eta + 2\tau)$. The change in the measure is one of the critical insights in this paper: while it sometimes leads to weaker branching vectors as compared to [1], the "starting value" in an application in the above proof of Theorem 1 is $2(k - |Z'|)$, not $3(k - |Z'|)$ as in [1]. Thus, to obtain the promised running time bound, we are fine with branching vectors of the form $(1, 2)$; that is, we are fine with branching steps in two directions, where in one direction the measure drops by at least one, and in the other direction by at least two. The change in the measure is similar to the one that happened in the work of Kociumaka and Pilipczuk for FEEDBACK VERTEX SET [11], as compared to a previous champion of Cao, Chen, and Liu [5].

We introduce now some definitions that will help us streamline later arguments. Let (G, W, R, k) be an instance of DIS-IFVS and let $F = V(G)\backslash W$. We say that $u \in F\backslash R$ is a *potential nice vertex* or *P-nice* if u is of degree 2 and exactly one of its neighbors is in W. For a vertex v in $G[F]$, we define the *nice degree* of v, denoted by $\mathrm{Ndeg}(v)$, as the number of P-nice neighbors of v. A *generalized degree* of v is $\mathrm{Gdeg}_W(v) = \mathrm{Ndeg}(v) + \mathrm{deg}_W(v)$. We say that $u \in F\backslash R$ is a *potential tent* or *P-tent* if $\mathrm{Gdeg}_W(u) = 2$ and $\mathrm{deg}(u) = 3$. For a vertex v in F, we define the *tent degree* of v, denoted by $\mathrm{Tdeg}(v)$, as the number of neighbors of v that are P-tents.

3.2   Reduction Rules for DIS-IFVS

Now we introduce some reduction rules for DIS-IFVS. We always apply the applicable reduction rule of the lowest number. First, let us introduce five reduction rules from [1].

Reduction Rule 1: Delete any vertex of degree at most one.

Reduction Rule 2: Let u, v be two adjacent vertices of degree two in $G\backslash W$ which are not nice vertices in F. Besides, u is adjacent to x while v is adjacent to y (x and y could be the same vertex). If neither u nor v is in R or both are in R, then delete one vertex in $\{u,v\}$ arbitrarily and connect the neighbors of the deleted vertex with a new edge. If exactly one of u and v is in R, say $v \in R$, then delete v and add an edge between its neighbors (i.e., an edge uy).

Reduction Rule 3: If $k < 0$ or $\mu < 0$, return that the input instance is a NO instance.

Reduction Rule 4: If there is a vertex $v \in R$ such that v has two neighbors in the same component of W, then return that the input instance is a NO instance.

Reduction Rule 5: If there is a vertex $v \in F\backslash R$ such that v has at least two neighbors in the same component of W, then remove v from G and add all vertices in $F \cap N(v)$ to R. In this case, k decreases by one.

It is not difficult to verify the safeness of Reduction Rules 1–5 as shown in [1]. But when analyzing Reduction Rules 1 and 5, we need to be careful since we use a different measure $\mu = k + \rho - (\eta + \tau)$. In Reduction Rule 1, if one deletes a neighbor w of a tent or a nice vertex v, then v stops being a tent or a nice vertex ($\eta + \tau$ could decrease by one), but also $\{w\}$ stops being a connected component of $G[W]$ (decreasing ρ by one). For Reduction Rule 5, it may happen that v is a tent or a nice vertex, and its deletion decreases $\eta + \tau$ by one. However, the removal of v also decreases k by one. Thus μ does not increase.

Now we introduce two new reduction rules.

Reduction Rule 6: If there is a vertex $v \in R$ such that $\mathrm{Gdeg}_W(v) \geq 1$ or $\mathrm{Tdeg}(v) \geq 1$, then remove v from R and add v to W.

Reduction Rule 7: If there is a vertex $v \in F\backslash R$ such that every neighbor $w \in N(v)\backslash(W \cup R)$ is of degree 2, and at least one such neighbor exists, then put $N(v)\backslash(W \cup R)$ into R.

We first show their safeness.

Claim 1. Reduction Rules 6 and 7 are safe.

Proof. The safeness of Reduction Rule 6 is straightforward. For the safeness of Reduction Rule 7, suppose that (G, W, R, k) is an input instance. Let v be the vertex satisfying the condition of Reduction Rule 7 and $(G, W, R \cup (N(v) \cap F), k)$ be the instance obtained after applying Reduction Rule 7. We claim that (G, W, R, k) is a YES instance if and only if $(G, W, R \cup (N(v) \cap F), k)$ is a YES instance. The "if" direction is straightforward, since we only increased the set R.

For the "only if" direction, let X be a solution of size at most k to the instance (G, W, R, k). If $X \cap N(v) = \emptyset$, then X is also a solution to $(G, W, R \cup (N(v) \cap F), k)$. Otherwise, we construct a vertex set $X' = (X \cup \{v\})\backslash(N(v) \cap F)$. Obviously $|X'| \leq k$. We will show that X' is a solution to $(G, W, R \cup (N(v) \cap$

$F), k)$. Clearly, it is disjoint with $W \cup R \cup N((v) \cap F)$ and independent, as it is disjoint with $N(v)$. To show that X' is a feedback vertex set in G, observe that since every vertex $w \in N(v)\backslash(W \cup R)$ is of degree 2, every cycle passing through w in G passes also through v. □

Since Reduction Rule 7 only moves vertices to R, its application does not change the measure; note that the neighbors of a vertex affected by Reduction Rule 7 can be neither a nice vertex nor a tent. However, the situation is not that easy for Reduction Rule 6, and we need to show that its application does not increase μ. To this end, we show a number of generic observations on how the measure μ changes if we modify a neighbor of a P-nice vertex or a P-tent.

Observation 1. *Let $v \in F$ be a vertex with a P-nice neighbor w. Consider the operation of moving v to W. Then, the vertex w becomes nice and η goes up at least by one.*

Observation 2. *Let $v \in F$ be a vertex with a P-tent neighbor w such that v is not P-nice. Consider the operation of putting v in a solution: deleting it from G and putting $N(v) \cap F$ into R. Then the application of reduction rules on w and its (possible) other neighbors in F decreases μ by at least one.*

Proof. The operation moves w to R and decreases its degree to 2. Since w is a P-tent and v is not a P-nice vertex, every neighbor $u \in (N(w) \cap F)\backslash\{v\}$ is a P-nice vertex. Consequently, Reduction Rule 2 reduces $(N[w] \cap F)\backslash\{v\}$ to a single vertex w', which is in R if $(N(w) \cap F)\backslash\{v\} \subseteq R$. Furthermore, $\deg(w') = \deg_W(w') = 2$. If w' has both neighbors in the same connected component of $G[W]$, then either Reduction Rule 4 rejects the instance or Reduction Rule 5 decreases k by one. Otherwise, if $w' \in R$, Reduction Rule 6 moves w' to W, decreasing ρ by one. If $w' \notin R$, then w' becomes a nice vertex, increasing η by one. Thus, in all cases, μ decreases by at least one. □

Observation 3. *Let $v \in F$ be a vertex with a P-tent neighbor w such that v is not P-nice. Consider the operation of moving v into W. Then the application of reduction rules on w and its (possible) other neighbors in F decrease μ by at least one.*

Proof. Since w is a P-tent and v is not P-nice, every neighbor $u \in (N(w) \cap F)\backslash\{v\}$ is P-nice. Consider such a vertex u; note that $u \in F\backslash R$ by the definition of P-nice. Reduction Rule 7 is applicable to w; this rule would move u to R and then Reduction Rule 6 would move u to W. Along this process, Reduction Rule 4 or 5 can be triggered on w, either rejecting the instance or decreasing k by one. Otherwise, if $w \in R$, Reduction Rule 6 moves w to W, decreasing ρ by two. Finally, in the last case we are left with $w \in F\backslash R$ with $\deg_W(w) = \deg(w) = 3$, that is, w becomes a tent and increases τ by one. Thus, in all cases, μ decreases by at least one. □

Armed with the above observations, we can now show that Reduction Rule 6 on its own does not increase the measure.

Claim 2. An application of Reduction Rule 6 does not increase the measure.

Proof. If v is a tent or a nice vertex, then η or τ decreases by one but ρ decreases by at least one because Reduction Rule 4 or 5 is not applicable. In this case, μ does not increase. If v is neither a tent nor a nice vertex and $\deg_W(v) \geq 1$, ρ does not increase, and η and τ do not decrease. In this case, μ does not increase.

We are left with the case $\deg_W(v) = 0$, and then ρ increases by one. If $\mathrm{Gdeg}_W(v) \geq 1$ but $\deg_W(v) = 0$, we have a P-nice neighbor w of v. Then, after v is moved to W, Observation 1 asserts that future application of reduction rules on w cause a measure decrease of at least one, offsetting the increase of ρ. Otherwise, $\mathrm{Tdeg}(v) \geq 1$, and we have a neighbor w of v that is a P-tent. Then, after v is moved to W, Observation 3 asserts that future application of reduction rules on w and its possible neighbors in F cause measure decrease of at least one. This finishes the proof. □

3.3 Branching for DIS-IFVS

Now we are ready to introduce the branching algorithm. We assume that all reduction rules have been applied exhaustively. As a branching pivot, we pick a vertex $v \in F$ that is neither a nice vertex nor a tent and satisfies one of the following three cases:

Case A: $\mathrm{Gdeg}_W(v) \geq 3$.
Case B: $\mathrm{Gdeg}_W(v) \geq 1$ and $\mathrm{Tdeg}(v) \geq 1$.
Case C: $\mathrm{Tdeg}(v) \geq 2$.

In case of more than one vertices of F satisfying one of the above cases, we prefer to pick a vertex v that satisfies an earlier case.

First, note that the nonapplicability of Reduction Rule 6 implies that the chosen branching pivot v does not lie in R.

No matter which case the chosen branching pivot v satisfies, we branch into two cases. In one case we include v into the solution: we delete v from the graph, include $N(v) \cap F$ into R, and decrease k by one. In the other case, we move v to W.

We now show that in each of the cases, the branching gives a branching vector $(1,2)$ or better with respect to the measure μ. That is, in one of the branches the measure drops by at least one, and in the other by at least two.

Case A: $\mathrm{Gdeg}_W(v) \geq 3$.

(i) Branch where v is deleted and all vertices in $N(v) \cap F$ are added to R. k decreases by 1, ρ stays the same, and η and ρ does not decrease as v is neither a nice vertex nor a tent. Thus, μ decreases by at least one.
(ii) Branch where v is moved from F to W. ρ decreases by $\deg_W(v) - 1$ (which may be -1 if $\deg_W(v) = 0$) and η increases by $\mathrm{Ndeg}(v)$. Since $\mathrm{Gdeg}_W(v) = \deg_W(v) + \mathrm{Ndeg}(v) \geq 3$ and τ does not decrease μ decreases by at least two.

Case B: $Gdeg_W(v) \geq 1$ and $Tdeg(v) \geq 1$.

(i) Branch where v is deleted and all vertices in $N(v) \cap F$ are added to R. First, k decreases by one. Furthermore, v has a P-tent neighbor w and Observation 2 asserts that future applications of reduction rules on w and its remaining neighbors in F decrease the measure by at least one. Thus, in total μ decreases by at least two.

(ii) Branch where v is moved from F to W. For every P-tent neighbor w of v, Observation 3 asserts that the application of reduction rules to w and its remaining neighbors in F cause a measure decrease of at least 1. If $deg_W(v) \geq 1$, then moving v to W does not increase ρ, and we are done. Otherwise, if $deg_W(v) = 0$, moving v to W increases ρ by 1 but the assumption $Gdeg_W(v) \geq 1$ implies that there also exists a P-nice neighbor w of v. For every such P-nice neighbor w of v, Observation 1 asserts that the future application of reduction rules on w and its remaining neighbors in F cause measure drop by at least 1. Consequently, in this case we also have a measure drop of at least 1.

Case C: $Tdeg(v) \geq 2$.

(i) Branch where v is deleted and all vertices in $N(v) \cap F$ are added to R. First, k decreases by one. Furthermore, for every P-tent neighbor w of v, Observation 2 asserts that the application of reduction rules on w and its remaining neighbors in F cause measure drop by at least one. Since $Tdeg(v) \geq 2$, together with the decrease of k we have a total measure decrease of at least 3.

(ii) Branch where v is moved from F to W. The move itself may increase ρ by one. For every P-tent neighbor w of v, Observation 3 asserts that the future application of reduction rules on w and its remaining neighbors in F cause measure drop by at least 1. Since $Tdeg(v) \geq 2$, in total we have a measure decrease by at least 1.

We are left with analysing what happens if no vertex of F satisfies any of the three cases for the choice of the branching pivot. As in [1], we rely on the following base case.

Lemma 3 ([1]). *Let (G, W, R, k) be an instance of DIS-IFVS where every vertex in $V(G) \backslash W$ is either a nice vertex or a tent. Then we can find an independent feedback vertex set $X \subseteq V(G) \backslash (W \cup R)$ for G of the minimum size in polynomial time.*

Lemma 3 follows from the observation by Cao et al. [5] and the fact that all nice vertices and tents form an independent set.

We show the following.

Lemma 4. *If no reduction rule can be applied and every vertex of F does not satisfy any of the cases for the choice of the branching pivot, then the remaining instance of DIS-IFVS can be solved in polynomial time.*

Proof. We claim that every vertex in F of the remaining graph G is either a tent or a nice vertex; the claim then follows by Lemma 3.

For contradiction, suppose that there is a connected component D of $G[F]$ that is not a singleton with a tent or a nice vertex. Since no vertex of D falls into Case A, $\text{Gdeg}_W(v) \leq 2$ for every $v \in D$; in particular, every leaf (a vertex in F that has only exactly one neighbor in F) $v \in D$ satisfies $\deg_W(v) \in \{1, 2\}$. Root the tree $G[D]$ at an arbitrary vertex, and consider a leaf $v \in D$ that is furthest from the root in $G[D]$ and, among such leaves, choose one maximizing $\deg_W(v)$. Note that $v \notin R$ as otherwise Reduction Rule 6 would move v to W.

First, assume $\deg_W(v) = 2$. Since v is a leaf of D and is not nice, v has exactly one neighbor $u \in D$, and v is a P-tent. Hence, $\text{Tdeg}(u) \geq 1$. If $\deg(u) = 2$, then Reduction Rule 7 applies to v if $u \notin R$ and once u is in R, then Reduction Rule 6 applies to u, making v a tent. Consequently, $\deg(u) \geq 3$. However, by the choice of v, $\deg_W(u) \geq 1$ or u is adjacent to another leaf v' of D. However, this implies that $\text{Gdeg}_W(u) \geq 1$ (if $\deg_W(u) \geq 1$ or v' exists and $\deg_W(v') = 1$) or $\text{Tdeg}(u) \geq 2$ (if v' exists and $\deg_W(v') = 2$), and Case B or C applies to u.

Second, assume $\deg_W(v) = 1$, and again let u be the unique neighbor of v in $G[D]$. If $\deg(u) = 2$, then Reduction Rule 2 is applicable. By the choice of v, every other leaf v' adjacent to u also satisfies $\deg_W(v') = 1$; that is, every child of u is P-nice as $u \notin R$. If $\text{Gdeg}_W(u) \geq 3$, then Case A applies to u. Hence, $\deg(u) = 3$ and $\text{Gdeg}_W(u) = 2$: u has a parent x in $G[D]$ and either one more child v' that is P-nice or a neighbor in W. In particular, u is a P-tent, and $\text{Tdeg}(x) \geq 1$.

If $\deg(x) = 2$, then Reduction Rule 7 would apply to u and move v to R, and consequently Reduction Rule 6 would move v to W. If $\text{Gdeg}_W(x) \geq 1$, then Case B applies to x. Hence, x has another child u' that is not P-nice. By the choice of v, the connected component of $G[D]\backslash\{x\}$ containing u' is a star with u' as a center. Furthermore, every child w of u' is P-nice (i.e., $\deg_W(w) = 1$). Since Case A is not applicable to u', we have $\text{Gdeg}_W(u') \leq 2$. If $\deg(u') = 2$, then either u' is P-nice (if $\deg_W(u') = 1$) or Reduction Rule 2 is applicable to u' and its child (if $\deg_W(u') = 0$). We infer that $\deg(u') = 3$ and $\text{Gdeg}_W(u') = 2$; in particular, u' is a P-tent. Hence, $\text{Tdeg}(x) \geq 2$ and case C applies to x. This completes the proof of the lemma.   □

Every step of the reduction rules and branching can be executed in polynomial time. In every case of branching, the branching vector is $(1, 2)$. Thus we get the following recurrence: $T(\mu) = T(\mu - 1) + T(\mu - 2)$. As a result, the running time of the algorithm for DIS-IFVS is $O^*(\varphi^{2k})$. This concludes the proof of Lemma 1 and thus of the whole Theorem 1.

4   Conclusion

In this paper, we presented a faster FPT algorithm for the INDEPENDENT FEEDBACK VERTEX SET problem by using a different measure, introducing some new reduction rules and improving the branching algorithm for the DISJOINT INDEPENDENT FEEDBACK VERTEX SET problem. Moreover, we introduce the notion of generalized degree and tent degree, which makes the reduction and branching more concise. The running time of our algorithm is $\mathcal{O}^*(3.619^k)$, which matches the running time of the current fastest FPT algorithm for the FEEDBACK VERTEX SET problem. As IFVS is a more general problem than FVS, any improvement for IFVS will lead to an improvement for the FPT algorithm of FVS. We conclude with re-iterating an open problem of [19]: does there exist a kernel of size $\mathcal{O}(k^2)$, as it is the case for FVS [21,22]?

References

1. Agrawal, A., Gupta, S., Saurabh, S., Sharma, R.: Improved algorithms and combinatorial bounds for independent feedback vertex set. In: 11th International Symposium on Parameterized and Exact Computation, IPEC 2016, 24–26 August 2016, Aarhus, Denmark. LIPIcs, vol. 63, pp. 2:1–2:14. Schloss Dagstuhl - Leibniz-Zentrum fuer Informatik (2016)
2. Floudas, C.A., Pardalos, P.M. (eds.): Encyclopedia of Optimization, 2nd edn. Springer, Heidelberg (2009). https://doi.org/10.1007/978-0-387-74759-0
3. Cygan, M., et al.: Parameterized Algorithms. Springer, Cham (2015). https://doi.org/10.1007/978-3-319-21275-3
4. Bodlaender, H.L.: On disjoint cycles. Int. J. Found. Comput. Sci. 5(1), 59–68 (1994). https://doi.org/10.1142/S0129054194000049
5. Cao, Y., Chen, J., Liu, Y.: On feedback vertex set new measure and new structures. In: Kaplan, H. (ed.) SWAT 2010. LNCS, vol. 6139, pp. 93–104. Springer, Heidelberg (2010). https://doi.org/10.1007/978-3-642-13731-0_10
6. Chen, J., Fomin, F.V., Liu, Y., Lu, S., Villanger, Y.: Improved algorithms for the feedback vertex set problems. In: Dehne, F., Sack, J.-R., Zeh, N. (eds.) WADS 2007. LNCS, vol. 4619, pp. 422–433. Springer, Heidelberg (2007). https://doi.org/10.1007/978-3-540-73951-7_37
7. Downey, R.G., Fellows, M.R.: Fixed parameter tractability and completeness. In: Complexity Theory: Current Research, Dagstuhl Workshop, 2–8 February 1992, pp. 191–225. Cambridge University Press (1992)
8. Downey, R.G., Fellows, M.R.: Parameterized Complexity. Monographs in Computer Science. Springer, New York (1999). https://doi.org/10.1007/978-1-4612-0515-9
9. Guo, J., Gramm, J., Hüffner, F., Niedermeier, R., Wernicke, S.: Compression-based fixed-parameter algorithms for feedback vertex set and edge bipartization. J. Comput. Syst. Sci. 72(8), 1386–1396 (2006). https://doi.org/10.1016/j.jcss.2006.02.001
10. Kanj, I., Pelsmajer, M., Schaefer, M.: Parameterized algorithms for feedback vertex set. In: Downey, R., Fellows, M., Dehne, F. (eds.) IWPEC 2004. LNCS, vol. 3162, pp. 235–247. Springer, Heidelberg (2004). https://doi.org/10.1007/978-3-540-28639-4_21
11. Kociumaka, T., Pilipczuk, M.: Faster deterministic feedback vertex set. Inf. Process. Lett. 114(10), 556–560 (2014). https://doi.org/10.1016/j.ipl.2014.05.001
12. Cygan, M., Nederlof, J., Pilipczuk, M., Pilipczuk, M., van Rooij, J.M.M., Wojtaszczyk, J.O.: Solving connectivity problems parameterized by treewidth in single exponential time. In: IEEE 52nd Annual Symposium on Foundations of Computer Science, FOCS 2011, Palm Springs, CA, USA, 22–25 October 2011, pp. 150–159. IEEE Computer Society (2011). https://doi.org/10.1109/FOCS.2011.23

13. Cygan, M., Pilipczuk, M., Pilipczuk, M., Wojtaszczyk, J.O.: Subset feedback vertex set is fixed-parameter tractable. SIAM J. Discrete Math. **27**(1), 290–309 (2013). https://doi.org/10.1137/110843071

14. Iwata, Y., Wahlström, M., Yoshida, Y.: Half-integrality, LP-branching, and FPT algorithms. SIAM J. Comput. **45**(4), 1377–1411 (2016). https://doi.org/10.1137/140962838

15. Lokshtanov, D., Ramanujan, M.S., Saurabh, S.: Linear time parameterized algorithms for subset feedback vertex set. ACM Trans. Algorithms **14**(1), 7:1–7:37 (2018). https://doi.org/10.1145/3155299

16. Cygan, M., Pilipczuk, M., Pilipczuk, M.: On group feedback vertex set parameterized by the size of the cutset. Algorithmica **74**(2), 630–642 (2016). https://doi.org/10.1007/s00453-014-9966-5

17. Guillemot, S.: FPT algorithms for path-transversal and cycle-transversal problems. Discrete Optim. **8**(1), 61–71 (2011). https://doi.org/10.1016/j.disopt.2010.05.003

18. Kratsch, S., Wahlström, M.: Representative sets and irrelevant vertices: new tools for kernelization. In: 53rd Annual IEEE Symposium on Foundations of Computer Science, FOCS 2012, New Brunswick, NJ, USA, 20–23 October 2012, pp. 450–459. IEEE Computer Society (2012). https://doi.org/10.1109/FOCS.2012.46

19. Misra, N., Philip, G., Raman, V., Saurabh, S., Sikdar, S.: FPT algorithms for connected feedback vertex set. J. Comb. Optim. **24**(2), 131–146 (2012). https://doi.org/10.1007/s10878-011-9394-2

20. Misra, N., Philip, G., Raman, V., Saurabh, S.: On parameterized independent feedback vertex set. Theor. Comput. Sci. **461**, 65–75 (2012). https://doi.org/10.1016/j.tcs.2012.02.012

21. Iwata, Y.: Linear-time kernelization for feedback vertex set. In: 44th International Colloquium on Automata, Languages, and Programming, ICALP 2017, 10–14 July 2017, Warsaw, Poland. LIPIcs, vol. 80, pp. 68:1–68:14. Schloss Dagstuhl - Leibniz-Zentrum fuer Informatik (2017). https://doi.org/10.4230/LIPIcs.ICALP.2017.68

22. Thomassé, S.: A $4k^2$ kernel for feedback vertex set. ACM Trans. Algorithms **6**(2), 32:1–32:8 (2010). https://doi.org/10.1145/1721837.1721848

Construction and Local Routing
for Angle-Monotone Graphs

Anna Lubiw[1] and Debajyoti Mondal[2(✉)]

[1] Cheriton School of Computer Science, University of Waterloo, Waterloo, Canada
alubiw@uwaterloo.ca
[2] Department of Computer Science, University of Saskatchewan, Saskatoon, Canada
dmondal@cs.usask.ca

Abstract. A geometric graph in the plane is *angle-monotone of width* γ if every pair of vertices is connected by an *angle-monotone path of width* γ, a path such that the angles of any two edges in the path differ by at most γ. Angle-monotone graphs have good spanning properties.

We prove that every point set in the plane admits an angle-monotone graph of width $90°$, hence with spanning ratio $\sqrt{2}$, and a subquadratic number of edges. This answers an open question posed by Dehkordi, Frati and Gudmundsson.

We show how to construct, for any point set of size n and any angle α, $0 < \alpha < 45°$, an angle-monotone graph of width $(90° + \alpha)$ with $O(\frac{n}{\alpha})$ edges. Furthermore, we give a local routing algorithm to find angle-monotone paths of width $(90° + \alpha)$ in these graphs. The *routing ratio*, which is the ratio of path length to Euclidean distance, is at most $1/\cos(45° + \frac{\alpha}{2})$, i.e., ranging from $\sqrt{2} \approx 1.414$ to 2.613. For the special case $\alpha = 30°$, we obtain the Θ_6-graph and our routing algorithm achieves the known routing ratio 2 while finding angle-monotone paths of width $120°$.

1 Introduction

The problem of constructing a geometric graph on a given set of points in the plane so that the graph is sparse yet has good spanning and/or routing properties has been very well-studied. The basic goal is to guarantee paths that are relatively short, and to be able to find such paths using local routing. Two fundamental concepts in this regard are *spanners* and *greedy graphs*. A geometric graph is a *t-spanner* if there is a path of stretch factor t between any two vertices, i.e., a path whose length is at most t times the Euclidean distance between the endpoints. A geometric graph is *greedy* if there is a path between every two vertices such that each intermediate vertex is closer to the destination than the previous vertex on the path. Greedy graphs permit *greedy routing* where a path from source to destination is found by the local rule of moving from the current

---

This work is partially supported in part by the Natural Sciences and Engineering Research Council of Canada (NSERC).

A. Brandstädt et al. (Eds.): WG 2018, LNCS 11159, pp. 356–368, 2018.
https://doi.org/10.1007/978-3-030-00256-5_29

vertex to any neighbor that is closer to the destination. However, greedy graphs are not necessarily t-spanners for any constant t.

The most desirable goal would be to construct sparse geometric graphs together with a local routing algorithm to find paths with bounded stretch factor that always get closer to the destination. This is the topic of our paper. There are two aspects to the goal: to construct sparse geometric graphs in which such paths exist, and to find the paths via a local routing algorithm.

Recently, Dehkordi et al. [9] introduced a class of graphs with good path properties: A graph is *angle-monotone* if there is a path between every two vertices that, after some rotation, is x- and y-monotone—equivalently, there is some 90° wedge such that the vector of every edge of the path lies in this wedge. This class was explored (and named) by Bonichon et al. [3]. Any angle-monotone path σ from s to t has the *self-approaching* property (see [1]) that a point moving along σ always gets closer to t.

The concept can be generalized to wedges of angles other than 90°—a path is *angle-monotone of width* γ ("generalized angle-monotone") if there is some wedge of angle γ such that the vector of every edge of the path lies in this wedge. Although graphs that are angle monotone of width greater than 90° are not necessarily self-approaching, they have good spanning properties. A graph that is angle-monotone of width $\gamma < 180°$ is a $(1/\cos\frac{\gamma}{2})$-spanner [3], thus a $\sqrt{2}$ spanner for $\gamma = 90°$ (the factor $\sqrt{2}$ is obvious based on the path being x- and y-monotone after some rotation).

Our specific goal in this paper is to construct sparse generalized angle-monotone graphs and design local routing algorithms to find generalized angle-monotone paths in them. There have been a few results on constructing angle-monotone graphs, but no previous results on local routing to find angle-monotone paths—except for some impossibility results.

Constructing Angle-Monotone Graphs. The best result on constructing planar angle-monotone graphs is due to Dehkordi et al. [9] who proved that any set of n points has a planar angle-monotone graph of width 90° using $O(n)$ Steiner points. They proved this by showing that a Gabriel triangulation is angle-monotone of width 90° (see [12] for a simpler proof), and then using the result that any point set can be augmented with $O(n)$ Steiner points to obtain a point set whose Delaunay triangulation is Gabriel. Without Steiner points, it is known that one cannot guarantee planar angle-monotone graphs for all point sets [3]. For the special case of n points in convex position, Dehkordi et al. [9] proved that there exists a (non-planar) angle-monotone graph with $O(n \log n)$ edges. In this paper we show that any point set has an angle-monotone graph with a subquadratic number of edges.

Turning to angle-monotone graphs of larger width, Bonichon et al. [3] showed that the half-Θ_6-graph on a set of n points, which is planar, is an angle-monotone graph of width 120°.

Local Routing on Angle-Monotone Graphs. A *k-local routing algorithm* finds a path one vertex at a time using only local information about the current vertex and its *k*-neighborhood plus the coordinates of the destination. The *routing ratio* of a local routing algorithm is the maximum stretch factor of any path found by the algorithm. The results mentioned in the previous two paragraphs imply that Gabriel graphs are $\sqrt{2}$-spanners, and half-Θ_6-graphs are 2-spanners (as was previously known [4, 8]). Are there local routing algorithms to find paths with good stretch factors, or paths that are angle-monotone in these classes of graphs? The answers are "yes" and "no", respectively. Bonichon et al. [3] gave a 1-local routing algorithm for Gabriel graphs that has routing ratio $(1 + \sqrt{2})$. On the other hand, they proved that no local routing algorithm can find angle-monotone paths in Gabriel graphs. Bose et al. [7] gave a 1-local routing algorithm for half-Θ_6-graphs that has routing ratio 2.887. They proved that this is the best ratio possible for any local routing algorithm, which implies that no local routing algorithm will find angle-monotone paths of width 120° in half-Θ_6-graphs. We construct a family of graphs together with a local routing algorithm that finds generalized angle-monotone paths.

Contributions. Our main results are as follows:

1. Given n points in the plane we construct an angle-monotone graph of width 90° with $O(\frac{n^2 \log \log n}{\log n})$ edges—a subquadratic number of edges. Since angle-monotone graphs are *increasing-chord graphs*, this answers Open Problem 4 from [9]. (We refer to [1, 2, 13–15] for results on self-approaching and increasing-chord graphs.)

2. Given n points in the plane and any α, $0 < \alpha < 45°$, we construct an angle-monotone graph of width $90° + \alpha$ with $O(\frac{n}{\alpha})$ edges. We give a 2-local routing algorithm for these graphs that finds angle-monotone paths of width $90° + \alpha$, thus of stretch factor $1/\cos(\frac{90° + \alpha}{2})$. In particular, for $\alpha = 30°$ our construction yields the [full] Θ_6-graph, and our local routing algorithm finds angle-monotone paths of width 120° and stretch factor 2. For this case, our algorithm is 1-local and very similar to the one of Bose et al. [7] that finds paths of stretch factor 2 in half-Θ_6-graphs, but our proof of correctness is simpler.

2   Angle-Monotone Graphs of Width 90°

In this section we show that any set of n points admits an angle-monotone graph of width 90° with $o(n^2)$ edges.

To achieve this, we will use the Erdős-Szekeres theorem [10] to partition the point set into subsets each with a logarithmic number of points in convex position. We will then construct an angle-monotone graph on each pair of subsets. Our construction is inspired by and builds upon a result in [9] that every 'one-sided convex point set' admits an increasing-chord graph with a linear number of edges. In fact, their proof yields an angle-monotone graph of width 90° (see the full version [11]). We use this in Lemma 4 below.

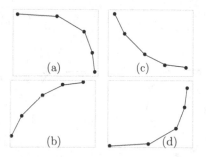 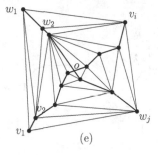

Fig. 1. (a) An $(x, -y)$-convex path. (b) An (x, y)-convex path. (c) An $(x, -y)$-concave path. (d) An (x, y)-concave path. (e) Illustration for Lemma 1.

We first introduce some preliminary definitions and notation. We will distinguish two types of x-monotone paths: an (x, y)-*monotone path* increases in both x- and y-coordinates, and an $(x, -y)$-*monotone path* increases in x-coordinate and decreases in y-coordinate. For each type of path we further distinguish convex and concave subtypes. Traversed in increasing x order, a convex path turns to the right, and a concave path turns to the left. Thus *an (x, y)-convex path* is an (x, y)-monotone path that turns to the right when traversed in increasing x order, and etc. for the other three types. See Figs. 1(a)–(d).

Lemma 1. *Let $P = (v_1, \ldots, v_i)$ be an $(x, -y)$-monotone path, and let $P' = (w_1, \ldots, w_j)$ be an (x, y)-monotone path. Then there exists an angle-monotone graph of width $90°$ and size $O(i + j)$ that spans P and P'.*

Proof. Assume without loss of generality that P and P' intersect, say at point o. (If necessary, we can add points $(-\infty, \infty)$ and $(\infty, -\infty)$ at the start and end of P respectively, and similarly for P'.) We will solve four subproblems for the points to the left of o, to the right of o, above o and below o, as illustrated in Fig. 1(e). Observe that any two points in $P \cup P'$ either lie in the same path, or in one of these half-spaces, so it suffices to find an angle-monotone graph of size $O(i + j)$ for each subproblem, and take the union.

Let $v_1, \ldots, v_{i'}$ and $w_1, \ldots, w_{j'}$ be the vertices to the left of the vertical line through o. We now construct an angle-monotone graph spanning these vertices as follows. Add an edge (v_1, w_1) and then move a vertical sweep-line ℓ from $(-\infty, 0)$ to o. Each time we encounter a new vertex q, we add the edges (q, v') and (q, w'), where v' (resp., w') is the rightmost vertex of P (resp., P') lying in the left-half plane of ℓ. We call v' and w' the *predecessor* of q in P and in P', respectively. The resulting graph H has size $O(i + j)$. We now show that H is an angle-monotone graph. For any pair of vertices a, b, if a, b belong to the same path, i.e., P or P', then they are already connected by an angle-monotone path. Otherwise, assume without loss of generality that $a \in P$, $b \in P'$, and b has a larger x-coordinate than a. Let b' be the predecessor of b in P. Follow the path P from a to b' and then take the edge (b', b). This is an $(x, -y)$-monotone path, and thus angle-monotone (equivalently, of width $90°$). □

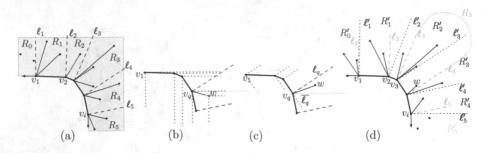

Fig. 2. (a)–(d) Illustration for Lemma 2.

Lemma 2. *Let $P = (v_1, \ldots, v_i)$ be an $(x, -y)$-convex path, and let R be the region (above P) bounded by P and the leftward and downward rays starting at v_1 and v_i, respectively. Then for any set W of j points in R, there exists a graph G of size $O(i + j)$ such that any pair of vertices $v \in P, w \in W$ is connected by an angle-monotone path of width $90°$.*

Proof. Let v_0 be any point on the leftward ray starting at v_1. For each q from 1 to i, let ℓ_q be the ray starting at v_q that lies perpendicular to $v_{q-1}v_q$ and enters region R. Since P is convex, the rays ℓ_q subdivide the region R into regions $R_0, R_1, \ldots, R_i$, e.g., see Fig. 2(a). For each point v_q, connect v_q to all the points in region $(R_q \cap W)$, e.g., see Fig. 2(b). Let G' be the resulting graph including the edges of P. We now claim that for any vertex v_t, $1 \leq t \leq q$ and for any $w \in (R_q \cap W)$ the path $v_t, \ldots, v_q, w$ is an angle-monotone path. If the y-coordinate of w is smaller than that of v_q, then this path is $(x, -y)$-monotone and hence angle-monotone, e.g., see Fig. 2(b). Otherwise, one can observe that all edges in the path have vectors that lie in the $90°$ clockwise wedge between ℓ_q and the line extending (v_{q-1}, v_q), e.g., see Fig. 2(c). Thus the path $v_t, \ldots, v_q, w$ is an angle-monotone path.

For each q from i to 1, we construct a graph G'' symmetrically by defining the perpendicular rays $\ell'_1, \ldots, \ell'_i$ and regions $R'_0, \ldots, R'_i$, as illustrated in Fig. 2(d). We construct the final graph G by taking the union of all the edges of G' and G''. It is straightforward to observe that G has at most $(i + 2j)$ edges.

To complete the proof, we must show that there is an angle-monotone path from any vertex v_t, $1 \leq t \leq i$ to any $w \in W$. Observe that R_q and R'_{q-1} intersect because P is convex. If $w \in (R_t \cup \cdots \cup R_i)$, then there is an angle-monotone path from v_t to w in G, and otherwise $w \in (R'_{t-1} \cup \cdots \cup R'_0)$ and there is an angle-monotone path from v_t to w in G''. □

Lemma 3. *Let $P = (v_1, \ldots, v_i)$ and $P' = (w_1, \ldots, w_j)$ be a pair of $(x, -y)$-convex (or, concave) paths. Then there exists an angle-monotone graph (spanning P and P') with width $90°$ and size $O(i + j)$.*

Proof. We prove the lemma assuming that P and P' are a pair of convex paths. The case when they are concave is symmetric. We consider two cases depending on whether P and P' intersect or not.

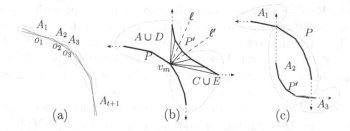

Fig. 3. (a)–(c) Illustration for Lemmas 3–4.

Case 1: First consider the case when P and P' do not intersect, and assume without loss of generality that P' lies above P. Since the vertices on P' are already connected by an angle-monotone path, we can apply Lemma 2 to obtain the required angle-monotone graph.

Case 2: Consider now the case when P and P' intersect. Let $o_1, \ldots, o_t$ be the points of intersections ordered from left to right, e.g., see Fig. 3(a). Let A_1 (resp., A_{t+1}) be the set of vertices of $(P \cup P')$ with x-coordinates smaller (resp., larger) than that of o_1 (resp., o_t). For every q, where $2 \leq q \leq t$, let A_q be the set of vertices of $(P \cup P')$ that lie to the left of o_q and to the right of o_{q-1}.

We process the sets $A_1, \ldots, A_{t+1}$ independently using Case 1, and let $G_{A_1}, \ldots, G_{A_{t+1}}$ be the resulting graphs. Compute the final graph G by taking the union of P, P', and $G_{A_1}, \ldots, G_{A_{t+1}}$. It is straightforward to verify that every pair of vertices in G is connected by an angle-monotone path. The number of edges in G is at most $\sum_{k=1}^{q} |A_k| \in O(i + j)$. $\qquad\square$

Lemma 4. *Let $P = (v_1, \ldots, v_i)$ be an $(x, -y)$-convex path, and let $P' = (w_1, \ldots, w_j)$ be an $(x, -y)$-concave path (or, vice versa). Then there exists an angle-monotone graph (spanning P and P') of width $90°$ and size $O(k \log k)$, where $k = \max\{i, j\}$.*

Proof. We extend P by adding leftward and downward rays starting at v_1 and v_i, respectively, e.g., see Fig. 3(b). We extend P' symmetrically. We now consider two cases depending on whether P, P' intersect or not.

Case A: If P' and P do not intersect, e.g., see Fig. 2(f), then P' lies above P. In this scenario we can find an angle-monotone graph of size $O(k)$ by applying Lemma 2.

Case B: If P and P' intersect, then they intersect in at most two points o_1, o_2, with o_1 to the left of o_2, e.g., see Fig. 3(c). The part to the left of o_1 and the part to the right of o_2 can be handled using Case A. In the middle we have a convex polygon, where the result of Dehkordi et al. [9] gives an angle-monotone graph of size $O(k \log k)$. See [11] for further details. $\qquad\square$

Theorem 1. *Let S be a point set with n points. Then there exists an angle-monotone graph (spanning S) of width $90°$ and size $O(\frac{n^2 \log \log n}{\log n})$ edges.*

Proof. By the Erdős-Szekeres theorem [10], every point set with n points contains a subset of $O(\log n)$ points in convex position. Urabe [16] observed that by repeatedly extracting such a convex set, one can partition a point set into $O(\frac{n}{\log n})$ convex polygons each of size $O(\log n)$. We partition each of these convex polygons into an (x, y)-convex path, an $(x, -y)$-convex path, an (x, y)-concave path, and an $(-x, -y)$-concave path.

For each pair of these paths, we apply Lemmas 1–4, as appropriate. Finally, we compute the required graph G by taking the union of all the $O(\frac{n^2}{\log^2 n})$ graphs. Since any pair of points in S either lie on the same path, or in one of these $O(\frac{n^2}{\log^2 n})$ graphs, they are connected by an angle-monotone path of width $90°$. Since the length of each path is at most $O(\log n)$, the size of G is $O(\frac{n^2}{\log^2 n}) \cdot O(\log n \log \log n) = O(\frac{n^2 \log \log n}{\log n})$. $\qquad\square$

Corollary 1. *Let S be a point set with t nested convex hulls. Then there exists an angle-monotone graph (spanning S) of width $90°$ with $O(t^2 n \log n)$ edges.*

Although the above construction of a subquadratic-size angle-monotone network with width $90°$ is somewhat involved, one can easily construct an angle-monotone graph with width $(90° + \alpha)$ and $O(\frac{n^{3/2}}{\alpha})$ edges, for any $0 < \alpha \le 90°$, as shown in [11].

3 Angle-Monotone Graphs of Width $(90° + \alpha)$

In this section we show how to construct, for any point set of size n and any angle α, $0 < \alpha < 45°$, an angle-monotone graph of width $(90° + \alpha)$ with $O(\frac{n}{\alpha})$ edges. We call these *layered 3-sweep graphs*. First, in Sect. 3.1, we introduce a *3-sweep graph* of a point set in which three lines are used to connect each point to three of its neighbors. The special case where the three lines form $60°$ wedges yields the half-Θ_6-graph. In Sect. 3.1, we analyze angle-monotonicity properties of 3-sweep graphs. Then, in Sect. 3.2, we define a *k-layer 3-sweep graph* as a union of k different 3-sweep graphs. We prove that a layered 3-sweep graph with an appropriate number of layers is an angle-monotone graph of width $(90° + \alpha)$ with $O(\frac{n}{\alpha})$ edges.

3.1 3-Sweep Graphs

Let $\triangle ABC$ be an acute triangle in $\mathbb{R}^2$ such that A, B, C appear in clockwise order on the perimeter of $\triangle ABC$, e.g., see Fig. 4(a). Let $\theta_a, \theta_b, \theta_c$ be the angles at A, B, C, respectively. For any point q let $W_{q,a}$ (the "a-wedge" of q) be the wedge with apex q such that the two sides of $W_{q,a}$ are parallel to AB and AC, i.e., $\triangle ABC$ can be translated such that A coincides with q and two sides of $\triangle ABC$ lie along the sides of $W_{q,a}$. Similarly, we define the wedges $W_{q,b}$ and $W_{q,c}$, e.g., see Fig. 4(b). The *a-nearest neighbor* of q in $W_{q,a}$ is defined to be the first point p that we hit (after q) while sweeping $W_{q,a}$ by a line L_{bc} parallel to

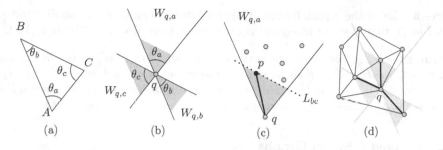

Fig. 4. (a) $\triangle ABC$. (b) $W_{q,a}, W_{q,b}, W_{q,c}$. (c) The a-nearest neighbor of q, where (q, p) is a θ_a-edge. (d) A 3-sweep graph.

BC (starting with the line through q). Figure 4(c) illustrates such an example. In the case of ties, we can pick arbitrarily as far as the results in this subsection are concerned. However, it is important that the local routing algorithm in Sect. 4 be able to find the a-nearest neighbor, so we break ties by choosing the most clockwise point. We call the edge (q, p) a θ_a-*edge*. We define b- and c-nearest neighbors and θ_b- and θ_c-edges analogously.

Given a set of points S, and three acute angles $\theta_a, \theta_b, \theta_c$ summing to $180°$, we define a *3-sweep graph* G on S with angles $\{\theta_a, \theta_b, \theta_c\}$ to be a geometric graph obtained by connecting every point $q \in S$ to its a-, b- and c-nearest neighbors, e.g., see Fig. 4(d). If $\theta_a = \theta_b = \theta_c = 60°$, then G is equivalent to the well known half-Θ_6-graph.

Bonichon et al. [4] proved that half-Θ_6-graphs are equivalent to Triangular Distance (TD) Delaunay triangulations, introduced by Chew [8]. A 3-sweep graph is also the same as the half-Θ_6-graph under a linear transformation (see [11]).

Both half-Θ_6 and 3-sweep graphs are special cases of *convex Delaunay graphs*, which were studied by Bose et al. [5]. They proved that every convex Delaunay graph is a t-spanner, but the value of t obtained from that proof is too large to be useful for our triangle T'—for details, see [11].

Every convex Delaunay graph is planar [5], and hence the following lemma is immediate. An independent proof of Lemma 5 is included in [11].

Lemma 5. *Every 3-sweep graph is planar.*

In the remainder of this section we analyze angle-monotonicity properties of 3-sweep graphs. We will show that for points q and t in a 3-sweep graph G with t in $W_{q,a}$ there is an angle-monotone path from q to t whose width depends on θ_a and on the position of t relative to the a-*path* of q. The a-*path* of q, denoted $P_{q,a}$, is defined to be the maximal path $q(= v_0), \ldots, v_k$ in G such that for each i from 1 to k, v_i is the a-nearest neighbor of v_{i-1}. We also define the *extended a-path* $\overline{P}_{q,a}$ to be the a-path $P_{q,a}$ together with $W_{v_k,a}$, which is empty of points since the a-path is maximal. We define [extended] b- and c-paths similarly.

Observe that if t is a vertex of $P_{q,a}$ then there is an angle-monotone path of width θ_a from q to t. The following lemma handles the case where $t \in W_{q,a}$, and

t does not lie on the a-path from q. The proof of the lemma is very similar to the proof in [3] that the half-Θ_6-graph is angle-monotone of width $120°$ (see [11]).

Lemma 6. *Let q, t be two vertices in G such that t lies in $W_{q,a}$. If t lies to the left (resp., right) of $\overline{P}_{q,a}$ then there is an angle-monotone path of width $(\theta_a + \theta_b)$ (resp., $(\theta_a + \theta_c)$) from q to t. Furthermore, the path consists of one subpath of the a-path of q followed by one subpath of the b-path (resp., c-path) of t.*

3.2 Layered 3-Sweep Graphs

In this subsection we define an angle-monotone graph of width $(90° + \alpha)$ for any angle α, $0 < \alpha < 45°$, such that $k = \frac{180}{\alpha}$ is an integer, and for any set S of n points. Our graph is defined as a k-layer 3-sweep graph.

Let $\triangle ABC$ be an acute triangle with A, B, C in clockwise order around the triangle, and with angles $\theta_a = 2\alpha$, $\theta_b = \theta_c = 90° - \alpha$. Orient $\triangle ABC$ so that the vertically upward ray starting at A bisects θ_a. Let G_1 be the 3-sweep graph of S with respect to the 3 lines through the sides of $\triangle ABC$.

We define G_i, $2 \leq i \leq k$ by successive rotations of $\triangle ABC$. Let $\triangle_i ABC$ be the triangle obtained by rotating $\triangle ABC$ clockwise around A with an angle of $\frac{i-1}{k} 360°$, and let G_i be the 3-sweep graph of S with respect to the three lines through the sides of $\triangle_i ABC$. The union of $G_1, \ldots, G_k$ is defined to be the k-*layer 3-sweep graph* H_k of S with respect to α.

Theorem 2. *Let H_k be a k-layer 3-sweep graph, with $k = \frac{180}{\alpha}$. Then H_k is an angle-monotone graph of width $(90° + \alpha)$ and the number of edges in H_k is $O(\frac{n}{\alpha})$.*

Proof. Let q and v be two points in S. Then v belongs to $W_{q,a}$ in some G_i, where $1 \leq i \leq k$. By Lemma 6, there exists an angle-monotone path of width $2\alpha + (90° - \alpha) = (90° + \alpha)$ between q and v in G_i, and hence also in H_k. By Lemma 5, each G_i is planar. Hence H_k has $O(nk) \in O(\frac{n}{\alpha})$ edges.   □

If $2\alpha = 60°$, then $k = 6$. Because of symmetries, $G_i = G_{i+2}$ so we really only have two 3-sweep graphs, and the resulting graph H_6 is the full-Θ_6-graph.

In the remainder of this section we compare k-layer 3-sweep graphs and full-Θ_k graphs, e.g., see Fig. 5(a). On the one hand, for $k > 6$, H_k may have up to 3 times as many edges as the Θ_k-graph though this is less if k is congruent to 2 mod 4 (see [11]). On the other hand, every H_k is an angle-monotone graph of width $(90° + \alpha)$, but it is not known whether Θ_k-graphs are angle-monotone with bounded width. For every $k = 4m + 4$, where m is a positive integer, one can construct a Θ_k graph of width approximately $(90° + 2\alpha)$. For example, if $k = 8$, then $2\alpha = 45°$, and H_8 is an angle-monotone graph of width $112.5°$. A Θ_8-graph may have comparatively large width, e.g., Fig. 5(b) shows a Θ_8-graph, where any angle-monotone path between u and v has width approximately $135°$.

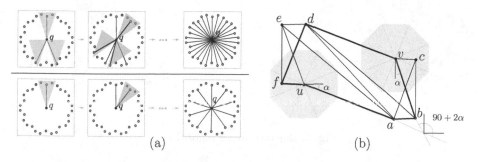

Fig. 5. (a) Illustration for the neighbors of q in (top) H_{10}, and (bottom) Θ_{10}. (b) An angle-monotone path between u and v of width approximately $(90° + 2\alpha) = 135°$ (inspired by an illustration in [6]).

4 Local Routing in Layered 3-Sweep Graphs

In this section we give a local routing algorithm for k-layer 3-sweep graphs. Specifically, our routing algorithm is *2-local*, meaning that at each step we assume knowledge of: the coordinates of the current vertex u, the coordinates of the target vertex, and the *2-neighborhood* of u, which consists of the neighbors of u and their neighbors. In the special case when $k = 6$, i.e., for full-Θ_6-graphs, we can restrict ourselves to 1-locality.

Theorem 3. *There is a 2-local routing algorithm that finds angle-monotone paths of width $90° + \alpha$ in any k-layer 3-sweep graph H_k, where $\alpha = 180°/k$. The algorithm has routing ratio $1/\cos(45° + \frac{\alpha}{2})$.*

Before giving the algorithm, we explain why we need 2-locality. Given a start vertex q and a target vertex t, we can find, based on the angle of line qt, which of the k 3-sweep graphs, say G_i has $t \in W_{q,a}$. Our routing algorithm will only use edges of G_i, so we need a way to tell if an edge of H_k belongs to G_i. Consider an edge from current vertex u to some vertex v. From their coordinates, we can decide whether v is in a *positive* wedge of u in G_i, i.e., one of $W_{u,a}, W_{u,b}$, or $W_{u,c}$. If so, then, by checking the other neighbors of u, we can detect if v is the unique a-, b-, or c-neighbor of u in that wedge in G_i. Otherwise, u is in a positive wedge of v, and, using 2-locality, we can check the neighbors of v to detect if u is the unique a-, b-, or c-neighbor of v in G_i.

For the special case of $\alpha = 30°$, H_k is the full-Θ_6-graph and our algorithm finds angle-monotone paths of width $120°$ and achieves routing ratio 2. In this case our algorithm, operating on a single 3-sweep graph, can be viewed as a slight variant of the algorithm of Bose et al. [7] for routing positively in a half-Θ_6-graph. Their algorithm achieves spanning ratio 2 but—as stated—includes a tie-breaking rule that prevents it from finding angle monotone paths of width $120°$. See [11] for further details. Our contribution is to simplify the statement of the algorithm, generalize to other angles, and give a much simpler proof of correctness using angle-monotonicity.

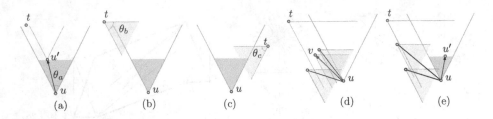

Fig. 6. Illustration for algorithm $\mathcal{A}$.

Algorithm $\mathcal{A}$ (Local Routing). Let H_k be a k-layer 3-sweep graph with angles $\theta_a = 2\alpha, \theta_b = \theta_c = 90° - \alpha$, and let q and t be two vertices in H_k. As discussed above, we can find out which 3-sweep graph, G_i, has t in $W_{q,a}$. We will route in G_i, using 2-locality to distinguish its edges as discussed above. For ease of description, orient the plane with $W_{q,a}$ pointing upward, centered on the vertical axis, so that edge BC of the reference triangle is horizontal. See Fig. 6. The general situation is that we have routed (forwarded the message) to some vertex u. Initially $u = q$. The algorithm stops when $u = t$.

- While t is an internal point of $W_{u,a}$, forward the message to u', where u' is the a-neighbor of u in $W_{u,a}$. See Fig. 6(a). Observe that u' is below or on the horizontal line through t.
- At this point, u either belongs to $W_{t,b}$ or $W_{t,c}$ (possibly lying on the boundary of the wedge). See Figs. 6(b)–(c). If u belongs to $W_{t,b}$, call routine $\mathcal{A}_L$, otherwise call routine $\mathcal{A}_R$.

Algorithm $\mathcal{A}_L$ (Left Routing). Invariant: $u \in W_{t,b}$. Until u reaches t do the following:

- Case 1. Forward the message to the first clockwise neighbor v of u in G_i such that $v \in W_{t,b}$ and $u \in W_{v,b}$, if such a vertex v exists. See Fig. 6(d).
- Case 2. If no such vertex v exists, then forward the message to vertex u', where u' is the a-neighbor of u in $W_{u,a}$. See Fig. 6(e).

Algorithm $\mathcal{A}_R$ (Right Routing). Invariant: $u \in W_{t,c}$. Symmetric to above.

We now prove that $\mathcal{A}$ finds an angle-monotone path of width $(90° + \alpha)$ from the source q to the destination t. Since we execute only one of $\mathcal{A}_L$ or $\mathcal{A}_R$ and they are symmetric, it suffices to consider the case where $\mathcal{A}_L$ is executed. The significant part of the proof is to show that the algorithm finds a path from q to t. The fact that the path is angle monotone of width $(90° + \alpha)$ follows immediately. In particular, the initial while loop of algorithm $\mathcal{A}$ uses only θ_a-edges, and algorithm $\mathcal{A}_L$ uses only θ_b- and θ_a-edges. Thus the path is angle monotone of width $(90° + \alpha)$. Note that the algorithm does not find a path with θ_a-edges appearing before θ_b-edges, as was guaranteed in Lemma 6.

In order to show that algorithm $\mathcal{A}$ finds a path from q to t we will show: (1) the invariant $u \in W_{t,b}$ holds for algorithm $\mathcal{A}_L$; (2) some measure improves at each routing step of the algorithm.

First consider the invariant $u \in W_{t,b}$. $W_{t,b}$ is bounded by two lines, ℓ and ℓ', where ℓ is the horizontal line through t. To show that $u \in W_{t,b}$, we must show that u is below, or on, ℓ, and to the right of, or on, ℓ'. When we first call $\mathcal{A}_L$, u is to the right of, or on, ℓ', and each step of $\mathcal{A}_L$ preserves this property—see Figs. 6(d) and (e). It remains to prove that u is below or on line ℓ. We will prove the stronger invariant that $P_{t,b}$ goes through or above u, i.e. that $P_{t,b}$ intersects the ray going vertically upward from u.

We begin by showing that this is true when we first call $\mathcal{A}_L$. If we call $\mathcal{A}_L$ because q is on ℓ', then $P_{t,b}$ must pass through or above q. The only other way to call $\mathcal{A}_L$ is because we just completed a step of the while loop of $\mathcal{A}$ where t was internal to $W_{u,a}$ but not internal to $W_{u',a}$, e.g., see Fig. 6(a). By Lemma 5, $P_{t,b}$ cannot cross the edge (u, u'). Hence it must pass above or through u'.

Now consider a general step of $\mathcal{A}_L$. We route from u to vertex w which is either vertex v in Case 1 (Fig. 6(d)) or vertex u' in Case 2 (Fig. 6(e)). Suppose (for a contradiction) that the path $P_{t,b}$ does not go through or above w. By induction we know that $P_{t,b}$ goes through or above u. By Lemma 5, $P_{t,b}$ cannot cross the edge (u, w). (This is where we use the assumption that (u, w) is an edge of G_i.) Thus $P_{t,b}$ must go through u and the other points of edge (u, w) must lie above the path. Let x be the vertex before u on $P_{t,b}$. Then $x \in W_{t,b}$ and $u \in W_{x,b}$. We now claim that the algorithm should have chosen x rather than w. First note that x is a candidate for vertex v in Case 1 of $\mathcal{A}_L$. Thus the algorithm would not have moved to Case 2. Next note that x comes before v in clockwise order around u, so the algorithm would have chosen x rather than v.

It remains to show that something improves at every step of the algorithm. Let d_a be the distance from u to the horizontal line through t. Let d_b be the distance from t to the line determined by the right boundary of $W_{u,a}$. In every iteration of the while loop of $\mathcal{A}$, d_a decreases and d_b does not increase. In Case 2 of $\mathcal{A}_L$, d_a decreases and d_b does not increase. Finally, in Case 1 of $\mathcal{A}_L$, d_b decreases and d_a does not increase. Thus $d_a + d_b$ strictly improves, and the algorithm must terminate. The path found by the algorithm is an angle-monotone path of width $90° + \frac{\theta_a}{2} = (90 + \alpha)$.

We show that for full-Θ_6-graphs, it suffices to know the 1-neighborhood of the current vertex. See [11] for further details.

Theorem 4. *There is a 1-local routing algorithm that finds angle-monotone paths of width $120°$ in any full-Θ_6-graph.*

5 Open Questions

1. (from [3]) What is $\gamma_{\min}$, the smallest γ such that every point set has a planar angle-monotone graph of width γ? It is known that $90° < \gamma_{\min} \leq 120°$.
2. We showed that every set of n points admits an angle-monotone graph of width $90°$ with $o(n^2)$ edges, but can a better bound be proved? $O(n \log n)$ edges? Even $O(n)$ is not ruled out.

3. Using Steiner points, we can construct angle-monotone graphs of width γ, for any given $\gamma > 0$, however, the size of the graph depends on some distance parameters of the point set (see [11]). What is the smallest γ such that every point set has an angle-monotone Steiner graph with width γ and $o(n^2)$ edges?

References

1. Alamdari, S., Chan, T.M., Grant, E., Lubiw, A., Pathak, V.: Self-approaching graphs. In: Didimo, W., Patrignani, M. (eds.) GD 2012. LNCS, vol. 7704, pp. 260–271. Springer, Heidelberg (2013). https://doi.org/10.1007/978-3-642-36763-2_23
2. Bahoo, Y., Durocher, S., Mehrpour, S., Mondal, D.: Exploring increasing-chord paths and trees. In: Proceedings of the 29th Canadian Conference on Computational Geometry. CCCG (2017)
3. Bonichon, N., Bose, P., Carmi, P., Kostitsyna, I., Lubiw, A., Verdonschot, S.: Gabriel triangulations and angle-monotone graphs: local routing and recognition. In: Hu, Y., Nöllenburg, M. (eds.) GD 2016. LNCS, vol. 9801, pp. 519–531. Springer, Cham (2016). https://doi.org/10.1007/978-3-319-50106-2_40
4. Bonichon, N., Gavoille, C., Hanusse, N., Ilcinkas, D.: Connections between theta-graphs, Delaunay triangulations, and orthogonal surfaces. In: Thilikos, D.M. (ed.) WG 2010. LNCS, vol. 6410, pp. 266–278. Springer, Heidelberg (2010). https://doi.org/10.1007/978-3-642-16926-7_25
5. Bose, P., Carmi, P., Collette, S., Smid, M.H.M.: On the stretch factor of convex Delaunay graphs. J. Comput. Geom. 1(1), 41–56 (2010)
6. Bose, P., Carufel, J.D., Morin, P., van Renssen, A., Verdonschot, S.: Towards tight bounds on theta-graphs: more is not always better. Theor. Comput. Sci. 616, 70–93 (2016)
7. Bose, P., Fagerberg, R., van Renssen, A., Verdonschot, S.: Optimal local routing on Delaunay triangulations defined by empty equilateral triangles. SIAM J. Comput. 44(6), 1626–1649 (2015)
8. Chew, L.P.: There is a planar graph almost as good as the complete graph. In: Proceedings of the 2nd Annual Symposium on Computational Geometry (SoCG), pp. 169–177 (1986)
9. Dehkordi, H.R., Frati, F., Gudmundsson, J.: Increasing-chord graphs on point sets. J. Graph Algorithms Appl. 19(2), 761–778 (2015)
10. Erdös, P., Szekeres, G.: A combinatorial theorem in geometry. Compos. Math. 2, 463–470 (1935)
11. Lubiw, A., Mondal, D.: Construction and local routing for angle-monotone graphs (2015). https://arxiv.org/abs/1801.06290
12. Lubiw, A., O'Rourke, J.: Angle-monotone paths in non-obtuse triangulations. In: Proceedings of the 29th Canadian Conference on Computational Geometry. CCCG (2017)
13. Mastakas, K., Symvonis, A.: On the construction of increasing-chord graphs on convex point sets. In: Proceedings of the 6th International Conference on Information, Intelligence, Systems and Applications (IISA), pp. 1–6. IEEE (2015)
14. Nöllenburg, M., Prutkin, R., Rutter, I.: On self-approaching and increasing-chord drawings of 3-connected planar graphs. J. Comput. Geom. 7(1), 47–69 (2016)
15. Rote, G.: Curves with increasing chords. Math. Proc. Camb. Philos. Soc. 115, 1–12 (1994)
16. Urabe, M.: On a partition into convex polygons. Discret. Appl. Math. 64(2), 179–191 (1996)

Characterization and Recognition of Tree 3-Spanner Admissible Directed Path Graphs of Diameter Three

B. S. Panda[1](✉) and Anita Das[2]

[1] Department of Mathematics, Indian Institute of Technology Delhi,
Hauz Khas, New Delhi 110016, India
`bspanda@maths.iitd.ac.in`
[2] Infosys Ltd., Bengaluru, India
`anitadas01@infosys.com`

Abstract. A spanning tree T of a graph G is a **tree t-spanner**, t an integer, if the distance between any two vertices in T is at most t times their distance in G. A graph that admits a tree t-spanner is called a **tree t-spanner admissible** graph. The problem of deciding whether a graph is tree t-spanner admissible is NP-complete for any fixed $t \geq 4$ and is linearly solvable for $t \leq 2$. The case $t = 3$ is still open and is conjectured to be NP-complete. In this paper, we present a structural characterization and a polynomial time recognition algorithm for tree 3-spanner admissible directed path graphs of diameter three.

Keywords: Tree spanners · Directed path graphs · NP-completeness

1 Introduction

A spanning tree T of a connected graph G is called a **tree t-spanner**, t an integer, if the distance between any two vertices in T is at most t times their distance in G. A graph that has a tree t-spanner is called a **tree t-spanner admissible graph**. Tree spanners are used as models for broadcast operations [19]. Tree spanners are also used in approximating bandwidth of graphs [21] and in biology [2] and has been widely studied in the literature (see [1–6, 12, 13, 17, 19–21]).

The problem of deciding whether an arbitrary graph is tree t-spanner admissible for any fixed $t \geq 4$ is shown to be NP-complete by Cai and Corneil [6] and remains NP-complete even for chordal graphs [3]. The tree t-spanner problem for $t \leq 2$ can be solved in linear time for arbitrary graphs [6]. However, the status of the case $t = 3$ is still open for arbitrary graphs and was conjectured to be NP-complete [6]. Motivated by this conjecture, researchers have investigated the tree 3-spanner problem for various special classes of graphs. The classes of graphs which admit tree 3-spanners are split graphs, co-graphs, and complements of bipartite graphs [5], interval graphs and permutation graphs [13], and

© Springer Nature Switzerland AG 2018
A. Brandstädt et al. (Eds.): WG 2018, LNCS 11159, pp. 369–381, 2018.
https://doi.org/10.1007/978-3-030-00256-5_30

very strongly chordal graphs [3]. The tree 3-spanner problem can be solved in polynomial time for planar graphs [8] and 2-sep directed path graphs [17].

Le and Le [12] claimed that directed path graphs always admit tree 3-spanners. However, Panda and Das [16] have disproved their result by showing that not all directed path graphs of diameter three admit tree 3-spanners. Hence it is interesting to characterize and recognize tree 3-spanner admissible directed path graphs of diameter three.

In this paper, we present a structural characterization and a polynomial time recognition algorithm for tree 3-spanner admissible directed path graphs of diameter three.

The rest of the paper is organized as follows. In Sect. 2, we present some pertinent definitions and preliminary results. In Sect. 3, we first prove some properties of directed path graphs of diameter three. We then characterize tree 3-spanner admissible directed path graphs of diameter three. An $O(n^3)$ time recognition algorithm for tree 3-spanner admissible directed path graphs of diameter three is given in Sect. 4, where n denotes the number of vertices of the input graph. Finally, Sect. 5 concludes the paper.

2   Preliminaries

All graphs considered in this paper are assumed to be connected. Let $G = (V, E)$ denote a graph and let n and m denote the number of vertices and number of edges of G, respectively. A set $S \subseteq V(G)$ is called a **clique** if $G[S]$, the induced subgraph of G on S, is a complete subgraph of G. A clique S is called a **maximal clique** if no proper superset of S is a clique of G. Throughout this paper, by a clique we mean a maximal clique unless otherwise mentioned. For a graph G, let $N_G(v) = \{w \in V | vw \in E\}$ be the set of neighbors of v. We will use $N(v)$ for $N_G(v)$ if the graph G is clear from the context. Let $N_G[v] = N_G(v) \cup \{v\}$. Let $d_G(x, y)$ denote the distance between x and y in G. A tree T having n vertices is called a **star** if it has a vertex v of degree $n - 1$. In this case, v is called the **star center** of T. If T has exactly two vertices, say x and y, of degree more than one, then T is called a **bi-star** and x and y are called the **bi-star centers** of T. A graph G is **chordal** if every cycle in G of length at least four has a chord, i.e., an edge joining two non-consecutive vertices of the cycle.

A directed graph T is a **directed tree** if the underlying graph T' of T, which is obtained from T by ignoring the directions of the edges in T, is a tree. A directed tree T is a **rooted directed tree** with root $r(T)$ if $r(T)$ is the only vertex in T of zero in-degree (the number of arcs entering $r(T)$). Let T be a rooted directed tree with root $r(T)$. If xy is an arc in T from x to y, then x is called the **parent** of y. All the vertices in the path from a vertex x to $r(T)$ are called **ancestors** of x. If x is an ancestor of y, then y is called a **descendant** of x. If x is an ancestor of y and $x \neq y$, then x is called a **proper ancestor** of y and y is called a **proper descendant** of x. The **least common ancestor** of x and y is the common ancestor z of x and y such that for any common ancestor z' of x and y, z' is an ancestor of z. The **depth** of a vertex x is the length of the path from r to x.

A vertex v of a graph G is called **simplicial** if $N[v]$ induces a clique. It is known [7] that every chordal graph G admits a simplicial vertex and if G is a non-complete chordal graph, then it admits two non-adjacent simplicial vertices. An ordering $\sigma = (v_1, v_2, \ldots, v_n)$ of $V(G)$ is called a **perfect elimination ordering** (PEO) of G if v_i is a simplicial vertex of $G[\{v_i, v_{i+1}, \ldots, v_n\}]$ for all i, $(1 \leq i \leq n)$. A graph G is chordal if and only if it has a PEO [9].

If $G - C$ is disconnected for a clique C and has components $H_i = (V_i, E_i)$, $1 \leq i \leq r$ and $r \geq 2$, then C is called a separating clique of G and $G_i = G[(V_i \cup C)]$ is called a **separated subgraph** of G with respect to C, where $1 \leq i \leq r$ and $r \geq 2$. Let $W(G_i) = \{v \in C |$ there is a $w \in V_i$ with $vw \in E(G_i)\}$. Cliques of G other than C which intersect C are called **relevant cliques** of G with respect to C. A relevant clique C_j of G_i for which $(C_j \cap C) = W(G_i)$ is called a **principal clique** of G_i.

The existence of a principal clique of every separated subgraph of a chordal graph is guaranteed by the following result due to Panda and Mohanty [18].

Lemma 1 ([18]). *Every separated subgraph G_i of a chordal graph with respect to the separating clique C has a principal clique.*

The following lemma is used in next section.

Lemma 2 ([16]). *If G_i is a separated subgraph of a chordal graph G with respect to C and C_j is any non-principal clique of G_i, then there exists a vertex $x \in C_j \setminus W(G_i)$ such that $xy \notin E(G)$ for all $y \in W(G_i) \setminus C_j$.*

In the following definitions, only relevant cliques are considered. Let C_1 and C_2 be two cliques of G. We say (1) C_1 and C_2 are **unattached**, denoted $C_1 | C_2$, if $C_1 \cap C_2 \cap C = \emptyset$; otherwise, they are **attached**; (2) C_1 **dominates** C_2, denoted $C_1 \geq C_2$, if $C_1 \cap C \supseteq C_2 \cap C$; (3) C_1 **properly dominates** C_2, denoted $C_1 > C_2$, if $C_1 \cap C \supset C_2 \cap C$; and (4) C_1 and C_2 are **antipodal**, denoted $C_1 \leftrightarrow C_2$, if they are attached and neither dominates the other.

Let G_1 and G_2 be two separated subgraphs of G with respect to C. We say (1) G_1 and G_2 are **unattached**, denoted $G_1 | G_2$, if $C_1 | C_2$ for every clique C_1 in G_1 and for every clique C_2 in G_2 (otherwise they are **attached**); (2) G_1 **dominates** G_2, denoted $G_1 \geq G_2$, if they are attached and for every clique C_1 in G_1, $C_1 \geq C_2$ for every cliques C_2 in G_2 or $C_1 | C_2$ for all cliques C_2 of G_2; (3) G_1 **properly dominates** G_2, denoted $G_1 > G_2$, if $G_1 \geq G_2$ but not $G_2 \geq G_1$; and (4) G_1 and G_2 are **antipodal**, denoted $G_1 \leftrightarrow G_2$, if they are attached and neither dominates the other.

The above concepts were introduced in [14].

Lemma 3 ([14]). *A collection of pairwise non-antipodal subgraphs of a chordal graph G can be arranged in such a way that $G_i > G_j$ implies $i < j$.*

Lemma 4. *If G is a tree 3-spanner admissible chordal graph and $\sigma = (v_1, v_2, \ldots, v_n)$ is a PEO of G, then $G_i = G[\{v_i, v_{i+1}, \ldots, v_n\}]$ is tree 3-spanner admissible for each i, $1 \leq i \leq n$.*

Proof. Let T be a tree 3-spanner of G. Now v_1 is a simplicial vertex of G. If $d_T(v_1) = 1$, then let $T' = T - v_1$. If $d_T(v_1) \geq 2$, then let $N_T(v_1) = \{w_1, w_2, \ldots, w_k\}$. Let $T' = (V', E')$, where $V' = V(T) \setminus \{v_1\}$, and $E(T') = (E(T) \setminus \{v_1w_1, v_1w_2, \ldots, v_1w_k\}) \cup \{w_1w_2, w_1w_3, \ldots, w_1w_k\}$, i.e., T' is obtained from T by removing the vertex v_1, hence removing the edges $v_1w_i, 1 \leq i \leq k$ and adding the edges $w_1w_i, 2 \leq i \leq k$. It is easy to see that T' is a tree 3-spanner of $G - v_1 = G[\{v_2, v_3, \ldots, v_n\}]$. Using the fact that v_i is a simplicial vertex of G_i, $2 \leq i \leq n$, inductively we can construct a tree 3-spanner of G_i for each i, $2 \leq i \leq n$, starting from a tree 3-spanner of $G_1 = G$. Hence the result.

Lemma 5. *Let G be a tree 3-spanner admissible chordal graph and let $S = \{G_1, G_2, \ldots, G_r\}$, $r \geq 2$ be the set of all separated subgraphs of G with respect to a separating clique C. Let $S_1 \subset S$. Then $\bigcup_{G_i \in S_1} G_i$ is tree 3-spanner admissible.*

Proof. Let $|V(\bigcup_{G_i \in S_1} G_i)| = l$ and let G_i be any separated subgraph. Let $|V(G_i)| = k$ and $|C| = s$. Let $k - s = j$. Since C is a clique of G_i and a non-complete chordal graph has two non-adjacent simplicial vertices, it is easy to construct a PEO $\alpha_1 = (v_1, v_2, \ldots, v_k)$ of G_i such that $\{v_{j+1}, v_{j+2}, \ldots, v_k\} = C$, i.e., the last s vertices of α_1 constitute C. Using this fact, it it easy to construct a PEO β of G such that the last l vertices of β constitute $V(\bigcup_{G_i \in S_1} G_i)$. So this fact and Lemma 4 imply that $\bigcup_{G_i \in S_1} G_i$ admits a tree 3-spanner.

The following important result will be used extensively in the later sections.

Lemma 6 ([4]). *Let T be a tree 3-spanner of a chordal graph G. For any clique C of G, one of the following conditions holds:*

(i) C induces a star in T.
(ii) Either C induces a bi-star in T or there is a vertex $v \notin C$ such that $C \cup \{v\}$ induces a bi-star in T.

Remark 1. Suppose that $T[C \cup \{x\}], x \notin C$ is a bi-star. If x is not a bi-star center of $T[C \cup \{x\}]$, then x must be a pendant vertex in $T[C \cup \{x\}]$ and hence $T[C]$ is a bi-star. So x and y must be bi-star centers of $T[C \cup \{x\}]$ for some $y \in N(x) \cap C$.

A graph $G = (V, E)$ is a **directed path graph** if there exists a rooted directed tree B and a family of directed paths $(\bar{v})_{v \in V}$ in B such that for all vertices v_1 and v_2, $v_1v_2 \in E$ if and only if $(\bar{v}_1) \cap (\bar{v}_2) \neq \emptyset$. The class of directed path graphs is properly contained in the class of chordal graphs. Directed path graphs were introduced by Gavril [10] and are characterized as follows:

Theorem 1 ([10]). *A graph $G = (V, E)$ is a directed path graph if and only if there exists a (rooted) directed tree B whose vertex set is the set of all cliques of G and such that, for each vertex v of G, the cliques containing v form a directed path $\bar{v}$ in B. If such a tree B exists, the family $(\bar{v})_{v \in V}$ is a representing family of directed paths on B for G.*

The tree B in the above theorem is called a **characteristic tree** of G. Directed path graphs are also known as **rooted directed vertex (RDV)** graphs and characteristic trees of directed path graphs are also called as **RDV clique trees** (see [14, 15]).

Let $G_i, 1 \leq i \leq r, r \geq 2$, be the separated subgraphs of G with respect to the separating clique C.

Panda [15] has given the following characterization of RDV graphs (same as directed path graphs).

Theorem 2 ([15]). *G is an RDV graph if and only if each G_i is RDV, and the G_i's can be two-colored such that no antipodal pairs have the same color, in one color every subgraph has an RDV clique tree rooted at C, in the other color no two relevant cliques are unattached, and every subgraph (with one possible exception) has an RDV clique tree rooted at a relevant clique. The exceptional subgraph, should it exist, is dominated by every other subgraph of the same color, and it has an RDV clique tree in which the vertex C has out degree zero.*

A k-sun, $k \geq 3$, is a graph with vertex set $\{v_1, v_2, \ldots, v_k\} \cup \{x_1, x_2, \ldots, x_k\}$ such that $\{v_1, v_2, \ldots, v_k\}$ forms a clique and $\{x_1, x_2, \ldots, x_k\}$ forms an independent set and x_i is adjacent to v_i and v_{i+1}, for $1 \leq i \leq k-1$ and x_k is adjacent to v_k and v_1. For any integer $k \geq 3$, a k**-planet** is obtained from the path of k vertices $v_1, v_2, \ldots, v_k$ and a triangle abc by adding edges $bv_i, 1 \leq i \leq k-1$ and $cv_i, 2 \leq i \leq k$. It is well known that a directed path graph is free from k-planet for $k = 3$ and for $k \geq 5$ and is free from k-sun for each $k \geq 3$.

A graph $G = (V, E)$ is called a **path graph** if there exists a tree T and a family of paths $(\bar{v})_{v \in V}$ in T such that $xy \in E$ for all vertices $x, y \in V$ if and only if $(\bar{x}) \cap (\bar{y}) \neq \emptyset$.

Path graphs are also known as **undirected vertex (UV)** graphs and the class of directed path graphs is a proper subclass of path graphs (see [14]).

Lemma 7 ([14]). *Let C be a clique in the UV graph G. If C is not a separator, then C is a leaf vertex (i.e., a vertex with degree 1) in any clique tree of G.*

The following result follows from the above lemma.

Lemma 8. *Let T be any RDV clique tree of a directed path graph G, and let T' be the tree obtained from T by ignoring the direction of the edges of T. If C is any non-separating clique of G, then C is a leaf vertex (i.e., a vertex with degree 1) in T'.*

Lemma 9 ([14]). *Let G be a UV graph having a clique tree T and let C be a separating clique of G. Let C' and C'' be any two relevant cliques of G with respect to C. If C' is present in the path from C to C'' in the clique tree T of G, then $C' \geq C''$.*

Let T be any RDV clique tree of a directed path graph G, and let T' be the tree obtained from T by ignoring the direction of the edges of T. Let C be a separating clique of G. Let $P = C, C', C''$ be a path in T'. By Lemma 9, $C' \geq C''$. If G_1 is a separated subgraph of G with respect to C containing the clique C', then $C \cap C' \supseteq C \cap C''$ for any relevant clique C'' of G_1. Hence C' is a principal clique of G_1.

The following result follows from the above discussion.

Lemma 10. *Let T be any RDV clique tree of a directed path graph G and T' be the tree obtained from T by ignoring the direction of the edges of T. Let $P = C, C', C''$ be a path in T'. If C is a separating clique in G and G_i is a separated subgraph containing the clique C', then C' is a principal clique of G_i.*

Let C be a separating clique of G. Let G_i, $1 \leq i \leq r$, $r \geq 2$, be the separated subgraphs of G with respect to C. A separated subgraph G_i is said to satisfy **property** P if $C_i \cap C_j \cap C \neq \emptyset$ for every pair of distinct relevant cliques C_i and C_j of G_i. A family $\{S_i\}_{i \in I}$ of subsets of a set S is said to satisfy **Helly Property** if $J \subseteq I$ and $S_i \cap S_j \neq \emptyset$ for all $i, j \in J$ imply $\bigcap_{j \in J} S_j \neq \emptyset$.

Lemma 11 ([16]). *If the separated subgraph G_i satisfies the property P, then the family $S = \{W(G_i) \cap C_j \mid C_j \neq C$ and C_j is a relevant clique of $G_i\}$ of subsets of $W(G_i)$ satisfies the Helly property.*

Lemma 12 ([16]). *Let G_i be a separated subgraph of G with respect to C such that all the cliques of G_i are relevant cliques. If G_i satisfies the property P, then there is an ordering $C_1, C_2, \ldots, C_k$ of the cliques of G_i other than C such that $C_j \cap W(G_i) \subseteq C_{j+1} \cap W(G_i)$ for all j, $1 \leq j \leq k-1$.*

3   Characterization

A separating clique C of G is called a **dominating separating clique** if it intersects every clique of G. The following lemma whose proof is omitted proves the existence of a dominating separating clique in every directed path graph of diameter three.

Lemma 13. *Every diameter three directed path graph admits a dominating separating clique.*

Let G be a directed path graph of diameter three and C be a dominating separating clique of G. Let G_i, $1 \leq i \leq r$, $r \geq 2$, be the separated subgraphs of G with respect to C.

Since G is a directed path graph, all the separated subgraphs of G with respect to C can be two colored satisfying the condition of Theorem 2. Without loss of generality, assume that each G_i has an RDV clique tree rooted at C if G_i is assigned color 1. Let $S_1 = \{G_i \mid G_i$ is assigned color 1$\}$ and $S_2 = \{G_i \mid G_i$ is assigned color 2$\}$.

If G is tree 3-spanner admissible, then by Lemma 5, each G_i is tree 3-spanner admissible. So the main idea behind our characterization of tree 3-spanner admissible directed path graph of diameter three is to find the conditions under which each G_i is tree 3-spanner admissible and to find the conditions under which G is tree 3-spanner admissible given that each G_i is tree 3-spanner admissible.

Suppose that G_i satisfies the property P. Let $C_1, C_2, \ldots, C_k, k \geq 1$ be the cliques of G_i other than C. So $(\cap_{i=1}^k (C \cap C_i)) \neq \emptyset$ by Lemma 11. Now each $x \in \cap_{i=1}^k (C \cap C_i)$ is a dominating vertex of G_i and $x \in W(G_i)$. If $W(G_i)$ contains a dominating vertex, say x, of G_i, then $x \in \cap_{i=1}^k (C \cap C_i)$. So G_i satisfies the property P in this case.

In view of the above discussion, we have the following result.

Lemma 14. *A separated subgraph G_i with respect to a dominating separating clique C satisfies the property P if and only if $W(G_i)$ contains a dominating vertex of G_i.*

The following lemma provides a sufficient condition for tree 3-spanner admissibility.

Lemma 15. *If each G_i, $1 \leq i \leq r, r \geq 2$, satisfies the property P, then G admits a tree 3-spanner.*

Proof. Since every separated subgraph G_i with respect to C satisfies the property P, G_i has a dominating vertex, say x_i, in C, $1 \leq i \leq r$, by Lemma 14. Let T_1 be a star having vertex set C. Let T be such that $E(T) = E(T_1) \cup (\bigcup_{i=1}^r \{x_i x | x \in (V(G_i) \setminus C)\})$, and $V(T) = V(G)$. It is easy to see that T is a tree 3-spanner of G. So G admits a tree 3-spanner. $\qquad\square$

Definition 1. *A pair of relevant cliques (C_i, C_i') of G_i is said to be a **violating pair** with respect to C if $C_i \cap C_i' \cap C = \emptyset$ and there exists a relevant clique C_k such that $C_i \cap C \subseteq C_k$, $(C_i \cap C_k) \setminus C \neq \emptyset$, $C_i' \cap C \subseteq C_k$, and $(C_i' \cap C_k) \setminus C \neq \emptyset$. In this case, C_k is called a **base clique** of the violating pair (C_i, C_i'). Each vertex of C_k which is present in either C_i or C_i' is called a **base vertex**. The sets $X(C_i, C_i') = C_i \cap C_k$ and $Y(C_i, C_i') = C_i' \cap C_k$ are called the set of base vertices of C_i and C_i', respectively. A clique C_i is called a **violating clique** with respect to C if there exists another clique C_i' such that (C_i, C_i') is a violating pair with respect to C.*

Lemma 16. *If G_i does not satisfy the property P, then it contains a violating pair of relevant cliques.*

Lemma 17. *If G_1, G_2, and G_3 are three mutually unattached separated subgraphs of G with respect to C such that each G_i, $1 \leq i \leq 3$, violates the property P, then G does not admit a tree 3-spanner.*

Recall that the set of all separated subgraphs of G with respect to the separating clique C can be two colored satisfying the condition of Theorem 2. Let G_i, $1 \leq i \leq r$, $r \geq 2$, be the separated subgraphs of G with respect to C. Let $S_1 = \{G_i | G_i$ is assigned color $1\}$ and $S_2 = \{G_i | G_i$ is assigned color $2\}$. Since no two relevant cliques in the color class two are unattached, no two separated subgraphs in S_2 are unattached. Also no two separated subgraphs in S_2 are antipodal and all the cliques of G_i other than C are relevant with respect to C for all $G_i \in S_2$. So by Lemma 3, there is an ordering of the separated subgraphs of S_2, say, $G'_1, G'_2, \ldots, G'_r$, such that G'_i dominates G'_j if and only if $i < j$. By Lemma 12, there is an ordering of the cliques of G'_i, say $C_1, C_2, \ldots, C_s$, such that $C_i \cap W(G_i) \subseteq C_{i+1} \cap W(G_i)$ for all $i, 1 \leq i \leq s-1$. If $G_i, G_j \in S_2$, and G_i dominates G_j, then each clique of G_i dominates every clique of G_j because no two cliques of color class two are unattached. So concatenating the ordering of the cliques of $G'_r, G'_{r-1}, \ldots, G'_1$ in this order, we get an ordering, say $C_1, C_2, \ldots, C_t$, of the cliques of the separated subgraphs of S_2 such that $C_i \cap C \subseteq C_{i+1} \cap C$ for all $i, 1 \leq i \leq t-1$. So we have the following result.

Lemma 18. *There is an ordering of the cliques of the separated subgraphs of S_2, say, $C_1, C_2, \ldots, C_t$, such that $C_i \cap C \subseteq C_{i+1} \cap C$ for all $i, 1 \leq i \leq t-1$.*

A separated subgraph G_i of S_1 is called a **maximal separated subgraph** if G_i is not dominated by any other separated subgraph of S_1. If a separated subgraph G_j of S_1 is dominated by a maximal separated subgraph G_i, then G_j is called a **neighbor separated subgraph** of G_i. Let $N(G_i)$ denote the set of all neighbor separated subgraphs of a maximal separated subgraph G_i and $N[G_i] = N(G_i) \cup \{G_i\}$. Let $A_{S_1} = \{G_i | G_i \in S_1$ and G_i is a maximal separated subgraph$\}$. Let $B_{S_1} = \{G_i | G_i \in A_{S_1}$ and there is some $G_j \in N[G_i]$ such that G_j violates the property $P\}$. Let $C_{S_1} = A_{S_1} \setminus B_{S_1}$.

Lemma 19. *Any two maximal separated subgraphs of S_1 are unattached.*

Proof. If possible suppose that $G_i, G_j \in S_1$ are two maximal separated subgraphs such that G_i and G_j are attached. Since $G_i, G_j \in S_1$, they are non-antipodal to each other. Since they are attached and are non-antipodal, one will dominate the other, say $G_i > G_j$. This contradicts the fact that G_j is a maximal separated subgraph. Hence the result.   □

Lemma 20. *If B_{S_1} contains more than two separated subgraphs, then G does not admit a tree 3-spanner.*

Proof. The proof follows from Lemmas 17 and 19.   □

Lemma 21. *Let $G_1, G_2, \ldots, G_r, r \geq 2$ be the separated subgraphs of a directed path graph with respect to a separating clique C. Let G_i be such that $W(G_i) \cap W(G_j) = \emptyset$ for $i \neq j$. If $G - (V(G_i) \setminus C)$ is tree 3-spanner admissible, then $G - (V(G_i) \setminus C)$ has a tree 3-spanner T such that $d_T(x, y) \leq 2$ for all $x, y \in W(G_i)$.*

Lemma 22 (Reduction Lemma). *If $G_k \in C_{S_1}$, then $\bigcup_{G_i \in S_1} G_i$ admits a tree 3-spanner if and only if $\bigcup_{G_i \in S_1} G_i - ((\bigcup_{G_j \in N[G_k]} V(G_j)) \setminus C)$ admits a tree 3-spanner.*

Let $S'_1 = S_1 \setminus \{G_j |$there is some $G_i \in C_{S_1}$ such that $G_j \in N[G_i]\}$. In view of the above reduction lemma, $\bigcup_{G_i \in S_1} G_i$ admits a tree 3-spanner if and only if $\bigcup_{G_i \in S'_1} G_i$ admits a tree 3-spanner. So we concentrate on S'_1 instead of S_1. Recall that $B_{S'_1} = \{G_i | G_i \in A_{S'_1}$ and there is some $G_j \in N[G_i]$ such that G_j violates property $P\}$. If $B_{S'_1}$ has no separated subgraphs, then by Lemma 15 $\bigcup_{G_i \in S'_1} G_i$

admits a tree 3-spanner. If $B_{S_1'}$ contains three or more separated graphs, then by Lemma 20, G does not admit a tree 3-spanner. So assume that $B_{S_1'}$ contains at most two separated graphs. We deal two cases, i.e., $|B_{S_1'}| = 1$ and $|B_{S_1'}| = 2$ separately.

3.1 $B_{S_1'}$ Contains Exactly Two Maximal Separated Subgraphs

Let G_i be a maximal separated subgraph and $VP(N[G_i]) = \{(C_i, C_i')|(C_i, C_i')$ is a violating pair of G_j such that $G_j \in N[G_i]\}$, $VC(N[G_i]) = \{C_i|$ there exists C_i' such that (C_i, C_i') is a violating pair of G_j with respect to C, where $G_j \in N[G_i]\}$, and $VC_V(N[G_i]) = \{x|x \in C_i \cap C,$ where $C_i \in VC(N[G_i])\}$.

Definition 2. *Let G_i be a maximal separated subgraph. We say that a subset S of $VC(N[G_i])$ satisfies* **property** P^* *if there exists a vertex $x \in W(G_i)$ such that for each violating clique $C_i \in S$, if there exists another violating clique C_i' in S such that (C_i, C_i') is a violating pair, then either $X(C_i, C_i') \subseteq N(x)$ or $Y(C_i, C_i') \subseteq N(x)$. Such a vertex x is called a* **root vertex** *of G_i with respect to S. Note that G_i can contain more than one root vertex with respect to S. A maximal separated subgraph G_i satisfies* **property** P_1 *if $VC(N[G_i])$ satisfies the property P^*. Define $R_{P_1}(VC(N[G_i])) = \{x \mid x$ is a root vertex of $G_i\}$.*

For a vertex $x \in V(G)$, let $N_G^i(x) = \{y|d_G(x, y) = i\}, i \geq 1$. So $N_G^1(x) = N_G(x)$. Since G is of diameter three, $N_{G_i}^3(x) = \emptyset$ for all $x \in W(G_i)$ and for each separated subgraph G_i.

Lemma 23. *A maximal separated subgraph $G_1 \in S_1'$ satisfies the property P_1 if and only if there exists a vertex x in C such that for each connected component D of $N_{G_1}^2(x)$ there exists a vertex $v_D \in N_{G_1}^1(x)$ such that v_D is adjacent to all the vertices of D.*

The following lemma shows that if a separated subgraph G_1 satisfies the property P_1, then it admits a tree 3-spanner T such that C induces a star in T.

Lemma 24. *If a maximal separated subgraph $G_1 \in S_1'$ satisfies the property P_1, then $H = \bigcup_{G_i \in N[G_1]} G_i$ admits a tree 3-spanner T such that C induces a star in T having star center at x, where $x \in R_{P_1}(VC(N[G_1]))$.*

Lemma 25. *Let G_1 and G_2 be two maximal separated subgraphs in S_1' and $G^* = \bigcup_{G_i \in N[G_1] \cup N[G_2]} G_i$. The following are true.*

(i) *If G^* admits a tree 3-spanner, then C induces a bi-star in every tree 3-spanner of G^*.*

(ii) *G^* admits a tree 3-spanner if and only if each of G_1 and G_2 satisfies the property P_1.*

(iii) *If G^* admits a tree 3-spanner, then C induces a bi-star in every tree 3-spanner of G^* such that one of the bi-star centers belongs to $R_{P_1}(VC(N[G_1]))$ and the other bi-star center belongs to $R_{P_1}(VC(N[G_2]))$.*

Let $S_1' = \{G_1, G_2, \ldots, G_r\}$, and $S_2 = \{G_1', G_2', \ldots, G_s'\}$. Let $G' = (\bigcup_{i=1}^{r} G_i) \cup (\bigcup_{i=1}^{s} G_i')$.

Definition 3. G' is said to satisfy **property** P_2 if the following conditions are true.

(i) S_1' contains exactly two maximal separated subgraphs, say G_1 and G_2, such that each of G_1 and G_2 satisfies the property P_1.

Let α be an ordering of the cliques of the separated subgraphs of S_2 satisfying the condition of Lemma 18. Let C_{min} be the clique having least index in α such that $C_{min} \cap W(G_1) \neq \emptyset$ and $C_{min} \cap W(G_2) \neq \emptyset$. Let C_{smin} be the clique having least index in the ordering α such that either $C_{smin} \cap W(G_1) \neq \emptyset$ or $C_{smin} \cap W(G_2) \neq \emptyset$. If C_{smin} exists, then without loss of generality $C_{smin} \cap W(G_1) \neq \emptyset$.

(ii) If C_{min} exists, then either $R_{P_1}(VC(N[G_1])) \cap C_{min} \neq \emptyset$ or $R_{P_1}(VC(N[G_2])) \cap C_{min} \neq \emptyset$.

(iii) If both C_{smin} and C_{min} exist and C_{min} is different from C_{smin}, then $R_{P_1}(VC(N[G_1])) \cap C_{min} \neq \emptyset$.

Lemma 26. Suppose that S_1' contains exactly two maximal separated subgraphs. Then G admits a tree 3-spanner if and only if $G' = (\bigcup_{i=1}^{r} G_i) \cup (\bigcup_{i=1}^{s} G_i')$ satisfies the property P_2.

3.2 $B_{S_1'}$ Contains Exactly One Maximal Separated Subgraph

Definition 4. A maximal separated subgraph G_i satisfies **property** P_3 if G_i violates the property P_1 and $VC(N[G_i])$ can be partitioned into two sets $VC1(N[G_i])$ and $VC2(N[G_i])$ such that each of $VC1(N[G_i])$ and $VC2(N[G_i])$ satisfies the property P^*.

Lemma 27. Let G_1 be the only maximal separated subgraph in S_1' and G_1 violates the property P_1. Let $G'' = \bigcup_{G_i \in N[G_1]} G_i$. The following are true.

(i) G'' admits a tree 3-spanner if and only if G_1 satisfies the property P_3.

(ii) If G'' admits a tree 3-spanner, then C induces a bi-star having one of the bi-star centers in $R_{P_1}(VC1(N[G_1]))$ and the other bi-star center in $R_{P_1}(VC2(N[G_1]))$ in every tree 3-spanner of G'.

Recall that $G' = (\bigcup_{i=1}^{r} G_i) \cup (\bigcup_{i=1}^{s} G_i')$.

Definition 5. G' satisfies **property** P_4 if the following conditions are true.

(i) S_1' contains exactly one maximal separated subgraph, say G_1, violating the property P. Suppose that G_1 violates the properties P_1 but satisfies the property P_3. Let $VC1(N[G_1])$ and $VC2(N[G_1])$ be a partition of $VC(N[G_1])$ with respect to which G_1 satisfies P_3. Let $VV_1(G_1) = VC1_V(N[G_1])$ and $VV_2(G_1) = VC2_V(N[G_1])$.

Let α be an ordering of the cliques of the separated subgraphs of S_2 satisfying the condition of Lemma 18. Let C_{min}' be the clique having least *index in α such that $C_{min} \cap VV_1(G_1) \neq \emptyset$ and $C'_{min} \cap VV_2(G_2) \neq \emptyset$. Let C'_{smin} be the clique having least index in the ordering α such that either $C'_{smin} \cap VV_1(G_1) \neq \emptyset$ or $C'_{smin} \cap VV_2(G_1) \neq \emptyset$. If C'_{smin} exists, without loss of generality, $C'_{smin} \cap VV_1(G_1) \neq \emptyset$.*

(ii) If C'_{min} exists, then either $R_{P_1}(VC1(N[G_1])) \cap C'_{min} \neq \emptyset$ or $R_{P_1}(VC2(N[G_1])) \cap C'_{min} \neq \emptyset$.

(iii) If both C'_{min} and C'_{smin} exists and C'_{smin} is different from C'_{min}, then $R_{P_1}(VC1(N[G_1])) \cap C'_{min} \neq \emptyset$.

Lemma 28. *Suppose that S'_1 contains exactly one maximal separated subgraph, say G_1. Let G_1 violate the property P_1. G admits a tree 3-spanner if and only if $G' = (\bigcup_{i=1}^{r} G_i) \cup (\bigcup_{i=1}^{s} G'_i)$ satisfies the property P_4.*

Lemma 29. *If S'_1 contains exactly one maximal separated subgraph, say G_1 and G_1 satisfies the property P_1, then G admits a tree 3-spanner.*

Proof. By Lemma 24, $H = \bigcup_{G_i \in N[G_1]} G_i$ admits a tree 3-spanner T such that C induces a star in T having star center at x, where $x \in R_{P_1}(VC(N[G_1]))$. Let $y \in C$ be a dominating vertex of $H_1 = \bigcup_{G_i \in S_2} G_i$. Let $x_j \in C$ be a dominating vertex for $G_j \in (S_1 \setminus S'_1)$. Let $T' = T \cup \{yz | z \in (V(H_1) \setminus C)\} \cup (\bigcup_{G_j \in (S_1 \setminus S'_1)} \{x_j x, x \in (V(G_j) \setminus C)\})$. Clearly T' is a tree 3-spanner of G. $\square$

3.3   Characterization Theorem

Next we present the characterization theorem whose proof follows from the above lemmas.

Theorem 3. *G admits a tree 3-spanner if and only if*

(i) S'_1 contains at most two maximal separated subgraphs.

(ii) If S'_1 contains exactly two maximal separated subgraphs, then G' satisfies the property P_2.

(iii) If S'_1 contains exactly one maximal separated subgraph, say G_1, then either G_1 satisfies the property P_1 or G' satisfies the property P_4.

Proof. **Necessity:** (i) follows from Lemma 17, (ii) follows from Lemma 26, and (iii) follows from Lemma 28.
Sufficiency: Follows from Lemmas 26, 28, and 29. $\square$

4   Recognition

In this section, we show that all the conditions of Theorem 3 can be tested in polynomial time which leads to a polynomial time recognition algorithm for tree 3-spanner admissible directed path graph of diameter three.

Due to space restriction, we omit the details. First of all find a dominating separating clique, say C, of G. Next we find all the separated subgraphs of G with respect to C and compute $S_1 = \{G_i | G_i$ is assigned color 1$\}$, $S_2 = \{G_i | G_i$

is assigned color 2}, S_1', and $H = (\bigcup_{G_i \in S_1'} G_i) \cup (\bigcup_{G_i \in S_2} G_i)$. We next find the number of maximal separated graphs, say k, of S_1'. If $k \geq 3$, then we declare that G does not admit a tree 3-spanner.

Suppose that $k = 2$. Let $G' = H$. We test whether G' satisfies the property P_2. If G' violates the property P_2, then declare that G does not admit a tree 3-spanner.

Next suppose that $k = 1$. We check whether G' satisfies the property P_4. If G' does not satisfy the property P_4, declare that G does not admit a tree 3-spanner; otherwise G admits a tree 3-spanner. It can be proved that all the above steps require at most $O(n^3)$ time.

The following theorem follows from the above discussion.

Theorem 4. *A tree 3-spanner admissible directed path graph of diameter three can be recognized in $O(n^3)$ time.*

5   Conclusion

We obtained a structural characterization of tree 3-spanner admissible directed path graphs of diameter three. Based on this characterization, we proposed an $O(n^3)$ time algorithm to recognize tree 3-spanner admissible directed path graphs of diameter three. It would be interesting to characterize and recognize tree 3-spanner admissible directed path graphs. This is an open problem.

References

1. Awerbuch, B., Baratz, A., Peleg, D.: Efficient broadcast and light-weighted spanners, Manuscript (1992)
2. Bandelt, H.J., Dress, A.: Reconstructing the shape of a tree from observed dissimilarity data. Adv. Appl. Math. **7**, 309–343 (1986)
3. Brandstädt, A., Chepoi, V., Dragan, F.F.: Distance approximating trees for chordal and dually chordal graphs. J Algorithms **30**, 166–184 (1999)
4. Brandstädt, A., Dragan, F.F., Le, H.O., Le, V.B.: Tree spanners on chordal graphs: complexity and algorithms. Theor. Comput. Sci **310**, 329–354 (2004)
5. Cai, L.: Tree spanners: spanning trees that approximate the distances, Ph.D. thesis, University of Toronto (1992)
6. Cai, L., Corneil, D.G.: Tree spanners. SIAM J. Discret. Math. **8**, 359–387 (1995)
7. Dirac, G.A.: On rigid circuit graphs. Abh. Math. Sem Univ. Hamburg **25**, 71–76 (1961)
8. Fekete, S.P., Kremer, J.: Tree spanners in planar graphs. Discret. Appl. Math. **108**, 85–103 (2001)
9. Fulkerson, D.R., Gross, O.A.: Incidence matrices and interval graphs. Pac. J. Math. **15**, 835–855 (1965)
10. Gavril, F.: A recognition algorithm for the intersection graphs of directed paths in directed trees. Discret. Math. **13**, 237–249 (1975)
11. Golumbic, M.C.: Algorithmic Graph Theory and Perfect Graphs. Annals of Discrete Mathematics, vol. 53. Elsevier, Amsterdam (2004)

12. Le, H.O., Le, V.B.: Optimal tree 3-spanners in directed path graphs. Networks **34**, 81–87 (1999)
13. Madanlal, M.S., Venkatesan, G., Rangan, C.P.: Tree 3-spanners on interval, permutation and regular bipartite graphs. Inform. Process. Lett. **59**, 97–102 (1996)
14. Monma, C.L., Wei, V.K.: Intersection graphs of paths in a tree. J. Combin. Theory Ser. B **41**, 141–181 (1986)
15. Panda, B.S.: The separator theorem for rooted directed vertex graphs. J. Combin. Theory Ser. B **81**, 156–162 (2001)
16. Panda, B.S., Das, A.: On tree 3-spanners in directed path graphs. Networks **50**, 203–210 (2007)
17. Panda, B.S., Das, A.: Tree 3-spanners in 2-sep directed path graphs: characterization, recognition, and construction. Discret. Appl. Math. **157**, 2153–2169 (2009)
18. Panda, B.S., Mohanty, S.P.: Intersection graphs of vertex disjoint paths in a tree. Discret. Math. **146**, 179–209 (1995)
19. Peleg, D.: Distributed Computing: A Locality-Sensitive Approach. SIAM Monograph on Discrete Mathematics and Applications. SIAM, Philadelphia (2000)
20. Peleg, D., Schaffer, A.A.: Graph spanners. J. Graph Theory **13**, 99–116 (1989)
21. Venkatesan, G., Rotics, U., Madanlal, M., Makowsky, J.A., Rangan, C.P.: Restrictions of minimum spanner problems. Inform. Comput. **136**, 143–164 (1997)

Author Index

Printed in the United States
By Bookmasters